HARCOURT BRACE JOVANOVICH COLLEGE OUTLINE SERIES

INTRODUCTORY ALGEBRA

Alan Wise

Department of Mathematics, University of San Diego

Harcourt Brace Jovanovich College Publishers

Fort Worth   Philadelphia   San Diego   New York   Orlando   Austin   San Antonio
Toronto   Montreal   London   Sydney   Tokyo

Requests for permission to make copies of any part of the work should be mailed to:

Permissions
Harcourt Brace Jovanovich, Publishers
Orlando, Florida 32887

Printed in the United States of America

Library of Congress Cataloging in Publication Data

Wise, Alan.
    Introductory Algebra.

    (Harcourt Brace Jovanovich college outline series)
    (Books for Professionals)
    Includes index.
    1. Algebra.   I. Title.   II. Series.   III. Series: Books for Professionals.
QA152.2.W57 1986   512.9   86-7630
ISBN 0-15-601524-2

*To my sister Diane and her family,
Nelson, Byron and Brett.*

PREFACE

The purpose of this book is to present a complete course in introductory algebra in the clear, concise form of an outline. This outline provides an in-depth review of the principles of introductory algebra for independent study, and contains essential supplementary material for mastering introductory algebra. Or, this outline can simply be used as a valuable, self-contained refresher course on the practical application of introductory algebra.

Although comprehensive enough to be used by itself for independent study, this outline is specifically designed to be used as a supplement to college and high school textbooks on the subject. But notice that the topics in this outline are more narrowly defined than the topics in many textbooks on the same subject. For instance, whereas applications of linear equations is included in the discussion of linear equations in some books and excluded in others, in this outline it is covered in its own separate chapter. This isolation not only helps you to find the specific topics that you need to study but also enables you to bypass those topics that you are already familiar with.

Notice, too, that regular features at the end of each chapter are specially designed to supplement your textbook and course work in introductory algebra.

RAISE YOUR GRADES This feature consists of a checkmarked list of open-ended thought questions to help you assimilate the material you have just studied. These thought questions invite you to compare concepts, interpret ideas, and examine the whys and wherefores of chapter material.

SUMMARY This feature consists of a brief restatement of the main ideas in each chapter, including definitions of key terms. Because it is presented in the efficient form of a numbered list, you can use it to refresh your memory quickly.

SOLVED PROBLEMS Each chapter of this outline contains a set of exercises and word problems and their step-by-step solutions. Undoubtedly the most valuable feature of this outline, these problems allow you to become proficient with the fundamental skills and applications of introductory algebra. Along with the sample midterm and final examinations, they also give you ample exposure to the kinds of questions that you are likely to encounter on a typical college exam. To make the most of these solved problems, try writing your own solutions first. Then compare your answers to the detailed solutions provided in the book. If you have trouble understanding a particular problem, or set of problems, use the example reference numbers to locate and review the appropriate instruction and parallel step-by-step examples that were developed earlier in the chapter.

SUPPLEMENTARY EXERCISES Each chapter of this outline concludes with a set of drill exercises and practical applications and their answers. The supplementary exercises are designed to help you master and retain all the newly discussed skills and concepts presented in the given chapter, and also contain example references.

Of course there are other features of this outline that you will find very helpful, too. One is the format itself, which serves both as a clear guide to important ideas and as a convenient structure upon which to organize your knowledge. A second is the attention devoted to methodology and the practical applications of introductory algebra. Yet a third is the careful listing of learning objectives in each section denoted by the uppercase letters A, B, C, D, or E.

We wish to thank the people who made this text possible, including Serena Hecker for painstakingly checking the solutions and answers to each exercise and word problem and Mel Friedman for his comments and suggestions during the development of this text.

San Diego, California   ALAN WISE

CONTENTS

1 ARITHMETIC REVIEW I: WHOLE NUMBERS, FRACTIONS, AND MIXED NUMBERS

THIS CHAPTER IS ABOUT

☑ **Identifying Numbers and Symbols**
☑ **Factoring Whole Numbers**
☑ **Identifying Fractions and Mixed Numbers**
☑ **Renaming Fractions and Mixed Numbers**
☑ **Comparing Fractions and Mixed Numbers**
☑ **Simplifying Fractions**
☑ **Computing with Fractions**

1-1. Identifying Numbers and Symbols

A. Identify whole numbers.

The numbers you use to count things are called the **counting numbers** or **natural numbers**.

EXAMPLE 1-1: List all the counting numbers.

Solution: The counting numbers are listed as 1, 2, 3, 4, 5, 6, 7, 8, 9, 10, 11, 12, · · ·
or just 1, 2, 3, · · ·

Note: The **ellipsis symbol** [· · ·] that follows the listed counting numbers indicates that the counting number pattern shown continues on forever.

When the number **zero** [0] is combined with the counting numbers, you get the **whole numbers**.

EXAMPLE 1-2: List all the whole numbers.

Solution: The whole numbers are listed as 0, 1, 2, 3, 4, 5, 6, 7, 8, 9, 10, 11, 12, · · ·
or just 0, 1, 2, 3, · · ·

B. Identify relation symbols.

To **compare** two whole numbers that are not equal, you identify which number is *greater* [larger] and/or which number is *less* [smaller]. **Relation symbols** are used to show how two numbers compare.

EXAMPLE 1-3: List the relation symbols used in arithmetic.

Solution: $<$ is read as "*is less than*" [9 is less than 10 or $9 < 10$].   } the three
    $>$ is read as "*is greater than*" [10 is greater than 9 or $10 > 9$].   } **basic relation**
    $=$ is read as "*is equal to*" [9 is equal to 9 or $9 = 9$].   } **symbols**
    $\neq$ is read as "*is not equal to*" [9 is not equal to 10 or $9 \neq 10$].
    $\approx$ is read as "*is approximately equal to*" [9.9 is approximately equal to 10 or $9.9 \approx 10$].
    $\leqslant$ is read as "*is less than or equal to*" [9 is less than or equal to 10 or $9 \leqslant 10$].
    $\geqslant$ is read as "*is greater than or equal to*" [10 is greater than or equal to 9 or $10 \geqslant 9$].

Note 1: The symbol = is called an **equality symbol.**

Note 2: The symbols <, >, ≠, ≤, ≥, and ≈ are called **inequality symbols.**

Note 3: Any two numbers can be compared using one and only one of the three basic relation symbols <, >, or =.

C. Identify operation symbols.

To join two or more amounts together, you **add.**

EXAMPLE 1-4: Identify the numbers and operation symbol used in the following **addition** problem: $2 + 3 = 5$

Solution: In $2 + 3 = 5$ or $\begin{matrix} 2 \\ +3 \\ \hline 5 \end{matrix}$: 2 and 3 are both called **addends.**
5 is called the **sum.**
+ is called an **addition symbol.**

Note: $2 + 3 = 5$ or $\begin{matrix} 2 \\ +3 \\ \hline 5 \end{matrix}$ is read as "two *plus* three is equal to five."

To take one amount away from another amount, you **subtract.**

EXAMPLE 1-5: Identify the numbers and operation symbol used in the following **subtraction** problem: $5 - 3 = 2$

Solution: In $5 - 3 = 2$ or $\begin{matrix} 5 \\ -3 \\ \hline 2 \end{matrix}$: 5 is called the **minuend.**
3 is called the **subtrahend.**
2 is called the **difference.**
− is called a **subtraction symbol.**

Note: $5 - 3 = 2$ or $\begin{matrix} 5 \\ -3 \\ \hline 2 \end{matrix}$ is read as "five *minus* three is equal to two."

Subtraction is related to addition.

EXAMPLE 1-6: Show how subtraction is related to addition.

Solution:

subtraction facts $\Bigg\langle$ 5 − 3 = 2 because 2 + 3 = 5
  5 − 2 = 3 because 3 + 2 = 5 $\Bigg\rangle$ related **addition facts**

To join two or more equal amounts together, you **multiply.**

EXAMPLE 1-7: Identify the numbers and operation symbol used in the following **multiplication** problem: $6 \times 2 = 12$

Solution: In $6 \times 2 = 12$ or $\begin{matrix} 2 \\ \times 6 \\ \hline 12 \end{matrix}$: 6 and 2 are both called **factors.**
6 is also called the **multiplicand.**
2 is also called the **multiplier.**
12 is called the **product.**
× is called a **multiplication symbol.**

Note: $6 \times 2 = 12$ or $\begin{matrix} 2 \\ \times 6 \\ \hline 12 \end{matrix}$ is read as "six *times* two is equal to twelve."

There are several different ways to write a multiplication problem like "6 times 2."

EXAMPLE 1-8: Write "6 times 2" in five equivalent ways.

Solution: 6 times 2 = 6 × 2 ⎫
           = 6 · 2 ⎪
           = 6(2) ⎬ all equivalent ways to write "6 times 2"
           = (6)2 ⎪
           = (6)(2) ⎭

Multiplication is related to addition.

EXAMPLE 1-9: Show how multiplication is related to addition.

number of addends
       value of each addend
      ↓   ↓

Solution: $6 \times 2 = 2 + 2 + 2 + 2 + 2 + 2$ ⟵ 6 repeated addends of 2

Note: Multiplication is just a short way to write two or more **repeated addends.**

To separate a given amount into two or more equal amounts, you **divide.**

EXAMPLE 1-10: Identify the numbers and operation symbol used in the following **division** problem: $9 \div 2 = 4 \text{ R}1$

Solution: In $9 \div 2 = 4 \text{ R}1$ or $2\overline{)9}$:

$$\begin{array}{r} 4\,\text{R}1 \\ 2\overline{)\,9\,} \\ -8 \\ \hline 1 \end{array}$$

9 is called the **dividend.**
2 is called the **divisor.**
4 is called the **quotient.**
1 is called the **remainder.**
4 R1 is called the **division answer.**
$)$ is called a **division box.**
$\div$ is called a **division symbol.**

Note: $9 \div 2 = 4 \text{ R}1$ or $2\overline{)9}^{\,4\,\text{R}1}$ is read as "nine *divided by* two is equal to four remainder one."

There are several different ways to write a division problem like "9 divided by 2."

EXAMPLE 1-11: Write "9 divided by 2" in four equivalent ways.

Solution: 9 divided by 2 = 9 ÷ 2 ⎫
                 = 9/2 ⎪
                 = $\frac{9}{2}$ ⎬ all equivalent ways of writing "9 divided by 2"
                 = $2\overline{)9}$ ⎭

Division is related to multiplication.

EXAMPLE 1-12: Show how division is related to multiplication.

Solution:

division facts ⟨ $12 \div 2 = 6$ because $6 \times 2 = 12$
                $12 \div 6 = 2$ because $2 \times 6 = 12$ ⟩ related **multiplication facts**

1-2. Factoring Whole Numbers

A. Identify prime and composite numbers.

A given nonzero whole number is a **whole-number factor** of a second whole number if the given nonzero whole number divides the second whole number *evenly* [with a remainder of zero].

To find all the whole-number factors of a given whole number, you can use division.

EXAMPLE 1-13: Find all the whole-number factors of 12.

Solution: The whole-number factors of 12 are 1, 2, 3, 4, 6, and 12 because:

whole-number factors of 12		not whole-number factors of 12	
$12 \div 1 = 12$ or 12 R0		$12 \div 5 = 2$ R2	
$12 \div 2 = 6$ or 6 R0		$12 \div 7 = 1$ R5	
$12 \div 3 = 4$ or 4 R0		$12 \div 8 = 1$ R4	
$12 \div 4 = 3$ or 3 R0	zero remainders	$12 \div 9 = 1$ R3	nonzero remainders
$12 \div 6 = 2$ or 2 R0		$12 \div 10 = 1$ R2	
$12 \div 12 = 1$ or 1 R0		$12 \div 11 = 1$ R1	

A whole number that has exactly two different whole-number factors is called a **prime number.**

Note: 12 is not a prime number because 12 has six different whole-number factors.

EXAMPLE 1-14: List the first ten prime numbers.

Solution: The first ten prime numbers are 2, 3, 5, 7, 11, 13, 17, 19, 23, and 29 because:

0 has more than two different whole-number factors [$0 \div 1 = 0, 0 \div 2 = 0, 0 \div 3 = 0, \cdots$ means every nonzero whole number is a whole-number factor of 0].

1 has only one different whole-number factor [$1 \div 1 = 1$ only means the only whole-number factor of 1 is 1].

2, 3, 5, 7, 11, 13, 17, 19, 23, and 29 are the first ten whole numbers that have exactly two different whole-number factors [2 has whole-number factors of 1 and 2; 3 has whole-number factors of 1 and 3; 5 has whole-number factors of 1 and 5; and so on].

A whole number that is greater than zero and has more than two different whole-number factors is called a **composite number.**

Note: 12 is a composite number because 12 has six different whole-number factors.

EXAMPLE 1-15: List the first ten composite numbers.

Solution: The first ten composite numbers are 4, 6, 8, 9, 10, 12, 14, 15, 16, and 18 because:

4, 6, 8, 9, 10, 12, 14, 15, 16, and 18 are the first ten whole numbers with more than two different whole-number factors [4 has whole-number factors of 1, 2, and 4; 6 has whole-number factors of 1, 2, 3, and 6; and so on].

Note 1: The whole numbers 0 and 1 are neither prime nor composite numbers.

Note 2: No prime number is a composite number, and vice versa. Every prime number has *exactly two* different whole-number factors and every composite number has *more than two* different whole-number factors.

Note 3: Every whole number is 0 or 1 or prime or composite.

B. **Factor composite numbers.**

To **factor a composite number,** you write it as a product of two or more counting numbers.

EXAMPLE 1-16: Factor 12 in as many different ways as possible using counting numbers.

Solution: $12 = 1 \cdot 12$
$\qquad\quad 12 = 2 \cdot 6$
$\qquad\quad 12 = 3 \cdot 4$ 12 can be factored in exactly four different ways using counting numbers
$\qquad\quad 12 = 2 \cdot 2 \cdot 3$

Note 1: The factorizations $1 \cdot 12$ and $12 \cdot 1$ are not considered to be different factorizations but **equivalent factorizations** because each contains the same factors [1 and 12].

Note 2: The factorization $2 \cdot 2 \cdot 3$ is called a **product of primes** because each factor is a prime number.

The following statement is one of the most important facts about the arithmetic of whole numbers.

The Fundamental Rule of Arithmetic

Every composite number can be factored as a product of primes in exactly one way, except for equivalent factorizations.

To **factor a composite number as a product of primes,** you can use division.

EXAMPLE 1-17: Factor 12 as a product of primes using division.

prime divisor

Solution: $12 \div 2 = 6$ ←—— composite quotient (continue the division process)

$\qquad\qquad\quad 6 \div 2 = 3$ ←—— prime quotient (Stop!)

$\qquad 12 = 2 \cdot 2 \cdot 3$ ←—— product of primes

Note 1: To factor a composite number as a product of primes when the quotient is a composite number, you continue the division process by dividing that quotient by another prime number until the quotient is a prime number.

Note 2: When the quotient is a prime number, you stop the division process and write each prime divisor and the prime quotient as the product of primes.

Note 3: The factorizations $2 \cdot 2 \cdot 3$, $2 \cdot 3 \cdot 2$, and $3 \cdot 2 \cdot 2$ are all considered equivalent factorizations because each contains two factors of 2 and one factor of 3.

1-3. Identifying Fractions and Mixed Numbers

A. **Identify fractions.**

If a and b are whole numbers $[b \neq 0]$, then $\dfrac{a}{b}$ is called a **common fraction** or just a **fraction.**

Note: Expressions like $\dfrac{5}{0}$ and $\dfrac{0}{0}$ are **not defined** as fractions because the denominator of a fraction can never be zero, *by definition.* Division by zero is always meaningless.

Every fraction has three distinct parts.

EXAMPLE 1-18: Name the three parts of the fraction $\frac{3}{4}$.

Solution: In the fraction $\frac{3}{4}$: 3 is called the **numerator.**
— is called a **fraction bar.**
4 is called the **denominator.**

When the numerator of a fraction is less than the denominator, the fraction is called a **proper fraction.** When the numerator of a fraction is greater than or equal to the denominator, the fraction is called an **improper fraction.**

EXAMPLE 1-19: Determine whether each of the following is a proper fraction or an improper fraction:
(a) $\frac{3}{4}$ (b) $\frac{4}{3}$ (c) $\frac{3}{3}$

Solution: (a) $\frac{3}{4}$ is a proper fraction because $3 < 4$ [the numerator 3 is less than the denominator 4].
(b) $\frac{4}{3}$ is an improper fraction because $4 > 3$ [the numerator 4 is greater than the denominator 3].
(c) $\frac{3}{3}$ is an improper fraction because $3 = 3$ [the numerator 3 is equal to the denominator 3].

Note: Every fraction is either a proper fraction or an improper fraction.

B. Identify mixed numbers.

The sum of a whole number and a proper fraction is called a **mixed number.**

EXAMPLE 1-20: Add 2 and $\frac{3}{4}$ to form a mixed number.

Solution: $2 + \frac{3}{4} = 2\frac{3}{4}$ ⟵ mixed number

whole number

proper fraction

Every mixed number has two distinct parts.

EXAMPLE 1-21: Name the two distinct parts of the mixed number $2\frac{3}{4}$.

Solution: In $2\frac{3}{4}$: 2 is called the **whole-number part.**
$\frac{3}{4}$ is called the **fraction part.**

1-4. Renaming Fractions and Mixed Numbers

A. Rename a fraction as an equivalent division problem and vice versa.

To **rename a fraction as an equivalent division problem,** you use

$$\frac{a}{b} = a \div b \quad \text{or} \quad b\overline{)a}$$

EXAMPLE 1-22: Rename $\frac{2}{3}$ as an equivalent division problem.

denominator

Solution: $\dfrac{2}{3} = 2 \div 3$ or $3\overline{)2}$ ⟵ equivalent division problems

numerator

To **rename a division problem as an equivalent fraction,** you use

$$a \div b = \frac{a}{b} \quad [b \neq 0]$$

EXAMPLE 1-23: Rename each of the following as an equivalent fraction: (a) $5 \div 2$ (b) $2\overline{)5}$

Solution: (a) $5 \div 2 = \dfrac{5 \leftarrow \text{dividend}}{2 \leftarrow \text{divisor}}$

(b) $2\overline{)5} = 5 \div 2 = \dfrac{5 \leftarrow \text{dividend}}{2 \leftarrow \text{divisor}}$

B. Rename a fraction with a denominator of 1 as an equal whole number and vice versa.

To **rename a fraction with a denominator of 1 as an equal whole number,** you use

$$\frac{a}{1} = a \div 1 = a$$

EXAMPLE 1-24: Rename each of the following as an equal whole number: (a) $\frac{5}{1}$ (b) $\frac{0}{1}$ (c) $\frac{1}{1}$

Solution: (a) $\frac{5}{1} = 5 \div 1 = 5$ or $\frac{5}{1} = 5$ (b) $\frac{0}{1} = 0 \div 1 = 0$ or $\frac{0}{1} = 0$
(c) $\frac{1}{1} = 1 \div 1 = 1$ or $\frac{1}{1} = 1$

Note: A fraction with a denominator of 1 is always equal to the whole number named in the numerator.

To **rename a whole number as an equal fraction with a denominator of 1,** you use

$$a = a \div 1 = \frac{a}{1}$$

EXAMPLE 1-25: Rename each of the following as an equal fraction with a denominator of 1: (a) 3 (b) 0 (c) 1

Solution: (a) $3 = 3 \div 1 = \frac{3}{1}$ or $3 = \frac{3}{1}$ (b) $0 = 0 \div 1 = \frac{0}{1}$ or $0 = \frac{0}{1}$
(c) $1 = 1 \div 1 = \frac{1}{1}$ or $1 = \frac{1}{1}$

Note: To rename a whole number as an equal fraction with a denominator of 1, always use the whole number as the numerator of the fraction.

C. Rename a mixed number as an equal improper fraction and vice versa.

To **rename a mixed number as an equal improper fraction,** you use

$$a\frac{b}{c} = \frac{c \cdot a + b}{c}$$

EXAMPLE 1-26: Rename $2\frac{3}{4}$ as an equal improper fraction:

Solution: $2\dfrac{3}{4} = \dfrac{4 \cdot 2 + 3}{4} = \dfrac{8 + 3}{4} = \dfrac{11}{4}$

To **rename an improper fraction as an equal whole or mixed number,** you divide the numerator by the denominator.

EXAMPLE 1-27: Rename each of the following as an equal mixed number or whole number: (a) $\frac{11}{4}$ (b) $\frac{12}{3}$

Solution: (a) $\frac{11}{4} = 2\frac{3}{4}$ because $\frac{11}{4} = 11 \div 4 = 4\overline{)11}$

(b) $\frac{12}{3} = 4$ because $\frac{12}{3} = 12 \div 3 = 4$

1-5. Comparing Fractions and Mixed Numbers

A. Compare fractions using <, >, or =.

To **compare two fractions** using <, >, or =, you compare their **cross products**.

EXAMPLE 1-28: Compare the following pairs of fractions using <, >, or =: (a) $\frac{3}{4}$ and $\frac{6}{8}$

(b) $\frac{6}{8}$ and $\frac{8}{10}$ (c) $\frac{8}{10}$ and $\frac{9}{12}$

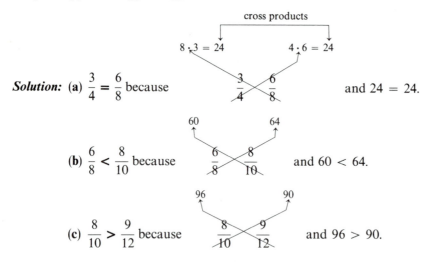

Solution: (a) $\frac{3}{4} = \frac{6}{8}$ because $\qquad$ and $24 = 24$.

(b) $\frac{6}{8} < \frac{8}{10}$ because $\qquad$ and $60 < 64$.

(c) $\frac{8}{10} > \frac{9}{12}$ because $\qquad$ and $96 > 90$.

B. Compare mixed numbers using <, >, or =.

To **compare two mixed numbers** with different whole-number parts using <, >, or =, you compare the different whole-number parts. To compare two mixed numbers with the same whole-number parts, you compare the fraction parts.

EXAMPLE 1-29: Compare the following pairs of mixed numbers using <, >, or =: (a) $2\frac{5}{8}$ and $3\frac{1}{8}$

(b) $1\frac{1}{2}$ and $1\frac{1}{3}$

Solution: (a) $2\frac{5}{8} < 3\frac{1}{8}$ because $2 < 3$.   Compare the different whole-number parts.

(b) $1\frac{1}{2} > 1\frac{1}{3}$ because $\qquad$ and $3 > 2$ Compare the fraction parts when the whole-number parts are the same.

1-6. Simplifying Fractions

A fraction for which the numerator and denominator do not share a **common prime factor** is called a **fraction in lowest terms**.

EXAMPLE 1-30: Determine whether each of the following fractions is in lowest terms: (a) $\frac{3}{4}$ (b) $\frac{4}{6}$

Solution: (a) $\frac{3}{4}$ is a fraction in lowest terms because 3 and 4 do not share a common prime factor:

$$\frac{3}{4} = \frac{3}{2 \cdot 2} \quad \text{no common prime factors}$$

(b) $\frac{4}{6}$ is not a fraction in lowest terms because 4 and 6 share a common prime factor of 2:

$$\frac{4}{6} = \frac{2 \cdot 2}{3 \cdot 2} \qquad \text{common prime factor of 2}$$

To **simplify a fraction** that is not in lowest terms, you first factor the numerator and denominator and then eliminate all the **common factors** using the following rule:

Fundamental Rule for Fractions

If a, b, and c are whole numbers [$b \neq 0$ and $c \neq 0$], then $\dfrac{a \cdot c}{b \cdot c} = \dfrac{a}{b}$.

Note: By the Fundamental Rule for Fractions, the value of a fraction will not change when you divide both the numerator and denominator by the same nonzero number.

EXAMPLE 1-31: Simplify the following fractions: **(a)** $\dfrac{4}{6}$ **(b)** $\dfrac{6}{18}$ **(c)** $\dfrac{36}{12}$ **(d)** $\dfrac{25}{20}$

Solution: **(a)** $\dfrac{4}{6} = \dfrac{\cancel{2} \cdot 2}{\cancel{2} \cdot 3}$ **(b)** $\dfrac{6}{18} = \dfrac{1 \cdot \cancel{6}}{3 \cdot \cancel{6}}$ or $\dfrac{6}{18} = \dfrac{\cancel{2} \cdot \cancel{3}}{\cancel{2} \cdot \cancel{3} \cdot 3}$

$\qquad = \dfrac{2}{3}$ $\qquad = \dfrac{1}{3}$ $\qquad = \dfrac{1}{3}$

(c) $\dfrac{36}{12} = \dfrac{3 \cdot \cancel{12}}{1 \cdot \cancel{12}}$ or $\dfrac{36}{12} = \dfrac{\cancel{2} \cdot \cancel{2} \cdot 3 \cdot \cancel{3}}{\cancel{2} \cdot \cancel{2} \cdot \cancel{3}}$ **(d)** $\dfrac{25}{20} = \dfrac{5 \cdot \cancel{5}}{4 \cdot \cancel{5}}$

$\qquad = \dfrac{3}{1}$ $\qquad = \dfrac{3}{1}$ $\qquad = \dfrac{5}{4}$ or $1\dfrac{1}{4}$

$\qquad = 3$ $\qquad = 3$

1-7. Computing with Fractions

A. Multiply with fractions.

To **multiply fractions,** you multiply the numerators and then multiply the denominators, as stated in the following rule:

Multiplication Rule for Fractions

If $\dfrac{a}{b}$ and $\dfrac{c}{d}$ are fractions, then $\dfrac{a}{b} \cdot \dfrac{c}{d} = \dfrac{a \cdot c}{b \cdot d}$.

Note: If you first eliminate all common factors before multiplying, the product will always be in lowest terms after multiplying.

EXAMPLE 1-32: Multiply **(a)** $\dfrac{5}{6} \times \dfrac{3}{4}$ **(b)** $\dfrac{2}{3} \cdot \dfrac{3}{5} \cdot \dfrac{15}{4}$.

Solution: **(a)** $\dfrac{5}{6} \times \dfrac{3}{4} = \dfrac{5}{2 \times \cancel{3}} \times \dfrac{\cancel{3}}{2 \times 2}$ **(b)** $\dfrac{2}{3} \cdot \dfrac{3}{5} \cdot \dfrac{15}{4} = \dfrac{\cancel{2}}{\cancel{3}} \cdot \dfrac{\cancel{3}}{\cancel{5}} \cdot \dfrac{\cancel{3} \cdot \cancel{5}}{\cancel{2} \cdot 2}$

$\qquad = \dfrac{5}{2} \times \dfrac{1}{4}$ $\qquad = \dfrac{1}{1} \cdot \dfrac{3}{1} \cdot \dfrac{1}{2}$

$\qquad = \dfrac{5 \times 1}{2 \times 4}$ $\qquad = \dfrac{1 \cdot 3 \cdot 1}{1 \cdot 1 \cdot 2}$

$\qquad = \dfrac{5}{8}$ $\qquad = \dfrac{3}{2}$ or $1\dfrac{1}{2}$

To **multiply fractions and whole numbers and/or mixed numbers,** you first rename each whole number and/or mixed number as a fraction.

EXAMPLE 1-33: Multiply (a) $4\left(\dfrac{2}{3}\right)$ (b) $\left(\dfrac{3}{4}\right)2\dfrac{2}{3}$.

Solution: (a) $4\left(\dfrac{2}{3}\right) = \dfrac{4}{1}\left(\dfrac{2}{3}\right)$ (b) $\left(\dfrac{3}{4}\right)2\dfrac{2}{3} = \left(\dfrac{3}{4}\right)\dfrac{8}{3}$

$$= \frac{4 \cdot 2}{1 \cdot 3} \qquad\qquad = \left(\frac{1}{4}\right)\frac{2 \times 4}{1}$$

$$= \frac{8}{3} \text{ or } 2\frac{2}{3} \qquad\qquad = \left(\frac{1}{1}\right)\frac{2}{1}$$

$$= \frac{2}{1}$$

$$= 2$$

B. Divide with fractions.

To divide fractions, you must know how to find the **reciprocal** of a fraction. The reciprocal of a fraction is defined as follows:

If $\dfrac{a}{b}$ is a fraction $[a \neq 0]$, then $\dfrac{b}{a}$ is called the **reciprocal** of $\dfrac{a}{b}$.

EXAMPLE 1-34: What is the reciprocal of $\frac{3}{4}$?

Solution: The reciprocal of $\dfrac{3}{4}$ is $\dfrac{4}{3}$.   Interchange the numerator and denominator.

Note: The product of a fraction and its reciprocal is always 1 $[\frac{3}{4} \times \frac{4}{3} = 1]$.

To **divide fractions,** you change to multiplication and write the reciprocal of the divisor as stated in the following rule:

Division Rule for Fractions

If $\dfrac{a}{b}$ and $\dfrac{c}{d}$ are fractions $[c \neq 0]$, then $\dfrac{a}{b} \div \dfrac{c}{d} = \dfrac{a}{b} \cdot \dfrac{d}{c}$.

EXAMPLE 1-35: Divide the following: (a) $\dfrac{3}{5} \div \dfrac{2}{3}$ (b) $\dfrac{3}{4} \div 8$ (c) $6 \div \dfrac{4}{3}$ (d) $2\dfrac{2}{5} \div 1\dfrac{1}{5}$

Solution: (a) $\dfrac{3}{5} \div \dfrac{2}{3} = \dfrac{3}{5} \cdot \dfrac{3}{2}$ Change to multiplication.

                         Write the reciprocal of the divisor.

$$ = \frac{9}{10} \qquad \text{Multiply fractions.}$$

(b) $\dfrac{3}{4} \div 8 = \dfrac{3}{4} \div \dfrac{8}{1}$ (c) $6 \div \dfrac{4}{3} = \dfrac{6}{1} \div \dfrac{4}{3}$ (d) $2\dfrac{2}{5} \div 1\dfrac{1}{5} = \dfrac{12}{5} \div \dfrac{6}{5}$

$$= \frac{3}{4} \cdot \frac{1}{8} \qquad\quad = \frac{6}{1} \cdot \frac{3}{4} \qquad\quad = \frac{12}{5} \cdot \frac{5}{6}$$

$$= \frac{3}{32} \qquad\quad = \frac{\cancel{2} \cdot 3}{1} \cdot \frac{3}{\cancel{2} \cdot 2} \qquad\quad = \frac{2 \cdot \cancel{6}}{1} \cdot \frac{1}{\cancel{6}}$$

$$\phantom{=\frac{3}{32}} \qquad\quad = \frac{9}{2} \text{ or } 4\frac{1}{2} \qquad\quad = \frac{2}{1} = 2$$

C. Add and subtract like fractions.

Fractions that have the same denominator are called **like fractions.** To **add like fractions** [or **subtract like fractions**] you add [or subtract] the numerators and write the same denominator, as stated in the following rules:

Addition and Subtraction Rules for Like Fractions

If $\dfrac{a}{c}$ and $\dfrac{b}{c}$ are like fractions, then $\dfrac{a}{c} + \dfrac{b}{c} = \dfrac{a+b}{c}$

$$\text{and} \quad \dfrac{a}{c} - \dfrac{b}{c} = \dfrac{a-b}{c}$$

EXAMPLE 1-36: Add the following like fractions: (a) $\dfrac{2}{8} + \dfrac{3}{8}$ (b) $\dfrac{3}{12} + \dfrac{5}{12}$

Solution: (a) $\dfrac{2}{8} + \dfrac{3}{8} = \dfrac{2+3}{8}$ (b) $\dfrac{3}{12} + \dfrac{5}{12} = \dfrac{3+5}{12}$

$$= \dfrac{5}{8} \qquad\qquad\qquad = \dfrac{8}{12}$$

$$= \dfrac{2 \times \cancel{4}}{3 \times \cancel{4}} \qquad \text{Simplify the sum when possible.}$$

$$= \dfrac{2}{3}$$

EXAMPLE 1-37: Subtract the following like fractions: (a) $\dfrac{5}{8} - \dfrac{2}{8}$ (b) $\dfrac{11}{12} - \dfrac{5}{12}$

Solution: (a) $\dfrac{5}{8} - \dfrac{2}{8} = \dfrac{5-2}{8}$ (b) $\dfrac{11}{12} - \dfrac{5}{12} = \dfrac{11-5}{12}$

$$= \dfrac{3}{8} \qquad\qquad\qquad = \dfrac{6}{12}$$

$$= \dfrac{1 \times \cancel{6}}{2 \times \cancel{6}} \qquad \text{Simplify the difference when possible.}$$

$$= \dfrac{1}{2}$$

Note: Remember to simplify the sum or difference whenever possible when adding or subtracting like fractions.

D. Find the least common denominator (LCD).

Fractions that have different denominators are called **unlike fractions.** To **add or subtract unlike fractions,** you must first find the **least common denominator** [**LCD**]. The LCD of two or more fractions is the smallest nonzero whole number that all of the denominators divide into evenly [with a remainder of zero]. To find the LCD of two or more unlike fractions, you can use the following method:

Factoring Method for Finding the LCD

1. Factor each denominator as a product of primes.
2. Identify the greatest number of times that each prime factor occurs in any single factorization from Step 1.
3. Write the LCD as the product of the factors found in Step 2.

Note 1: When all the denominators divide the larger denominator evenly, the LCD is the larger denominator.

Note 2: When no two of the denominators share a common prime factor, the LCD is the product of all the denominators.

EXAMPLE 1-38: Find the LCD for (**a**) $\frac{3}{10}$ and $\frac{7}{12}$ (**b**) $\frac{3}{4}$ and $\frac{5}{8}$ (**c**) $\frac{2}{3}$ and $\frac{3}{4}$ (**d**) $\frac{3}{5}, \frac{5}{6}, \frac{1}{8}$.

Solution: (**a**) The LCD for $\frac{3}{10}$ and $\frac{7}{12}$ is 60 because

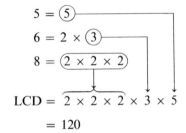

$$10 = 2 \times \textcircled{5}$$

Factor each denominator as a product of primes.

$$12 = \textcircled{2 \times 2} \times \textcircled{3}$$

Circle the greatest number of times that each different prime factor occurs in any **single** factorization.

$$\text{LCD} = 2 \times 2 \times 3 \times 5$$

Write the LCD as a product of the circled factors.

$$= 60$$

(**b**) The LCD for $\frac{3}{4}$ and $\frac{5}{8}$ is 8 because the smaller denominator 4 divides the larger denominator 8 evenly.

(**c**) The LCD for $\frac{2}{3}$ and $\frac{3}{4}$ is 12, the product of the denominators [$3 \times 4 = 12$], because the two denominators do not share a common prime factor.

(**d**) The LCD for $\frac{3}{5}, \frac{5}{6}$, and $\frac{1}{8}$ is 120 because

$$5 = \textcircled{5}$$

Use the factoring method.

$$6 = 2 \times \textcircled{3}$$

$$8 = \textcircled{2 \times 2 \times 2}$$

$$\text{LCD} = 2 \times 2 \times 2 \times 3 \times 5$$

$$= 120$$

Note: In Example 1-38(**d**), the factoring method must be used to find the LCD because neither 5 nor 6 divides 8 evenly and 6 and 8 share a common prime factor of 2.

E. Add and subtract unlike fractions.

To **add or subtract unlike fractions,** you can use the following rules:

Addition and Subtraction Rules for Unlike Fractions

1. Find the LCD of all the unlike fractions.
2. Build up the unlike fractions to get like fractions using the LCD.
3. Add or subtract the like fractions from Step 2.
4. Simplify the sum or difference from Step 3 when possible.

EXAMPLE 1-39: Add the following unlike fractions: (**a**) $\dfrac{1}{2} + \dfrac{1}{3}$ (**b**) $\dfrac{11}{15} + \dfrac{3}{5} + \dfrac{1}{6}$

Solution: (**a**) The LCD for $\dfrac{1}{2}$ and $\dfrac{1}{3}$ is 6.   Find the LCD.

building factor

$$\frac{1}{2} = \frac{1 \cdot 3}{2 \cdot 3} = \frac{3}{6}$$

Build up to get like fractions using the LCD.

$$\frac{1}{3} = \frac{1 \cdot 2}{3 \cdot 2} = \frac{2}{6} \quad \text{LCD}$$

$$\frac{1}{2} + \frac{1}{3} = \frac{3}{6} + \frac{2}{6}$$

Substitute the like fractions for the original unlike fractions.

$$= \frac{3 + 2}{6}$$

Add the like fractions.

$$= \frac{5}{6}$$

(b) The LCD for $\dfrac{11}{15}, \dfrac{3}{5},$ and $\dfrac{1}{6}$ is 30.   Use the factoring method.

$$\dfrac{11}{15} = \dfrac{11 \cdot 2}{15 \cdot 2} = \dfrac{22}{30} \longleftarrow$$ Build up to get like fractions using the LCD.

$$\dfrac{3}{5} = \dfrac{3 \cdot 6}{5 \cdot 6} = \dfrac{18}{30} \longleftarrow \left.\right\} \; \text{LCD}$$

$$\dfrac{1}{6} = \dfrac{1 \cdot 5}{6 \cdot 5} = \dfrac{5}{30} \longleftarrow$$

$$\dfrac{11}{15} + \dfrac{3}{5} + \dfrac{1}{6} = \dfrac{22}{30} + \dfrac{18}{30} + \dfrac{5}{30}$$ Substitute the like fractions for the original unlike fractions.

$$= \dfrac{22 + 18 + 5}{30}$$ Add the like fractions.

$$= \dfrac{45}{30}$$

$$= \dfrac{3 \times \cancel{15}}{2 \times \cancel{15}}$$ Simplify when possible.

$$= \dfrac{3}{2} \text{ or } 1\dfrac{1}{2}$$

EXAMPLE 1-40: Subtract the following: $\dfrac{11}{12} - \dfrac{1}{6}$

Solution: The LCD for $\dfrac{11}{12}$ and $\dfrac{1}{6}$ is 12.   Find the LCD.

$$\dfrac{11}{12} \; \xrightarrow{\text{same}} \; \dfrac{11}{12} \longleftarrow$$

$$\dfrac{1}{6} = \dfrac{1 \cdot 2}{6 \cdot 2} = \dfrac{2}{12} \longleftarrow \left.\right\} \; \text{LCD}$$ Build up to get like fractions using the LCD.

$$\dfrac{11}{12} - \dfrac{1}{6} = \dfrac{11}{12} - \dfrac{2}{12}$$ Substitute the like fractions for the original unlike fractions.

$$= \dfrac{11 - 2}{12}$$ Subtract the unlike fractions.

$$= \dfrac{9}{12}$$

$$= \dfrac{\cancel{3} \times 3}{\cancel{3} \times 4}$$ Simplify when possible.

$$= \dfrac{3}{4}$$

RAISE YOUR GRADES

Can you . . . ?

☑ identify counting numbers [natural numbers] and whole numbers
☑ identify relation symbols
☑ identify operation symbols

☑ identify prime and composite numbers
☑ factor a composite number in as many different ways as possible
☑ factor a composite number as a product of primes
☑ identify fractions and mixed numbers
☑ rename a fraction as an equivalent division problem and vice versa
☑ rename a fraction with a denominator of one as a whole number and vice versa
☑ rename a mixed number as an equal improper fraction and vice versa
☑ compare whole numbers, fractions, and mixed numbers using $<$, $>$, or $=$
☑ simplify a fraction by renaming it in lowest terms
☑ find the least common denominator [LCD] of two or more fractions
☑ add, subtract, multiply, and divide with like fractions
☑ add, subtract, multiply, and divide with unlike fractions

SUMMARY

1. The counting numbers [natural numbers] are 1, 2, 3, $\cdots$.
2. The whole numbers are 0, 1, 2, 3, $\cdots$.
3. To compare two whole numbers that are not equal, you identify which number is greater [larger] and/or which number is less [smaller].
4. The relation symbols used to show how two numbers compare are $<$ [is less than], $>$ [is greater than], $=$ [is equal to], $\neq$ [is not equal to], $\leqslant$ [is less than or equal to], $\geqslant$ [is greater than or equal to], and $\approx$ [is approximately equal to].
5. The symbol $=$ is called an equality symbol and the symbols $<$, $>$, $\neq$, $\leqslant$, $\geqslant$, and $\approx$ are called inequality symbols.
6. Any two numbers can be compared using exactly one and only one of the three basic relation symbols $<$, $>$, or $=$.
7. To write "2 plus 3," you write $2 + 3$.
8. To write "5 minus 3," you write $5 - 3$.
9. To write "6 times 2," you write 6×2, $6 \cdot 2$, $6(2)$, $(6)2$, or $(6)(2)$.
10. To write "9 divided by 2," you write $9 \div 2$, $9/2$, $\frac{9}{2}$, or $2\overline{)9}$.
11. A given nonzero whole number is a whole-number factor of a second whole number if the given nonzero whole number divides the second whole number evenly [with a zero remainder].
12. A whole number that has exactly two different whole-number factors is called a prime number.
13. The two different whole-number factors of a prime number are always the whole number 1 and the prime itself.
14. A whole number that is greater than zero with more than two different whole-number factors is called a composite number.
15. Every whole number is 0, 1, prime, or composite.
16. To factor a composite number, you write it as a product of two or more nonzero whole numbers.
17. **The Fundamental Rule of Arithmetic:** Every composite number can be factored as a product of primes in exactly one way, except for equivalent factorizations.
18. If a and b are whole numbers [$b \neq 0$], then $\frac{a}{b}$ is called a common fraction, or just a fraction.
19. Fractions that have a zero in the denominator are considered not defined, since any number that is divided by zero is not defined.
20. In the fraction $\frac{a}{b}$, a is called the numerator, b is called the denominator, and $-$ is called the fraction bar.
21. When the numerator of a fraction is less than the denominator, the fraction is called a proper fraction.
22. When the numerator of a fraction is greater than or equal to the denominator, the fraction is called an improper fraction.

23. The sum of a whole number and a proper fraction is called a mixed number $\left[a + \dfrac{b}{c} = a\dfrac{b}{c}\right]$.

24. In the mixed number $a\dfrac{b}{c}$, a is called the whole number part and $\dfrac{b}{c}$ is called the fraction part.

25. To rename a fraction as an equivalent division problem, you use $\dfrac{a}{b} = a \div b$.

26. To rename a division problem as an equivalent fraction, you use $a \div b = \dfrac{a}{b}\,[b \neq 0]$.

27. To rename a fraction that has a denominator of 1 as an equal whole number, you use $\dfrac{a}{1} = a$.

28. To rename a whole number as an equal fraction with a denominator of 1, you use $a = \dfrac{a}{1}$.

29. To rename a mixed number as an equal improper fraction, you use $a\dfrac{b}{c} = \dfrac{c \cdot a + b}{c}$.

30. To rename an improper fraction as an equal whole or mixed number, you divide the numerator by the denominator.
31. To compare two fractions using $<$, $>$, or $=$, you compare their cross products.
32. To compare two mixed numbers with different whole-number parts using $<$, $>$, or $=$, you compare the different whole-number parts. To compare two mixed numbers with the same whole-number parts, you compare the fraction parts.
33. A fraction for which the numerator and denominator do not share a common prime factor is called a fraction in lowest terms.
34. To simplify a fraction that is not in lowest terms, you first factor the numerator and denominator and then eliminate all common factors using the **Fundamental Rule for Fractions:** If a, b, and c are whole numbers $[b \neq 0$ and $c \neq 0]$, then $\dfrac{a \cdot c}{b \cdot c} = \dfrac{a}{b}$.

35. To multiply with fractions, you multiply the numerators and then multiply the denominators as stated in the **Multiplication Rule for Fractions:** If $\dfrac{a}{b}$ and $\dfrac{c}{d}$ are fractions, then $\dfrac{a}{b} \cdot \dfrac{c}{d} = \dfrac{a \cdot c}{b \cdot d}$.

36. If you first eliminate all common factors before multiplying, the product will always be in lowest terms after multiplying.
37. If $\dfrac{a}{b}$ is a fraction $[a \neq 0]$, then $\dfrac{b}{a}$ is called the reciprocal of $\dfrac{a}{b}$.
38. To divide with fractions, you change to multiplication and write the reciprocal of the divisor as stated in the **Division Rule for Fractions:** If $\dfrac{a}{b}$ and $\dfrac{c}{d}$ are fractions $[c \neq 0]$, then $\dfrac{a}{b} \div \dfrac{c}{d} = \dfrac{a}{b} \cdot \dfrac{d}{c}$.

39. Fractions that have the same denominator are called like fractions.
40. To add or subtract like fractions, you add or subtract the numerators and write the same denominator, as stated in the **Addition and Subtraction Rules for Like Fractions:**

If $\dfrac{a}{c}$ and $\dfrac{b}{c}$ are like fractions, then $\dfrac{a}{c} + \dfrac{b}{c} = \dfrac{a + b}{c}$

and $\dfrac{a}{c} - \dfrac{b}{c} = \dfrac{a - b}{c}$.

41. Fractions with different denominators are called unlike fractions.
42. To add or subtract unlike fractions, you must first find the least common denominator (LCD).
43. The LCD of two or more fractions is the smallest nonzero whole number that all of the denominators divide into evenly.
44. To find the LCD of two or more fractions, you can use the **Factoring Method for Finding the LCD:**
    (a) Factor each denominator as a product of primes.
    (b) Identify the greatest number of times that each prime factor occurs in any single factorization from Step (a).
    (c) Write the LCD as the product of the factors found in Step (b).

45. When all the denominators divide the larger denominator evenly, the LCD is the larger denominator.

46. When no two of the denominators share a common prime factor, the LCD is the product of all the denominators.

47. To add or subtract unlike fractions, you can use the **Addition and Subtraction Rules for Unlike Fractions:**

(a) Find the LCD.

(b) Build up the unlike fractions to get like fractions using the LCD.

(c) Add or subtract the like fractions.

(d) Simplify the sum or difference when possible.

SOLVED PROBLEMS

PROBLEM 1-1   List the relation symbols used in arithmetic.

Solution: Recall that each relation symbol is either an equality symbol or an inequality symbol [see Example 1-3]:

Equality Symbol	Inequality Symbols	
$=$ [is equal to]	$<$ [is less than] $\leqslant$ [is less than or equal to] $>$ [is greater than]	
	$\neq$ [is not equal to] $\geqslant$ [is greater than or equal to] $\approx$ [is approximately equal to]	

PROBLEM 1-2   Identify each number and operation symbol in (a) $3 + 4 = 7$ (b) $9 - 5 = 4$ (c) $2 \times 3 = 6$ (d) $7 \div 2 = 3\,R1$

Solution: Recall that the four basic arithmetic operations are addition, subtraction, multiplication, and division:

(a) In $3 + 4 = 7$,   3 and 4 are both called addends.

7 is called the sum.

$+$ is called an addition symbol [see Example 1-4].

(b) In $9 - 5 = 4$,   9 is called the minuend.

5 is called the subtrahend.

4 is called the difference.

$-$ is called a subtraction symbol [see Example 1-5].

(c) In $2 \times 3 = 6$,   2 and 3 are both called factors.

2 is also called the multiplicand.

3 is also called the multiplier.

6 is called the product.

$\times$ is called a multiplication symbol [see Example 1-7].

(d) In $7 \div 2 = 3\,R1$, 7 is called the dividend.

2 is called the divisor.

3 is called the quotient.

1 is called the remainder.

3 R1 is called the division answer.

$\div$ is called a division symbol [see Example 1-10].

PROBLEM 1-3   Write "5 times 4" in five equivalent ways.

Solution: 5 times $4 = 5 \times 4 = 5 \cdot 4 = 5(4) = (5)4 = (5)(4)$ [See Example 1-8.]

PROBLEM 1-4   Write "7 divided by 3" in four equivalent ways.

Solution: 7 divided by $3 = 7 \div 3 = 7/3 = \frac{7}{3} = 3\overline{)7}$ [See Example 1-11.]

PROBLEM 1-5 Circle each prime number and cross out each composite number in the following group of numbers:

0	1	2	3	4	5	6	7	8	9	10	11	12	13	14	15	16	17	18	19
20	21	22	23	24	25	26	27	28	29	30	31	32	33	34	35	36	37	38	39
40	41	42	43	44	45	46	47	48	49	50	51	52	53	54	55	56	57	58	59
60	61	62	63	64	65	66	67	68	69	70	71	72	73	74	75	76	77	78	79
80	81	82	83	84	85	86	87	88	89	90	91	92	93	94	95	96	97	98	99	100

Solution: Recall that a whole number with exactly two different whole-number factors is called a prime number, and a whole number greater than zero with more than two different whole number factors is called a composite number [see Examples 1-14 and 1-15]:

0	1	②	③	4	⑤	6	⑦	8	9	10	⑪	12	⑬	14	15	16	⑰	18	⑲
20	21	22	㉓	24	25	26	27	28	㉙	30	㉛	32	33	34	35	36	㊲	38	39
40	㊶	42	㊸	44	45	46	㊿	48	49	50	51	52	㊾	54	55	56	57	58	㊾
60	㊻	62	63	64	65	66	㊿	68	69	70	㊹	72	㊺	74	75	76	77	78	㊾
80	81	82	㊴	84	85	86	87	88	㊾	90	91	92	93	94	95	96	㊾	98	99	100

PROBLEM 1-6 Factor each whole number in as many different ways as possible using two counting numbers: **(a)** 2 **(b)** 6 **(c)** 18 **(d)** 36

Solution: Recall that to factor a number, you write it as the product of two or more nonzero numbers [see Example 1-16]:

(a) $2 = 1 \times 2$

(b) $6 = 1 \times 6$
$6 = 2 \times 3$

(c) $18 = 1 \times 18$
$18 = 2 \times 9$
$18 = 3 \times 6$

(d) $36 = 1 \times 36$
$36 = 2 \times 18$
$36 = 3 \times 12$
$36 = 4 \times 9$
$36 = 6 \times 6$

PROBLEM 1-7 Factor each composite number as a product of primes: **(a)** 10 **(b)** 18 **(c)** 60

Solution: Recall the Fundamental Rule of Arithmetic: Every composite number can be factored as a product of primes in exactly one way, except for equivalent factorizations [see Example 1-17].

(a) $10 \div 2 = 5$
$10 = 2 \times 5$

(b) $18 \div 2 = 9$
$9 \div 3 = 3$
$18 = 2 \times 3 \times 3$

(c) $60 \div 2 = 30$
$30 \div 2 = 15$
$15 \div 3 = 5$
$60 = 2 \times 2 \times 3 \times 5$

PROBLEM 1-8 Rename each fraction as an equivalent division problem: **(a)** $\frac{3}{4}$ **(b)** $\frac{5}{2}$ **(c)** $\frac{4}{9}$

Solution: Recall that to rename a fraction as an equivalent division problem, you use $\frac{a}{b} = a \div b$ or $b\overline{)a}$ [see Example 1-22]:

(a) $\frac{3}{4} = 3 \div 4$ or $4\overline{)3}$ **(b)** $\frac{5}{2} = 5 \div 2$ or $2\overline{)5}$ **(c)** $\frac{4}{9} = 4 \div 9$ or $9\overline{)4}$

PROBLEM 1-9 Rename each division problem as an equivalent fraction: **(a)** $7 \div 8$ **(b)** $2\overline{)5}$ **(c)** 3/2

Solution: Recall that to rename a division problem as an equivalent fraction, you use $a \div b = \dfrac{a}{b}$ $[b \neq 0]$ [see Example 1-23]:

(a) $7 \div 8 = \frac{7}{8}$ (b) $2\overline{)5} = 5 \div 2 = \frac{5}{2}$ (c) $3/2 = 3 \div 2 = \frac{3}{2}$

PROBLEM 1-10   Rename each whole number with a denominator of 1 as an equal whole number:
(a) $\frac{3}{1}$ (b) $\frac{1}{1}$ (c) $\frac{0}{1}$

Solution: Recall that to rename a fraction with a denominator of 1 as an equal whole number, you use $\dfrac{a}{1} = a \div 1 = a$ [see Example 1-24]:

(a) $\frac{3}{1} = 3 \div 1 = 3$ or $\frac{3}{1} = 3$ (b) $\frac{1}{1} = 1 \div 1 = 1$ or $\frac{1}{1} = 1$ (c) $\frac{0}{1} = 0 \div 1 = 0$ or $\frac{0}{1} = 0$

PROBLEM 1-11   Rename each whole number as an equal fraction with a denominator of 1: (a) 5
(b) 1   (c) 0

Solution: Recall that to rename a whole number as an equal fraction with a denominator of 1, you use $a = a \div 1 = \dfrac{a}{1}$ [see Example 1-25]:

(a) $5 = 5 \div 1 = \frac{5}{1}$ or $5 = \frac{5}{1}$ (b) $1 = 1 \div 1 = \frac{1}{1}$ or $1 = \frac{1}{1}$ (c) $0 = 0 \div 1 = \frac{0}{1}$ or $0 = \frac{0}{1}$

PROBLEM 1-12   Rename each mixed number as an equal improper fraction: (a) $1\frac{3}{4}$ (b) $2\frac{1}{2}$
(c) $5\frac{3}{4}$

Solution: Recall that to rename a mixed number as an equal improper fraction, you see $a\dfrac{b}{c} = \dfrac{c \cdot a + b}{c}$
[see Example 1-26]:

(a) $1\dfrac{3}{4} = \dfrac{4 \cdot 1 + 3}{4} = \dfrac{4 + 3}{4} = \dfrac{7}{4}$ (b) $2\dfrac{1}{2} = \dfrac{2 \cdot 2 + 1}{2} = \dfrac{4 + 1}{2} = \dfrac{5}{2}$ (c) $5\dfrac{3}{4} = \dfrac{4 \cdot 5 + 3}{4} = \dfrac{23}{4}$

PROBLEM 1-13   Rename each improper fraction as either an equal whole number or mixed number: (a) $\frac{3}{2}$ (b) $\frac{20}{5}$ (c) $\frac{8}{3}$

Solution: Recall that to rename an improper fraction as either an equal whole number or mixed number, you divide the numerator by the denominator [see Example 1-27]:

(a) $\frac{3}{2} = 1\frac{1}{2}$ because $2\overline{)3}$, with quotient 1 ← remainder, 2 ← divisor, -2, 1

(b) $\frac{20}{5} = 4$ because $5\overline{)20}$, $\dfrac{4}{-20}$, 0

(c) $\frac{8}{3} = 2\frac{2}{3}$ because $3\overline{)8}$, 2 R2, -6, 2

PROBLEM 1-14   Compare fractions using $<$, $>$, or $=$: (a) $\frac{12}{18}$ and $\frac{10}{15}$ (b) $\frac{5}{8}$ and $\frac{12}{16}$
(c) $\frac{18}{20}$ and $\frac{20}{24}$

Solution: Recall that to compare two fractions using $<$, $>$, or $=$, you compare their cross products [see Example 1-28]:

(a) $\frac{12}{18} = \frac{10}{15}$ because $15 \times 12 = 180$ and $18 \times 10 = 180$ and $180 = 180$
(b) $\frac{5}{8} < \frac{12}{16}$ because $16 \times 5 = 80$ and $8 \times 12 = 96$ and $80 < 96$
(c) $\frac{18}{20} > \frac{20}{24}$ because $24 \times 18 = 432$ and $20 \times 20 = 400$ and $432 > 400$

PROBLEM 1-15   Compare mixed numbers using $<$, $>$, or $=$: (a) $5\frac{1}{4}$ and $4\frac{3}{4}$ (b) $2\frac{3}{4}$ and $2\frac{9}{12}$
(c) $1\frac{11}{16}$ and $1\frac{3}{4}$

Solution: Recall that to compare two mixed numbers using $<$, $>$, or $=$ with different whole-number parts, you compare the different whole-number parts. To compare two mixed numbers with the same whole-number parts, you compare the fraction parts [see Example 1-29]:

(a) $5\frac{1}{4} > 4\frac{3}{4}$ because $5 > 4$ (b) $2\frac{3}{4} = 2\frac{9}{12}$ because $\frac{3}{4} = \frac{9}{12}$ (c) $1\frac{11}{16} < 1\frac{3}{4}$ because $\frac{11}{16} < \frac{3}{4}$

PROBLEM 1-16 Simplify each proper fraction: (a) $\frac{2}{4}$ (b) $\frac{4}{6}$ (c) $\frac{3}{12}$ (d) $\frac{5}{15}$ (e) $\frac{6}{8}$ (f) $\frac{4}{20}$ (g) $\frac{3}{24}$ (h) $\frac{4}{10}$

Solution: Recall that to simplify a proper fraction that is not in lowest terms, you reduce the fraction to lowest terms using the Fundamental Rule for Fractions, which states: If a, b, and c are whole numbers [$b \neq 0$ and $c \neq 0$], then $\dfrac{a \cdot c}{b \cdot c} = \dfrac{a}{b}$ [see Example 1-31]:

(a) $\dfrac{2}{4} = \dfrac{1 \times \cancel{2}}{2 \times \cancel{2}} = \dfrac{1}{2}$ (b) $\dfrac{4}{6} = \dfrac{\cancel{2} \times 2}{\cancel{2} \times 3} = \dfrac{2}{3}$ (c) $\dfrac{3}{12} = \dfrac{1 \times \cancel{3}}{\cancel{3} \times 4} = \dfrac{1}{4}$ (d) $\dfrac{5}{15} = \dfrac{1 \times \cancel{5}}{3 \times \cancel{5}} = \dfrac{1}{3}$

(e) $\dfrac{6}{8} = \dfrac{\cancel{2} \times 3}{\cancel{2} \times 4} = \dfrac{3}{4}$ (f) $\dfrac{4}{20} = \dfrac{1 \times \cancel{4}}{\cancel{4} \times 5} = \dfrac{1}{5}$ (g) $\dfrac{3}{24} = \dfrac{1 \times \cancel{3}}{\cancel{3} \times 8} = \dfrac{1}{8}$ (h) $\dfrac{4}{10} = \dfrac{\cancel{2} \times 2}{\cancel{2} \times 5} = \dfrac{2}{5}$

PROBLEM 1-17 Multiply with fractions: (a) $\frac{2}{3} \cdot \frac{1}{5}$ (b) $\frac{2}{3} \times \frac{3}{4} \times \frac{5}{6}$ (c) $\frac{5}{8}(4)$ (d) $(2\frac{1}{4})\frac{5}{6}$

Solution: Recall that to multiply with fractions, you first rename any whole or mixed numbers as equal fractions, then eliminate all common factors, then multiply all the numerators and then multiply all the denominators [see Examples 1-32 and 1-33]:

(a) $\dfrac{2}{3} \cdot \dfrac{1}{5} = \dfrac{2 \cdot 1}{3 \cdot 5} = \dfrac{2}{15}$ (b) $\dfrac{2}{3} \times \dfrac{\cancel{3}}{4} \times \dfrac{5}{6} = \dfrac{\cancel{2}}{1} \times \dfrac{1}{\cancel{2} \times 2} \times \dfrac{5}{6} = \dfrac{1}{1} \times \dfrac{1}{2} \times \dfrac{5}{6} = \dfrac{1 \times 1 \times 5}{1 \times 2 \times 6} = \dfrac{5}{12}$

(c) $\dfrac{5}{8}(4) = \dfrac{5}{8} \cdot \dfrac{4}{1} = \dfrac{5}{2 \cdot \cancel{4}} \cdot \dfrac{\cancel{4}}{1} = \dfrac{5}{2} \cdot \dfrac{1}{1} = \dfrac{5 \cdot 1}{2 \cdot 1} = \dfrac{5}{2}$ or $2\dfrac{1}{2}$

(d) $\left(2\dfrac{1}{4}\right)\dfrac{5}{6} = \dfrac{9}{4} \cdot \dfrac{5}{6} = \dfrac{\cancel{3} \cdot 3}{4} \cdot \dfrac{5}{2 \cdot \cancel{3}} = \dfrac{3}{4} \cdot \dfrac{5}{2} = \dfrac{3 \cdot 5}{4 \cdot 2} = \dfrac{15}{8}$ or $1\dfrac{7}{8}$

PROBLEM 1-18 Divide with fractions: (a) $\frac{3}{4} \div \frac{2}{3}$ (b) $\frac{2}{3} \div \frac{5}{6}$ (c) $\frac{2}{3} \div 4$ (d) $2\frac{1}{2} \div \frac{1}{4}$

Solution: Recall that to divide with fractions, you first rename any whole or mixed numbers as fractions, then change to multiplication and write the reciprocal of the divisor, and then multiply fractions [see Example 1-35]:

(a) $\frac{3}{4} \div \frac{2}{3} = \frac{3}{4} \cdot \frac{3}{2} = \frac{9}{8}$ or $1\frac{1}{8}$ (b) $\frac{2}{3} \div \frac{5}{6} = \frac{2}{3} \cdot \frac{6}{5} = \frac{4}{5}$ (c) $\frac{2}{3} \div 4 = \frac{2}{3} \div \frac{4}{1} = \frac{2}{3} \cdot \frac{1}{4} = \frac{1}{6}$
(d) $2\frac{1}{2} \div \frac{1}{4} = \frac{5}{2} \div \frac{1}{4} = \frac{5}{2} \cdot \frac{4}{1} = 10$

PROBLEM 1-19 Add or subtract like fractions: (a) $\frac{2}{5} + \frac{1}{5}$ (b) $\frac{3}{10} + \frac{1}{10}$ (c) $\frac{4}{5} - \frac{3}{5}$ (d) $\frac{7}{8} - \frac{3}{8}$

Solution: Recall that to add or subtract like fractions, you add or subtract the numerators, write the same denominator, and then simplify the sum or difference when possible [see Examples 1-36 and 1-37]:

(a) $\dfrac{2}{5} + \dfrac{1}{5} = \dfrac{2 + 1}{5} = \dfrac{3}{5}$ (b) $\dfrac{3}{10} + \dfrac{1}{10} = \dfrac{3 + 1}{10} = \dfrac{4}{10} = \dfrac{2 \cdot 2}{2 \cdot 5} = \dfrac{2}{5}$

(c) $\dfrac{4}{5} - \dfrac{3}{5} = \dfrac{4 - 3}{5} = \dfrac{1}{5}$ (d) $\dfrac{7}{8} - \dfrac{3}{8} = \dfrac{7 - 3}{8} = \dfrac{4}{8} = \dfrac{1 \cdot 4}{2 \cdot 4} = \dfrac{1}{2}$

PROBLEM 1-20 Find the LCD for: (a) $\frac{1}{2}$ and $\frac{3}{4}$ (b) $\frac{1}{2}$ and $\frac{2}{3}$ (c) $\frac{3}{4}$ and $\frac{5}{6}$ (d) $\frac{1}{12}$, $\frac{3}{8}$, and $\frac{5}{9}$

Solution: Recall that the least common denominator [LCD] of two or more fractions is the smallest nonzero whole number that all of the denominators divide into evenly [see Example 1-38]:

(a) The LCD of $\frac{1}{2}$ and $\frac{3}{4}$ is 4 because 2 divides the larger denominator 4.

(b) The LCD of $\frac{1}{2}$ and $\frac{2}{3}$ is 6 (2×3) because 2 and 3 do not share a common prime factor.

(c) The LCD of $\frac{3}{4}$ and $\frac{5}{6}$ is 12 because $4 = \boxed{2 \times 2}$ and $6 = 2 \times \boxed{3}$ means: LCD $= 2 \times 2 \times 3 = 12$.

(d) The LCD of $\frac{1}{12}$, $\frac{3}{8}$, and $\frac{5}{9}$ is 72 because $12 = 2 \times 2 \times 3$, $8 = \boxed{2 \times 2 \times 2}$, and $9 = \boxed{3 \times 3}$ means: LCD $= 2 \times 2 \times 2 \times 3 \times 3 = 72$.

PROBLEM 1-21   Add or subtract unlike fractions: **(a)** $\frac{1}{3} + \frac{1}{4}$ **(b)** $\frac{3}{4} + \frac{1}{6} + \frac{5}{8}$ **(c)** $\frac{4}{5} - \frac{3}{10}$

Solution: Recall that to add or subtract unlike fractions, you first find the LCD of all the unlike fractions, then build up the fractions to get like fractions using the LCD, and then add or subtract the like fractions, simplifying the answer when possible [see Examples 1-39 and 1-40]:

(a) The LCD for $\dfrac{1}{3}$ and $\dfrac{1}{4}$ is 12: $\dfrac{1}{3} + \dfrac{1}{4} = \dfrac{1 \cdot 4}{3 \cdot 4} + \dfrac{1 \cdot 3}{4 \cdot 3} = \dfrac{4}{12} + \dfrac{3}{12} = \dfrac{7}{12}$

(b) The LCD for $\dfrac{3}{4}, \dfrac{1}{6}$ and $\dfrac{5}{8}$ is 24: $\dfrac{3}{4} + \dfrac{1}{6} + \dfrac{5}{8} = \dfrac{3 \cdot 6}{4 \cdot 6} + \dfrac{1 \cdot 4}{6 \cdot 4} + \dfrac{5 \cdot 3}{8 \cdot 3} = \dfrac{18}{24} + \dfrac{4}{24} + \dfrac{15}{24} = \dfrac{37}{24}$ or $1\dfrac{13}{24}$

(c) The LCD for $\dfrac{4}{5}$ and $\dfrac{3}{10}$ is 10: $\dfrac{4}{5} - \dfrac{3}{10} = \dfrac{4 \cdot 2}{5 \cdot 2} - \dfrac{3}{10} = \dfrac{8}{10} - \dfrac{3}{10} = \dfrac{5}{10} = \dfrac{1}{2}$

Supplementary Exercises

PROBLEM 1-22   Factor each composite number as a product of primes: **(a)** 4   **(b)** 6   **(c)** 8
(d) 9   **(e)** 14   **(f)** 15   **(g)** 16   **(h)** 20   **(i)** 21   **(j)** 22   **(k)** 24   **(l)** 25   **(m)** 26
(n) 27   **(o)** 28   **(p)** 30   **(q)** 32   **(r)** 33   **(s)** 34   **(t)** 35   **(u)** 36   **(v)** 40
(w) 42   **(x)** 48   **(y)** 56   **(z)** 100

PROBLEM 1-23   Rename each whole or mixed number as an equal fraction: **(a)** 2   **(b)** 5
(c) $3\frac{1}{2}$ **(d)** $1\frac{2}{3}$ **(e)** $2\frac{3}{4}$ **(f)** $4\frac{3}{5}$ **(g)** 6   **(h)** 0   **(i)** $2\frac{1}{3}$ **(j)** $5\frac{1}{4}$ **(k)** $3\frac{1}{5}$ **(l)** $2\frac{7}{10}$
(m) $1\frac{1}{2}$ **(n)** $4\frac{1}{3}$ **(o)** 1   **(p)** 12   **(q)** $2\frac{5}{6}$ **(r)** $1\frac{1}{10}$ **(s)** $3\frac{3}{8}$ **(t)** $4\frac{2}{5}$ **(u)** $1\frac{1}{8}$ **(v)** $2\frac{7}{8}$
(w) 3   **(x)** 100   **(y)** $2\frac{4}{5}$ **(z)** $3\frac{7}{10}$

PROBLEM 1-24   Rename each improper fraction as an equal whole or mixed number: **(a)** $\frac{5}{1}$
(b) $\frac{3}{1}$ **(c)** $\frac{5}{2}$ **(d)** $\frac{5}{4}$ **(e)** $\frac{8}{5}$ **(f)** $\frac{5}{3}$ **(g)** $\frac{11}{6}$ **(h)** $\frac{13}{8}$ **(i)** $\frac{8}{2}$ **(j)** $\frac{10}{5}$ **(k)** $\frac{17}{9}$
(l) $\frac{13}{10}$ **(m)** $\frac{7}{2}$ **(n)** $\frac{20}{3}$ **(o)** $\frac{1}{1}$ **(p)** $\frac{6}{6}$ **(q)** $\frac{11}{4}$ **(r)** $\frac{10}{3}$ **(s)** $\frac{25}{8}$ **(t)** $\frac{24}{5}$ **(u)** $\frac{18}{2}$
(v) $\frac{24}{8}$ **(w)** $\frac{37}{6}$ **(x)** $\frac{51}{8}$ **(y)** $\frac{100}{10}$ **(z)** $\frac{25}{1}$

PROBLEM 1-25   Compare fractions and mixed numbers using $<$, $>$, or $=$: **(a)** $\frac{1}{2}$ and $\frac{1}{3}$
(b) $\frac{1}{4}$ and $\frac{1}{2}$ **(c)** $\frac{4}{8}$ and $\frac{1}{2}$ **(d)** $\frac{1}{3}$ and $\frac{1}{4}$ **(e)** $\frac{2}{3}$ and $\frac{3}{4}$ **(f)** $\frac{3}{4}$ and $\frac{1}{2}$
(g) $1\frac{1}{2}$ and $2\frac{1}{5}$ **(h)** $3\frac{1}{3}$ and $2\frac{7}{8}$ **(i)** $2\frac{4}{5}$ and $2\frac{1}{2}$ **(j)** $3\frac{3}{4}$ and $3\frac{3}{4}$ **(k)** $5\frac{6}{12}$ and $5\frac{1}{2}$
(l) $\frac{3}{8}$ and $\frac{2}{6}$ **(m)** $\frac{5}{6}$ and $\frac{4}{5}$ **(n)** $\frac{5}{8}$ and $\frac{3}{5}$ **(o)** $\frac{2}{5}$ and $\frac{3}{4}$ **(p)** $4\frac{3}{5}$ and $2\frac{7}{8}$ **(q)** $5\frac{1}{2}$ and $4\frac{1}{2}$
(r) $2\frac{3}{4}$ and $2\frac{7}{8}$ **(s)** $1\frac{7}{10}$ and $1\frac{7}{8}$ **(t)** $4\frac{5}{8}$ and $4\frac{5}{9}$ **(u)** $\frac{3}{10}$ and $\frac{9}{30}$ **(v)** $\frac{15}{20}$ and $\frac{3}{4}$
(w) $2\frac{1}{9}$ and $1\frac{8}{9}$ **(x)** $1\frac{3}{4}$ and $1\frac{7}{8}$ **(y)** $\frac{2}{5}$ and $\frac{2}{3}$ **(z)** $\frac{4}{12}$ and $\frac{1}{3}$

PROBLEM 1-26   Simplify: **(a)** $\frac{2}{6}$ **(b)** $\frac{4}{8}$ **(c)** $\frac{9}{12}$ **(d)** $\frac{4}{6}$ **(e)** $\frac{8}{10}$ **(f)** $\frac{3}{12}$ **(g)** $\frac{6}{10}$
(h) $\frac{9}{18}$ **(i)** $\frac{12}{18}$ **(j)** $\frac{20}{25}$ **(k)** $\frac{2}{4}$ **(l)** $\frac{2}{8}$ **(m)** $\frac{2}{10}$ **(n)** $\frac{6}{8}$ **(o)** $\frac{6}{12}$ **(p)** $\frac{4}{10}$ **(q)** $\frac{2}{12}$
(r) $\frac{4}{16}$ **(s)** $\frac{8}{12}$ **(t)** $\frac{12}{16}$ **(u)** $\frac{4}{12}$ **(v)** $\frac{10}{20}$ **(w)** $\frac{10}{12}$ **(x)** $\frac{5}{15}$ **(y)** $\frac{18}{20}$ **(z)** $\frac{20}{24}$

PROBLEM 1-27   Multiply with fractions and mixed numbers: **(a)** $\frac{3}{4} \times \frac{2}{3}$ **(b)** $\frac{1}{2} \cdot \frac{1}{4}$ **(c)** $\frac{2}{3}(2\frac{1}{4})$
(d) $(2\frac{1}{2})\frac{4}{5}$ **(e)** $(1\frac{3}{4})(1\frac{1}{7})$ **(f)** $1\frac{1}{2} \times 2\frac{1}{2}$ **(g)** $\frac{5}{8} \cdot \frac{1}{4}$ **(h)** $\frac{5}{6}(\frac{2}{3})$ **(i)** $(\frac{3}{8})1\frac{1}{3}$ **(j)** $(2\frac{1}{4})(\frac{1}{3})$
(k) $1\frac{1}{2} \times \frac{3}{4} \times 2\frac{1}{3} \times 4$ **(l)** $\frac{5}{8} \cdot 2\frac{1}{2} \cdot \frac{3}{10} \cdot 2$ **(m)** $(1\frac{1}{2})(1\frac{1}{3})(1\frac{1}{4})(1\frac{1}{5})$ **(n)** $\frac{3}{8} \times 1\frac{1}{3} \times \frac{4}{5} \times 1\frac{7}{10} \times 0 \times 5\frac{7}{10}$
(o) $\frac{1}{2} \cdot 1\frac{1}{2} \cdot \frac{3}{4} \cdot 1\frac{1}{3} \cdot \frac{4}{5} \cdot 1\frac{1}{4} \cdot \frac{5}{6} \cdot 1\frac{1}{5}$ **(p)** $\frac{2}{3}(\frac{1}{5})$ **(q)** $(\frac{1}{2})\frac{3}{4}$ **(r)** $(\frac{2}{5})(\frac{5}{9})$ **(s)** $\frac{2}{3} \times \frac{9}{2}$ **(t)** $\frac{3}{8} \cdot \frac{3}{4}$
(u) $\frac{5}{3}(2)$ **(v)** $(\frac{5}{8})8$ **(w)** $\frac{2}{3} \times \frac{3}{4} \times \frac{5}{6}$ **(x)** $\frac{16}{5} \cdot 4 \cdot \frac{1}{10} \cdot \frac{5}{4}$ **(y)** $(1\frac{1}{3})(12)(\frac{1}{2})$ **(z)** $\frac{3}{10}(3\frac{3}{4})2$

PROBLEM 1-28   Divide with fractions and mixed numbers: **(a)** $\frac{3}{4} \div \frac{1}{2}$ **(b)** $\frac{2}{3} \div \frac{1}{2}$ **(c)** $\frac{5}{8} \div \frac{5}{8}$
(d) $\frac{0}{2} \div \frac{7}{10}$ **(e)** $\frac{1}{3} \div 2$ **(f)** $3 \div \frac{2}{5}$ **(g)** $3\frac{1}{2} \div \frac{1}{2}$ **(h)** $\frac{1}{4} \div 2\frac{1}{2}$ **(i)** $1\frac{1}{2} \div 3$ **(j)** $2 \div 3\frac{3}{4}$
(k) $1\frac{1}{5} \div 1\frac{1}{3}$ **(l)** $2\frac{1}{2} \div 2\frac{1}{4}$ **(m)** $\frac{7}{12} \div \frac{5}{12}$ **(n)** $1\frac{3}{4} \div 1\frac{3}{4}$ **(o)** $\frac{5}{18} \div 1\frac{1}{9}$ **(p)** $\frac{3}{4} \div \frac{2}{3}$
(q) $\frac{3}{8} \div \frac{2}{3}$ **(r)** $\frac{2}{3} \div \frac{5}{6}$ **(s)** $\frac{1}{6} \div \frac{2}{5}$ **(t)** $\frac{1}{2} \div \frac{1}{3}$ **(u)** $\frac{1}{3} \div \frac{1}{2}$ **(v)** $\frac{3}{5} \div \frac{3}{5}$ **(w)** $\frac{0}{3} \div \frac{7}{9}$
(x) $\frac{5}{6} \div 3$ **(y)** $2\frac{1}{2} \div \frac{1}{4}$ **(z)** $0 \div 5\frac{3}{4}$

PROBLEM 1-29   Add and subtract with fractions and mixed numbers: **(a)** $\frac{1}{8} + \frac{5}{8}$ **(b)** $\frac{1}{5} + \frac{2}{5}$
(c) $\frac{1}{2} + \frac{5}{6}$ **(d)** $\frac{3}{8} + \frac{1}{2}$ **(e)** $\frac{2}{15} + \frac{3}{5} + \frac{1}{3}$ **(f)** $\frac{7}{12} + \frac{5}{6} + \frac{1}{2}$ **(g)** $\frac{5}{6} + \frac{7}{8} + \frac{9}{10}$ **(h)** $\frac{1}{8} + \frac{3}{16} + \frac{1}{4} + \frac{1}{2}$
(i) $\frac{5}{6} + \frac{1}{2} + \frac{5}{12} + \frac{1}{3}$ **(j)** $\frac{3}{10} + \frac{1}{12} + \frac{2}{15} + \frac{5}{18}$ **(k)** $\frac{11}{20} + \frac{13}{15} + \frac{7}{8} + \frac{7}{9}$ **(l)** $\frac{5}{12} + \frac{11}{12} + \frac{7}{12}$
(m) $\frac{3}{5} + \frac{4}{5} + \frac{2}{5}$ **(n)** $\frac{1}{6} + \frac{5}{6} + \frac{5}{6} + \frac{5}{6}$ **(o)** $\frac{2}{3} - \frac{1}{3}$ **(p)** $\frac{7}{8} - \frac{1}{8}$ **(q)** $\frac{11}{12} - \frac{7}{12}$ **(r)** $\frac{1}{2} - \frac{1}{2}$
(s) $\frac{9}{10} - \frac{3}{10}$ **(t)** $\frac{3}{4} - \frac{1}{2}$ **(u)** $\frac{1}{2} - \frac{1}{6}$ **(v)** $\frac{5}{8} - \frac{1}{4}$ **(w)** $\frac{1}{4} - \frac{1}{6}$ **(x)** $\frac{5}{6} - \frac{5}{8}$ **(y)** $\frac{7}{10} - \frac{1}{6}$
(z) $\frac{5}{12} - \frac{3}{20}$

Answers to Supplementary Exercises

(1-22)   **(a)** $2 \cdot 2$ **(b)** $2 \cdot 3$ **(c)** $2 \cdot 2 \cdot 2$ **(d)** $3 \cdot 3$ **(e)** $2 \cdot 7$ **(f)** $3 \cdot 5$ **(g)** $2 \cdot 2 \cdot 2 \cdot 2$
    (h) $2 \cdot 2 \cdot 5$ **(i)** $3 \cdot 7$ **(j)** $2 \cdot 11$ **(k)** $2 \cdot 2 \cdot 2 \cdot 3$ **(l)** $5 \cdot 5$ **(m)** $2 \cdot 13$
    (n) $3 \cdot 3 \cdot 3$ **(o)** $2 \cdot 2 \cdot 7$ **(p)** $2 \cdot 3 \cdot 5$ **(q)** $2 \cdot 2 \cdot 2 \cdot 2 \cdot 2$ **(r)** $3 \cdot 11$ **(s)** $2 \cdot 17$
    (t) $5 \cdot 7$ **(u)** $2 \cdot 2 \cdot 3 \cdot 3$ **(v)** $2 \cdot 2 \cdot 2 \cdot 5$ **(w)** $2 \cdot 3 \cdot 7$ **(x)** $2 \cdot 2 \cdot 2 \cdot 2 \cdot 3$
    (y) $2 \cdot 2 \cdot 2 \cdot 7$ **(z)** $2 \cdot 2 \cdot 5 \cdot 5$

(1-23)   **(a)** $\frac{2}{1}$ **(b)** $\frac{5}{1}$ **(c)** $\frac{7}{2}$ **(d)** $\frac{5}{3}$ **(e)** $\frac{11}{4}$ **(f)** $\frac{23}{5}$ **(g)** $\frac{6}{1}$ **(h)** $\frac{0}{1}$ **(i)** $\frac{7}{3}$ **(j)** $\frac{21}{4}$
    (k) $\frac{16}{5}$ **(l)** $\frac{27}{10}$ **(m)** $\frac{3}{2}$ **(n)** $\frac{13}{3}$ **(o)** $\frac{1}{1}$ **(p)** $\frac{12}{1}$ **(q)** $\frac{17}{6}$ **(r)** $\frac{11}{10}$ **(s)** $\frac{29}{8}$
    (t) $\frac{22}{5}$ **(u)** $\frac{9}{8}$ **(v)** $\frac{23}{8}$ **(w)** $\frac{3}{1}$ **(x)** $\frac{100}{1}$ **(y)** $\frac{14}{5}$ **(z)** $\frac{37}{10}$

(1-24)   **(a)** 5 **(b)** 3 **(c)** $1\frac{1}{2}$ **(d)** $1\frac{1}{4}$ **(e)** $1\frac{3}{5}$ **(f)** $1\frac{2}{3}$ **(g)** $1\frac{5}{6}$ **(h)** $1\frac{5}{8}$ **(i)** 4
    (j) 2 **(k)** $1\frac{8}{9}$ **(l)** $1\frac{3}{10}$ **(m)** $3\frac{1}{2}$ **(n)** $6\frac{2}{3}$ **(o)** 1 **(p)** 1 **(q)** $2\frac{3}{4}$ **(r)** $3\frac{1}{3}$
    (s) $3\frac{1}{8}$ **(t)** $4\frac{4}{5}$ **(u)** 9 **(v)** 3 **(w)** $6\frac{1}{6}$ **(x)** $6\frac{3}{8}$ **(y)** 10 **(z)** 25

(1-25)   **(a)** $>$ **(b)** $<$ **(c)** $=$ **(d)** $>$ **(e)** $<$ **(f)** $>$ **(g)** $<$ **(h)** $>$ **(i)** $>$
    (j) $<$ **(k)** $=$ **(l)** $>$ **(m)** $>$ **(n)** $>$ **(o)** $<$ **(p)** $>$ **(q)** $>$ **(r)** $<$
    (s) $<$ **(t)** $>$ **(u)** $=$ **(v)** $=$ **(w)** $>$ **(x)** $<$ **(y)** $<$ **(z)** $=$

(1-26)   **(a)** $\frac{1}{3}$ **(b)** $\frac{1}{2}$ **(c)** $\frac{3}{4}$ **(d)** $\frac{2}{3}$ **(e)** $\frac{4}{5}$ **(f)** $\frac{1}{4}$ **(g)** $\frac{3}{5}$ **(h)** $\frac{1}{2}$ **(i)** $\frac{2}{3}$ **(j)** $\frac{4}{5}$
    (k) $\frac{1}{2}$ **(l)** $\frac{1}{4}$ **(m)** $\frac{1}{5}$ **(n)** $\frac{3}{4}$ **(o)** $\frac{1}{2}$ **(p)** $\frac{2}{5}$ **(q)** $\frac{1}{6}$ **(r)** $\frac{1}{4}$ **(s)** $\frac{2}{3}$ **(t)** $\frac{3}{4}$
    (u) $\frac{1}{3}$ **(v)** $\frac{1}{2}$ **(w)** $\frac{5}{6}$ **(x)** $\frac{1}{3}$ **(y)** $\frac{9}{10}$ **(z)** $\frac{5}{6}$

(1-27)   **(a)** $\frac{1}{2}$ **(b)** $\frac{1}{8}$ **(c)** $\frac{3}{2}$ or $1\frac{1}{2}$ **(d)** 2 **(e)** 2 **(f)** $\frac{15}{4}$ or $3\frac{3}{4}$ **(g)** $\frac{5}{32}$ **(h)** $\frac{5}{9}$ **(i)** $\frac{1}{2}$
    (j) $\frac{3}{4}$ **(k)** $\frac{21}{2}$ or $10\frac{1}{2}$ **(l)** $\frac{15}{16}$ **(m)** 3 **(n)** 0 **(o)** $\frac{3}{4}$ **(p)** $\frac{2}{15}$ **(q)** $\frac{3}{8}$ **(r)** $\frac{2}{9}$
    (s) 3 **(t)** 2 **(u)** $\frac{10}{3}$ or $3\frac{1}{3}$ **(v)** 5 **(w)** $\frac{5}{12}$ **(x)** $\frac{8}{5}$ or $1\frac{3}{5}$ **(y)** 8 **(z)** $\frac{9}{4}$ or $2\frac{1}{4}$

(1-28) (a) $\frac{3}{2}$ or $1\frac{1}{2}$ (b) $\frac{4}{3}$ or $1\frac{1}{3}$ (c) 1 (d) 0 (e) $\frac{1}{6}$ (f) $\frac{15}{2}$ or $7\frac{1}{2}$ (g) 7 (h) $\frac{1}{10}$

(i) $\frac{1}{2}$ (j) $\frac{8}{15}$ (k) $\frac{9}{10}$ (l) $\frac{10}{9}$ or $1\frac{1}{9}$ (m) $\frac{7}{5}$ or $1\frac{2}{5}$ (n) 1 (o) $\frac{1}{4}$

(p) $\frac{9}{8}$ or $1\frac{1}{8}$ (q) $\frac{9}{16}$ (r) $\frac{4}{5}$ (s) $\frac{5}{12}$ (t) $\frac{3}{2}$ or $1\frac{1}{2}$ (u) $\frac{2}{3}$ (v) 1 (w) 0

(x) $\frac{5}{18}$ (y) 10 (z) 0

(1-29) (a) $\frac{3}{4}$ (b) $\frac{3}{5}$ (c) $\frac{4}{3}$ or $1\frac{1}{3}$ (d) $\frac{7}{8}$ (e) $\frac{16}{15}$ or $1\frac{1}{15}$ (f) $\frac{23}{12}$ or $1\frac{11}{12}$

(g) $\frac{313}{120}$ or $2\frac{73}{120}$ (h) $\frac{17}{16}$ or $1\frac{1}{16}$ (i) $\frac{25}{12}$ or $2\frac{1}{12}$ (j) $\frac{143}{180}$ (k) $\frac{221}{72}$ or $3\frac{5}{72}$

(l) $\frac{23}{12}$ or $1\frac{11}{12}$ (m) $\frac{9}{5}$ or $1\frac{4}{5}$ (n) $\frac{8}{3}$ or $2\frac{2}{3}$ (o) $\frac{1}{3}$ (p) $\frac{3}{4}$ (q) $\frac{1}{3}$ (r) 0 (s) $\frac{3}{5}$

(t) $\frac{1}{4}$ (u) $\frac{1}{3}$ (v) $\frac{3}{8}$ (w) $\frac{1}{12}$ (x) $\frac{5}{24}$ (y) $\frac{8}{15}$ (z) $\frac{4}{15}$

2 ARITHMETIC REVIEW II: DECIMALS, PERCENTS, SQUARES, AND SQUARE ROOTS

THIS CHAPTER IS ABOUT

☑ **Identifying Decimals**
☑ **Comparing and Rounding Decimals**
☑ **Renaming with Decimals and Whole Numbers**
☑ **Computing with Decimals**
☑ **Renaming with Decimals and Fractions**
☑ **Renaming with Percents**
☑ **Finding Squares and Square Roots**

2-1. Identifying Decimals

A. Identify decimal numbers.

The numbers used to measure things accurately are called **decimal numbers** or just **decimals.** Every decimal has three distinct parts.

EXAMPLE 2-1: Name the three parts of the decimal 3.25.

Solution: In the decimal 3.25: 3 is called the **whole-number part.**
. is called the **decimal point.**
25 is called the **decimal-fraction part.**

B. Identify decimal fractions.

When the whole-number part of a decimal is zero, the decimal is called a **decimal fraction.**

EXAMPLE 2-2: Determine if each of the following is a decimal fraction: **(a)** 3.25 **(b)** 15.0 **(c)** 0.375

Solution: **(a)** 3.25 is not a decimal fraction because the whole-number part [3] is not zero.
(b) 15.0 is not a decimal fraction because the whole-number part [15] is not zero.
(c) 0.375 is a decimal fraction because the whole-number part [0] is zero.

Caution: Always write zero for the whole-number part in a decimal fraction. A decimal fraction like 0.375 should *not* be written as .375.

C. Identify terminating decimals.

A decimal that ends in a given place is called a **terminating decimal.** For example, the decimal 3.25 is a terminating decimal because it ends in the hundredths place. A decimal that does not end in any given place is called a **nonterminating decimal.** For example, 0.222 · · · is a nonterminating decimal because the ellipsis symbol [· · ·] indicates that the 2s continue on forever.

EXAMPLE 2-3: Which one of the following is a nonterminating decimal? **(a)** 0.3   **(b)** 0.33
(c) 0.333   **(d)** 0.333···

Solution: **(a)** 0.3 is a terminating decimal because it ends in the tenths place.
        (b) 0.33 is a terminating decimal because it ends in the hundredths place.
        (c) 0.333 is a terminating decimal because it ends in the thousandths place.
        (d) 0.333··· is a nonterminating decimal because the ellipsis symbol [···] indicates that
        the 3s continue on forever.

D. Identify repeating decimals.

A nonterminating decimal that repeats one or more digits forever (like 0.333···) is called a **repeating decimal.**

EXAMPLE 2-4: Which of the following is a repeating decimal? **(a)** 1.234···   **(b)** 27.2727···

Solution: **(a)** 1.234··· is not a repeating decimal because the counting pattern shown goes on
        forever: 1.2345678910111213···
        (b) 27.2727··· is a repeating decimal because the digits 2 and 7 repeat forever in the same
        way.

Repeating decimals [like 0.666··· and 27.2727···] are usually written with **bar notation** instead of ellipsis notation. To write a repeating decimal using bar notation, you write each digit that repeats in the decimal-fraction part under the bar.

EXAMPLE 2-5: Write the following decimals using bar notation: **(a)** 0.666···   **(b)** 27.2727···
(c) 0.8333···

Solution: **(a)** $0.666··· = 0.\overline{6}$
        (b) $27.2727··· = 27.\overline{27}$ ⟷ bar notation
        (c) $0.8333··· = 0.8\overline{3}$

simplest form

Note: $0.666··· = 0.\overline{6} = 0.6\overline{6} = 0.66\overline{6}$, and so on.

Caution: 27.2727··· cannot be written as $\overline{27}$ ⟵ wrong
        because $\overline{27} = 272727···$
        but $27.\overline{27} = 27.2727···$

Caution: Do not write 0.8333··· as $0.\overline{83}$ ⟵ wrong
        because $0.\overline{83} = 0.838383···$
        but $0.8\overline{3} = 0.8333···$

2-2. Comparing and Rounding Decimals

A. Identify the correct digit in a decimal given the digit's place value.

To help identify the correct **digit** in a decimal given the digit's **place-value,** you can use a **place-value chart.**

EXAMPLE 2-6: In 1.732051, draw a line under the digit that is in the thousandths place.

Solution:

Decimal Place-Value Chart [from millions to millionths]

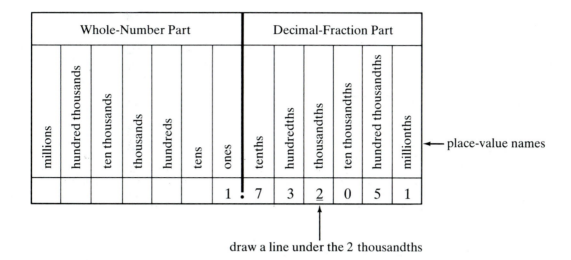

draw a line under the 2 thousandths

B. Compare decimals.

To **compare two decimals** using <, >, or =, you will find it helpful to write one decimal under the other while aligning the decimal points and like values.

EXAMPLE 2-7: Compare the following decimals: 0.09 and 0.1

Solution: 0.09 is less than 0.1 or 0.09 < 0.1 because 0 tenths < 1 tenth.

	Whole-Number Part							Decimal-Fraction Part					
millions	hundred thousands	ten thousands	thousands	hundreds	tens	ones		tenths	hundredths	thousandths	ten thousandths	hundred thousandths	millionths
						0	.	0	9				
						0		1					

same   different [0 tenths < 1 tenth]

Note: The inequality symbols < and > should *always* point at the smaller number.

C. Round decimals.

When an exact answer is not needed for a problem, you may want to make the computation easier by **rounding decimals** before computing. The place to which you round decimals will depend on how accurate your answer should be.

If the digit to the right of the digit to be rounded is 5 or more, then you **round up** as follows:

1. Increase the digit to be rounded by 1.
2. Replace the digits to the right of the rounded digit with zeros in the whole-number part.
3. Omit the digits to the right of the rounded digit in the decimal-fraction part.

If the digit to the right of the digit to be rounded is less than 5, then you **round down** as follows:

1. Leave the digit to be rounded the same.
2. Replace the digits to the right of the rounded digit with zeros in the whole-number part.
3. Omit the digits to the right of the rounded digit in the decimal fraction part.

EXAMPLE 2-8: Round 3.14159 to the nearest thousandth.

thousandths

Solution: $3.14159 = 3.14\underline{1}59$ Draw a line under the digit to be rounded.

$= 3.14\underline{1}59$ Is the digit to the right 5 or more? *Yes:* $5 = 5$

≈ 3.142 Then round up: Increase 1 by 1 to get 2.
Omit 5 and 9.

Note: To the nearest thousandth, $3.14159 \approx 3.142$.

EXAMPLE 2-9: Round 963.25 to the nearest ten.

Solution: $963.25 = 9\underline{6}3.25$ Draw a line under the digit to be rounded.

$= 9\underline{6}3.25$ Is the digit to the right 5 or more? *No:* $3 < 5$

≈ 960 Then round down: Leave the 6 the same.
Replace 3 with a zero.
Omit 2 and 5.

Note: To the nearest ten, $963.25 \approx 960$.

2-3. Renaming with Decimals and Whole Numbers

Any number of zeros can be written on the right-hand end of the decimal-fraction part of a decimal without changing the value of that decimal

$$1.5 = 1.5\underbrace{000 \cdots 0}$$

any number of zeros

A. Rename a whole number as an equal decimal.

To **rename a given whole number as an equal decimal,** you use the given whole number for the whole-number part of the decimal and write one or more zeros for the decimal-fraction part.

EXAMPLE 2-10: Rename 8 as an equal decimal.

whole number
decimal

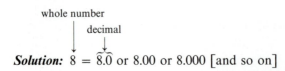

Solution: $8 = 8.0$ or 8.00 or 8.000 [and so on]

B. Rename certain decimals as equal whole numbers.

To **rename a decimal as an equal whole number when the decimal-fraction part is zero,** you just write the whole-number part of the decimal.

EXAMPLE 2-11: Rename 5.0 as an equal whole number.

decimal   whole-number part

Solution: $\overset{\downarrow}{5.\overset{\frown}{0}}$ = $\overset{\downarrow}{5}$

C. Rename a decimal as an equal decimal with a given number of decimal places.

Each digit in the decimal-fraction part of a decimal counts as 1 **decimal place.** The number of decimal places in a given decimal is the number of digits that are in the decimal-fraction part.

EXAMPLE 2-12: Find the number of decimal places in **(a)** 0.70   **(b)** 21.5   **(c)** 0.001   **(d)** 5

Solution: **(a)** 0.70 has 2 decimal places because there are 2 digits in the decimal-fraction part.
(b) 21.5 has 1 decimal place because there is 1 digit in the decimal-fraction part.
(c) 0.001 has 3 decimal places because there are 3 digits in the decimal-fraction part.
(d) 5 has 0 decimal places because 5 has no decimal-fraction part.

Note: Every whole number has 0 decimal places.

To **rename a decimal as an equal decimal containing more decimal places,** you write the necessary number of zeros on the right-hand end of the decimal-fraction part.

EXAMPLE 2-13: Rename 1.5 and 3 as equal decimals with **(a)** 1 decimal place
(b) 2 decimal places
(c) 3 decimal places

Solution: **(a)** 1.5 = 1.5   ⟵ 1 decimal place   3 = 3.0   ⟵ 1 decimal place
(b) 1.5 = 1.50   ⟵ 2 decimal places   3 = 3.00   ⟵ 2 decimal places
(c) 1.5 = 1.500   ⟵ 3 decimal places   3 = 3.000   ⟵ 3 decimal places

2-4. Computing with Decimals

A. Add and subtract with decimals.

To **add with decimals** or **subtract with decimals,** you can use the following rules:

Addition and Subtraction Rules for Decimals

1. Rename any whole numbers as decimals.
2. Rename, if necessary, so that each decimal has the same number of decimal places.
3. Write vertical addition or subtraction form while aligning the decimal points and like values.
4. Add or subtract as you would for whole numbers.
5. Write the decimal point in the sum or difference directly below the other decimal points.

EXAMPLE 2-14: Add the following decimals: **(a)** 0.3 + 0.4   **(b)** 0.5 + 2.73 + 18.2 + 1.125
(c) 258 + 63.49

Solution: **(a)**
$$\begin{array}{r} 0.3 \\ +0.4 \\ \hline 0.7 \end{array}$$

(b)
$$\begin{array}{r} \overset{1\ \ 1}{0.5\,0\,0} \longleftarrow 0.5 \\ 2.7\,3\,0 \longleftarrow 2.73 \\ 1\,8.2\,0\,0 \longleftarrow 18.2 \\ +\ 1.1\,2\,5 \\ \hline 2\,2.5\,5\,5 \end{array}$$

(c)
$$\begin{array}{r} \overset{1\ \ 1}{2\,5\,8.0\,0} \longleftarrow 258 \\ +\ 6\,3.4\,9 \\ \hline 3\,2\,1.4\,9 \end{array}$$

EXAMPLE 2-15: Subtract the following decimals: **(a)** 0.7 − 0.3   **(b)** 25.375 − 4.8
(c) 45.2 − 9.125   **(d)** 5 − 0.008

Solution:

(a) $\begin{array}{r} 0.7 \\ -0.3 \\ \hline 0.4 \end{array}$

(b) $\begin{array}{r} 2\overset{4}{\cancel{5}}.\overset{1}{3}\,7\,5 \\ -\;\;4.8\,0\,0 \\ \hline 2\,0.5\,7\,5 \end{array}$

(c) $\begin{array}{r} \overset{3}{\cancel{4}}\,{}^{1}5.\overset{1}{2}\,\overset{9}{\cancel{0}}\,{}^{1}0 \\ -\;\;9.1\,2\,5 \\ \hline 3\,6.0\,7\,5 \end{array}$ ⟵ 45.2

(d) $\begin{array}{r} \overset{4}{\cancel{5}}.\overset{9}{\cancel{0}}\,\overset{9}{\cancel{0}}\,{}^{1}0 \\ -0.0\,0\,8 \\ \hline 4.9\,9\,2 \end{array}$ ⟵ 5

B. Multiply with decimals.

To **multiply with decimals,** you can use the following rules:

Multiplication Rules for Decimals

1. Write vertical multiplication form while aligning the right-hand digits.
2. Multiply as you would for whole numbers.
3. Put the same number of decimal places in the product as the total number of decimal places contained in both factors.

EXAMPLE 2-16: Multiply the following decimals: **(a)** 3×0.4 **(b)** $25(3.49)$ **(c)** $(5.14)23.6$ **(d)** $(0.02)(0.03)$

Solution: (a) $\begin{array}{r} 0.4 \\ \times\;\;\; 3 \\ \hline 1.2 \end{array}$ ⟵ 1 decimal place / ⟵ 0 decimal places / ⟵ 1 [1 + 0] decimal places

(b) $\begin{array}{r} \overset{1}{}\overset{2\,4}{3.49} \\ \times\;\;\; 25 \\ \hline 1745 \\ 698 \\ \hline 87.25 \end{array}$ ⟵ 2 decimal places / ⟵ 0 decimal places / ⟵ 2 [2 + 0] decimal places

(c) $\begin{array}{r} \overset{1\,3}{} \\ \overset{1\,2}{23.6} \\ \times\,5.14 \\ \hline 944 \\ 236 \\ 1180 \\ \hline 121.304 \end{array}$ ⟵ 1 decimal place / ⟵ 2 decimal places / ⟵ 3 [1 + 2] decimal places

(d) $\begin{array}{r} 0.03 \\ \times\,0.02 \\ \hline 0.0006 \end{array}$ ⟵ 2 decimal places / ⟵ 2 decimal places / ⟵ 4 [2 + 2] decimal places

write zeros as needed

The whole numbers 10, 100, 1000, $\cdots$ are called **powers of 10.** To **multiply by a power of 10,** you can use the following rule:

Multiply by Powers of 10 Rule

To multiply by a power of 10, you move the decimal point one decimal place to the right for each zero in the power of 10.

EXAMPLE 2-17: Multiply 0.75 by the following powers of 10 **(a)** 10 **(b)** 100 **(c)** 1000.

Solution:
(a) $10 \times 0.75 = 07.5 = 7.5$ One zero in 10 means 1 decimal place to the right.
(b) $100 \times 0.75 = 075. = 75$ Two zeros in 100 means 2 decimal places to the right.
(c) $1000 \times 0.75 = 0750. = 750$ Three zeros in 1000 means 3 decimal places to the right.

C. Divide with decimals.

To **divide with decimals,** you can use the following rules:

Division Rules for Decimals

1. Write division box form: $a \div b = b\,)\overline{a}$
2. If the divisor is a decimal, you must make the divisor a whole number by multiplying both the divisor and dividend by the same power of 10.
3. Divide as you would for whole numbers.
4. Put the same number of decimal places in the quotient as in the dividend from Step 2.

EXAMPLE 2-18: Divide the following decimals: **(a)** $3 \overline{)1.2}$ **(b)** $4.5 \div 0.2$ **(c)** $45/0.15$

(d) $\dfrac{0.0001}{0.002}$

Solution: **(a)**

$$
3 \overline{)1.2} \quad\begin{array}{l}\text{0.4} \leftarrow \text{same number of}\\ \leftarrow \text{decimal places}\end{array}
$$

$$
\begin{array}{r}
0.4\\
3 \overline{)\ 1.2}\\
-1\,2\\
\hline
0
\end{array}
$$

(b) $4.5 \div 0.2 = 0.2 \overline{)4.5} = 2_{\wedge} \overline{)45}_{\wedge}$ ← whole-number divisor

write the decimal point in the answer directly above the **caret** [∧]

$$
\begin{array}{r}
22.5\\
2_{\wedge} \overline{)\ \ 45_{\wedge}0}\\
-4\\
\hline
\emptyset 5\\
-\ 4\\
\hline
10\\
-10\\
\hline
0
\end{array}
$$

(c) $45/0.15 = 0.15 \overline{)45} = 0.15 \overline{)45.00} = 15_{\wedge} \overline{)4500}_{\wedge}$

$\times 100 \quad \times 100$

$$
\begin{array}{r}
300\\
15_{\wedge} \overline{)\ 4500}_{\wedge}\\
-45\\
\hline
\emptyset\emptyset\emptyset
\end{array}
$$

(d) $\dfrac{0.0001}{0.002} = 0.002 \overline{)0.0001} = 2_{\wedge} \overline{)0_{\wedge}1}$

$\times 1000 \quad \times 1000$

write zeros as needed

$$
\begin{array}{r}
0.05\\
2_{\wedge} \overline{)0_{\wedge}10}\\
-10\\
\hline
0
\end{array}
$$

Recall: The whole numbers 10, 100, 1000, $\cdots$ are called powers of 10.

To **divide by a power of 10,** you can use the following rule:

Divide by a Power of 10 Rule
To divide by a power of 10, you move the decimal point one decimal place to the left for each zero in the power of 10.

EXAMPLE 2-19: Divide 3.8 by the following powers of 10: **(a)** 10   **(b)** 100   **(c)** 1000

Solution:
(a) $3.8 \div 10 = .38 = 0.38$ One zero in 10 means 1 decimal place to the left.
(b) $3.8 \div 100 = .038 = 0.038$ Two zeros in 100 means 2 decimal places to the left.
(c) $3.8 \div 1000 = .0038 = 0.0038$ Three zeros in 1000 means 3 decimal places to the left.

2-5. Renaming with Decimals and Fractions

A. Rename a decimal as an equal fraction.
To **rename a decimal as an equal fraction,** you can use the value of the decimal-fraction part to write the correct denominator.

EXAMPLE 2-20: Rename each decimal as an equal fraction in simplest form: **(a)** 0.5   **(b)** 3.25
(c) 0.001

$$
\begin{array}{c}
\text{value} \quad \text{fraction simplest form} \\
\downarrow \qquad \downarrow \qquad \downarrow
\end{array}
$$

Solution: **(a)** $0.5 \quad = \quad 5 \text{ tenths} \quad = \frac{5}{10} = \frac{1}{2}$

(b) $3.25 = 325 \text{ hundredths} = \frac{325}{100} = \frac{13}{4}$ or $3\frac{1}{4}$

(c) $0.001 = 1 \text{ thousandth} = \frac{1}{1000}$

Note: A value of tenths, hundredths, thousandths, $\cdots$ means the denominator of the proper fraction is 10, 100, 1000, $\cdots$, respectively.

B. Rename a fraction as an equal decimal.

To **rename a fraction as an equal decimal,** you divide the numerator by the denominator to get a decimal quotient.

EXAMPLE 2-21: Rename each fraction as an equal decimal: **(a)** $\frac{3}{4}$ **(b)** $\frac{1}{3}$

Solution: **(a)** $\frac{3}{4} = 3 \div 4$ or $4\overline{)3}$:

$$
\begin{array}{r}
0.75 \longleftarrow \tfrac{3}{4} \text{ as an equal decimal} \\
4\overline{)\,3.00} \\
-28 \\
\hline
20 \\
-20 \\
\hline
0 \longleftarrow \text{Stop! (remainder is zero)}
\end{array}
$$

(b) $\frac{1}{3} = 1 \div 3$ or $3\overline{)1}$:

$$
\begin{array}{r}
0.333\cdots = 0.\overline{3} \longleftarrow \tfrac{1}{3} \text{ as an equal decimal} \\
3\overline{)\,1.000} \\
-9 \\
\hline
10 \\
-9 \\
\hline
10 \\
-9 \\
\hline
1
\end{array}
$$

repeating 1s

2-6. Renaming with Percents

A. Rename using the percent symbol %.

To rename part of a whole, you can use a number value, decimal, fraction, or **percent.**

EXAMPLE 2-22: Write part of a whole as a number value, decimal, fraction, and percent.

Solution:

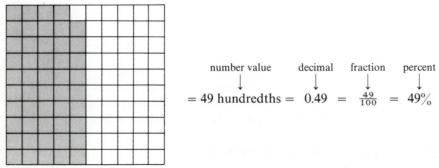

$$
\begin{array}{ccccc}
\text{number value} & & \text{decimal} & \text{fraction} & \text{percent} \\
\downarrow & & \downarrow & \downarrow & \downarrow
\end{array}
$$

$$
= 49 \text{ hundredths} = 0.49 = \frac{49}{100} = 49\%
$$

The **percent symbol** % stands for the word *percent.* Read 49% as "forty-nine percent." Percent means "hundredths," or "out of one hundred," or "per hundred," or "$\times \frac{1}{100}$," or "$\div 100$."

EXAMPLE 2-23: Rename 50% in five equivalent ways.

Solution: 50% = 50 hundredths

= 50 out of one hundred

= 50 per hundred $\left.\right\}$ all equivalent ways of writing 50%

= $50 \times \frac{1}{100}$

= $50 \div 100$

B. Rename a percent as an equal decimal.

To **rename a percent as an equal decimal,** you can use: % means "÷100."

EXAMPLE 2-24: Rename 50% as an equal decimal using: % means "÷100."

Solution: 50% = 50 ÷ 100 ⟵ % means "÷100"

= 0.5   50 ÷ 100 = 0.5

To rename a percent as an equal decimal, you can also use the following shortcut.

Shortcut 2-1: To rename a percent as an equal decimal:
1. Move the decimal point two decimal places to the **left** [divide by 100].
2. Eliminate the percent symbol [%].

EXAMPLE 2-25: Rename 125% as an equal decimal using Shortcut 2-1.

Solution: 125% = 1.25%   Move the decimal point two places to the left.

= 1.25   Eliminate the percent symbol.

Check: Does 125% = 1.25 using % means "÷100"? *Yes:* 125% = 125 ÷ 100 = 1.25

C. Rename a percent as an equal fraction.

To **rename a percent as an equal fraction,** you can use: % means "÷100."

EXAMPLE 2-26: Rename 50% as an equal fraction using: % means "÷100."

Solution: 50% = 50 ÷ 100 ⟵ % means "÷100"

$$= \frac{50}{100} \qquad 50 \div 100 = \frac{50}{100}$$

$$= \frac{1}{2} \qquad \frac{50}{100} = \frac{1 \cdot 50}{2 \cdot 50} = \frac{1}{2}$$

Note: 50% = 0.5 or $\frac{1}{2}$ is correct because $0.5 = \frac{5}{10} = \frac{1}{2}$.

To rename a percent as an equal fraction, you will usually find it easier to use: % means "$\times \frac{1}{100}$."

EXAMPLE 2-27: Rename the following percents as equal fractions using: % means "$\times \frac{1}{100}$:" **(a)** $33\frac{1}{3}\%$ **(b)** 62.5%

Solution: **(a)** $33\frac{1}{3}\% = 33\frac{1}{3} \times \frac{1}{100}$ ⟵ % means "$\times \frac{1}{100}$"

$$= \frac{100}{3} \times \frac{1}{100} \qquad 33\frac{1}{3} = \frac{100}{3}$$

$$= \frac{1}{3} \qquad \frac{\cancel{100}}{3} \times \frac{1}{\cancel{100}} = \frac{1}{3}$$

(b) $62.5\% = 62.5 \times \dfrac{1}{100}$ ←— % means "$\times \dfrac{1}{100}$"

$$= \dfrac{625}{10} \times \dfrac{1}{100} \qquad 62.5 = 625 \text{ tenths} = \dfrac{625}{10}$$

$$= \dfrac{\cancel{5} \times \cancel{5} \times \cancel{5} \times 5}{2 \times \cancel{5}} \times \dfrac{1}{4 \times \cancel{5} \times \cancel{5}}$$

$$= \dfrac{5}{2} \times \dfrac{1}{4}$$

$$= \dfrac{5}{8}$$

Note: Every percent can be renamed as either an equal decimal or as an equal fraction.

D. **Rename a decimal as an equal percent.**

Recall from Shortcut 2-1 that to rename a percent as an equal decimal, you move the decimal point two places to the left and eliminate the percent symbol. Conversely, to **rename a decimal as an equal percent,** you do the opposite of each step in Shortcut 2-1.

Shortcut 2-2: To rename a decimal as an equal percent:

1. Move the decimal point two decimal places to the **right** [multiply by 100].
2. Draw the percent symbol [%].

EXAMPLE 2-28: Rename 0.25 as an equal percent using Shortcut 2-2.

Solution: $0.25 = 025.\%$ Move the decimal point two places to the right.
                               Draw the percent symbol [%].

         $= 25\%$ $025. = 25$

Check: Does $25\% = 0.25$ using: % means "$\div 100$"? *Yes:* $25\% = 25 \div 100 = 0.25$

E. **Rename a fraction as an equal percent.**

To **rename a fraction as an equal percent,** you can first rename the fraction as a decimal and then as a percent.

EXAMPLE 2-29: Rename $\frac{5}{8}$ as an equal percent by first renaming as a decimal.

Solution: $\dfrac{5}{8} = 0.625$ ←— $8 \overline{)\, 5.000}$ with quotient 0.625

        $= 62.5\%$ $0.625 = 62.5\%$

Check: Does $62.5\% = \frac{5}{8}$ using: % means "$\div 100$"?

$$Yes:\ 62.5\% = 62.5 \div 100 = 0.625 = \dfrac{625}{1000} = \dfrac{5 \cdot \cancel{125}}{8 \cdot \cancel{125}} = \dfrac{5}{8}$$

Note: To check $62.5\% = \frac{5}{8}$, you first divide by 100 [$\div 100$] and eliminate the percent symbol [%]: $62.5\% = 62.5 \div 100 = \frac{5}{8}$.

You can also rename a fraction as an equal percent by performing the steps in Shortcut 2-3:

Shortcut 2-3: To rename a fraction as an equal percent:
1. Multiply by 100.
2. Draw the percent symbol [%].

EXAMPLE 2-30: Rename $\frac{2}{3}$ as an equal percent using Shortcut 2-3:

Solution: $\frac{2}{3} = \left(100 \times \frac{2}{3}\right)\%$ Multiply by 100.
Draw the percent symbol [%].

$= \frac{200}{3}\%$ $100 \times \frac{2}{3} = \frac{100}{1} \times \frac{2}{3} = \frac{200}{3}$

$= 66\frac{2}{3}\%$ $3 \overline{\smash{)}200} \; ^{66\frac{2}{3}}$

Note: Every decimal and fraction can be renamed as an equal percent.

2-7. Finding Squares and Square Roots

A. Write exponential notation.

A short way to write the product of repeated factors is **exponential notation.**

To write a product of repeated factors as exponential notation, you first write the repeated factor and then write the number of repeated factors.

EXAMPLE 2-31: Write $2 \cdot 2 \cdot 2 \cdot 2$ as exponential notation.

repeated factor
number of repeated factors

Solution: $2 \cdot 2 \cdot 2 \cdot 2 = 2^4$ } $\longleftarrow$ exponential notation   *Think:* 2 is repeated as a factor 4 times.

Note: In 2^4, 2 is called the **base** and 4 is called the **exponent** or **power.** The exponential notation 2^4 is read as "two to the **fourth power**" or "two **raised** to the fourth power."

B. Evaluate exponential notation.

To **evaluate exponential notation,** you first write the exponential notation as a product of repeated factors and then multiply to get a whole number, fraction, or decimal product.

EXAMPLE 2-32: Evaluate the following exponential notation: **(a)** 3^2 **(b)** 2^3

Solution: **(a)** In 3^2, the base 3 has a power [exponent] of 2.
A power [exponent] of 2 means the base 3 is used as a repeated factor 2 times.

$3^2 = 3 \cdot 3$ Write 2 repeated factors of 3.

$= 9$ Multiply.

(b) In 2^3, the base 2 has a power [exponent] of 3.
A power [exponent] of 3 means the base 2 is used as a repeated factor 3 times.

$2^3 = 2 \cdot 2 \cdot 2$ Write 3 repeated factors of 2.

$= 8$ Multiply.

Note: The exponential notation 3^2 is read as "three to the **second power**" or "three **squared**."
The exponential notation 2^3 is read as "two to the **third power**" or "two **cubed**."

C. Find the square of a given number.

If $a = b^2$, then a is called the **square** of b. To **find the square of a given number,** you multiply the given number times itself.

EXAMPLE 2-33: Find the square of each of the following numbers: **(a)** 3   **(b)** $\frac{3}{4}$ **(c)** 0.1

Solution: **(a)** The square of 3 is 9 because $3^2 = 3 \cdot 3 = 9$.

(b) The square of $\frac{3}{4}$ is $\frac{9}{16}$ because $\left(\frac{3}{4}\right)^2 = \frac{3}{4} \times \frac{3}{4} = \frac{9}{16}$.

(c) The square of 0.1 is 0.01 because $0.1^2 = (0.1)(0.1) = 0.01$.

You should know all the **whole-number squares** up to 100 from memory.

EXAMPLE 2-34: Name all the whole-number squares up to 100.

Solution: 1 is a whole-number square because $1^2 = 1$.
4 is a whole-number square because $2^2 = 4$.
9 is a whole-number square because $3^2 = 9$.
16 is a whole-number square because $4^2 = 16$.
25 is a whole-number square because $5^2 = 25$.
36 is a whole-number square because $6^2 = 36$.
49 is a whole-number square because $7^2 = 49$.
64 is a whole-number square because $8^2 = 64$.
81 is a whole-number square because $9^2 = 81$.
100 is a whole-number square because $10^2 = 100$.

D. Find a square root of a known square.

If $a = b^2$, then b is called a **square root** of a. To **find a square root of a given number,** you find another number that, when squared, will equal the given number.

EXAMPLE 2-35: Find a square root of **(a)** 9   **(b)** $\frac{9}{16}$ **(c)** 0.01.

Solution: **(a)** A square root of 9 is 3 because $9 = 3 \cdot 3 = 3^2$ [see Example 2-33 **(a)**].

(b) A square root of $\frac{9}{16}$ is $\frac{3}{4}$ because $\frac{9}{16} = \frac{3 \times 3}{4 \times 4} = \left(\frac{3}{4}\right)^2$ [see Example 2-33 **(b)**].

(c) A square root of 0.01 is 0.1 because $0.01 = (0.1)(0.1) = 0.1^2$ [see Example 2-33 **(c)**].

"A square root of 9" can be written with symbols as $\sqrt{9}$, where $\sqrt{}$ is called the **square root symbol** or **radical sign**, 9 is called the **square** or **radicand**, and $\sqrt{9}$ is called a **radical**.

E. Find a square or square root using a table.

To find the square of any whole number from 0 to 100, you can use Appendix Table 2 found in the back of this book, as shown in part in the following example.

EXAMPLE 2-36: Find 47^2 using Appendix Table 2.

Solution:

From Appendix Table 2

Number N	Square N^2	Square Root $\sqrt{N}$
35	1225	5.916
36	1296	6
37	1369	6.083
38	1444	6.164
39	1521	6.245
40	1600	6.325
41	1681	6.403
42	1764	6.481
43	1849	6.557
44	1936	6.633
45	2025	6.708
46	2116	6.782
47 →	2209	6.856
48	2304	6.928
49	2401	7

Place your left finger on the base 47 in the "Number N" column.

Place your right finger at the top of the "Square N^2" column.

Move your left finger straight across and your right finger straight down until they meet at $47^2 = 2209$.

Note: $47^2 = 47 \cdot 47 = 2209$

To find a square root of any whole number from 0 to 100, you can use Appendix Table 2 as shown in part in the following example.

EXAMPLE 2-37: Find $\sqrt{47}$ using Appendix Table 2.

Solution:

From Appendix Table 2

Number N	Square N^2	Square Root $\sqrt{N}$
35	1225	5.916
36	1296	6
37	1369	6.083
38	1444	6.164
39	1521	6.245
40	1600	6.325
41	1681	6.403
42	1764	6.481
43	1849	6.557
44	1936	6.633
45	2025	6.708
46	2116	6.782
47 →	2209 →	6.856
48	2304	6.928
49	2401	7

Place your left finger on the radicand 47 in the "Number N" column.

Place your right finger at the top of the "Square Root $\sqrt{N}$" column.

Move your left finger straight across and your right finger straight down until they meet at $\sqrt{47} \approx 6.856$.

Note: $\sqrt{47} \approx 6.856$ because $(6.856)(6.856) = 47.004736 \approx 47$.

F. Find a square or square root using a calculator.

To find the square of any 4-digit whole number or decimal, you can use a standard 8-digit display calculator. To find the square of any 5-digit whole number or decimal, you can use a standard 10-digit display calculator.

EXAMPLE 2-38: Find **(a)** 276^2 **(b)** 915.8^2 using a calculator.

	Press	Display	Interpret
Solution: **(a)**	2 7 6	276	276
	×	276	276 ×
	2 7 6	276	276 × 276
	=	76176	276 × 276 = 76,176
(b)	9 1 5 . 8	915.8	915.8
	×	915.8	915.8 ×
	9 1 5 . 8	915.8	915.8 × 915.8
	=	838689.64	915.8 × 915.8 = 838,689.64

To find or approximate a square root of any 8-digit whole number or decimal, you can use a standard 8-digit calculator. To find or approximate a square root of any 10-digit whole number or decimal, you can use a standard 10-digit calculator.

EXAMPLE 2-39: Approximate **(a)** $\sqrt{785273}$ **(b)** $\sqrt{14{,}906.385}$ using a calculator.

	Press	Display	Interpret
Solution: **(a)**	7 8 5 2 7 3	785273	785,273
	√	886.15630	$\sqrt{785{,}273} \approx 886.15630$
(b)	1 4 9 0 6 . 3 8 5	14906.385	14,906.385
	√	122.09171	$\sqrt{14{,}906.385} \approx 122.09171$

To find or approximate a square root of a fraction using a calculator, you first divide the numerator by the denominator and then press the square root key as shown in Example 2-40.

EXAMPLE 2-40: Approximate $\sqrt{\frac{17}{32}}$ using a calculator.

	Press	Display	Interpret
Solution:	1 7	17	17
	÷	17	17 ÷
	3 2	32	17 ÷ 32
	=	0.53125	17 ÷ 32 = 0.53125
	√	0.7288690	$\sqrt{0.53125} \approx 0.7288690$

Note: $\sqrt{\frac{17}{32}} \approx 0.7$ [nearest tenth], 0.73 [nearest hundredth], 0.729 [nearest thousandth].

RAISE YOUR GRADES

Can you . . . ?

☑ identify decimal numbers, decimal fractions, terminating decimals, and repeating decimals
☑ identify the correct digit in a decimal given the digit's place value
☑ compare decimals using <, >, or =
☑ round decimals
☑ rename a decimal that has only zeros in the decimal-fraction part as an equal whole number and vice versa
☑ rename a decimal as an equal decimal with a given number of decimal places
☑ add, subtract, multiply and divide with decimals
☑ multiply or divide by powers of 10

☑ rename a decimal as an equal fraction and vice versa
☑ rename a percent in five equivalent ways
☑ rename a percent as an equal decimal or fraction and vice versa
☑ write repeated factors as exponential notation
☑ evaluate exponential notation
☑ find the square of a given number using a table or a calculator
☑ find a square root of a known square using a table or a calculator

SUMMARY

1. The numbers used to measure things accurately are called decimal numbers or just decimals.
2. In the decimal 3.25, 3 is called the whole-number part, 25 is called the decimal-fraction part, and . is called the decimal point.
3. A decimal like 0.75 that has zero as its whole-number part is called a decimal fraction.
4. The decimal 3.25 is called a terminating decimal because it ends in a given place—the hundredths place.
5. A decimal that does not end in any given place is called a nonterminating decimal.
6. A nonterminating decimal (like $0.666\cdots$ or $0.\overline{27}$) that repeats one or more digits forever is called a repeating decimal.
7. To help identify the correct digit in a decimal given the digit's place value, you can use a place-value chart.
8. To compare two decimals using $<$, $>$, or $=$, you will find it helpful to write one decimal under the other while aligning the decimal points and like values.
9. If the digit to the right of the digit to be rounded is 5 or more, then you round up as follows:
   (a) Increase the digit to be rounded by 1.
   (b) Replace the digits to the right of the rounded digit with zeros in the whole-number part.
   (c) Omit the digits to the right of the rounded digit in the decimal-fraction part.
10. If the digit to the right of the digit to be rounded is less than 5, then you round down as follows:
   1. Leave the digit to be rounded the same.
   2. Replace the digits to the right of the rounded digit with zeros in the whole-number part.
   3. Omit the digits to the right of the rounded digit in the decimal-fraction part.
11. Any number of zeros can be written on the right-hand end of the decimal-fraction part of a decimal without changing the value of that decimal.
12. To rename a given whole number as an equal decimal, you use the given whole number as the whole-number part of the decimal and write one or more zeros for the decimal-fraction part.
13. When the decimal-fraction part of a decimal is zero, you can rename the decimal as an equal whole number by just writing the whole-number part of the decimal.
14. Each digit in the decimal-fraction part of a decimal counts as 1 decimal place.
15. To rename a decimal as an equal decimal containing more decimal places, you write the necessary number of zeros on the right-hand end of the decimal-fraction part.
16. To add with decimals or subtract with decimals, you can use the following rules:
   Addition and Subtraction Rules for Decimals
   (a) Rename any whole numbers as decimals.
   (b) Rename, if necessary, so that each decimal has the same number of decimal places.
   (c) Write vertical addition or subtraction form while aligning the decimal points and like values.
   (d) Add or subtract as you would for whole numbers.
   (e) Write the decimal point in the sum or difference directly below the other decimal points.
17. To multiply with decimals, you can use the following rules:
   Multiplication Rules for Decimals
   (a) Write vertical multiplication form while aligning the right-hand digits.
   (b) Multiply as you would for whole numbers.
   (c) Put the same number of decimal places in the product as the total number of decimal places contained in both factors.

18. The whole numbers 10, 100, 1000, ⋯ are called powers of 10.
19. **Multiply by Powers of 10 Rule**

    To multiply by a power of 10, you move the decimal point one decimal place to the right for each zero in the power of 10.
20. To divide with decimals, you can use the following rules:

    Division Rules for Decimals

    (a) Write division box form: $a \div b = b \overline{)a}$

    (b) If the divisor is a decimal, you must make the divisor a whole number by multiplying both the divisor and dividend by the same power of 10.

    (c) Divide as you would for whole numbers.

    (d) Put the same number of decimal places in the quotient as in the dividend from Step (b).
21. **Divide by a Power of 10 Rule**

    To divide by a power of 10, you move the decimal point one decimal place to the left for each zero in the power of 10.
22. To rename a decimal as an equal fraction, you can use the value of the decimal-fraction part to write the correct denominator.
23. To rename a fraction as an equal decimal, you divide the numerator by the denominator to get a decimal quotient.
24. The symbol % stands for the word *percent*.
25. Percent means "hundredths," or "out of one hundred," or "per hundred," or "$\times \frac{1}{100}$," or "$\div 100$."
26. To rename a percent as an equal decimal:

    (a) Move the decimal point two places to the **left** [divide by 100].

    (b) Eliminate the percent symbol [X].
27. To rename a percent as an equal fraction, you can use: % means "$\times \frac{1}{100}$."
28. To rename a decimal as an equal percent:

    (a) Move the decimal point two places to the **right** [multiply by 100].

    (b) Write the percent symbol [%].
29. To rename a fraction as an equal percent:

    (a) Multiply by 100.

    (b) Draw the percent symbol [%].
30. In the exponential notation 2^3, 2 is called the base, 3 is called the exponent or power.
31. To write repeated factors like $2 \times 2 \times 2$ as exponential notation, you use the repeated factor (2) as the base and the number of repeated factors (3) as the power [$2 \times 2 \times 2 = 2^3$].
32. To evaluate exponential notation, you first write the exponential notation as a product of repeated factors and then multiply to get a whole number, fraction, or decimal product.
33. If $a = b^2$, then a is called the square of b and b is called a square root of a.
34. To find the square of a given number, you multiply the given number times itself.
35. To find a square root of a given number, you find another number that, when squared, will equal the given number.
36. "A square root of 9" can be written with symbols as $\sqrt{9}$, where $\sqrt{}$ is called the square root symbol or radical sign, 9 is called the square or radicand, and $\sqrt{9}$ is called a radical.
37. To find the square or a square root of any whole number from 0 to 100, you can use Appendix Table 2 found in the back of this book.
38. To find the square or a square root of a whole number, decimal, or fraction, you can usually use a standard 8- or 10-digit calculator.

SOLVED PROBLEMS

PROBLEM 2-1   Identify each decimal as either terminating or repeating: **(a)** 0.3   **(b)** 0.333
(c) $0.\overline{3}$ **(d)** 0.333 ⋯

Solution: Recall that a terminating decimal ends in a given place and a repeating decimal repeats one or more digits forever [see Examples 2.3 and 2.4]:

(a) 0.3 is a terminating decimal because it ends in the tenths place.

(b) 0.333 is a terminating decimal because it ends in the thousandths place.

(c) $0.\overline{3}$ is a repeating decimal because it repeats 3s forever.
(d) $0.333\cdots$ is a repeating decimal because it repeats 3s forever.

PROBLEM 2-2 Rename each repeating decimal using bar notation in simplest form:

(a) $0.333\cdots$ (b) $0.666\cdots$ (c) $33.333\cdots$ (d) $66.666\cdots$ (e) $0.1666\cdots$ (f) $0.8333\cdots$
(g) $0.181818\cdots$ (h) $0.090909\cdots$ (i) $0.142857142857142857142857\cdots$

Solution: Recall that to write a repeating decimal using bar notation, you write each digit that repeats in the decimal-fraction part under the bar [see Example 2.5]:

(a) $0.333\cdots = 0.\overline{3}$ (b) $0.666\cdots = 0.\overline{6}$ (c) $33.333\cdots = 33.\overline{3}$ (d) $66.666\cdots = 66.\overline{6}$
(e) $0.1666\cdots = 0.1\overline{6}$ (f) $0.8333\cdots = 0.8\overline{3}$ (g) $0.181818\cdots = 0.\overline{18}$
(h) $0.090909\cdots = 0.\overline{09}$ (i) $0.142857142857142857\cdots = 0.\overline{142857}$

PROBLEM 2-3 Draw a line under the digit in the (a) tenths place   (b) tens place
(c) hundredths place   (d) thousandths place in the number 5413.067982.

Solution: Recall that to help identify the correct digit in a decimal given the digit's place value, you can use a place-value chart [see Example 2-6]:

(a) 5413.0̲67982   (b) 54̲13.067982   (c) 5413.06̲7982   (d) 5413.067̲982

PROBLEM 2-4 Compare each pair of decimals using $<$, $>$, or $=$: (a) 3.14 and 3.141
(b) 0.5 and 0.49   (c) 1.2780 and 1.278

Solution: Recall that to compare two decimals using $<$, $>$, or $=$, you may find it helpful to write one decimal under the other while aligning the decimal points and like values [see Example 2-7]:

(a) $3.14 < 3.141$ because $3.14 = 3.140$, and 0 thousandths $<$ 1 thousandth
(b) $0.5 > 0.49$ because 5 tenths $>$ 4 tenths   (c) $1.2780 = 1.278$

PROBLEM 2-5 Round 349.615 to the nearest (a) ten   (b) tenth   (c) whole number
(d) hundredth

Solution: Recall that if the digit to the right of the digit to be rounded is 5 or more [less than 5], then you round up [round down] as follows:

1. Increase the digit to be rounded by 1. [Leave the digit to be rounded the same.]
2. Replace the digits to the right of the rounded digit with zeros in the whole-number part.
3. Omit the digits to the right of the rounded digit in the decimal-fraction part.

Recall that if the digit to the right of the digit to be rounded is less than 5, then you round down by leaving the digit to be rounded the same, and continue with Steps 2 and 3 [see Examples 2-8 and 2-9]:

(a) $34\overset{\frown}{9}.615 \approx 350$ (b) $349.\overset{\frown}{6}15 \approx 349.6$ (c) $34\overset{\frown}{9}.615 \approx 350$ $(49 + 1 = 50)$
(d) $349.6\overset{\frown}{1}5 \approx 349.62$

PROBLEM 2-6 Rename each whole number as an equal decimal: (a) 0   (b) 1   (c) 25

Solution: Recall that to rename a given whole number as an equal decimal, you use the given whole number for the whole-number part of the decimal and write one or more zeros for the decimal-fraction part [see Example 2-10]:

(a) $0 = 0.0$ or 0.00 or 0.000 or 0.0000 [and so on]
(b) $1 = 1.0$ or 1.00 or 1.000 or 1.0000 [and so on]
(c) $25 = 25.0$ or 25.00 or 25.000 or 25.0000 [and so on]

PROBLEM 2-7 Rename each decimal as an equal whole number: (a) 2.0   (b) 14.00   (c) 1.000

Solution: Recall that when the decimal-fraction part of a decimal is zero, you can rename the decimal as an equal whole number by just writing the whole-number part of the decimal [see Example 2-11]:

(a) $2.0 = 2$ (b) $14.00 = 14$ (c) $1.000 = 1$

PROBLEM 2-8   Rename both 39 and 2.3 as equal decimals with (a) 1 decimal place
(b) 2 decimal places   (c) 3 decimal places

Solution: Recall that any number of zeros can be written on the right-hand end of the decimal-fraction part of a decimal without changing the value of the decimal. The number of decimal places in a decimal is the number of digits that are in the decimal-fraction part.

(a)	**(b)**	**(c)**

39 = 39.0 [1 decimal place] = 39.00 [2 decimal places] = 39.000 [3 decimal places]
2.3 = 2.3 [1 decimal place] = 2.30 [2 decimal places] = 2.300 [3 decimal places]

PROBLEM 2-9   Add or subtract decimals as indicated: (a) 2.5 + 15 + 0.75   (b) 7.8 − 5
(c) 4 − 3.125

Solution: Recall that to add or subtract with decimals, you can use the following rules:

Addition and Subtraction Rules for Decimals
1. Rename any whole numbers as decimals.
2. Rename, if necessary, so that each decimal has the same number of decimal places.
3. Write vertical addition or subtraction form while aligning the decimal points and like values.
4. Add or subtract as you would for whole numbers.
5. Write the decimal point in the sum or difference directly below the other decimal points [see Examples 2-14 and 2-15]:

(a)
$$\begin{array}{r} 1 \\ 2.50 \longleftarrow 2.5 \\ 15.00 \longleftarrow 15 \\ +\ 0.75 \\ \hline 18.25 \end{array}$$

(b)
$$\begin{array}{r} 7.8 \\ -5.0 \longleftarrow 5 \\ \hline 2.8 \end{array}$$

(c)
$$\begin{array}{r} 3\ 9\ 9\ 1 \\ 4.0\ 0\ 0 \longleftarrow 4 \\ -3.1\ 2\ 5 \\ \hline 0.8\ 7\ 5 \end{array}$$

PROBLEM 2-10   Multiply with decimals: (a) 1.5 × 4   (b) (0.2)(0.3)(0.4)   (c) 3.04(0.25)

Solution: Recall that to multiply with decimals, you can use the following rules:

Multiplication Rules for Decimals
1. Write vertical multiplication form while aligning the right-hand digits.
2. Multiply as you would for whole numbers.
3. Put the same number of decimal places in the product as the total number of decimal places contained in both factors [see Example 2-16]:

(a)
$$\begin{array}{r} 1.5 \longleftarrow \text{1 decimal place} \\ \times\ 4 \longleftarrow \text{0 decimal places} \\ \hline 6.0 \longleftarrow \text{1 [1 + 0] decimal place} \end{array}$$

(b)
$$\begin{array}{r} 0.2 \longleftarrow \text{1 decimal place} \\ \times\ 0.3 \longleftarrow \text{1 decimal place} \\ \hline 0.06 \longleftarrow \text{2 [1 + 1] decimal places} \\ \times\ 0.4 \longleftarrow \text{1 decimal place} \\ \hline 0.024 \longleftarrow \text{3 [2 + 1] decimal places} \end{array}$$

(c)
$$\begin{array}{r} 2 \\ 3.04 \longleftarrow \text{2 decimal places} \\ \times 0.25 \longleftarrow \text{2 decimal places} \\ \hline 1520 \\ 608 \\ \hline 0.7600 \longleftarrow \text{4 [2 + 2] decimal places} \end{array}$$

PROBLEM 2-11   Multiply by powers of 10: (a) 10(3.5)   (b) 3.141 × 100   (c) (1000)0.2

Solution: Recall that to multiply by a power of 10, you move the decimal point one decimal place to the right for each zero in the power of 10 [see Example 2-17]:

(a) 10(3.5) = 35. = 35   (b) 3.141 × 100 = 314.1   (c) (1000)0.2 = 0200. = 200

PROBLEM 2-12   Divide with decimals: (a) 2.18 ÷ 2   (b) 0.75/2.5   (c) $\dfrac{8}{0.4}$

Solution: Recall that to divide with decimals, you can use the following rules:

Division Rules for Decimals

1. Write division box form: $a \div b = b\overline{)a}$
2. If the divisor is a decimal, you must make the divisor a whole number by multiplying both the divisor and dividend by the same power of 10.
3. Divide as you would for whole numbers.
4. Put the same number of decimal places in the quotient as in the dividend from Step 2 [see Example 2-18]:

(a) $2.18 \div 2 = 2\overline{)2.18}$
$$\begin{array}{r} 1.09 \\ 2\,\overline{)\ 2.18} \\ -2 \\ \hline 0\ 18 \\ -\ \ 18 \\ \hline 0 \end{array}$$

(b) $0.75/2.5 = 2.5\overline{)0.75} = 25\wedge\overline{)\ 7\wedge5}$
$$\begin{array}{r} 0.3 \\ 25\wedge\,\overline{)\ 7\wedge5} \\ -7\ 5 \\ \hline 0 \end{array}$$

(c) $\dfrac{8}{0.4} = 0.4\overline{)8} = 4\wedge\overline{)\ 80\wedge}$
$$\begin{array}{r} 20 \\ 4\wedge\,\overline{)\ 80\wedge} \\ -8 \\ \hline 00 \end{array}$$

PROBLEM 2-13 Divide by powers of 10: (a) $32.5 \div 10$ (b) $0.5/100$ (c) $\dfrac{12.5}{1000}$

Solution: Recall that to divide by a power of 10, you move the decimal point one decimal place to the left for each zero in the power of 10 [see Example 2-19]:

(a) $32.5 \div 10 = 3.25$ (b) $0.5/100 = 0.005$ (c) $\dfrac{12.5}{1000} = 0.0125$

PROBLEM 2-14 Rename each decimal as an equal fraction in simplest form:

(a) 0.4 (b) 0.75 (c) 3.125

Solution: Recall that to rename a decimal as an equal fraction, you can use the value of the decimal-fraction part to write the correct denominator [see Example 2-20]:

(a) $0.4 = 4$ tenths $= \frac{4}{10} = \frac{2}{5}$ (b) $0.75 = 75$ hundredths $= \frac{75}{100} = \frac{3}{4}$
(c) $3.125 = 3125$ thousandths $= \frac{3125}{1000} = \frac{25}{8}$ or $3\frac{1}{8}$

PROBLEM 2-15 Rename each fraction as an equal decimal: (a) $\frac{1}{2}$ (b) $\frac{5}{6}$

Solution: Recall that to rename a fraction as an equal decimal, you divide the numerator by the denominator to get a decimal quotient [see Example 2-21]:

(a) $\frac{1}{2} = 1 \div 2$ or $2\overline{)1}$:
$$\begin{array}{r} 0.5 \longleftarrow \tfrac{1}{2}\text{ as an equal decimal} \\ 2\,\overline{)\ 1.0} \\ -1\ 0 \\ \hline 0 \end{array}$$

(b) $\frac{5}{6} = 5 \div 6$ or $6\overline{)5}$:
$$\begin{array}{r} 0.8333\cdots = 0.8\overline{3} \longleftarrow \tfrac{5}{6}\text{ as an equal decimal} \\ 6\,\overline{)\ 5.0000} \\ -4\ 8 \\ \hline 20 \\ -18 \\ \hline 20 \\ -18 \\ \hline 20 \\ -18 \\ \hline 2 \end{array}$$

PROBLEM 2-16 Rename 25% using (a) hundredths   (b) out of one hundred
(c) per hundred   (d) $\times \frac{1}{100}$ (e) $\div 100$

Solution: Recall that percent (%) means "hundredths," or "out of one hundred," or "per hundred," or "$\times \frac{1}{100}$," or "$\div 100$:" [see Example 2-23]:

(a) 25% = 25 hundredths   (b) 25% = 25 out of one hundred   (c) 25% = 25 per hundred
(d) 25% = $25 \times \frac{1}{100}$ (e) 25% = $25 \div 100$

PROBLEM 2-17 Rename each percent as an equal decimal: (a) 23%   (b) 5%   (c) 175%
(d) 0.1%

Solution: Recall that to rename a percent as an equal decimal, you:

1. Move the decimal point two decimal places to the **left** [divide by 100].
2. Eliminate the percent symbol (%) [see Example 2-25]:

(a) 23% = .23% = 0.23   (b) 5% = .05% = 0.05   (c) 175% = 1.75% = 1.75
(d) 0.1% = .001% = 0.001

PROBLEM 2-18 Rename each percent as an equal fraction: (a) 25%   (b) $66\frac{2}{3}$%   (c) 37.5%

Solution: Recall that to rename a percent as an equal fraction, you can use: % means "$\times \frac{1}{100}$" [see Example 2-27]:

(a) $25\% = 25 \times \dfrac{1}{100} = \dfrac{25}{1} \times \dfrac{1}{4 \times 25} = \dfrac{1}{1} \times \dfrac{1}{4} = \dfrac{1}{4}$

(b) $66\frac{2}{3}\% = 66\frac{2}{3} \times \dfrac{1}{100} = \dfrac{200}{3} \times \dfrac{1}{100} = \dfrac{2 \times 100}{3} \times \dfrac{1}{100} = \dfrac{2}{3} \times \dfrac{1}{1} = \dfrac{2}{3}$

(c) $37.5\% = 37.5 \times \dfrac{1}{100} = \dfrac{375}{10} \times \dfrac{1}{100} = \dfrac{3 \times 5 \times 5 \times 5}{2 \times 5} \times \dfrac{1}{2 \times 2 \times 5 \times 5} = \dfrac{3}{2} \times \dfrac{1}{4} = \dfrac{3}{8}$

PROBLEM 2-19 Rename each decimal as an equal percent: (a) 0.13   (b) 0.07   (c) 1.25
(d) 0.005

Solution: Recall that to rename a decimal as an equal percent, you:

1. Move the decimal point two decimal places to the **right** [multiply by 100].
2. Draw the percent symbol (%) [see Example 2-28]:

(a) 0.13 = 013.% = 13%   (b) 0.07 = 007.% = 7%   (c) 1.25 = 125.% = 125%
(d) 0.005 = 000.5% = 0.5%

PROBLEM 2-20 Rename each fraction as an equal percent: (a) $\frac{1}{8}$ (b) $\frac{1}{3}$

Solution: Recall that to rename a fraction as an equal percent, you:

1. Multiply the fraction by 100.
2. Draw the percent symbol (%) [see Example 2-30]:

(a) $\dfrac{1}{8} = \left(100 \times \dfrac{1}{8}\right)\% = \dfrac{100}{1} \times \dfrac{1}{8}\% = \dfrac{4 \times 25}{1} \times \dfrac{1}{2 \times 4}\% = \dfrac{25}{2}\% = 12\frac{1}{2}\%$ or 12.5%

(b) $\dfrac{1}{3} = \left(100 \times \dfrac{1}{3}\right)\% = \dfrac{100}{1} \times \dfrac{1}{3}\% = \dfrac{100}{3}\% = 33\frac{1}{3}\%$

PROBLEM 2-21 Write each product of repeated factors as exponential notation:

(a) $\frac{3}{4} \cdot \frac{3}{4}$ (b) $2 \cdot 2 \cdot 2$ (c) $(0.5)(0.5)(0.5)(0.5)$

Solution: Recall that to write a product of repeated factors as exponential notation, you first write the repeated factor as the base and then write the number of repeated factors as the power (exponent) [see Example 2-31]:

(a) $\frac{3}{4} \cdot \frac{3}{4} = \left(\frac{3}{4}\right)^2$ *Think:* $\frac{3}{4}$ is repeated as a factor 2 times.
(b) $2 \cdot 2 \cdot 2 = 2^3$ *Think:* 2 is repeated as a factor 3 times.
(c) $(0.5)(0.5)(0.5)(0.5) = 0.5^4$ *Think:* 0.5 is repeated as a factor 4 times.

PROBLEM 2-22   Evaluate **(a)** 5^2 **(b)** 2^5 **(c)** $\frac{3^2}{4}$ **(d)** $\left(\frac{3}{4}\right)^2$ **(e)** 1.2^2 **(f)** 0.2^3

Solution: Recall that to evaluate exponential notation, you first write the exponential notation as a product of repeated factors and then multiply to get a whole number, fraction, or decimal product [see Example 2-32]:

(a) $5^2 = 5 \cdot 5 = 25$ **(b)** $2^5 = 2 \cdot 2 \cdot 2 \cdot 2 \cdot 2 = 4 \cdot 2 \cdot 2 \cdot 2 = 8 \cdot 2 \cdot 2 = 16 \cdot 2 = 32$

(c) $\dfrac{3^2}{4} = \dfrac{3 \cdot 3}{4} = \dfrac{9}{4}$ or $2\frac{1}{4}$ **(d)** $\left(\dfrac{3}{4}\right)^2 = \dfrac{3}{4} \cdot \dfrac{3}{4} = \dfrac{3 \cdot 3}{4 \cdot 4} = \dfrac{9}{16}$

(e) $1.2^2 = (1.2)(1.2) = 1.44$ **(f)** $0.2^3 = (0.2)(0.2)(0.2) = (0.04)(0.2) = 0.008$

PROBLEM 2-23   Find the square of each number: **(a)** 53   **(b)** 25.38

Solution: Recall that if $a = b^2$, then a is called the square of b. To find the square of a given number, you multiply the number times itself. You can find a square by using paper and pencil, a calculator, or Appendix Table 2 [see Examples 2-33, 2-34, 2-36, and 2-38]:

(a) $53^2 = 53 \times 53 = 2809$ ⟵ using paper and pencil, a calculator, or Appendix Table 2
(b) $25.38^2 = (25.38)(25.38) = 644.1444$ ⟵ using paper and pencil or a calculator

PROBLEM 2-24   Find a square root of each number: **(a)** 80   **(b)** 0.5   **(c)** $\frac{1}{4}$

Solution: Recall that if $a = b^2$, then b is called a square root of a. To find a square root of a given number, you find another number that, when squared, will equal the given number. You can find a square root by using paper and pencil, a calculator, or Appendix Table 2 [see Examples 2-35, 2-37, 2-39, and 2-40,]:

(a) $\sqrt{80} \approx 8.944$ using Appendix Table 2
(b) $\sqrt{0.5} \approx 0.7071067$ ⟵ using a calculator
(c) $\sqrt{\frac{1}{4}} = \sqrt{0.25} = 0.5$ ⟵ using a calculator or paper and pencil

Supplementary Exercises

PROBLEM 2-25   Draw a line under the digit in each decimal that has the given value:

(a) thousandths: 6582.197034   **(b)** tens: 3075.192864   **(c)** tenths: 4132.709865
(d) hundredths: 9827.651430   **(e)** hundreds: 7043.928165   **(f)** millionths: 5897.264013
(g) ten thousandths: 1087.649235   **(h)** hundred thousandths: 2540.738169
(i) hundredths: 5813.490267   **(j)** millionths: 4871.032965   **(k)** thousandths: 1026.589736
(l) tenths: 6105.478329   **(m)** hundred thousandths: 7580.613294
(n) hundreds: 4259.836107   **(o)** thousands: 2768.105934   **(p)** ten thousandths: 8315.069274

PROBLEM 2-26   Compare decimals using $<$, $>$, or $=$:

(a) 2.35 and 2.53   **(b)** 0.01 and 0.001   **(c)** 3.1 and 3.10   **(d)** 1.01 and 1.011
(e) 34.89 and 34.90   **(f)** 26.78 and 25.79   **(g)** 0.0101 and 0.01   **(h)** 47 and 47.1

(i) 0.9 and 1.0   **(j)** 0.05 and 0.5   **(k)** 1.03 and 0.4   **(l)** 25.0 and 25   **(m)** 49 and 48.9
(n) 0.625 and 0.75   **(o)** 3.14 and 3.141   **(p)** 2.25 and 3.95

PROBLEM 2-27   Round each decimal to the given place: **(a)** 5.8361 to the nearest hundredth
(b) 3.14 to the nearest whole number   **(c)** 0.81539 to the nearest thousandth
(d) 2.7506 to the nearest tenth   **(e)** 0.815649 to the nearest ten thousandth
(f) 0.2185 to the nearest hundredth   **(g)** 1.851436 to the nearest hundred thousandth
(h) 139.0005 to the nearest ten   **(i)** 0.25 to the nearest tenth
(j) 0.138 to the nearest hundredth   **(k)** 1.82539 to the nearest thousandth
(l) 0.238504 to the nearest ten thousandth   **(m)** 0.851246 to the nearest hundred thousandth
(n) 0.834 to the nearest tenth   **(o)** 4582.1675 to the nearest hundred
(p) 3487.139 to the nearest whole number   **(q)** 8213.92 to the nearest thousand
(r) 8056.2 to the nearest ten

PROBLEM 2-28   Rename each whole number as an equal decimal: **(a)** 4   **(b)** 9   **(c)** 0
(d) 7   **(e)** 15   **(f)** 18   **(g)** 135   **(h)** 1   **(i)** 5   **(j)** 26   **(k)** 10   **(l)** 11   **(m)** 12
(n) 25   **(o)** 100   **(p)** 1000

PROBLEM 2-29   Rename each decimal as an equal whole number: **(a)** 5.0   **(b)** 1.0   **(c)** 0.00
(d) 4.00   **(e)** 15.00   **(f)** 7.000   **(g)** 8.0   **(h)** 3.0000   **(i)** 2.0   **(j)** 5.000   **(k)** 1.00
(l) 0.0000   **(m)** 20.0   **(n)** 89.00   **(o)** 125.000   **(p)** 10.0000

PROBLEM 2-30   Rename each number as an equal decimal with the given number of decimal places:

(a) 2 with 1 decimal place   **(b)** 4 with 5 decimal places   **(c)** 8 with 3 decimal places
(d) 1 with 4 decimal places   **(e)** 1.5 with 2 decimal places   **(f)** 0.9 with 3 decimal places
(g) 10.25 with 4 decimal places   **(h)** 6.09 with 5 decimal places   **(i)** 3 with 1 decimal place
(j) 5 with 2 decimal places   **(k)** 2 with 3 decimal places   **(l)** 9 with 4 decimal places
(m) 2.9 with 2 decimal places   **(n)** 0.1 with 3 decimal places
(o) 10.37 with 5 decimal places   **(p)** 5.07 with 4 decimal places

PROBLEM 2-31   Add with decimals:

(a) 1.125 + 3.375   **(b)** 4 + 8.25   **(c)** 5.1 + 16   **(d)** 0.915 + 0.6
(e) 1.4 + 18.35   **(f)** 123 + 0.123   **(g)** 14.58 + 8 + 0.06 + 12.1 + 125
(h) 15.00 + 0.008 + 0.9 + 18 + 1.01 + 5.8953   **(i)** 0.5 + 0.8   **(j)** 8.4 + 7.9
(k) 12.49 + 5.89   **(l)** 16.785 + 0.792   **(m)** 0.85 + 10.6   **(n)** 51.6 + 8.925
(o) 15.49 + 8.79 + 8.49 + 5.99 + 25.69   **(p)** 0.812 + 0.05 + 0.1 + 0.725 + 0.3495 + 0.9
(q) 3 + 0.1   **(r)** 0.5 + 2   **(s)** 8 + 1.2   **(t)** 3.5 + 1   **(u)** 4 + 1.25   **(v)** 7.49 + 12
(w) 5 + 0.25   **(x)** 0.125 + 3   **(y)** 90 + 127.39   **(z)** 25.185 + 689

PROBLEM 2-32   Subtract with decimals:

(a) 8.451 − 0.079   **(b)** 6.85 − 3   **(c)** 5 − 2.84   **(d)** 6.3 − 5.859   **(e)** 7.285 − 6.9
(f) 18.49 − 9.99   **(g)** 8 − 0.1   **(h)** 17.5 − 9   **(i)** 0.9 − 0.1   **(j)** 8.3 − 3.7
(k) 258.79 − 168.85   **(l)** 4.1857 − 2.3462   **(m)** 4.15 − 0.9   **(n)** 18.125 − 5.3
(o) 2.1 − 1.89   **(p)** 35.75 − 0.3652   **(q)** 5.25 − 3   **(r)** 9.5 − 6   **(s)** 28.49 − 15
(t) 629.029 − 587   **(u)** 5 − 0.1   **(v)** 3 − 1.85   **(w)** 23 − 7.49   **(x)** 500 − 8.205

PROBLEM 2-33   Multiply with decimals:

(a) 3 × 0.2   **(b)** 0.1 × 4   **(c)** 6 × 0.8   **(d)** 0.9 × 7   **(e)** 2 × 0.5   **(f)** 0.4 × 5
(g) 3.45 × 3   **(h)** 2 × 0.75   **(i)** 0.3 × 0.6   **(j)** 0.5 × 0.2   **(k)** 3.49 × 0.9
(l) 0.1 × 1.25   **(m)** 5.25 × 0.35   **(n)** 2.15 × 8.03   **(o)** 2.03 × 0.125   **(p)** 5.621 × 2.12
(q) 0.3 × 0.1   **(r)** 0.4 × 0.02   **(s)** 0.06 × 0.01   **(t)** 0.003 × 0.03   **(u)** 10 × 2.5
(v) 100 × 2.5   **(w)** 1000 × 2.5   **(x)** 10,000 × 2.5   **(y)** 100,000 × 0.005
(z) 10,000,000 × 0.000002

PROBLEM 2-34 Divide with decimals:

(a) $1.0 \div 5$ (b) $0.18 \div 3$ (c) $0.03 \div 2$ (d) $0.0504 \div 6$ (e) $1.4 \div 20$ (f) $22.42 \div 3$
(g) $11.4 \div 9$ (h) $77.5 \div 31$ (i) $0.18 \div 0.2$ (j) $0.36 \div 0.3$ (k) $1.08 \div 0.4$
(l) $7.875 \div 2.1$ (m) $0.6603 \div 0.31$ (n) $1.1137 \div 0.43$ (o) $0.045 \div 0.135$
(p) $13.5 \div 2.025$ (q) $0.04 \div 2$ (r) $0.015 \div 3$ (s) $0.0008 \div 8$ (t) $0.0001 \div 5$
(u) $2.8 \div 10$ (v) $2.8 \div 100$ (w) $2.8 \div 1000$ (x) $2.8 \div 10,000$ (y) $5857 \div 100,000$
(z) $258,973 \div 1,000,000$

PROBLEM 2-35 Divide and round the quotient to the given place value:

(a) $5 \div 0.3$; nearest tenth   (b) $0.1 \div 0.6$; nearest tenth
(c) $1.3 \div 0.07$; nearest hundredth   (d) $2 \div 0.09$; nearest hundredth
(e) $3.1 \div 1.2$; nearest thousandth   (f) $0.1 \div 0.15$; nearest thousandth
(g) $4 \div 0.021$; nearest whole number   (h) $35.9 \div 2.4$; nearest whole number
(i) $2.1395 \div 16.8$; nearest tenth   (j) $5.03 \div 0.825$; nearest hundredth

PROBLEM 2-36 Rename each decimal fraction as an equal fraction in simplest form: (a) 0.3
(b) 0.9   (c) 0.2   (d) 0.6   (e) 0.25   (f) 0.05   (g) 0.375   (h) 0.1875   (i) 0.1
(j) 0.7   (k) 0.4   (l) 0.8   (m) 0.75   (n) 0.01   (o) 0.125   (p) 0.025   (q) 0.0625
(r) 0.4375   (s) 2.9   (t) 1.3   (u) 3.75   (v) 6.50   (w) 9.125   (x) 7.375   (y) 5.0625
(z) 8.4000

PROBLEM 2-37 Rename each fraction as an equal decimal: (a) $\frac{1}{2}$ (b) $\frac{1}{4}$ (c) $\frac{1}{5}$ (d) $\frac{4}{5}$
(e) $\frac{1}{8}$ (f) $\frac{5}{8}$ (g) $\frac{1}{10}$ (h) $\frac{7}{10}$ (i) $\frac{1}{16}$ (j) $\frac{5}{32}$ (k) $\frac{1}{3}$ (l) $\frac{1}{6}$ (m) $\frac{1}{9}$ (n) $\frac{5}{9}$ (o) $\frac{1}{11}$
(p) $\frac{7}{11}$ (q) $\frac{1}{12}$ (r) $\frac{5}{12}$ (s) $\frac{1}{7}$ (t) $\frac{3}{7}$ (u) $\frac{29}{10}$ (v) $\frac{43}{10}$ (w) $\frac{1}{100}$ (x) $\frac{23}{100}$ (y) $\frac{859}{100}$
(z) $\frac{1}{1000}$

PROBLEM 2-38 Rename each percent as a decimal: (a) 25%   (b) 50%   (c) 75%   (d) 90%
(e) 10%   (f) 20%   (g) 12.5%   (h) 37.5%   (i) 30%   (j) 40%   (k) $62\frac{1}{2}\%$ (l) 87.5%
(m) 60%   (n) 80%   (o) $6\frac{1}{4}\%$ (p) 18.75%   (q) 100%   (r) 125%   (s) 31.25%
(t) $43\frac{3}{4}\%$ (u) 250%   (v) 1000%   (w) 70%   (x) 90%   (y) 15%   (z) 35%

PROBLEM 2-39 Rename each percent as a fraction: (a) $16\frac{2}{3}\%$ (b) $83\frac{1}{3}\%$ (c) $33\frac{1}{3}\%$
(d) $66\frac{2}{3}\%$ (e) $8\frac{1}{3}\%$ (f) $41\frac{2}{3}\%$ (g) $58\frac{1}{3}\%$ (h) $91\frac{2}{3}\%$ (i) 1%   (j) 25%   (k) 50%
(l) 75%   (m) 10%   (n) 20%   (o) 30%   (p) 40%   (q) 60%   (r) 70%   (s) 80%
(t) 90%   (u) 5%   (v) 4%   (w) 15%   (x) 35%   (y) 100%   (z) 1000%

PROBLEM 2-40 Rename each decimal as a percent: (a) 0.1   (b) 0.2   (c) 0.625   (d) 0.875
(e) 0.5   (f) 0.6   (g) 0.3125   (h) 0.4375   (i) 0.9   (j) 0.25   (k) 0.75   (l) 0.05
(m) 0.125   (n) 0.375   (o) 0.3   (p) 0.4   (q) 0.0625   (r) 0.1875   (s) 0.7   (t) 0.8
(u) 0.5625   (v) 0.6875   (w) 1   (x) 1.5   (y) 2.5   (z)10

PROBLEM 2-41 Rename each fraction as a percent: (a) $\frac{1}{2}$ (b) $\frac{1}{4}$ (c) $\frac{1}{5}$ (d) $\frac{1}{8}$
(e) $\frac{1}{10}$ (f) $\frac{1}{12}$ (g) $\frac{1}{16}$ (h) $\frac{3}{4}$ (i) $\frac{2}{5}$ (j) $\frac{3}{8}$ (k) $\frac{3}{10}$ (l) $\frac{5}{12}$ (m) $\frac{3}{16}$ (n) $\frac{3}{5}$
(o) $\frac{5}{8}$ (p) $\frac{7}{10}$ (q) $\frac{7}{12}$ (r) $\frac{5}{16}$ (s) $\frac{4}{5}$ (t) $\frac{7}{8}$ (u) $\frac{9}{10}$ (v) $\frac{7}{16}$ (w) $\frac{1}{3}$ (x) $\frac{2}{3}$
(y) $\frac{1}{6}$ (z) $\frac{5}{6}$

PROBLEM 2-42 Write each group of repeated factors as exponential notation: (a) 8×8
(b) 2.3×2.3 (c) $\frac{9}{10} \times \frac{9}{10}$ (d) $7 \times 7 \times 7$ (e) $0.1 \times 0.1 \times 0.1$ (f) $\frac{3}{4} \times \frac{3}{4} \times \frac{3}{4}$

(g) $10 \times 10 \times 10 \times 10$ (h) $24.75 \times 24.75 \times 24.75 \times 24.75$ (i) $\frac{2}{3} \times \frac{2}{3} \times \frac{2}{3}$ (j) $\dfrac{6 \times 6}{5}$

(k) $\dfrac{2}{5 \times 5 \times 5}$ (l) $\dfrac{9 \times 9 \times 9}{3 \times 3}$ (m) $5 \times 5 \times 4$ (n) $3 \times 6 \times 6 \times 6 \times 6$ (o) 0×0

(p) $1 \times 1 \times 1$ (q) $2 \times 3 \times 2$ (r) $5 \times 4 \times 4 \times 5 \times 4 \times 5 \times 5$ (s) $\frac{1}{2} \times \frac{1}{2} \times \frac{1}{2}$

(t) $6 \times 6 \times 6 \times 6$ **(u)** $1.5 \times 1.5 \times 1.5 \times 1.5$ **(v)** $\frac{2}{3} \times \frac{2}{3} \times \frac{2}{3} \times \frac{2}{3}$ **(w)** $\dfrac{8 \times 8}{3}$

(x) $\dfrac{2 \times 2}{9 \times 9 \times 9}$ **(y)** $\dfrac{2 \times 2}{3 \times 3 \times 3}$ **(z)** $10 \times 10 \times 10 \times 10 \times 10 \times 10$

PROBLEM 2-43 Evaluate each exponential notation: **(a)** 9^2 **(b)** 0.2^2 **(c)** $(\frac{3}{4})^2$
(d) 3^3 **(e)** 0.2^3 **(f)** $(\frac{1}{2})^3$ **(g)** 10^2 **(h)** 10^4 **(i)** 10^6 **(j)** 10^{11} **(k)** 0^5 **(l)** 1^8
(m) $\frac{3^2}{2}$ **(n)** $(\frac{3}{2})^2$ **(o)** 5^2 **(p)** $(\frac{2}{3})^2$ **(q)** 0^2 **(r)** 25.3^2 **(s)** 10^3 **(t)** $(\frac{4}{5})^2$ **(u)** 0^3
(v) $\frac{18}{2^3}$ **(w)** $\frac{3^2}{2^4}$ **(x)** $\frac{5^2}{4}$ **(y)** $(\frac{5}{4})^2$ **(z)** $\frac{5}{4^2}$

PROBLEM 2-44 Find each square using paper and pencil, Appendix Table 2, or a calculator:

(a) 0^2 **(b)** 1^2 **(c)** 2^2 **(d)** 3^2 **(e)** 4^2 **(f)** 5^2 **(g)** 6^2 **(h)** 7^2 **(i)** 8^2 **(j)** 9^2
(k) 10^2 **(l)** 11^2 **(m)** 12^2 **(n)** 13^2 **(o)** 14^2 **(p)** 15^2 **(q)** 225^2 **(r)** $(\frac{1}{3})^2$
(s) $(\frac{1}{4})^2$ **(t)** $(\frac{2}{5})^2$ **(u)** $(\frac{9}{10})^2$ **(v)** $(\frac{9}{100})^2$ **(w)** 0.1^2 **(x)** 2.5^2 **(y)** 0.25^2 **(z)** 75.85^2

PROBLEM 2-45 Find or approximate each square root. Round to the nearest thousandth when necessary: **(a)** $\sqrt{0}$ **(b)** $\sqrt{1}$ **(c)** $\sqrt{2}$ **(d)** $\sqrt{3}$ **(e)** $\sqrt{4}$ **(f)** $\sqrt{5}$ **(g)** $\sqrt{6}$ **(h)** $\sqrt{7}$
(i) $\sqrt{8}$ **(j)** $\sqrt{9}$ **(k)** $\sqrt{10}$ **(l)** $\sqrt{11}$ **(m)** $\sqrt{12}$ **(n)** $\sqrt{13}$ **(o)** $\sqrt{14}$ **(p)** $\sqrt{15}$
(q) $\sqrt{225}$ **(r)** $\sqrt{\frac{1}{3}}$ **(s)** $\sqrt{\frac{1}{4}}$ **(t)** $\sqrt{\frac{2}{5}}$ **(u)** $\sqrt{\frac{9}{10}}$ **(v)** $\sqrt{\frac{9}{100}}$ **(w)** $\sqrt{0.1}$ **(x)** $\sqrt{2.5}$
(y) $\sqrt{0.25}$ **(z)** $\sqrt{75.85}$

Answers to Supplementary Exercises

(2-25)   **(a)** 6582.19$\underline{7}$034   **(b)** 30$\underline{7}$5.192864   **(c)** 4132.$\underline{7}$09865   **(d)** 9827.6$\underline{5}$1430
     (e) $\underline{7}$043.928165   **(f)** 5897.26401$\underline{3}$ **(g)** 1087.649$\underline{2}$35   **(h)** 2540.738$\underline{1}$69
     (i) 5813.4$\underline{9}$0267   **(j)** 4871.03296$\underline{5}$ **(k)** 1026.589$\underline{7}$36   **(l)** 6105.4$\underline{7}$8329
     (m) 7580.6132$\underline{9}$4   **(n)** 4$\underline{2}$59.836107   **(o)** $\underline{2}$768.105934   **(p)** 8315.069$\underline{2}$74

(2-26)   **(a)** $2.35 < 2.53$ **(b)** $0.01 > 0.001$ **(c)** $3.1 = 3.10$ **(d)** $1.01 < 1.011$
     (e) $34.89 < 34.90$ **(f)** $26.78 > 25.79$ **(g)** $0.0101 > 0.01$ **(h)** $47 < 47.1$
     (i) $0.9 < 1.0$ **(j)** $0.05 < 0.5$ **(k)** $1.03 > 0.4$ **(l)** $25.0 = 25$
     (m) $49 > 48.9$ **(n)** $0.625 < 0.75$ **(o)** $3.14 < 3.141$ **(p)** $2.25 < 3.95$

(2-27)   **(a)** 5.84   **(b)** 3   **(c)** 0.815   **(d)** 2.8   **(e)** 0.8156   **(f)** 0.22   **(g)** 1.85144
     (h) 140   **(i)** 0.3   **(j)** 0.14   **(k)** 1.825   **(l)** 0.2385   **(m)** 0.85125   **(n)** 0.8
     (o) 4600   **(p)** 3487   **(q)** 8000   **(r)** 8060

(2-28)   **(a)** 4.0   **(b)** 9.0   **(c)** 0.0   **(d)** 7.0   **(e)** 15.0   **(f)** 18.0   **(g)** 135.0   **(h)** 1.0
     (i) 5.0   **(j)** 26.0   **(k)** 10.0   **(l)** 11.0   **(m)** 12.0   **(n)** 25.0   **(o)** 100.0
     (p) 1000.0

(2-29)   **(a)** 5   **(b)** 1   **(c)** 0   **(d)** 4   **(e)** 15   **(f)** 7   **(g)** 8   **(h)** 3   **(i)** 2   **(j)** 5
     (k) 1   **(l)** 0   **(m)** 20   **(n)** 89   **(o)** 125   **(p)** 10

(2-30)   **(a)** 2.0   **(b)** 4.00000   **(c)** 8.000   **(d)** 1.0000   **(e)** 1.50   **(f)** 0.900   **(g)** 10.2500
     (h) 6.09000   **(i)** 3.0   **(j)** 5.00   **(k)** 2.000   **(l)** 9.0000   **(m)** 2.90   **(n)** 0.100
     (o) 10.37000   **(p)** 5.0700

(2-31)   **(a)** 4.5   **(b)** 12.25   **(c)** 21.1   **(d)** 1.515   **(e)** 19.75   **(f)** 123.123   **(g)** 159.74
     (h) 40.8133   **(i)** 1.3   **(j)** 16.3   **(k)** 18.38   **(l)** 17.577   **(m)** 11.45   **(n)** 60.525
     (o) 64.45   **(p)** 2.9365   **(q)** 3.1   **(r)** 2.5   **(s)** 9.2   **(t)** 4.5   **(u)** 5.25
     (v) 19.49   **(w)** 5.25   **(x)** 3.125   **(y)** 217.39   **(z)** 714.185

(2-32) (a) 8.372 (b) 3.85 (c) 2.16 (d) 0.441 (e) 0.385 (f) 8.5 (g) 7.9
(h) 8.5 (i) 0.8 (j) 4.6 (k) 89.94 (l) 1.8395 (m) 3.25 (n) 12.825
(o) 0.21 (p) 35.3848 (q) 2.25 (r) 3.5 (s) 13.49 (t) 42.029 (u) 4.9
(v) 1.15 (w) 15.51 (x) 491.795

(2-33) (a) 0.6 (b) 0.4 (c) 4.8 (d) 6.3 (e) 1 (f) 2 (g) 10.35 (h) 1.5
(i) 0.18 (j) 0.1 (k) 3.141 (l) 0.125 (m) 1.8375 (n) 17.2645 (o) 0.25375
(p) 11.91652 (q) 0.03 (r) 0.008 (s) 0.0006 (t) 0.00009 (u) 25 (v) 250
(w) 2500 (x) 25,000 (y) 500 (z) 20

(2-34) (a) 0.2 (b) 0.06 (c) 0.015 (d) 0.0084 (e) 0.07 (f) $7.47\overline{3}$ (g) $1.2\overline{6}$
(h) 2.5 (i) 0.9 (j) 1.2 (k) 2.7 (l) 3.75 (m) 2.13 (n) 2.59 (o) $0.\overline{3}$
(p) $6.\overline{6}$ (q) 0.02 (r) 0.005 (s) 0.0001 (t) 0.00002 (u) 0.28 (v) 0.028
(w) 0.0028 (x) 0.00028 (y) 0.05857 (z) 0.258973

(2-35) (a) 16.7 (b) 0.2 (c) 18.57 (d) 22.22 (e) 2.583 (f) 0.667 (g) 190
(h) 15 (i) 0.1 (j) 6.10

(2-36) (a) $\frac{3}{10}$ (b) $\frac{9}{10}$ (c) $\frac{1}{5}$ (d) $\frac{3}{5}$ (e) $\frac{1}{4}$ (f) $\frac{1}{20}$ (g) $\frac{3}{8}$ (h) $\frac{3}{16}$ (i) $\frac{1}{10}$
(j) $\frac{7}{10}$ (k) $\frac{2}{5}$ (l) $\frac{4}{5}$ (m) $\frac{3}{4}$ (n) $\frac{1}{100}$ (o) $\frac{1}{8}$ (p) $\frac{1}{40}$ (q) $\frac{1}{16}$ (r) $\frac{7}{16}$
(s) $\frac{29}{10}$ or $2\frac{9}{10}$ (t) $\frac{13}{10}$ or $1\frac{3}{10}$ (u) $\frac{15}{4}$ or $3\frac{3}{4}$ (v) $\frac{13}{2}$ or $6\frac{1}{2}$ (w) $\frac{73}{8}$ or $9\frac{1}{8}$
(x) $\frac{59}{8}$ or $7\frac{3}{8}$ (y) $\frac{81}{16}$ or $5\frac{1}{16}$ (z) $\frac{42}{5}$ or $8\frac{2}{5}$

(2-37) (a) 0.5 (b) 0.25 (c) 0.2 (d) 0.8 (e) 0.125 (f) 0.625 (g) 0.1
(h) 0.7 (i) 0.0625 (j) 0.15625 (k) $0.\overline{3}$ (l) $0.1\overline{6}$ (m) $0.\overline{1}$ (n) $0.\overline{5}$
(o) $0.\overline{09}$ (p) $0.\overline{63}$ (q) $0.08\overline{3}$ (r) $0.41\overline{6}$ (s) $0.\overline{142857}$ (t) $0.\overline{428571}$
(u) 2.9 (v) 4.3 (w) 0.01 (x) 0.23 (y) 8.59 (z) 0.001

(2-38) (a) 0.25 (b) 0.5 (c) 0.75 (d) 0.9 (e) 0.1 (f) 0.2 (g) 0.125 (h) 0.375
(i) 0.3 (j) 0.4 (k) 0.625 (l) 0.875 (m) 0.6 (n) 0.8 (o) 0.0625
(p) 0.1875 (q) 1 (r) 1.25 (s) 0.3125 (t) 0.4375 (u) 2.5 (v) 10 (w) 0.7
(x) 0.9 (y) 0.15 (z) 0.35

(2-39) (a) $\frac{1}{6}$ (b) $\frac{5}{6}$ (c) $\frac{1}{3}$ (d) $\frac{2}{3}$ (e) $\frac{1}{12}$ (f) $\frac{5}{12}$ (g) $\frac{7}{12}$ (h) $\frac{11}{12}$ (i) $\frac{1}{100}$
(j) $\frac{1}{4}$ (k) $\frac{1}{2}$ (l) $\frac{3}{4}$ (m) $\frac{1}{10}$ (n) $\frac{1}{5}$ (o) $\frac{3}{10}$ (p) $\frac{2}{5}$ (q) $\frac{3}{5}$ (r) $\frac{7}{10}$
(s) $\frac{4}{5}$ (t) $\frac{9}{10}$ (u) $\frac{1}{20}$ (v) $\frac{1}{25}$ (w) $\frac{3}{20}$ (x) $\frac{7}{20}$ (y) $\frac{1}{1}$ or 1 (z) $\frac{10}{1}$ or 10

(2-40) (a) 10% (b) 20% (c) 62.5% or $62\frac{1}{2}$% (d) 87.5% or $87\frac{1}{2}$% (e) 50% (f) 60%
(g) 31.25% or $31\frac{1}{4}$% (h) 43.75% or $43\frac{3}{4}$% (i) 90% (j) 25% (k) 75% (l) 5%
(m) 12.5% or $12\frac{1}{2}$% (n) 37.5% or $37\frac{1}{2}$% (o) 30% (p) 40% (q) 6.25% or $6\frac{1}{4}$%
(r) 18.75% or $18\frac{3}{4}$% (s) 70% (t) 80% (u) 56.25% or $56\frac{1}{4}$% (v) 68.75% or $68\frac{3}{4}$%
(w) 100% (x) 150% (y) 250% (z) 1000%

(2-41) (a) 50% (b) 25% (c) 20% (d) $12\frac{1}{2}$% or 12.5% (e) 10% (f) $8\frac{1}{3}$%
(g) $6\frac{1}{4}$% or 6.25% (h) 75% (i) 40% (j) $37\frac{1}{2}$% or 37.5% (k) 30% (l) $41\frac{2}{3}$%
(m) $18\frac{3}{4}$% or 18.75% (n) 60% (o) $62\frac{1}{2}$% or 62.5% (p) 70% (q) $58\frac{1}{3}$%
(r) $31\frac{1}{4}$% or 31.25% (s) 80% (t) $87\frac{1}{2}$% or 87.5% (u) 90% (v) $43\frac{3}{4}$% or 43.75%
(w) $33\frac{1}{3}$% (x) $66\frac{2}{3}$% (y) $16\frac{2}{3}$% (z) $83\frac{1}{3}$%

(2-42) (a) 8^2 (b) 2.3^2 (c) $(\frac{9}{10})^2$ (d) 7^3 (e) 0.1^3 (f) $(\frac{3}{4})^3$ (g) 10^4 (h) 24.75^4
(i) $(\frac{2}{3})^3$ (j) $\frac{6^2}{5}$ (k) $\frac{2}{5^3}$ (l) $\frac{9^3}{3^2}$ (m) $5^2 \times 4$ (n) 3×6^4 (o) 0^2 (p) 1^3
(q) $2^2 \times 3$ or 3×2^2 (r) $5^4 \times 4^3$ or $4^3 \times 5^4$ (s) $(\frac{1}{2})^3$ (t) 6^4 (u) 1.5^4
(v) $(\frac{2}{3})^4$ (w) $\frac{8^2}{3}$ (x) $\frac{2^2}{9^3}$ (y) $\frac{2^2}{3^3}$ (z) 10^6

(2-43)   **(a)** 81   **(b)** 0.04   **(c)** $\frac{9}{16}$ **(d)** 27   **(e)** 0.008   **(f)** $\frac{1}{8}$ **(g)** 100   **(h)** 10,000
(i) 1,000,000   **(j)** 100,000,000,000   **(k)** 0   **(l)** 1   **(m)** $\frac{9}{2}$ **(n)** $\frac{9}{4}$ **(o)** 25
(p) $\frac{4}{9}$ **(q)** 0   **(r)** 640.09   **(s)** 1000   **(t)** $\frac{16}{25}$ **(u)** 0   **(v)** $\frac{18}{8}$ or $\frac{9}{4}$
(w) $\frac{9}{16}$ **(x)** $\frac{25}{4}$ **(y)** $\frac{25}{16}$ **(z)** $\frac{5}{16}$

(2-44)   **(a)** 0   **(b)** 1   **(c)** 4   **(d)** 9   **(e)** 16   **(f)** 25   **(g)** 36   **(h)** 49   **(i)** 64
(j) 81   **(k)** 100   **(l)** 121   **(m)** 144   **(n)** 169   **(o)** 196   **(p)** 225   **(q)** 50,625
(r) $\frac{1}{9}$ **(s)** $\frac{1}{16}$ **(t)** $\frac{4}{25}$ **(u)** $\frac{81}{100}$ **(v)** $\frac{81}{10,000}$ **(w)** 0.01   **(x)** 6.25   **(y)** 0.0625
(z) 5753.2225

(2-45)   **(a)** 0   **(b)** 1   **(c)** 1.414   **(d)** 1.732   **(e)** 2   **(f)** 2.236   **(g)** 2.449   **(h)** 2.646
(i) 2.828   **(j)** 3   **(k)** 3.162   **(l)** 3.317   **(m)** 3.464   **(n)** 3.606   **(o)** 3.742
(p) 3.873   **(q)** 15   **(r)** 0.577   **(s)** 0.5   **(t)** 0.632   **(u)** 0.949   **(v)** 0.3
(w) 0.316   **(x)** 1.581   **(y)** 0.5   **(z)** 8.709

3 REAL NUMBERS

THIS CHAPTER IS ABOUT

☑ **Identifying Real Numbers**
☑ **Comparing Real Numbers**
☑ **Finding Absolute Values**
☑ **Computing with Integers**
☑ **Computing with Rational Numbers**
☑ **Using the Order of Operations**

3-1. Identifying Real Numbers

A. Identify positive and negative numbers.

Recall: The whole numbers are listed as 0, 1, 2, 3, $\cdots$

To graph whole numbers, you can use a **number line.**

EXAMPLE 3-1: Graph whole numbers on a number line.

Solution:

On a number line (**a**) the point that represents zero is called the **origin.**
(**b**) all points to the right of the origin represent **positive numbers.**
(**c**) all points to the left of the origin represent **negative numbers.**

Note: To write a negative number, you use a **negative sign** $(-)$.

EXAMPLE 3-2: Write some negative numbers.

Solution: $-1, -\frac{3}{4}, -1.5, -\sqrt{6}, -\pi$ ⟵ negative numbers

Note: To write a positive number, you do *not* have to write a **positive sign** $(+)$, although you may include one.

EXAMPLE 3-3: Write some positive numbers.

Solution: $+1$ or 1, $+\frac{3}{4}$ or $\frac{3}{4}$, $+1.5$ or 1.5, $+\sqrt{6}$ or $\sqrt{6}$, $+\pi$ or π ⟵ positive numbers

To **graph positive and negative numbers,** you can use a number line.

EXAMPLE 3-4: Graph some positive and negative numbers.

Solution

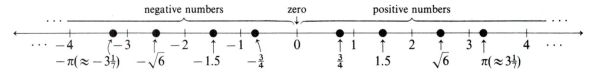

B. **Identify integers.**

Note: On a number line, the points representing
(a) $-\frac{3}{4}$ and $\frac{3}{4}$ are the same distance from the origin.
(b) -1 and 1 are the same distance from the origin.
(c) -1.5 and 1.5 are the same distance from the origin.
(d) $-\sqrt{6}$ and $\sqrt{6}$ are the same distance from the origin.
(e) $-\pi$ and π are the same distance from the origin.

Two numbers with **unlike signs** $(+, -)$ that are the same distance from the origin (0) on a number line are called **opposites.** To write **the opposite of a number,** you just change the number's sign to the **opposite sign.**

EXAMPLE 3-5: Write the opposite of each of the following numbers:
(a) $+4$ (b) $\frac{1}{2}$ (c) -2.3 (d) 0

Solution
(a) The opposite of $+4$ is -4 because $+4$ and -4 have opposite signs and they are both the same distance from 0 on a number line.
(b) The opposite of $\frac{1}{2}$ is $-\frac{1}{2}$ because $\frac{1}{2}$ and $-\frac{1}{2}$ have opposite signs $(\frac{1}{2} = +\frac{1}{2})$ and they are both the same distance from 0 on a number line.
(c) The opposite of -2.3 is $+2.3$ (or just 2.3) because -2.3 and $+2.3$ have opposite signs and they are both the same distance from 0 on a number line.
(d) The opposite of 0 is **0** because 0 can be written as two numbers with opposite signs $(0 = +0 = -0)$ and 0 is at the origin.

Note: The sum of two opposites is always zero: $4 + (-4) = 0$

The whole numbers $(0, 1, 2, 3, \cdots)$ and their opposites $(\cdots, -3, -2, -1, 0)$ are called **integers.**

EXAMPLE 3-6: List the set of integers.

Solution: $\cdots, -3, -2, -1, 0, 1, 2, 3, \cdots \longleftarrow$ integers

Note 1: In Example 3-6, the number collection
(a) $1, 2, 3, \cdots$ is called the **positive integers.**
(b) $\cdots, -3, -2, -1$ is called the **negative integers.**

Note 2: The ellipsis notation used in Example 3-6 indicates that the positive integers continue on forever in a positive counting pattern, and that the negative integers continue on forever in a negative counting pattern.

EXAMPLE 3-7: List the (a) first ten positive integers   (b) first ten negative integers.

Solution: $1, \quad 2, \quad 3, \quad 4, \quad 5, \quad 6, \quad 7, \quad 8, \quad 9, \quad 10 \longleftarrow$ first ten positive integers
$-10, -9, -8, -7, -6, -5, -4, -3, -2, -1 \longleftarrow$ first ten negative integers

Note: The negative numbers $-\frac{3}{4}$, -1.5, $-\sqrt{6}$, and $-\pi$ are not integers because $\frac{3}{4}$, 1.5, $\sqrt{6}$, and π are not whole numbers. *Only* the whole numbers and their opposites are integers.

C. **Identify rational numbers.**

When you draw a number line, you will usually need to graph several integers as reference points.

EXAMPLE 3-8: Draw a number line graphing several integers as reference points.

same distance from the origin (0)

Solution:

If a and b are integers ($b \neq 0$), then the collection of all numbers that can be written in the fractional form $\frac{a}{b}$ are called **rational numbers.**

Note: Integers and whole numbers are rational numbers.

EXAMPLE 3-9: Identify which of the following numbers are rational numbers: (a) $\frac{1}{2}$ (b) $\frac{-3}{4}$

(c) $\frac{2}{-5}$ (d) $1\frac{1}{2}$ (e) 5   (f) -2 (g) 0.1   (h) -1.3 (i) $0.\overline{3}$ (j) $\sqrt{2}$ (k) π

Solution

(a) $\frac{1}{2}$ is a rational number because 1 and 2 are integers.

(b) $\frac{-3}{4}$ is a rational number because -3 and 4 are integers.

(c) $\frac{2}{-5}$ is a rational number because 2 and -5 are integers.

(d) $1\frac{1}{2}$ is a rational number because $1\frac{1}{2}$ can be written as $\frac{3}{2}$, and 3 and 2 are integers.

(e) 5 is a rational number because 5 can be written as $\frac{5}{1}$, and 5 and 1 are integers.

(f) -2 is a rational number because -2 can be written as $\frac{-2}{1}$, and -2 and 1 are integers.

(g) 0.1 is a rational number because 0.1 can be written as $\frac{1}{10}$, and 1 and 10 are integers.

(h) -1.3 is a rational number because -1.3 can be written as $\frac{-13}{10}$, and -13 and 10 are integers.

(i) $0.\overline{3}$ is a rational number because $0.\overline{3}$ can be written as $\frac{1}{3}$, and 1 and 3 are integers.

(j) $\sqrt{2}$ is not a rational number because $\sqrt{2}$ cannot be written in the form $\frac{a}{b}$ where a and b are integers: $\sqrt{2} = 1.41421356237\cdots$ and is not a terminating or repeating decimal.

(k) π is not a rational number because π cannot be written in the form $\frac{a}{b}$ where a and b are integers: $\pi = 3.1415926535\cdots$ and is not a terminating or repeating decimal.

Note: If the decimal form of a number does *not* terminate or repeat, then that number is *not* a rational number. If the decimal form of a number does terminate or repeat, then that number is a rational number.

D. Identify irrational numbers.

Numbers whose decimal form does not terminate and does not repeat are called **irrational numbers.**

EXAMPLE 3-10: List some irrational numbers.

Solution: $\sqrt{2}, -\sqrt{3}, \sqrt{5}, -\sqrt{6}, \pi, -\pi$ ◄——— irrational numbers

Note: The positive or negative square root of any whole number that is not a square is an irrational number ($\cdots, -\sqrt{5}, -\sqrt{3}, -\sqrt{2}, \sqrt{2}, \sqrt{3}, \sqrt{5}, \cdots$).

E. Identify real numbers.

The collection of all rational and irrational numbers is called the set of **real numbers.**

Note 1: The real numbers include whole numbers, integers, fractions, and decimals.

Note 2: The only numbers that will be used in this text are real numbers.

EXAMPLE 3-11: Draw a diagram that shows the relationships between real numbers, irrational numbers, rational numbers, integers, and whole numbers.

Solution:

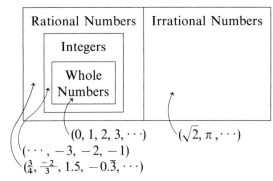

Note: Because every real number can be represented by a point on a number line, a number line is sometimes called a **real number line.** Every point on a number line represents a real number; that is, a number line is a pictorial representation of the collection of all real numbers.

3-2. Comparing Real Numbers

To compare two real numbers a and b using a number line, you write
(a) $a < b$ if a is to the **left** of b;
(b) $a > b$ if a is to the **right** of b;
(c) $a = b$ if a and b are represented by the **same** point.

EXAMPLE 3-12: Compare each of the following pairs of real numbers using $<$, $>$, or $=$:
(a) $\frac{3}{4}$ and 0.7 (b) -5 and 2 (c) -3 and -4 (d) $-\frac{1}{3}$ and $-0.\overline{3}$

Solution

(a) $\frac{3}{4} > 0.7$ because $0.7 = \frac{7}{10}$, and $\frac{3}{4} > \frac{7}{10}$ (overbrace: cross products) ($10 \times 3 = 30 > 28 = 4 \times 7$).
(b) $-5 < 2$ because every negative number is to the left of every positive number on a number line.
(c) $-3 > -4$ because -3 is to the right of -4 on a number line.
(d) $-\frac{1}{3} = -0.\overline{3}$ because $0.\overline{3} = \frac{1}{3}$.

3-3. Finding Absolute Values

The distance between a given real number and the origin (0) on a number line is called its **absolute value.** The **absolute value symbol** is $| \ |$. Read $|-5|$ as "the absolute value of negative five."

EXAMPLE 3-13: Find each absolute value: **(a)** $|+5|$ **(b)** $|-5|$ **(c)** $|\frac{3}{4}|$ **(d)** $|0|$

Solution
(a) $|+5| = 5$ because $+5$ (or just 5) is 5 units from 0 on a number line.
(b) $|-5| = 5$ because -5 is 5 units from 0 on a number line.
(c) $|\frac{3}{4}| = \frac{3}{4}$ because $\frac{3}{4}$ is $\frac{3}{4}$ of a unit from 0 on a number line.
(d) $|0| = 0$ because 0 is 0 units from 0 on a number line.

Note: The absolute value of any real number is *never* negative.

3-4. Computing with Integers

A. Add integers.
To add two or more integers with like signs, you use the following Addition Rules for Two or More Real Numbers with Like Signs:

Addition Rules for Two or More Real Numbers with Like Signs
1. Find the sum of the absolute values of the addends.
2. Write the same like sign on the sum.

Note: You can use this rule for all integers because all integers are real numbers.

EXAMPLE 3-14: Add the following integers with like signs: **(a)** $+5 + (+4)$
(b) $-4 + (-3)$ **(c)** $8 + 5 + 9 + 7$

Solution:

absolute values

(a) $+5 + (+4) = ?\ 9$ *Think:* $5 + 4 = 9$ ⟵ sum of absolute values

$= +9$ or 9 *Think:* Because both addends are positive, the sum is positive.

absolute values

(b) $-4 + (-3) = ?\ 7$ *Think:* $4 + 3 = 7$ ⟵ sum of absolute values

$= -7$ *Think:* Because both addends are negative, the sum is negative.

(c) $8 + 5 + 9 + 7 = ?\ 29$ *Think:* $8 + 5 + 9 + 7 = 29$

$= +29$ or 29 *Think:* No sign on a number means it is positive.

To add two integers with unlike signs, you can use the following Addition Rules for Two Real Numbers with Unlike Signs:

Addition Rules for Two Real Numbers with Unlike Signs
1. Find the difference between the absolute values of the addends.
2. Write the sign of the number with the larger absolute value on the sum.

Note: You can use this rule for all integers because all integers are real numbers.

EXAMPLE 3-15: Add the following integers with unlike signs: **(a)** $-3 + 7$ **(b)** $2 + (-8)$
(c) $+6 + (-6)$

Solution

larger absolute value

(a) $-3 + 7 = ?\ 4$ *Think:* $7 - 3 = 4$ ⟵ difference between absolute values

$= +4$ or 4 *Think:* Because 7 has the larger absolute value, the sum is positive.

larger absolute value
↓

(b) $2 + (-8) = ?\ 6$ *Think:* $8 - 2 = 6$ ⟵ difference between absolute values

            $= -6$ *Think:* Because -8 has the larger absolute value, the sum is negative.

(c) $+6 + (-6) = 0$ *Think:* The sum of two numbers that are opposites is always 0.

Note: The sum of any given real number and zero is always the given real number.

EXAMPLE 3-16: Add the following numbers:
(a) $0 + 7$ **(b)** $0 + (-8)$ **(c)** $5 + 0$ **(d)** $-2 + 0$

Solution: **(a)** $0 + 7 = 7$ **(b)** $0 + (-8) = -8$ **(c)** $5 + 0 = 5$ **(d)** $-2 + 0 = -2$

B. Subtract integers.

To subtract two integers, you can use the following Subtraction Rules for Two Real Numbers:

Subtraction Rules for Two Real Numbers
1. Change subtraction to addition and write the opposite of the subtrahend.
2. Follow the rules for adding real numbers.
3. Check by adding the proposed difference to the original subtrahend to see if you get the original minuend.

Note: You can use these rules for all integers because all integers are real numbers.

EXAMPLE 3-17: Subtract the following integers:
(a) $5 - 8$ **(b)** $-2 - 7$ **(c)** $6 - (-4)$ **(d)** $-3 - (-9)$

Solution

change to +

(a) $5 - 8 = 5 + (-8)$ Add the opposite of the subtrahend.

opposite of 8

         $= -3$ Add real numbers with unlike signs [see Example 3-15].

    Check: $-3 + 8 = +5$ ⟵ -3 checks

change to +

(b) $-2 - 7 = -2 + (-7)$ Add the opposite of the subtrahend.

opposite of 7

         $= -9$ Add real numbers with like signs [see Example 3-14].

    Check: $-9 + 7 = -2$ ⟵ -9 checks

change to +

(c) $6 - (-4) = 6 + (+4)$ Add the opposite of the subtrahend.

opposite of -4

         $= +10$ or 10 Add real numbers with like signs.

    Check: $10 + (-4) = 6$ ⟵ 10 checks

change to +

(d) $-3 - (-9) = -3 + 9$ Add the opposite of the subtrahend.

opposite of -9

$$= 6$$ Add real numbers with unlike signs.

Check: $6 + (-9) = -3$ ⟵ 6 checks

C. Multiply integers.

Recall: There are several different ways to write the multiplication problem "3 times 4."

EXAMPLE 3-18: Write "3 times 4" in five different ways.

Solution: 3 times 4 $= 3 \times 4$

$$= 3 \cdot 4$$

$$= 3(4)$$ all equivalent ways of writing "3 times 4"

$$= (3)4$$

$$= (3)(4)$$

To multiply two integers, you can use the following Multiplication Rules for Two Real Numbers:

Multiplication Rules for Two Real Numbers
1. Find the product of the absolute values.
2. Make the product
  (a) positive if the two factors have like signs;
  (b) negative if the two factors have unlike signs;
  (c) zero if either one of the factors is zero.

EXAMPLE 3-19: Multiply the following pairs of integers: **(a)** $9 \times (-7)$ **(b)** $-5 \cdot 2$
(c) $(9)(3)$ **(d)** $(-2)0$ **(e)** $-4(-6)$

Solution

absolute values

(a) $9 \times (-7) = \ ? \ 63$ *Think:* $9 \times 7 = 63$ ⟵ product of the absolute values

$$= -63$$ The factors 9 and -7 have unlike signs so the product is negative.

(b) $-5 \cdot 2 = \ ? \ 10$ *Think:* $5 \times 2 = 10$

$$= -10$$ The factors -5 and 2 have unlike signs so the product is negative.

(c) $(9)(3) = \ ? \ 27$ *Think:* $9 \times 3 = 27$

$$= +27 \text{ or } 27$$ The factors 9 and 3 have like signs so the product is positive.

(d) $(-2)0 = 0$ *Think:* When one of the factors is zero the product is always zero.

(e) $-4(-6) = \ ? \ 24$ *Think:* $4 \times 6 = 24$

$$= +24 \text{ or } 24$$ *Think:* The factors -4 and -6 have like signs so the product is positive.

Note: The product of two negative numbers is always positive!

To multiply more than two integers you can use the following Multiplication Rules for Two or More Real Numbers:

Multiplication Rules for Two or More Real Numbers
1. Find the product of all the absolute values.
2. Make the product
   (a) positive if there are an **even number** $(0, 2, 4, \cdots)$ of negative factors;
   (b) negative if there are an **odd number** $(1, 3, 5, \cdots)$ of negative factors;
   (c) zero if one or more factors are zero.

Note: You can use these rules for all integers because all integers are real numbers.

EXAMPLE 3-20: Multiply the following groups of integers: (a) $5 \cdot 3 \cdot 2$ (b) $4(-1)3$
(c) $-5(-2)2$ (d) $-1(-6)(-3)$ (e) $2(-3)(-1)(0)(-3)(-2)$

Solution
(a) $5 \cdot 3 \cdot 2 = ?\ 30$ *Think:* $5 \times 3 \times 2 = 30$ ⟵ product of the absolute values

$= +30$ or 30 There is an even number (0) of negative signs so the product is positive.

(b) $4(-1)3 = ?\ 12$ *Think:* $4 \times 1 \times 3 = 12$

$= -12$ There is an odd number (1) of negative signs so the product is negative.

(c) $-5(-2)2 = ?\ 20$ *Think:* $5 \times 2 \times 2 = 20$

$= +20$ or 20 There is an even number (2) of negative signs so the product is positive.

(d) $-1(-6)(-3) = ?\ 18$ *Think:* $1 \times 6 \times 3 = 18$

$= -18$ There is an odd number (3) of negative signs so the product is negative.

(e) $2(-3)(-1)(0)(-3)(-2) = 0$ *Think:* When one or more factors are zero, the product is always zero.

D. **Divide integers.**

Recall: There are several different ways to write the division problem "8 divided by 3."

EXAMPLE 3-21: Write "8 divided by 3" in six different ways.

Solution: 8 divided by $3 = 8 \div 3$
$= 8/3$
$= \frac{8}{3}$
$= 8 \cdot \frac{1}{3}$ } all equivalent ways of writing "8 divided by 3"
$= \frac{1}{3} \cdot 8$
$= 3\overline{)8}$

To divide two integers, you can use the following Division Rules for Two Real Numbers:

Division Rules for Two Real Numbers
1. Find the quotient of the absolute values.
2. Use the sign rules for multiplication to write the correct sign on the quotient.
3. Check by multiplying the proposed quotient by the original divisor to see if you get the original dividend.

Note: You can use these rules for all integers because all integers are real numbers.

Recall: Zero divided by any nonzero number is always zero.
       Division by zero is not defined.

EXAMPLE 3-22: Divide the following pairs of integers: **(a)** $15 \div 3$ **(b)** Divide -10 by 2. **(c)** $\frac{-8}{-2}$ **(b)** $-3\overline{)-18}$ **(e)** $\frac{0}{3}$ **(f)** $\frac{3}{0}$ **(g)** $\frac{0}{0}$

Solution

(a) $15 \div 3 = +5 \text{ or } 5$ Use the sign rules for multiplication.

   Check: $5 \times 3 = 15 \longleftarrow$ 5 checks

(b) Divide -10 by $2 = -10 \div 2$ Find the quotient of the absolute values.

        $= -5$ Use the sign rules for multiplication.

   Check: $-5 \times 2 = -10 \longleftarrow -5$ checks

(c) $\frac{-8}{-2} = 8 \div (-2)$ Find the quotient of the absolute values.

        $= -4$ Use the sign rules for multiplication.

   Check: $-4(-2) = +8 \text{ or } 8 \longleftarrow -4$ checks

(d) $-3\overline{)-18} = -18 \div (-3)$ Find the quotient of the absolute values.

        $= +6 \text{ or } 6$ Use the sign rules for multiplication.

   Check: $6(-3) = -18 \longleftarrow$ 6 checks

(e) $\frac{0}{3} = 0$ Zero divided by any nonzero real number is always zero.

   Check: $0 \times 3 = 0 \longleftarrow$ 0 checks

(f) $\frac{3}{0}$ is not defined.

(g) $\frac{0}{0}$ is not defined.

3-5. Computing with Rational Numbers

A. Rename rational numbers.

Recall: If a and b are integers ($b \neq 0$), then any number that can be written in the form $\frac{a}{b}$ is called a rational number.

Every rational number in fraction form has three signs: the sign of the fraction, the sign of the numerator, and the sign of the denominator.

EXAMPLE 3-23: Identify the three signs for each of the following rational numbers:
(a) $\frac{3}{4}$ **(b)** $-\frac{-3}{4}$

Solution

If exactly two of a fraction's three signs are changed, you will always get an equal fraction.

The following are all equivalent ways to write a fraction with form $\frac{a}{b}$.

$$\frac{a}{b} = -\frac{-a}{b} = -\frac{a}{-b} = \frac{-a}{-b}$$

EXAMPLE 3-24: Rename $\dfrac{3}{4}$ in three equivalent ways using sign changes.

Solution: $\dfrac{3}{4} = -\dfrac{-3}{4}$ *Think:* $+\dfrac{+3}{+4} = -\dfrac{-3}{+4}$ ⟵ two sign changes

$\dfrac{3}{4} = -\dfrac{3}{-4}$ *Think:* $+\dfrac{+3}{+4} = -\dfrac{+3}{-4}$ ⟵ two sign changes

$\dfrac{3}{4} = \dfrac{-3}{-4}$ *Think:* $+\dfrac{+3}{+4} = +\dfrac{-3}{-4}$ ⟵ two sign changes

Note: The fractions $\dfrac{3}{4}$, $-\dfrac{-3}{4}$, $-\dfrac{3}{-4}$, and $\dfrac{-3}{-4}$ are equal fractions.

Caution: To get an equal fraction using sign changes, you must change exactly two of a fraction's three signs.

The following are all equivalent ways to write a fraction with form $-\dfrac{a}{b}$:

$$-\frac{a}{b} = \frac{-a}{b} = \frac{a}{-b} = -\frac{-a}{-b}$$

EXAMPLE 3-25: Rename $-\dfrac{3}{4}$ in three equivalent ways using sign changes.

Solution: $-\dfrac{3}{4} = \dfrac{-3}{4}$ *Think:* $-\dfrac{+3}{+4} = +\dfrac{-3}{+4}$ ⟵ two sign changes

$-\dfrac{3}{4} = \dfrac{3}{-4}$ *Think:* $-\dfrac{+3}{+4} = +\dfrac{+3}{-4}$ ⟵ two sign changes

$-\dfrac{3}{4} = -\dfrac{-3}{-4}$ *Think:* $-\dfrac{+3}{+4} = -\dfrac{-3}{-4}$ ⟵ two sign changes

Note: The fractions $-\dfrac{3}{4}$, $\dfrac{-3}{4}$, $\dfrac{3}{-4}$, and $-\dfrac{-3}{-4}$ are equal fractions.

B. Simplify rational numbers.

A rational number in fraction form for which the numerator and denominator do not share a common integer factor other than 1 or -1 is called a **rational number in lowest terms.**

EXAMPLE 3-26: Which of the following rational numbers are in lowest terms?

(a) $\dfrac{3}{4}$ (b) $\dfrac{-5}{2}$ (c) $\dfrac{-6}{-10}$

Solution

(a) $\dfrac{3}{4}$ is a rational number in lowest terms because the only common integer factors of 3 and 4 are 1 and -1: $\dfrac{3}{4} = \dfrac{1(3)}{1(4)}$ and $\dfrac{3}{4} = \dfrac{-3}{-4} = \dfrac{-1(3)}{-1(4)}$.

(b) $\dfrac{-5}{2}$ is a rational number in lowest terms because the only common integer factors of -5 and 2 are 1 and -1: $\dfrac{-5}{2} = \dfrac{1(-5)}{1(2)}$ and $\dfrac{-5}{2} = \dfrac{5}{-2} = \dfrac{-1(-5)}{-1(2)}$.

(c) $\dfrac{-6}{-10}$ is not a rational number in lowest terms because 6 and 10 share a common integer factor

of 2 or -2: $\dfrac{-6}{-10} = \dfrac{2(-3)}{2(-5)}$ or $\dfrac{-6}{-10} = \dfrac{-2(3)}{-2(5)}$

To **simplify a rational number in fraction form,** you write the rational number using as few negative signs as possible and then **reduce the rational number in fraction form** to lowest terms using the following rule:

Fundamental Rule for Rational Numbers

$$\text{If } a, b, \text{ and c are integers } (b \neq 0 \text{ and } c \neq 0), \text{ then } \frac{a \cdot c}{b \cdot c} = \frac{a}{b}.$$

Note: By the Fundamental Rule for Rational Numbers, the value of a rational number in fraction form will not change when you divide both the numerator and denominator by the same nonzero number: $\dfrac{a \cdot \cancel{c}}{b \cdot \cancel{c}} = \dfrac{a}{b}$

EXAMPLE 3-27: Simplify the following fractions: **(a)** $\dfrac{6}{10}$ **(b)** $-\dfrac{6}{3}$ **(c)** $\dfrac{-12}{8}$ · **(d)** $-\dfrac{-10}{-25}$

Solution:

(a) $\dfrac{6}{10} = \dfrac{2 \cdot 3}{2 \cdot 5}$ Factor both the numerator and denominator.

$= \dfrac{\cancel{2} \cdot 3}{\cancel{2} \cdot 5}$ Use the Fundamental Rule for Rational Numbers.

$= \dfrac{3}{5}$ ⟵ simplest form

(b) $-\dfrac{6}{3} = -\dfrac{2 \cdot \cancel{3}}{1 \cdot \cancel{3}}$ Use the Fundamental Rule for Rational Numbers.

$= -\dfrac{2}{1}$

$= -2$ ⟵ simplest form

(c) $\dfrac{-12}{8} = -\dfrac{12}{8}$

$= -\dfrac{2 \cdot 2 \cdot 3}{2 \cdot 2 \cdot 2}$ or $-\dfrac{\cancel{4} \cdot 3}{\cancel{4} \cdot 2}$ Use the Fundamental Rule for Rational Numbers.

$= -\dfrac{3}{2}$

⟩ simplest form

or $-1\dfrac{1}{2}$

(d) $-\dfrac{-10}{-25} = -\dfrac{10}{25}$ Write as few negative signs as possible: $-\dfrac{-10}{-25} = -\dfrac{+10}{+25}$

$= -\dfrac{2 \cdot \cancel{5}}{5 \cdot \cancel{5}}$ Use the Fundamental Rule for Rational Numbers.

$= -\dfrac{2}{5}$ ⟵ simplest form

C. **Add rational numbers.**

To add rational numbers, you use the Addition Rules for Real Numbers with Like or Unlike Signs given in Section 3-4, part A.

EXAMPLE 3-28: Add the following rational numbers: **(a)** $\frac{1}{4} + 0.5$ **(b)** $-2.5 + 1.75$
(c) $\frac{3}{4} + (-\frac{1}{2})$ **(d)** $-3.2 + (-1\frac{1}{3})$

Solution

(a) Rename to get fractions   or   Rename to get decimals

$$\frac{1}{4} + 0.5 = \frac{1}{4} + \frac{5}{10} \qquad\qquad \frac{1}{4} + 0.5 = 0.25 + 0.5$$

$$= \frac{1}{4} + \frac{1}{2} \qquad\qquad\qquad = 0.75$$

$$= \frac{1}{4} + \frac{2}{4}$$

$$= \frac{3}{4}$$

            larger absolute value
            ↓

(b) $-2.5 + 1.75 = ?\,0.75$ *Think:* $2.5 - 1.75 = 0.75$ ⟵ difference between absolute values

$$= -0.75$$

Think: Because -2.5 has the larger absolute value, the sum is negative.

            larger absolute value
            ↓

(c) $\frac{3}{4} + \left(-\frac{1}{2}\right) = ?\,\frac{1}{4}$ *Think:* $\frac{3}{4} - \frac{1}{2} = \frac{1}{4}$ ⟵ difference between the absolute values

$$= +\frac{1}{4} \text{ or } \frac{1}{4}$$

Think: Because $\frac{3}{4}$ has the larger absolute value, the sum is positive.

(d) $-3.2 + \left(-1\frac{1}{3}\right) = -3\frac{2}{10} + \left(-1\frac{1}{3}\right)$ Rename 3.2 as a mixed number $\left(3\frac{2}{10}\right)$.

$$= -3\frac{1}{5} + \left(-1\frac{1}{3}\right) \qquad \text{Simplify } 3\frac{2}{10} \text{ to } 3\frac{1}{5}.$$

$$= -\frac{16}{5} + \left(-\frac{4}{3}\right) \qquad \text{Rename as fractions.}$$

$$= -\frac{48}{15} + \left(-\frac{20}{15}\right) \qquad \text{Use the LCD 15 to get like fractions.}$$

$$= ?\,\frac{68}{15} \qquad\qquad\qquad \textit{Think: } \frac{48}{15} + \frac{20}{15} = \frac{68}{15}$$

$$= -\frac{68}{15} \qquad\qquad\qquad \textit{Think: } \text{Because both addends are negative, the sum is negative.}$$

D. Subtract rational numbers.

To subtract rational numbers, you use the Subtraction Rules for Real Numbers given in Section 3-4, part B.

EXAMPLE 3-29: Subtract the following rational numbers: $-4.5 - 2\frac{3}{4}$

Solution: You can subtract by first renaming as decimals, or first renaming as fractions.

Rename to get decimals

$$-4.5 - 2\frac{3}{4} = -4.5 - 2.75 \qquad \text{Rename } 2\frac{3}{4} \text{ as } 2.75.$$

$$= -4.5 + (-2.75) \qquad \text{Add the opposite of the subtrahend.}$$

$$= -7.25$$

Rename to get fractions

$$-4.5 - 2\frac{3}{4} = -4\frac{5}{10} - 2\frac{3}{4} \qquad \text{Rename } -4.5 \text{ as } -4\frac{5}{10}.$$

$$= -4\frac{5}{10} + \left(-2\frac{3}{4}\right) \qquad \text{Add the opposite of the subtrahend.}$$

$$= -4\frac{1}{2} + \left(-2\frac{3}{4}\right) \qquad \text{Rename } 4\frac{5}{10} \text{ as } 4\frac{1}{2}.$$

$$= -4\frac{2}{4} + \left(-2\frac{3}{4}\right) \qquad \text{Use the LCD 4 to get like fractions.}$$

$$= -6\frac{5}{4} \qquad \text{Add rational numbers.}$$

$$= -7\frac{1}{4} \qquad \text{Rename } \frac{5}{4} \text{ as } 1\frac{1}{4}.$$

E. Multiply rational numbers.

To multiply rational numbers, you use the Multiplication Rules for Real Numbers given in Section 3-4, part C.

EXAMPLE 3-30: Multiply the following rational numbers: **(a)** $\frac{1}{2}(0.25)$ **(b)** $-\frac{1}{4}\cdot\frac{2}{3}$

(c) $1.5(-2.3)$ **(d)** $-0.5\left(-\frac{2}{3}\right)$

Solution

(a) Rename to get fractions or Rename to get decimals

$$\frac{1}{2}(0.25) = \frac{1}{2}\cdot\frac{1}{4} \qquad\qquad \frac{1}{2}(0.25) = 0.5(0.25)$$

$$= \frac{1}{8} \qquad\qquad\qquad\quad = 0.125$$

(b) $-\frac{1}{4}\cdot\frac{2}{3} = -\frac{1}{2\cdot\cancel{2}}\cdot\frac{\cancel{2}}{3}$ Eliminate common factors.

$$= ?\frac{1}{6} \qquad \textit{Think: } \frac{1}{2}\cdot\frac{1}{3} = \frac{1}{6} \longleftarrow \text{product of the absolute values}$$

$$= -\frac{1}{6} \qquad \textit{Think: } \text{Because the factors have unlike signs, the product is negative.}$$

(c) $1.5(-2.3) = ? \, 3.45$ *Think:* $1.5(2.3) = 3.45 \longleftarrow$ product of the absolute values

$\qquad\qquad = -3.45$ *Think:* Because the fractions have unlike signs, the product is negative.

(d) $-0.5\left(-\dfrac{2}{3}\right) = -\dfrac{1}{2}\left(-\dfrac{2}{3}\right)$ Rename 0.5 as $\dfrac{1}{2}$.

$\qquad\qquad = \dfrac{-1(-2)}{2(3)}$ Multiply fractions.

$\qquad\qquad = ? \, \dfrac{1}{3}$ *Think:* $\dfrac{1(\cancel{2})}{\cancel{2}(3)} = \dfrac{1}{3} \longleftarrow$ product of the absolute values

$\qquad\qquad = +\dfrac{1}{3} \text{ or } \dfrac{1}{3}$ *Think:* Because both factors are negative, the product is positive.

F. Divide rational numbers.

To divide rational numbers, you use the Division Rules for Real Numbers in Section 3-4, part D.

EXAMPLE 3-31: Divide the following rational numbers: $\dfrac{3}{4} \div (-0.25)$

Solution: You can divide by first renaming as decimals, or first renaming as fractions.

Rename to get decimals

$\dfrac{3}{4} \div (-0.25) = 0.75 \div (-0.25)$ Rename $\dfrac{3}{4}$ as 0.75.

$\qquad\qquad\qquad = ? \, 3$ Divide decimals.

$\qquad\qquad\qquad = -3$ *Think:* Because the decimals have unlike signs, the quotient is negative.

Rename to get fractions

$\dfrac{3}{4} \div (-0.25) = \dfrac{3}{4} \div \left(-\dfrac{1}{4}\right)$ Rename (-0.25) as $\left(-\dfrac{1}{4}\right)$.

$\qquad\qquad\qquad = \dfrac{3}{4} \cdot \left(-\dfrac{4}{1}\right)$ Multiply by the reciprocal of the divisor.

$\qquad\qquad\qquad = -\dfrac{3}{1}$ Multiply fractions and simplify.

$\qquad\qquad\qquad = -3$

Check: $-3(-0.25) = +0.75 \text{ or } 0.75 \longleftarrow$ -3 checks

3-6. Using the Order of Operations

To avoid getting a wrong answer when computing with real numbers, you must use the following rules:

Order of Operations
- First, **clear grouping symbols** such as ( ), [ ], { }, —, and $\sqrt{}$ by performing the operations inside of them.
- Then, evaluate each power notation and radical.
- Next, multiply and divide in order from left to right.
- Last, add and subtract in order from left to right.

Note: The saying *Please Excuse My Dear Aunt Sally* can help you remember the Order of Operations: **P**arentheses (grouping symbols), **E**xponents (and radicals), **M**ultiply and **D**ivide, **A**dd and **S**ubtract.

EXAMPLE 3-32: Evaluate the following expressions: **(a)** $3 + 5 \cdot 2$ **(b)** $(1 - 5)^2 - 2\sqrt{9} + \dfrac{12}{4}$

(c) $\dfrac{5 - \sqrt{5^2 - 4(6)(-4)}}{2(6)}$

Solution

(a) $3 + 5 \cdot 2 = 3 + 10$ *Think:* Multiply first.

           $= 13$ Then add.

(b) $(1 - 5)^2 - 2\sqrt{9} + \frac{12}{4} = (-4)^2 - 2\sqrt{9} + 3$ *Think:* Clear grouping symbols first.

           $= 16 - 2 \cdot 3 + 3$ Evaluate each power notation and radical.

           $= 16 - 6 + 3$ Multiply and divide in order from left to right.

           $= 10 + 3$ Add and subtract in order from left to right.

           $= 13$

(c) $\dfrac{5 - \sqrt{5^2 - 4(6)(-4)}}{2(6)} = \dfrac{5 - \sqrt{25 - 4(6)(-4)}}{2(6)}$ *Think:* $5^2 = 25$

           $= \dfrac{5 - \sqrt{25 - 24(-4)}}{2(6)}$ $4(6) = 24$

           $= \dfrac{5 - \sqrt{25 - (-96)}}{2(6)}$ $24(-4) = -96$

           $= \dfrac{5 - \sqrt{25 + 96}}{2(6)}$ $25 - (-96) = 25 + (+96)$

           $= \dfrac{5 - \sqrt{121}}{2(6)}$ $25 + 96 = 121$

           $= \dfrac{5 - 11}{2(6)}$ $\sqrt{121} = 11$ because $11 \cdot 11 = 121$

           $= \dfrac{-6}{2(6)}$ $5 - 11 = -6$

           $= \dfrac{-1(\cancel{6})}{2(\cancel{6})}$ $-6 = -1(6)$

           $= \dfrac{-1}{2}$ or $-\dfrac{1}{2}$ $\dfrac{a \cdot c}{b \cdot c} = \dfrac{a}{b}$

Caution: To avoid getting a wrong answer when there are no grouping symbols present, always multiply and divide *before* adding and subtracting.

EXAMPLE 3-33: Evaluate the following expressions: **(a)** $8 + 6 \div 2$ **(b)** $18 - 2 \cdot 3$

Solution

Correct Method	Wrong Method
(a) $8 + 6 \div 2 = 8 + 3$ Divide first.	$8 + 6 \div 2 = 14 \div 2$ No! Divide first.
$= 11 \longleftarrow$ correct answer	$= 7 \longleftarrow$ wrong answer
(b) $18 - 2 \cdot 3 = 18 - 6$ Multiply first.	$18 - 2 \cdot 3 = 16 \cdot 3$ No! Multiply first.
$= 12 \longleftarrow$ correct answer	$= 48 \longleftarrow$ wrong answer

Caution: Be sure to work from left to right when adding, subtracting, multiplying, or dividing.

EXAMPLE 3-34: Evaluate the following expressions: **(a)** $3 - 5 + 7$ **(b)** $12 \div 2 \times 3$

Solution

Correct Method	Wrong Method
(a) $3 - 5 + 7 = -2 + 7$ Work in order from left to right.	$3 - 5 + 7 = 3 - 12$ No! Never add and subtract from right to left.
$= 5$ ⟵ correct answer	$= -9$ ⟵ wrong answer
(b) $12 \div 2 \times 3 = 6 \times 3$ Work in order from left to right.	$12 \div 2 \times 3 = 12 \div 6$ No! Never multiply and divide from right to left.
$= 18$ ⟵ correct answer	$= 2$ ⟵ wrong answer

RAISE YOUR GRADES

Can you . . . ?

☑ identify positive and negative numbers
☑ graph positive and negative numbers
☑ write the opposite of any positive or negative number
☑ list the positive integers, negative integers, and integers
☑ identify rational numbers, irrational numbers, and real numbers
☑ draw a diagram showing the relationships between real numbers, rational numbers, irrational numbers, integers, and whole numbers
☑ compare any two real numbers using $<$, $>$, or $=$
☑ find the absolute value of any given real number
☑ add, subtract, multiply, and divide integers
☑ identify when division is not defined
☑ identify the three signs of a rational number in fraction form
☑ write rational numbers in the form $\frac{a}{b}$ in three equivalent ways using sign changes
☑ simplify rational numbers in fraction form
☑ add, subtract, multiply, and divide rational numbers
☑ evaluate an expression using the Order of Operations

SUMMARY

1. On a number line
   (a) the point that represents zero is called the origin.
   (b) all points to the right of the origin represent positive numbers.
   (c) all points to the left of the origin represent negative numbers.
2. To write a negative number, you use a negative sign $(-)$.
3. To write a positive number, you do not need to write a positive sign $(+)$.
4. Two numbers with unlike signs $(+, -)$ that are the same distance from the origin (0) on a number line are called opposites.
5. To find the opposite of a real number, you just change the number's sign to the opposite sign.
6. The sum of two opposites is always zero.
7. The whole numbers $(0, 1, 2, 3, \cdots)$ and their opposites $(\cdots, -3, -2, -1, 0)$ are called integers $(\cdots, -3, -2, -1, 0, 1, 2, 3, \cdots)$.
8. The positive integers are $1, 2, 3, \cdots$

9. The negative integers are $\cdots, -3, -2, -1$.

10. If a and b are integers $(b \neq 0)$, then the collection of all numbers that can be written in fractional form $\dfrac{a}{b}$ are called rational numbers.

11. Numbers whose decimal form does not terminate and does not repeat are called irrational numbers.

12. The collection of all rational and irrational numbers is called the set of real numbers.

13. A diagram showing the relationships between real numbers, rational numbers, irrational numbers, integers, and whole numbers can be drawn as follows:

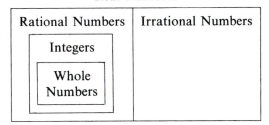

Real Numbers

14. Every real number is represented by a point on a number line and every point on a number line represents a real number. That is, a real number line is a pictorial representation of the collection of all real numbers.

15. To compare two real numbers a and b on a number line, you write
   (a) $a < b$ if a is to the left of b.
   (b) $a > b$ if a is to the right of b.
   (c) $a = b$ if a and b are represented by the same point.

16. The distance between a given real number and the origin (0) on a number line is its absolute value.

17. The absolute value of any real number is never negative.

18. Addition Rules for Two or More Real Numbers with Like Signs
   (a) Find the sum of the absolute values.
   (b) Write the same like sign on the sum.

19. Addition Rules for Two Real Numbers with Unlike Signs
   (a) Find the difference between the absolute values.
   (b) Write the sign of the number with the larger absolute value on the sum.

20. Subtraction Rules for Two Real Numbers
   (a) Change subtraction to addition by writing the opposite of the subtrahend.
   (b) Follow the rules for adding real numbers.
   (c) Check by adding the proposed difference to the original subtrahend to see if you get the original minuend.

21. A multiplication problem like "3 times 4" can be written as 3×4, $3 \cdot 4$, $3(4)$, $(3)4$, or $(3)(4)$.

22. Multiplication Rules for Two or More Real Numbers
   (a) Find the product of the absolute values.
   (b) Make the product
      (a) positive if there are an even number $(0, 2, 4, \cdots)$ of negative factors.
      (b) negative if there are an odd number $(1, 3, 5, \cdots)$ of negative factors.
      (c) zero if one or more factors are zero.

23. A division problem like "8 divided by 3" can be written as $8 \div 3$, $8/3$, $\frac{8}{3}$, $8 \cdot \frac{1}{3}$, $\frac{1}{3} \cdot 8$, or $3\overline{)8}$.

24. Division Rules for Two Real Numbers
   (a) Find the quotient of the absolute values.
   (b) Use the sign rules for multiplication to write the correct sign on the quotient.
   (c) Check by multiplying the proposed quotient by the original divisor to see if you get the original dividend.

25. Every rational number in fraction form has three signs: the sign of the fraction, the sign of the numerator, and the sign of the denominator.

26. The following are all equivalent ways to write a fraction with form $\dfrac{a}{b}$:

$$\frac{a}{b} = -\frac{-a}{b} = -\frac{a}{-b} = \frac{-a}{-b}$$

27. *Caution:* To get an equal fraction using **sign** changes, you must change exactly two of a fraction's three signs.

28. The following are all equivalent ways to write a fraction with form $-\dfrac{a}{b}$:

$$-\frac{a}{b} = \frac{-a}{b} = \frac{a}{-b} = -\frac{-a}{-b}$$

29. A rational number in fraction form for which the numerator and denominator do not share a common integer factor other than 1 or -1 is called a rational number in lowest terms.

30. To simplify a rational number in fraction form, you write the rational number using as few negative signs as possible and then reduce the rational number in fraction form to lowest terms using the Fundamental Rule for Rational Numbers.

31. **Fundamental Rule for Rational Numbers**

If a, b, and c are integers ($b \neq 0$ and $c \neq 0$), then $\dfrac{a \cdot c}{b \cdot c} = \dfrac{a}{b}$.

32. To avoid getting a wrong answer when computing with real numbers, you must use the Order of Operations rules in the order given below.

33. **Order of Operations**
    (a) Perform operations inside grouping symbols such as ( ), [ ], { }, —, and $\sqrt{}$.
    (b) Evaluate power notation and radicals.
    (c) Multiply and divide in order from left to right.
    (d) Add and subtract in order from left to right.

34. The saying "Please Excuse My Dear Aunt Sally" can help you remember the Order of Operations rules: **P**arentheses (enclosure symbols), **E**xponents (and radicals), **M**ultiply and **D**ivide, **A**dd and **S**ubtract.

35. *Caution:* To avoid getting a wrong answer when there are no grouping symbols present, always multiply and divide *before* adding and subtracting.

36. *Caution:* Be sure to work from left to right when adding, subtracting, multiplying, or dividing.

SOLVED PROBLEMS

PROBLEM 3-1 Write the opposite of each real number: (a) $+8$ (b) $-\frac{2}{3}$ (c) 0.6 (d) 0

Solution: Recall that to find the opposite of a real number, you just change the number's sign to the opposite sign [see Example 3-5]:
(a) The opposite of $+8$ is -8. (b) The opposite of $-\frac{2}{3}$ is $\frac{2}{3}$. (c) The opposite of 0.6 is -0.6.
(d) The opposite of 0 is 0.

PROBLEM 3-2 Identify which of the following numbers are rational numbers and which are irrational numbers: (a) 0 (b) 2 (c) -3 (d) $\frac{3}{4}$ (e) $\frac{-5}{3}$ (f) 2.3 (g) -0.25 (h) $0.\overline{6}$ (i) $\sqrt{9}$ (j) $-\sqrt{10}$ (k) π

Solution: Recall that the rational numbers include all the whole numbers, integers, fractions, terminating decimals, and repeating decimals. The irrational numbers include all numbers whose decimal form does not terminate and does not repeat [see Examples 3-9 and 3-10]:
(a)–(i) The numbers $0, 2, -3, \frac{3}{4}, \frac{-5}{3}, 2.3, -0.25, 0.\overline{6}$, and $\sqrt{9}$ (because $\sqrt{9} = 3$) are all rational numbers.
(j) $-\sqrt{10}$ is an irrational number because $-\sqrt{10} = -3.162277\cdots$, and does not terminate or repeat. (k) π is an irrational number because $\pi = 3.1415926535\cdots$, and does not terminate or repeat.

PROBLEM 3-3 Compare each pair of real numbers using $<$, $>$, or $=$: (a) $2, 5$ (b) $2, -5$
(c) $-2, 5$ (d) $-2, -5$ (e) $\frac{8}{12}, \frac{12}{18}$ (f) $0, \frac{1}{2}$ (g) $0, -\frac{3}{4}$ (h) $-\frac{2}{3}, -0.\overline{6}$

Solution: Recall that to compare two real numbers a and b using a number line, you write $a < b$ if a is to the left of b, $a > b$ if a is to the right of b, and $a = b$ if a and b represent the same point [see Example 3-12]:

(a) $2 < 5$ because 2 is to the left of 5 on a number line.

(b) $2 > -5$ because every positive number is to the right of every negative number.

(c) $-2 < 5$ because every negative number is to the left of every positive number.

(d) $-2 > -5$ because -2 is to the right of -5 on a number line.

(e) $\dfrac{8}{12} = \dfrac{12}{18}$ because $\dfrac{8}{12} = \dfrac{2 \cdot \cancel{4}}{3 \cdot \cancel{4}} = \dfrac{2}{3}$ and $\dfrac{12}{18} = \dfrac{2 \cdot \cancel{6}}{3 \cdot \cancel{6}} = \dfrac{2}{3}$.

(f) $0 < \frac{1}{2}$ because zero is to the left of every positive number.

(g) $0 > -\frac{3}{4}$ because zero is to the right of every negative number.

(h) $-\frac{2}{3} = -0.\overline{6}$ because $2 \div 3 = 0.666 \cdots = 0.\overline{6}$

PROBLEM 3-4 Find each absolute value: **(a)** $|+2|$ **(b)** $\left|-\frac{2}{3}\right|$ **(c)** $|2.75|$ **(d)** $|0|$

Solution: Recall that the distance that a given real number is from the origin (0) on a number line is the given real number's absolute value [see Example 3-13]:

(a) $|+2| = 2$ because 2 is 2 units from 0 on a number line.

(b) $\left|-\frac{2}{3}\right| = \frac{2}{3}$ because $-\frac{2}{3}$ is $\frac{2}{3}$ of a unit from 0 on a number line.

(c) $|2.75| = 2.75$ because 2.75 is 2.75 units from 0 on a number line.

(d) $|0| = 0$ because 0 is 0 units from 0 on a number line.

PROBLEM 3-5 Add integers with like signs: **(a)** $+5 + (+8)$ **(b)** $-6 + (-3)$ **(c)** $4 + 7$ **(d)** $+8 + (+7) + (+2)$ **(e)** $-5 + (-4) + (-7) + (-8)$

Solution: Recall that to add two or more real numbers with like signs, you first find the sum of the absolute values and then write the same like sign on the sum [see Example 3-14]:

(a) $+5 + (+8) = +(5 + 8) = +13$ or 13 (b) $-6 + (-3) = -(6 + 3) = -9$

(c) $4 + 7 = 11$ or $+11$ (d) $+8 + (+7) + (+2) = +(8 + 7 + 2) = +17$ or 17

(e) $-5 + (-4) + (-7) + (-8) = -(5 + 4 + 7 + 8) = -24$

PROBLEM 3-6 Add integers with unlike signs: **(a)** $+8 + (-6)$ **(b)** $2 + (-9)$ **(c)** $-7 + (+8)$ **(d)** $-8 + 1$ **(e)** $0 + 6$ **(f)** $0 + (-7)$ **(g)** $+9 + 0$ **(h)** $-4 + 0$

Solution: Recall that to add two integers with unlike signs, you first find the difference between the absolute values and then write the sign of the number with the larger absolute value on the sum [see Examples 3-15 and 3-16]:

(a) $+8 + (-6) = +(8 - 6) = +2$ or 2 (b) $2 + (-9) = -(9 - 2) = -7$

(c) $-7 + (+8) = +(8 - 7) = +1$ or 1 (d) $-8 + 1 = -(8 - 1) = -7$

(e) $0 + 6 = 6$ (f) $0 + (-7) = -7$ (g) $+9 + 0 = +9$ or 9 (h) $-4 + 0 = -4$

PROBLEM 3-7 Subtract integers: **(a)** $+5 - (+2)$ **(b)** $6 - 8$ **(c)** $-6 - (+4)$ **(d)** $-2 - 9$ **(e)** $+8 - (-3)$ **(f)** $5 - (-8)$ **(g)** $-5 - (-2)$ **(h)** $-3 - (-7)$

Solution: Recall that to subtract two integers, you first change subtraction to addition while writing the opposite of the subtrahend and then follow the rules for adding real numbers [see Example 3-17]:

(a) $+5 - (+2) = +5 + (-2) = +3$ or 3 (b) $6 - 8 = 6 + (-8) = -2$

(c) $-6 - (+4) = -6 + (-4) = -10$ (d) $-2 - 9 = -2 + (-9) = -11$

(e) $+8 - (-3) = +8 + (+3) = +11$ or 11   **(f)** $5 - (-8) = 5 + (+8) = +13$ or 13
(g) $-5 - (-2) = -5 + (+2) = -3$ **(h)** $-3 - (-7) = -3 + (+7) = +4$ or 4

PROBLEM 3-8 Multiply integers: **(a)** $5 \cdot 2$ **(b)** $+4(+3)$ **(c)** $(-2)3$ **(d)** $5(-3)$
(e) $+7(-4)$ **(f)** $(-5)(+6)$ **(g)** $-2(-8)$ **(h)** $(-3)(-2)(-5)$ **(i)** $5(-1)2(-1)(-2)(-1)$

Solution: Recall that to multiply two or more real numbers, you first find the product of all the absolute values and then make the product positive if there are an even number of negative factors; negative if there are an odd number of negative factors; zero if one or more factors are zero [see Examples 3-19 and 3-20]:

even number (0) of negative factors   odd number (1) of negative factors
positive product   negative product

(a) $5 \cdot 2 = 10$ **(b)** $+4(+3) = +12$ or 12   **(c)** $(-2)3 = -6$ **(d)** $5(-3) = -15$
(e) $+7(-4) = -28$ **(f)** $(-5)(+6) = -30$ **(g)** $-2(-8) = +16$ or 16
(h) $(-3)(-2)(-5) = -30$ **(i)** $5(-1)2(-1)(-2)(-1) = +20$ or 20

PROBLEM 3-9 Divide integers: **(a)** $12 \div 3$ **(b)** $-18 \div 2$ **(c)** $20 \div (-5)$
(d) $-10 \div (-2)$ **(e)** Divide $+8$ by $+2$ **(f)** $\frac{+15}{-3}$ **(g)** $-6\overline{)-12}$ **(h)** $0 \div (-4)$
(i) $-4 \div 0$ **(j)** $0 \div 0$

Solution: Recall that to divide two integers, you first find the quotient of the absolute values and then use the sign rules for multiplication to write the correct sign on the quotient [see Example 3-22]:

two positive numbers   one negative number
positive quotient   negative quotient

(a) $12 \div 3 = 4$ **(b)** $-18 \div 2 = -9$

two negative numbers
positive quotient

(c) $20 \div (-5) = -4$ **(d)** $-10 \div (-2) = +5$ or 5
(e) Divide $+8$ by $+2 = +8 \div (+2) = +4$ or 4   **(f)** $\frac{+15}{-3} = +15 \div (-3) = -5$
(g) $-6\overline{)-12} = -12 \div (-6) = +2$ or 2   **(h)** $0 \div (-4) = 0$
(i) $-4 \div 0$ is not defined.   **(j)** $0 \div 0$ is not defined.

PROBLEM 3-10 Rename each fraction in three equivalent ways using sign changes:

(a) $\frac{1}{2}$ **(b)** $-\frac{-2}{3}$ **(c)** $-\frac{5}{-8}$ **(d)** $\frac{-1}{-4}$ **(e)** $-\frac{3}{5}$ **(f)** $\frac{-5}{6}$ **(g)** $\frac{2}{-9}$ **(h)** $-\frac{-1}{-3}$

Solution: Recall that if exactly two of a fraction's three signs are changed, you will always get an equal fraction [see Examples 3-24 and 3-25]:

(a) $\frac{1}{2} = -\frac{-1}{2} = -\frac{1}{-2} = \frac{-1}{-2}$ **(b)** $-\frac{-2}{3} = -\frac{2}{-3} = \frac{-2}{-3} = \frac{2}{3}$

(c) $-\frac{5}{-8} = -\frac{-5}{8} = \frac{-5}{-8} = \frac{5}{8}$ **(d)** $\frac{-1}{-4} = \frac{1}{4} = -\frac{-1}{4} = -\frac{1}{-4}$

(e) $-\frac{3}{5} = \frac{-3}{5} = \frac{3}{-5} = -\frac{-3}{-5}$ **(f)** $\frac{-5}{6} = -\frac{5}{6} = \frac{5}{-6} = -\frac{-5}{-6}$

(g) $\frac{2}{-9} = -\frac{2}{9} = \frac{-2}{9} = -\frac{-2}{-9}$ **(h)** $-\frac{-1}{-3} = -\frac{1}{3} = \frac{-1}{3} = \frac{1}{-3}$

PROBLEM 3-11 Simplify each rational number: **(a)** $\frac{12}{18}$ **(b)** $\frac{-10}{-12}$ **(c)** $-\frac{42}{-14}$

(d) $\frac{-22}{-8}$ **(e)** $-\frac{12}{3}$ **(f)** $\frac{-15}{10}$ **(g)** $\frac{24}{-36}$ **(h)** $-\frac{-12}{-30}$

Solution: Recall the Fundamental Rule for Rational Numbers: If a, b, and c are integers ($b \neq 0$ and $c \neq 0$), then $\dfrac{a \cdot c}{b \cdot c} = \dfrac{a}{b}$ [see Example 3-27]:

(a) $\dfrac{12}{18} = \dfrac{2 \cdot \cancel{6}}{3 \cdot \cancel{6}} = \dfrac{2}{3}$
(b) $-\dfrac{-10}{12} = \dfrac{10}{12} = \dfrac{\cancel{2} \cdot 5}{\cancel{2} \cdot 6} = \dfrac{5}{6}$
(c) $-\dfrac{42}{-14} = \dfrac{42}{14} = \dfrac{3 \cdot \cancel{14}}{1 \cdot \cancel{14}} = \dfrac{3}{1} = 3$

(d) $\dfrac{-22}{-8} = \dfrac{22}{8} = \dfrac{\cancel{2} \cdot 11}{\cancel{2} \cdot 4} = \dfrac{11}{4} = 2\dfrac{3}{4}$
(e) $-\dfrac{12}{3} = -\dfrac{\cancel{3} \cdot 4}{1 \cdot \cancel{3}} = -\dfrac{4}{1} = -4$

(f) $\dfrac{-15}{10} = -\dfrac{15}{10} = -\dfrac{3 \cdot \cancel{5}}{2 \cdot \cancel{5}} = -\dfrac{3}{2}$ or $-1\dfrac{1}{2}$
(g) $\dfrac{24}{-36} = -\dfrac{24}{36} = -\dfrac{2 \cdot \cancel{12}}{3 \cdot \cancel{12}} = -\dfrac{2}{3}$

(h) $-\dfrac{-12}{-30} = -\dfrac{12}{30} = -\dfrac{2 \cdot \cancel{6}}{5 \cdot \cancel{6}} = -\dfrac{2}{5}$

PROBLEM 3-12 Add rational numbers: (a) $3 + \frac{3}{8}$ (b) $-0.5 + 0.2$ (c) $0.75 + (-\frac{3}{8})$
(d) $-1\frac{1}{2} + (-1.25)$ (e) $6 + \frac{1}{2} + 0.2 + (-3\frac{1}{4}) + (-2.75) + (-1)$

Solution: Recall that to add rational numbers, you use the Addition Rules for Real Numbers with Like (or Unlike) Signs given in Section 3-4, part A [see Example 3-28]:

(a) $3 + \frac{3}{8} = 3\frac{3}{8}$ or 3.375 (b) $-0.5 + 0.2 = -0.3$ or $-\frac{3}{10}$
(c) $0.75 + (-\frac{3}{8}) = 0.75 + (-0.375) = 0.375$ or $\frac{3}{8}$
(d) $-1\frac{1}{2} + (-1.25) = -1\frac{1}{2} + (-1\frac{1}{4}) = -2\frac{3}{4}$ or -2.75
(e) $6 + \frac{1}{2} + 0.2 + (-3\frac{1}{4}) + (-2.75) + (-1) = 6 + 0.5 + 0.2 + (-3.25) + (-2.75) + (-1)$

$$= 6.7 + (-7)$$
$$= -0.3 \text{ or } -\tfrac{3}{10}$$

PROBLEM 3-13 Subtract rational numbers: (a) $5 - \frac{3}{4}$ (b) $1.3 - 2.9$ (c) $0.25 - (-\frac{4}{5})$
(d) $-2\frac{1}{2} - (-3.75)$

Solution: Recall that to subtract rational numbers, you use the Subtraction Rules for Real Numbers given in Section 3-4, part B [see Example 3-29]:

(a) $5 - \frac{3}{4} = 4\frac{4}{4} - \frac{3}{4} = 4\frac{1}{4}$ or 4.25 (b) $1.3 - 2.9 = -1.6$ or $-1\frac{3}{5}$
(c) $0.25 - (-\frac{4}{5}) = 0.25 + (+\frac{4}{5}) = 0.25 + 0.8 = 1.05$ or $1\frac{1}{20}$
(d) $-2\frac{1}{2} - (-3.75) = -2\frac{1}{2} + (+3.75) = -2.5 + 3.75 = 1.25$ or $1\frac{1}{4}$

PROBLEM 3-14 Multiply rational numbers: (a) $\frac{1}{2}(-\frac{3}{4})$ (b) $-0.5(2.6)$ (c) $\frac{1}{4}(0.25)$
(d) $-1.25(-2\frac{3}{4})$ (e) $-3(0.5)(-\frac{1}{2})(-2.5)(-2)(0)(-8)(-1.75)$

Solution: Recall that to multiply rational numbers, you use the Multiplication Rules for Real Numbers given in Section 3-4, part C [see Example 3-30]:

(a) $\frac{1}{2}(-\frac{3}{4}) = -\frac{3}{8}$ or -0.375 (b) $-0.5(2.6) = -1.3$ or $-1\frac{3}{10}$
(c) $\frac{1}{4}(0.25) = \frac{1}{4} \cdot \frac{1}{4} = \frac{1}{16}$ or 0.0625 (d) $-1.25(-2\frac{3}{4}) = -1.25(-2.75) = 3.4375$ or $3\frac{7}{16}$
(e) $-3(0.5)(-\frac{1}{2})(-2.5)(-2)(0)(-8)(-1.75) = 0$

PROBLEM 3-15 Divide rational numbers: (a) $-\frac{2}{3} \div \frac{3}{4}$ (b) $1.2 \div (-0.5)$ (c) $\frac{3}{4} \div 0.25$
(d) $-2.5 \div (-1\frac{1}{2})$ (e) $0 \div 0.5$ (f) $-3.6 \div 0$

Solution: Recall that to divide rational numbers, you use the Division Rules for Real Numbers given in Section 3-4, part D [see Example 3-31]:

(a) $-\frac{2}{3} \div \frac{3}{4} = -\frac{2}{3} \cdot \frac{4}{3} = -\frac{8}{9}$ or $-0.\overline{8}$ (b) $1.2 \div (-0.5) = -2.4$ or $-2\frac{2}{5}$
(c) $\frac{3}{4} \div 0.25 = \frac{3}{4} \div \frac{1}{4} = \frac{3}{\cancel{4}} \cdot \frac{\cancel{4}}{1} = \frac{3}{1} = 3$ (d) $-2.5 \div (-1\frac{1}{2}) = -2.5 \div (-1.5) = 1.\overline{6}$ or $1\frac{2}{3}$
(e) $0 \div 0.5 = 0$ (f) $-3.6 \div 0$ is not defined

PROBLEM 3-16 Evaluate using the Order of Operations: **(a)** $8 - 6 + 2$ **(b)** $8 \div 2 \cdot 4$

(c) $9 - 3 \cdot 5$ **(d)** $9 + 3 \div 4$ **(e)** $4 + 6(-2)$ **(f)** $\dfrac{4 + 6}{2}$ **(g)** $10 \div (2 + 3)$

(h) $5 - \dfrac{8}{1 + 3} \cdot 3 + 6$ **(i)** $2 + [6 - (-4)(-2 + 8)] \div 6$

(j) $-3 + \{-2[16 + (-4 - (-3))] \div (-5)\}(-2)$ **(k)** $3 \cdot 2^3$ **(l)** $(-5 \cdot 2)^2$
(m) $-2 \cdot 3^2 + 10$ **(n)** $3^2 \cdot 2 \div 3\sqrt{16}$ **(o)** $-\sqrt{5^2 - 4^2}$ **(p)** $3^2(1 - 2) + 5\sqrt{4}$

(q) $2^3(3 - 7) \div 16\sqrt{9}$ **(r)** $\dfrac{11 + \sqrt{(-11)^2 - 4(6)(-10)}}{2(6)}$

Solution: Recall that to evaluate using the Order of Operations, you:
1. Perform operations inside grouping symbols such as ( ), [ ], { }, —, and $\sqrt{}$.
2. Evaluate each power notation and radical.
3. Multiply or divide in order from left to right.
4. Add or subtract in order from left to right.
   [See Examples 3-32, 3-33, and 3-34]:

(a) $8 - 6 + 2 = 2 + 2 = 4$ **(b)** $8 \div 2 \cdot 4 = 4 \cdot 4 = 16$ **(c)** $9 - 3 \cdot 5 = 9 - 15 = -6$

(d) $9 + 3 \div 4 = 9 + \dfrac{3}{4} = 9\dfrac{3}{4}$ **(e)** $4 + 6(-2) = 4 + (-12) = -8$ **(f)** $\dfrac{4 + 6}{2} = \dfrac{10}{2} = 5$

(g) $10 \div (2 + 3) = 10 \div 5 = 2$

(h) $5 - \dfrac{8}{1 + 3} \cdot 3 + 6 = 5 - \dfrac{8}{4} \cdot 3 + 6 = 5 - 2 \cdot 3 + 6 = 5 - 6 + 6 = -1 + 6 = 5$

(i) $2 + [6 - (-4)(-2 + 8)] \div 6 = 2 + [6 - (-4)(6)] \div 6$

$\qquad\qquad = 2 + [6 - (-24)] \div 6$

$\qquad\qquad = 2 + 30 \div 6$

$\qquad\qquad = 2 + 5$

$\qquad\qquad = 7$

(j) $-3 + \{-2[16 + (-4 - (-3))] \div (-5)\}(-2) = -3 + \{-2[16 + (-1)] \div (-5)\}(-2)$

$\qquad\qquad\qquad = -3 + \{-2[15] \div (-5)\}(-2)$

$\qquad\qquad\qquad = -3 + \{-30 \div (-5)\}(-2)$

$\qquad\qquad\qquad = -3 + \{6\}(-2)$

$\qquad\qquad\qquad = -3 + (-12)$

$\qquad\qquad\qquad = -15$

(k) $3 \cdot 2^3 = 3 \cdot 8 = 24$ **(l)** $(-5 \cdot 2)^2 = (-10)^2 = (-10)(-10) = 100$
(m) $-2 \cdot 3^2 + 10 = -2 \cdot 9 + 10 = -18 + 10 = -8$
(n) $3^2 \cdot 2 \div 3\sqrt{16} = 9 \cdot 2 \div 3(4) = 18 \div 3(4) = 6(4) = 24$
(o) $-\sqrt{5^2 - 4^2} = -\sqrt{25 - 16} = -\sqrt{9} = -(3) = -3$
(p) $3^2(1 - 2) + 5\sqrt{4} = 3^2(-1) + 5\sqrt{4} = 9(-1) + 5(2) = -9 + 5(2) = -9 + 10 = 1$
(q) $2^3(3 - 7) \div 16\sqrt{9} = 2^3(-4) \div 16\sqrt{9} = 8(-4) \div 16(3) = -32 \div 16(3) = -2(3) = -6$

(r) $\dfrac{11 + \sqrt{(-11)^2 - 4(6)(-10)}}{2(6)} = \dfrac{11 + \sqrt{121 - 4(6)(-10)}}{2(6)}$

$$= \dfrac{11 + \sqrt{121 - 24(-10)}}{2(6)}$$

$$= \dfrac{11 + \sqrt{121 + 240}}{2(6)}$$

$$= \dfrac{11 + \sqrt{361}}{2(6)}$$

$$= \dfrac{11 + 19}{2(6)}$$

$$= \dfrac{30}{2(6)}$$

$$= \dfrac{5(\cancel{6})}{2(\cancel{6})}$$

$$= \dfrac{5}{2} \text{ or } 2\dfrac{1}{2} \text{ or } 2.5$$

Supplementary Exercises

PROBLEM 3-17 Write the opposite of each real number: **(a)** 6   **(b)** -8 **(c)** $+9$
(d) 0   **(e)** 1   **(f)** -1 **(g)** $\frac{3}{4}$ **(h)** $-\frac{2}{3}$ **(i)** $+\frac{1}{8}$ **(j)** 1.25   **(k)** -2.5 **(l)** $+0.75$
(m) $0.\overline{6}$ **(n)** $-0.\overline{3}$ **(o)** $\sqrt{2}$ **(p)** $-\sqrt{5}$ **(q)** $-\pi$ **(r)** π **(s)** $2\frac{1}{2}$ **(t)** $-3\frac{3}{4}$
(u) $0.1\overline{6}$ **(v)** $-0.8\overline{3}$ **(w)** 2^3 **(x)** -3^2 **(y)** $(-4)^2$ **(z)** $-(-5)^3$

PROBLEM 3-18 Compare each pair of real numbers using $<$, $>$, or $=$: **(a)** 5, 8
(b) $-3, 4$ **(c)** 5, -9 **(d)** $-6, -8$ **(e)** 0, 7   **(f)** 0, -5 **(g)** $\frac{1}{2}, \frac{1}{3}$ **(h)** $-\frac{3}{4}, \frac{1}{2}$
(i) $-\frac{15}{20}, -\frac{2}{3}$ **(j)** $\frac{0}{-5}, 0$ **(k)** 1.5, 1.6   **(l)** $0.\overline{3}, -\frac{1}{3}$ **(m)** $-0.75, -\frac{3}{4}$ **(n)** $\frac{15}{16}, \frac{31}{32}$
(o) 0.023, 0.032   **(p)** $-0.12, -0.21$ **(q)** $-1.0101, 1.0101$ **(r)** $\frac{7}{16}, 0.4375$ **(s)** $1\frac{1}{2}, \frac{3}{2}$
(t) $2.5, -2\frac{1}{2}$ **(u)** $-1.75, -1\frac{1}{4}$ **(v)** $\frac{300}{7}, 42\frac{6}{7}$ **(w)** $-\frac{18}{40}, -\frac{12}{32}$ **(x)** $-0.0001, -0.000101$
(y) $2.125, \frac{17}{8}$ **(z)** $\pi, 3.14$

PROBLEM 3-19 Find each absolute value: **(a)** $|2|$ **(b)** $|-9|$ **(c)** $|+7|$ **(d)** $|0|$ **(e)** $|1|$
(f) $|-1|$ **(g)** $|\frac{1}{2}|$ **(h)** $|-\frac{2}{3}|$ **(i)** $|+\frac{1}{8}|$ **(j)** $|1.25|$ **(k)** $|2.5|$ **(l)** $|0.75|$ **(m)** $|0.\overline{6}|$
(n) $|0.\overline{3}|$ **(o)** $|\sqrt{2}|$ **(p)** $|-\sqrt{5}|$ **(q)** $|-\pi|$ **(r)** $|\pi|$ **(s)** $|2\frac{1}{2}|$ **(t)** $|-3\frac{3}{4}|$ **(u)** $|0.1\overline{6}|$
(v) $|0.8\overline{3}|$ **(w)** $|2^3|$ **(x)** $|-3^2|$ **(y)** $|(-4)^2|$ **(z)** $|-(-5)^3|$

PROBLEM 3-20 Compute with integers: **(a)** $+4 + (+2)$ **(b)** $7 + (-7)$ **(c)** $4 - (-7)$
(d) $-6 + 5$ **(e)** $-8 + (-9)$ **(f)** $-8 - 6$ **(g)** $-7 - (-8)$ **(h)** $2 - 9$
(i) $2 + 3 + 5$ **(j)** $-3 + 5 - 4$ **(k)** $4 - (-3) + 5$ **(l)** $-1 - 3 + (-4)$
(m) $2 - 3 + 1 + (-2)$ **(n)** $-2 - 3 + (-4) - (-3)$ **(o)** 4×8 **(p)** $36 \div 6$ **(q)** $-6(7)$
(r) $-21 \div 3$ **(s)** $8(-3)$ **(t)** $\frac{18}{6}$ **(u)** $-7(-7)$ **(v)** $\frac{-35}{-7}$ **(w)** $5 \cdot 3 \cdot 2$
(x) $50 \div 5 \div (-2)$ **(y)** $0 \times 8 \div (-6)$ **(z)** $3(-2)(-6) \div 4 \div 0$

PROBLEM 3-21 Simplify each rational number: (a) $\frac{24}{40}$ (b) $-\frac{20}{50}$ (c) $\frac{16}{-80}$

(d) $\frac{-75}{-90}$ (e) $-\frac{-20}{120}$ (f) $-\frac{84}{-120}$ (g) $-\frac{8}{10}$ (h) $-\frac{-3}{-9}$ (i) $\frac{3}{3}$ (j) $\frac{-12}{12}$

(k) $\frac{9}{-3}$ (l) $\frac{-18}{-2}$ (m) $-\frac{10}{-6}$ (n) $-\frac{-14}{4}$ (o) $-\frac{48}{24}$ (p) $-\frac{-56}{-32}$ (q) $\frac{256}{32}$

(r) $\frac{-125}{40}$ (s) $\frac{15}{-25}$ (t) $\frac{-20}{-45}$ (u) $-\frac{12}{18}$ (v) $-\frac{-9}{27}$ (w) $\frac{20}{-1}$ (x) $-\frac{-90}{1}$

(y) $-\frac{0}{-8}$ (z) $-\frac{-6}{0}$

PROBLEM 3-22 Compute with rational numbers: (a) $0.5 + 0.9$ (b) $1\frac{1}{5} + (-0.9)$
(c) $-0.8 + 2\frac{3}{5}$ (d) $-3\frac{1}{2} + (-2.1)$ (e) $\frac{1}{2} + \frac{7}{16}$ (f) $\frac{-3}{4} + 0.75$ (g) $-3\frac{1}{2} + 2.5$
(h) $-5\frac{1}{8} + (-7.25)$ (i) $0.2 - 0.6$ (j) $-0.8 - 1\frac{1}{2}$ (k) $-\frac{3}{8} - (-0.25)$ (l) $4\frac{3}{16} - (2.125)$
(m) $0.5(0.2)$ (n) $\frac{1}{2} \cdot \frac{3}{4}$ (o) $-4.8(-\frac{3}{10})$ (p) $-\frac{1}{4}(-0.\overline{3})$ (q) $0.8(-0.7)$ (r) $-\frac{3}{8} \cdot \frac{2}{3}$
(s) $0.7 \div 2\frac{2}{5}$ (t) $\frac{3}{8} \div 0.5$ (u) $-1.44 \div (-4.8)$ (v) $-\frac{1}{12} \div (-0.25)$ (w) $-4.5 \div 1.8$
(x) $\frac{1}{4} \div 1.8$ (y) $-6.125 \div (-1\frac{5}{16})$ (z) $-8\frac{3}{4} \div 2.5$

PROBLEM 3-23 Evaluate using the Order of Operations: (a) $4 \cdot 6(-3)$ (b) $\frac{-4}{-6} + 3$
(c) $4 - 6 \div (-3)$ (d) $-4(-6) - 3$ (e) $[4 + (-6)](-3)$ (f) $(-4 - 6) \div 3$
(g) $-5[-7 + 2(-3)]$ (h) $-20 \div [10 - (-3)(-2)]$ (i) $7[-4 - (-10) \div 2] + 2$
(j) $5 - [12 + (-2)(-4)] \div (-5)$ (k) $-6 + \{-4[2 - (\frac{-20}{10} - 2)] - (-15)\} \div 9$
(l) $4 - \{3 + [-2 + (-8 + \frac{-24}{3}) \div (-4)](-2)\}2$ (m) $4 \cdot 2^3$ (n) $3(-5^2)$ (o) $5(1 - 2)^2$

(p) $\frac{(4-1)^2}{6}$ (q) $4(5)^2 + 8(5) - 6$ (r) $(5 - 1)^2 + \frac{8}{2} - (3 + 5)^2$

(s) $2 - [5(6 - 3^2)^2 + 3] \div 8$ (t) $\sqrt{\frac{2(-4)}{-5(40)}}$ (u) $\frac{-13 + \sqrt{13^2 - 4(6)(6)}}{2(6)}$

(v) $\sqrt{8^2 - 15}$ (w) $-\sqrt{49} + 1^8$ (x) $\sqrt{3^2 + 4^2}$ (y) $\sqrt{5^2 - 3^2}$ (z) $-2(3^2 - \sqrt{9})^2$

Answers to Supplementary Exercises

(3-17) (a) -6 (b) 8 (c) -9 (d) 0 (e) -1 (f) 1 (g) $-\frac{3}{4}$ (h) $\frac{2}{3}$ (i) $-\frac{1}{8}$
(j) -1.25 (k) 2.5 (l) -0.75 (m) $-0.\overline{6}$ (n) $0.\overline{3}$ (o) $-\sqrt{2}$ (p) $\sqrt{5}$ (q) π
(r) $-\pi$ (s) $-2\frac{1}{2}$ (t) $3\frac{3}{4}$ (u) $-0.1\overline{6}$ (v) $0.8\overline{3}$ (w) -2^3 (x) 3^2 (y) $-(-4)^2$
(z) $(-5)^3$

(3-18) (a) $5 < 8$ (b) $-3 < 4$ (c) $5 > -9$ (d) $-6 > -8$ (e) $0 < 7$ (f) $0 > -5$
(g) $\frac{1}{2} > \frac{1}{3}$ (h) $-\frac{3}{4} < \frac{1}{2}$ (i) $-\frac{15}{20} < -\frac{2}{3}$ (j) $\frac{-0}{-5} = 0$ (k) $1.5 < 1.6$ (l) $0.\overline{3} > -\frac{1}{3}$
(m) $-0.75 = -\frac{3}{4}$ (n) $\frac{15}{16} < \frac{31}{32}$ (o) $0.023 < 0.032$ (p) $-0.12 > -0.21$
(q) $-1.0101 < 1.0101$ (r) $\frac{7}{16} = 0.4375$ (s) $1\frac{1}{2} = \frac{3}{2}$ (t) $2.5 > -2\frac{1}{2}$ (u) $-1.75 < -1\frac{1}{4}$
(v) $\frac{300}{7} = 42\frac{6}{7}$ (w) $-\frac{18}{40} < -\frac{12}{32}$ (x) $-0.0001 > -0.000101$ (y) $2.125 = \frac{17}{8}$
(z) $\pi > 3.14$

(3-19) (a) 2 (b) 9 (c) 7 (d) 0 (e) 1 (f) 1 (g) $\frac{1}{2}$ (h) $\frac{2}{3}$ (i) $\frac{1}{8}$ (j) 1.25
(k) 2.5 (l) 0.75 (m) $0.\overline{6}$ (n) $0.\overline{3}$ (o) $\sqrt{2}$ (p) $\sqrt{5}$ (q) π (r) π (s) $2\frac{1}{2}$
(t) $3\frac{3}{4}$ (u) $0.1\overline{6}$ (v) $0.8\overline{3}$ (w) 2^3 (x) 3^2 (y) 4^2 (z) 5^3

(3-20) **(a)** 6 **(b)** 0 **(c)** 11 **(d)** -1 **(e)** -17 **(f)** -14 **(g)** 1 **(h)** -7 **(i)** 10 **(j)** -2 **(k)** 12 **(l)** -8 **(m)** -2 **(n)** -6 **(o)** 32 **(p)** 6 **(q)** -42 **(r)** -7 **(s)** -24 **(t)** 3 **(u)** 49 **(v)** 5 **(w)** 30 **(x)** -5 **(y)** 0 **(z)** not defined

(3-21) **(a)** $\frac{3}{5}$ **(b)** $-\frac{2}{5}$ **(c)** $-\frac{1}{5}$ **(d)** $\frac{5}{6}$ **(e)** $\frac{1}{6}$ **(f)** $\frac{7}{10}$ **(g)** $-\frac{4}{5}$ **(h)** $-\frac{1}{3}$ **(i)** 1 **(j)** -1 **(k)** -3 **(l)** 9 **(m)** $\frac{5}{3}$ or $1\frac{2}{3}$ **(n)** $\frac{7}{2}$ or $3\frac{1}{2}$ **(o)** -2 **(p)** $-\frac{7}{4}$ or $-1\frac{3}{4}$ **(q)** 8 **(r)** $-\frac{25}{8}$ or $-3\frac{1}{8}$ **(s)** $-\frac{3}{5}$ **(t)** $\frac{4}{9}$ **(u)** $-\frac{2}{3}$ **(v)** $\frac{1}{3}$ **(w)** -20 **(x)** 90 **(y)** 0 **(z)** not defined

(3-22) **(a)** 1.4 or $1\frac{2}{5}$ **(b)** 0.3 or $\frac{3}{10}$ **(c)** 1.8 or $1\frac{4}{5}$ **(d)** -5.6 or $-5\frac{3}{5}$ **(e)** $\frac{15}{16}$ or 0.9375 **(f)** 0 **(g)** -1 **(h)** -12.375 or $-12\frac{3}{8}$ **(i)** -0.4 or $-\frac{2}{5}$ **(j)** -2.3 or $-2\frac{3}{10}$ **(k)** -0.125 or $-\frac{1}{8}$ **(l)** 2.0625 or $2\frac{1}{16}$ **(m)** 0.1 or $\frac{1}{10}$ **(n)** $\frac{3}{8}$ or 0.375 **(o)** 1.44 or $1\frac{11}{25}$ **(p)** $\frac{1}{12}$ or $0.08\overline{3}$ **(q)** -0.56 or $-\frac{14}{25}$ **(r)** $-\frac{1}{4}$ or -0.25 **(s)** $0.291\overline{6}$ or $\frac{7}{24}$ **(t)** $\frac{3}{4}$ or 0.75 **(u)** 0.3 or $\frac{3}{10}$ **(v)** $\frac{1}{3}$ or $0.\overline{3}$ **(w)** -2.5 or $-2\frac{1}{2}$ **(x)** $\frac{5}{36}$ or 0.138 **(y)** $4\frac{2}{3}$ or $4.\overline{6}$ **(z)** $-3\frac{1}{2}$ or -3.5

(3-23) **(a)** -72 **(b)** $\frac{11}{3}$ or $3\frac{2}{3}$ **(c)** 6 **(d)** 21 **(e)** 6 **(f)** $-\frac{10}{3}$ or $-3\frac{1}{3}$ **(g)** 65 **(h)** -5 **(i)** 9 **(j)** 9 **(k)** -7 **(l)** -60 **(m)** 32 **(n)** -75 **(o)** 5 **(p)** $\frac{3}{2}$ or $1\frac{1}{2}$ **(q)** 134 **(r)** -44 **(s)** -4 **(t)** $\frac{1}{5}$ **(u)** $-\frac{2}{3}$ **(v)** 7 **(w)** -6 **(x)** 5 **(y)** 4 **(z)** -72

4 PROPERTIES AND EXPRESSIONS

THIS CHAPTER IS ABOUT

☑ **Identifying Properties**
☑ **Using the Distributive Properties**
☑ **Clearing Parentheses**
☑ **Combining Like Terms**
☑ **Evaluating Expressions**
☑ **Evaluating Formulas**

4-1. Identifying Properties

A. Write indicated products and quotients that involve variables.

It is often necessary in algebra to write **indicated products** of real numbers and/or **variables** (letters).

Recall: There are five correct ways to write the indicated product of two real numbers. For example, the real number expression "2 times 5" can be written as

$$2 \times 5 \qquad 2 \cdot 5 \qquad 2(5) \qquad (2)5 \qquad (2)(5)$$

There are also five correct ways to write the indicated product of variables and real numbers, or the indicated product of variables and variables.

EXAMPLE 4-1: Write five correct ways to indicate the products of the following expressions:
(a) 2 times x **(b)** -3 times y **(c)** m times n

Solution
(a) 2 times x $= 2 \cdot x$ $= 2(x)$ $= (2)x$ $= (2)(x)$ $= 2x$

(b) -3 times $y = -3 \cdot y = -3(y) = (-3)y = (-3)(y) = -3y$ ⟶ preferred form

(c) m times n $= m \cdot n$ $= m(n)$ $= (m)n$ $= (m)(n)$ $= mn$

Note: When a variable is used as a factor, the traditional "times" symbol $\times$ is never used because it looks too much like the letter x.

It is often necessary in algebra to write **indicated quotients** of real numbers and/or variables.

Recall: There are six correct ways to write the indicated quotient of two real numbers. For example, the real number expression "8 divided by -3" can be written as

$$8 \div (-3) \qquad 8/(-3) \qquad 8 \cdot \frac{1}{-3} \qquad \frac{1}{-3} \cdot 8 \qquad \frac{8}{-3} \qquad -3\overline{)8}$$

There are five correct ways to write the indicated quotient of variables and real numbers, or the indicated quotient of variables and variables.

EXAMPLE 4-2: Write five correct ways to indicate the quotients of the following expressions:
(a) x divided by -3 **(b)** 8 divided by y **(c)** h divided by k

Solution

(a) x divided by $-3 = x \div (-3) = x/(-3) = x \cdot \dfrac{1}{-3} = \dfrac{1}{-3} \cdot x = \dfrac{x}{-3}$

(b) 8 divided by $y \quad = 8 \div y \quad = 8/y \quad = 8 \cdot \dfrac{1}{y} \quad = \dfrac{1}{y} \cdot 8 \quad = \dfrac{8}{y} \quad \longleftarrow$ preferred form

(c) h divided by $k \quad = h \div k \quad = h/k \quad = h \cdot \dfrac{1}{k} \quad = \dfrac{1}{k} \cdot h \quad = \dfrac{h}{k}$

Note: When variables are used in algebra, the traditional division symbol $\div$ and the slash division symbol /, although acceptable, are seldom used to write an indicated quotient. The preferred form of division notation is $\dfrac{a}{b}$.

EXAMPLE 4-3: Write the following expressions using the preferred form of division notation:
(a) x divided by $(x + 3)$ **(b)** $(y - 2)$ divided by 6   **(c)** $(m + n)$ divided by $(m - n)$

Solution

(a) x divided by $(x + 3)$ $= \dfrac{x}{x + 3}$

(b) $(y - 2)$ divided by 6 $= \dfrac{y - 2}{6} \quad \longleftarrow$ preferred division form

(c) $(m + n)$ divided by $(m - n) = \dfrac{m + n}{m - n}$

Note: If you choose to use a form other than the preferred form of division notation, you must use parentheses around the divisor to avoid any errors in computation, as in the expressions $x \div (x + 3)$ or $x/(x + 3)$.

Caution: Never write an expression like x divided by $(x + 3)$ as $x \div x + 3$ or $x/x + 3$. Omitting the parentheses will cause you to get a wrong answer when you simplify the expression.

B. Identify the commutative properties.

The set of real numbers follows certain rules called **properties.**

When you change the **order** of real number addends, the sum of the addends does not change.

EXAMPLE 4-4: Show that the sum does not change when the order of the addends is changed in $5 + 2$.

Solution: different order $\quad \begin{array}{l} 5 + 2 = 7 \\ 2 + 5 = 7 \end{array}$ same sum

Example 4-4 illustrates the following property:

Commutative Property of Addition
If a and b are real numbers, then $a + b = b + a$.

When you change the order of real number factors, the product of the factors does not change.

EXAMPLE 4-5: Show that the product does not change when the order of the factors is changed in $2 \cdot 3$.

Solution: different order $\quad \begin{array}{l} 2 \cdot 3 = 6 \\ 3 \cdot 2 = 6 \end{array}$ same product

Example 4-5 illustrates the following property:

Commutative Property of Multiplication
If a and b are real numbers, then $ab = ba$.

The Commutative Property of Addition and the Commutative Property of Multiplication are jointly called the **commutative properties.**

Caution: Subtraction of real numbers is not commutative.

EXAMPLE 4-6: Show that the difference changes when the order of the minuend and subtrahend are changed in $5 - 2$.

Solution: different order $\begin{array}{l} 5 - 2 = 3 \\ 2 - 5 = -3 \end{array}$ different differences

Caution: Division of real numbers is not commutative.

EXAMPLE 4-7: Show that the quotient changes when the order of the dividend and divisor are changed in $2 \div 1$.

Solution: different order $\begin{array}{l} 2 \div 1 = 2 \\ 1 \div 2 = \frac{1}{2} \end{array}$ different quotients

C. Identify the associative properties.

When you change the **grouping** of real number addends, the sum of the addends does not change.

EXAMPLE 4-8: Show that the sum does not change when the grouping of the addends is changed in $(2 + 3) + 4$.

Solution: different groupings $\begin{array}{l} (2 + 3) + 4 = 5 + 4 = 9 \\ 2 + (3 + 4) = 2 + 7 = 9 \end{array}$ same sum

Note: Use the Order of Operations first to clear grouping symbols.

Example 4-8 illustrates the following property:

Associative Property of Addition
If a, b, and c are real numbers, then $(a + b) + c = a + (b + c)$.

When you change the grouping of real number factors, the product of the factors does not change.

EXAMPLE 4-9: Show that the product does not change when the grouping of the factors is changed in $(2 \cdot 3)4$.

Solution: different groupings $\begin{array}{l} (2 \cdot 3)4 = 6 \cdot 4 = 24 \\ 2(3 \cdot 4) = 2 \cdot 12 = 24 \end{array}$ same product

Note: Use the Order of Operations first to clear grouping symbols.

Example 4-9 illustrates the following property:

Associative Property of Multiplication
If a, b, and c are real numbers, then $(ab)c = a(bc)$.

The Associative Property of Addition and the Associative Property of Multiplication are jointly called the **associative properties.**

Caution: Subtraction of real numbers is not associative.

EXAMPLE 4-10: Show that the difference changes when the grouping is changed in $(9 - 5) - 3$.

Solution: different groupings $\longrightarrow$ $(9 - 5) - 3 = 4 - 3 = 1 \longleftarrow$ different answers
$\longrightarrow$ $9 - (5 - 3) = 9 - 2 = 7 \longleftarrow$

Note: Use the Order of Operations first to clear grouping symbols.

Caution: Division of real numbers is not associative.

EXAMPLE 4-11: Show that the quotient changes when the grouping is changed in $(24 \div 6) \div 2$.

Solution: different groupings $\longrightarrow$ $(24 \div 6) \div 2 = 4 \div 2 = 2 \longleftarrow$ different answers
$\longrightarrow$ $24 \div (6 \div 2) = 24 \div 3 = 8 \longleftarrow$

D. Identify the distributive properties.

To multiply a sum by a real number, you can first add and then multiply, or you can first multiply and then add.

EXAMPLE 4-12: Compute $2(4 + 3)$ by **(a)** first adding and then multiplying, **(b)** first multiplying and then adding.

Solution

(a) First add, then multiply: $2(4 + 3) = 2 \cdot 7 = 14 \longleftarrow$ same answer
(b) First multiply, then add: $2(4 + 3) = 2 \cdot 4 + 2 \cdot 3 = 8 + 6 = 14 \longleftarrow$

Example 4-12, part **(b)** illustrates the following property:

Distributive Property of Multiplication Over Addition
If a, b, and c are real numbers, then $a(b + c) = ab + ac$
and $(b + c)a = ba + ca$.

To multiply a difference by a real number, you can first subtract and then multiply, or you can first multiply and then subtract.

EXAMPLE 4-13: Compute $4(5 - 2)$ by **(a)** first subtracting and then multiplying, **(b)** first multiplying and then subtracting.

Solution: **(a)** First subtract, then multiply: $4(5 - 2) = 4 \cdot 3 = 12 \longleftarrow$ same answer
(b) First multiply, then subtract: $4(5 - 2) = 4 \cdot 5 - 4 \cdot 2 = 20 - 8 = 12 \longleftarrow$

Example 4-13, part **b** illustrates the following property:

Distributive Property of Multiplication Over Subtraction
If a, b, and c are real numbers, then $a(b - c) = ab - ac$
and $(b - c)a = ba - ca$.

Note: The Distributive Property of Multiplication Over Addition and the Distributive Property of Multiplication Over Subtraction are jointly called the **distributive properties.**

E. Identify properties involving zero or one.

There are several important properties of real numbers that involve zero.

EXAMPLE 4-14: List the important properties of real numbers that involve zero.

Solution: Properties Involving Zero
If a and b are real numbers, then:

$a + 0 = 0 + a = a$ **Identity Property for Addition**
Because $a + 0 = a$ for every real number a, 0 is called the *identity element for addition.*

$$a + (-a) = -a + a = 0$$

Additive Inverse Property
Because $a + (-a) = 0$ for every real number a, the symbol $-a$ is called the *opposite* of a or the *additive inverse* of a.

$$a - 0 = a + 0 = a$$
$$0 - a = 0 + (-a) = -a$$
$$a - b = 0 \text{ means } a = b$$

$$a \cdot 0 = 0 \cdot a = 0$$

Zero-Factor Property
The Zero-Factor Property states that a zero factor always gives a zero product.

$$ab = 0 \text{ means } a = 0 \text{ or } b = 0$$

Zero-Product Property
The Zero-Product Property states that the only way for the product of two real numbers to be zero is for one or both of the real numbers to be zero.

$$0 \div a = \frac{0}{a} = 0 \ (a \neq 0)$$

Zero-Dividend Property
The Zero-Dividend Property states that zero divided by any nonzero real number equals zero.

$$a \div 0 \text{ and } \frac{a}{0} \text{ are not defined.}$$

Zero-Divisor Property
The Zero-Divisor Property states that dividing by zero is not defined.

There are several important properties of real numbers that involve the number one.

EXAMPLE 4-15: List the important properties of real numbers that involve one.

Solution:
If a and b are real numbers, then:

Properties Involving One

$$a \cdot 1 = 1 \cdot a = a$$

Identity Property for Multiplication
Because $a \cdot 1 = a$ for every real number a, one is called the *identity element for multiplication.*

$$a \cdot \frac{1}{a} = \frac{1}{a} \cdot a = 1 \ (a \neq 0)$$

Multiplicative Inverse Property
Because $a \cdot \dfrac{1}{a} = 1$ for every nonzero real number a, $\dfrac{1}{a} (a \neq 0)$ is called the reciprocal of a or the *multiplicative inverse* of a.

$$a \div 1 = \frac{a}{1} = a$$

$$a \div b = \frac{a}{b} = 1 \text{ means } a = b$$
$$(a \neq 0 \text{ and } b \neq 0)$$

Unit-Fraction Property
The Unit-Fraction Property states that the only way for the quotient of two real numbers to equal one is for the two real numbers to be the same.

F. Identify properties involving negative signs.

There are several important properties of real numbers that involve negative signs.

EXAMPLE 4-16: List the important properties of real numbers that involve negative signs.

Solution: **Properties Involving Negative Signs**
If a and b are real numbers, then:

$$-a = -1 \cdot a = a(-1)$$
Read $-a$ as "the opposite of a."

$$-(-a) = a$$
Read $-(-a)$ as "the opposite of the opposite of a."

$$ab = -a(-b)$$

$$-(ab) = -a(b) = a(-b) = -(-a)(-b)$$

G. Identify the Subtraction and Division Properties.

There are two important properties of real numbers that involve subtraction and division. You have already been introduced to each of these properties in previous sections of this book.

The subtraction property of real numbers states that to subtract two real numbers, you change subtraction to addition while writing the opposite of the subtrahend.

EXAMPLE 4-17: State the subtraction property algebraically.

Solution: **Subtraction Property**
If a and b are real numbers, then $a - b = a + (-b)$.

The **division property** of real numbers states that to divide fractions, you change division to multiplication while writing the reciprocal of the divisor.

EXAMPLE 4-18: State the division property algebraically.

Solution: **Division Property**
If a and b are real numbers, then $\dfrac{a}{b} = a \cdot \dfrac{1}{b}$ or $\dfrac{1}{b} \cdot a$.

4-2. Using the Distributive Properties

A. Compute using the distributive properties.

Caution: To compute with real numbers using the distributive properties, you must multiply each number inside the parentheses separately by the number outside the parentheses.

EXAMPLE 4-19: Evaluate the following expressions using the distributive properties: **(a)** $(3 + 1)5$
(b) $3(7 - 2)$

Solution

(a) Correct Method

$$(3 + 1)5 = 3 \cdot 5 + 1 \cdot 5$$
$$= 15 + 5$$
$$= 20 \longleftarrow \text{correct answer}$$

Wrong Method

$$(3 + 1)5 = 3 + 1 \cdot 5 \quad \text{No! (Multiply both 3 and 1 by 5.)}$$
$$= 3 + 5$$
$$= 8 \longleftarrow \text{wrong answer}$$

(b) Correct Method

$$3(7 - 2) = 3 \cdot 7 - 3 \cdot 2$$
$$= 21 - 6$$
$$= 15 \longleftarrow \text{correct answer}$$

Wrong Method

$$3(7 - 2) = 3 \cdot 7 - 2 \quad \text{No! (Multiply both 7 and 2 by 3.)}$$
$$= 21 - 2$$
$$= 19 \longleftarrow \text{wrong answer}$$

B. Compute without using the distributive properties.

Sometimes it is easier to compute with real numbers without using the distributive properties. To compute without using the distributive properties, you use the Order of Operations.

EXAMPLE 4-20: Evaluate the following without using the distributive properties:
(a) $(3 + 1)5$ **(b)** $3(7 - 2)$

Solution

(a) $(3 + 1)5 = 4 \cdot 5$ Use the Order of Operations by first adding inside the parentheses.

$$= 20 \longleftarrow \text{same answer as found in Example 4-19}$$
using a distributive property

(b) $3(7 - 2) = 3 \cdot 5$ Use the Order of Operations by first subtracting inside the parentheses.

$\qquad\qquad = 15 \longleftarrow$ same answer as found in Example 4-19
$\qquad\qquad\qquad\qquad\qquad$ using a distributive property

14-3. Clearing Parentheses

A. Identify algebraic expressions.

A number, variable, or the sum, difference, product, quotient, or square root of numbers and variables is called an **algebraic expression.**

EXAMPLE 4-21: List several different algebraic expressions.

Solution: The following are all algebraic expressions:

(a) 3   **(b)** x **(c)** $2y$ **(d)** $\dfrac{abc}{5}$ **(e)** $\dfrac{u}{v}$ **(f)** $2w + 3$ **(g)** $6z - \sqrt{z}$

(h) $m^2 - 2mn + n^2$

B. Identify terms.

In an algebraic expression that does not contain grouping symbols, the addition symbols separate the **terms.**

EXAMPLE 4-22: Identify the terms of each algebraic expression listed in Example 4-21.

Solution
(a) The number 3 is called a **constant term.**
(b) The variable x is called a **letter term.**
(c) Terms with only products and/or quotients of numbers and variables like $2y$, $\dfrac{abc}{5}$, and $\dfrac{u}{v}$ are called **general terms.**
(d) $2w + 3$ has two terms ($2w$ and 3) which are separated by an addition sign
(e) $6z - \sqrt{z}$ has two terms ($6z$ and $-\sqrt{z}$) because $6z - \sqrt{z} = 6z + (-\sqrt{z})$.
(f) $m^2 - 2mn + n^2$ has three terms m^2, $-2mn$, and n^2.

Caution: To identify the terms of algebraic expressions like $-2(x + 3)$ or $\dfrac{w - 2}{5}$, you must first **clear grouping symbols.** Often this can be done by using the distributive properties.

C. Clear parentheses using the distributive properties.

Recall that to **clear parentheses** by evaluating an indicated product like $(3 + 1)5$ or $3(7 - 2)$, it is almost always easier to compute without using the distributive properties [see Example 4-20].

However, to clear parentheses in an algebraic expression like $-2(x + 3)$ or $(m - n)3$, you must use the distributive properties because it is otherwise impossible to add different terms like x and 3 or to subtract different terms like m and n.

EXAMPLE 4-23: Clear parentheses and simplify the following expressions: **(a)** $-2(x + 3)$
(b) $(m - n)3$ **(c)** $5(2 + y)$ **(d)** $(-3r - 2s)(-4)$

Solution
(a) $-2(x + 3) = -2 \cdot x + (-2)3$ **(b)** $(m - n)3 = m \cdot 3 - n \cdot 3$

$\qquad\qquad\quad = -2x + (-6)$ $= 3 \cdot m - 3 \cdot n$

$\qquad\qquad\quad = -2x - 6$ $= 3m - 3n$

(c) $5(2 + y) = 5 \cdot 2 + 5 \cdot y$
$$= 10 + 5y$$
$$\text{or } 5y + 10$$

(d) $(-3r - 2s)(-4) = -3r(-4) - 2s(-4)$
$$= -3(-4)r - 2(-4)s$$
$$= +12r - (-8s)$$
$$= 12r + (+8s)$$
$$= 12r + 8s$$

Note: The terms of **(a)** $-2(x + 3)$ are $-2x$ and -6 because $-2(x + 3) = -2x - 6$.

(b) $(m - n)3$ are $3m$ and $-3n$ because $(m - n)3 = 3m - 3n$.

(c) $5(2 + y)$ are $5y$ and 10 because $5(2 + y) = 5y + 10$.

(d) $(-3r - 2s)(-4)$ are $12r$ and $8s$ because $(-3r - 2s)(-4) = 12r + 8s$.

When parentheses have a positive sign (or no sign) in front of them, you can clear parentheses by just writing the same algebraic expression that is inside of the parentheses.

EXAMPLE 4-24: Clear parentheses in $+(3a + 2)$.

Solution: $+(3a + 2) = 3a + 2$ because $+(3a + 2) = +1(3a + 2)$
$$= (+1)3a + (+1)2$$
$$= 3a + 2$$

Note: The terms of $+(3a + 2)$ are $3a$ and 2 because $+(3a + 2) = 3a + 2$.

When parentheses have a negative sign in front of them, you can clear parentheses by just writing the opposite of each term inside the parentheses.

EXAMPLE 4-25: Clear parentheses in $-(7b - 3)$.

Solution: $-(7b - 3) = -7b + 3$ because $-(7b - 3) = -1(7b - 3)$
$$= (-1)7b - (-1)3$$
$$= -7b - (-3)$$
$$= -7b + (+3)$$
$$= -7b + 3$$

Note: The terms of $-(7b - 3)$ are $-7b$ and 3 because $-(7b - 3) = -7b + 3$.

4-4. Combining Like Terms

A. Identify the numerical coefficient and the literal part of a term.

In a term, the number that multiplies the variable(s) is called the **numerical coefficient**.

EXAMPLE 4-26: Identify the numerical coefficient in each of the following: **(a)** $2y$ **(b)** x
(c) $-w^2$ **(d)** $\dfrac{abc}{5}$

Solution
(a) In $2y$, the numerical coefficient is 2.
(b) In x, the numerical coefficient is 1 because $x = 1x$.

(c) In $-w^2$, the numerical coefficient is -1 because $-w^2 = -1w^2$.

(d) In $\dfrac{abc}{5}$, the numerical coefficient is $\dfrac{1}{5}$ because $\dfrac{abc}{5} = \dfrac{1}{5}abc$.

The part of a term that is not the numerical coefficient is called the **literal part** of the term.

EXAMPLE 4-27: Identify the literal part in each of the following:

(a) $2y$ **(b)** x **(c)** $-w^2$ **(d)** $\dfrac{abc}{5}$

Solution
(a) In $2y$, the literal part is the variable y.
(b) In x, the literal part is the variable x.
(c) In $-w^2$, the literal part is the variable w^2.

(d) In $\dfrac{abc}{5}$, the literal part is the product of variables abc.

B. Identify like terms.

Terms that have the same literal part are called **like terms**. Terms that have different literal parts are called **unlike terms**.

EXAMPLE 4-28: Identify the like terms in each of the following: **(a)** $2x + 5 - x$

(b) $y - 5y + 3x$ **(c)** $\dfrac{mn}{2} - 3 + 5mn$ **(d)** $5w + \dfrac{3}{w}$ **(e)** $-5y + 3x$ **(f)** $mn + 2m$

(g) $2x^2 + 3x$

Solution
(a) In $2x + 5 - x$, the like terms are $2x$ and $-x$ because they have the same literal part x.
(b) In $y - 5y + 3x$, the like terms are y and $-5y$ because they have the same literal part y.

(c) In $\dfrac{mn}{2} - 3 + 5mn$, the like terms are $\dfrac{mn}{2}$ and $5mn$ because they have the same literal part mn.

(d) In $5w + \dfrac{3}{w}$, there are no like terms because the literal parts of $5w$ and $\dfrac{3}{w}$ are different (w and $\dfrac{1}{w}$, respectively).

(e) In $-5y + 3x$, there are no like terms because the literal parts of $-5y$ and $3x$ are different (y and x, respectively).

(f) In $mn + 2m$, there are no like terms because the literal parts of mn and $2m$ are different (mn and m, respectively).

(g) In $2x^2 + 3x$, there are no like terms because the literal parts of $2x^2$ and $3x$ are different (x^2 and x, respectively).

C. Combine like terms.

To **combine like terms** in an algebraic expression, you use the distributive properties to add or subtract the numerical coefficients of like terms and then write the same like literal part on the sum or difference.

EXAMPLE 4-29: Combine like terms in the following expressions: **(a)** $2w + 3w$ **(b)** $8ab - 5ab$

(c) $2x + 5 - x$ **(d)** $y - 5y + 3x$ **(e)** $\dfrac{mn}{2} - 3 + 5mn$ **(f)** $8u - 5v - 6u + v$

Solution

(a) $2w + 3w = (2 + 3)w$ Combine like terms using the distributive property of multiplication over addition: $ac + bc = (a + b)c$

$\qquad\qquad = 5w$

(b) $8ab - 5ab = (8 - 5)ab$ Combine the like terms using the distributive property of multiplication over subtraction: $ac - bc = (a - b)c$

$\qquad\qquad = 3ab$

like terms

(c) $2x + 5 - x = \overbrace{2x - x} + 5$ Group like terms.

$= 2x - 1x + 5$ *Think:* $-x = -1x$

$= (2 - 1)x + 5$ Combine like terms.

$= 1x + 5$

$= x + 5$ Simplify.

(d) $y - 5y + 3x = 1y - 5y + 3x$ *Think:* $y = 1y$

$= (1 - 5)y + 3x$ Combine like terms.

$= -4y + 3x$

$= 3x - 4y$ Write terms in alphabetical order.

like terms

(e) $\dfrac{mn}{2} - 3 + 5mn = \overbrace{\dfrac{mn}{2} + 5mn} - 3$ Group like terms.

$= \dfrac{1}{2}mn + 5mn - 3$ *Think:* $\dfrac{mn}{2} = \dfrac{1}{2}mn$

$= \left(\dfrac{1}{2} + 5\right)mn - 3$ Combine like terms.

$= 5\dfrac{1}{2}mn - 3$ or $\dfrac{11}{2}mn - 3$ or $\dfrac{11mn}{2} - 3$

(f) $8u - 5v - 6u + v = (8u - 6u) + (-5v + 1v)$ Group like terms together.

$= (8 - 6)u + (-5 + 1)v$ Combine like terms.

$= 2u + (-4)v$

$= 2u - 4v$ Simplify.

Note 1: In $3x - 4y$, $3x$ and $-4y$ cannot be combined because $3x$ and $-4y$ are not like terms.

Note 2: In $2u - 4v$, $2u$ and $-4v$ cannot be combined because $2u$ and $-4v$ are not like terms.

Caution: Only like terms can be combined.

D. Clear parentheses and combine like terms.

To clear parentheses and combine like terms, you first combine any like terms inside the parentheses, then clear the parentheses, and then combine any like terms that remain.

EXAMPLE 4-30: Clear parentheses and combine like terms in the following expressions:
(a) $2(x + 5) - 5x$ **(b)** $-(2y - 3) + (3y - 5)$ **(c)** $3(2a - b + 5a) - 8b$
(d) $6m - 2(m + n)$

Solution
(a) $2(x + 5) - 5x = \mathbf{2} \cdot x + \mathbf{2} \cdot 5 - 5x$ Clear parentheses.

$= 2x + 10 - 5x$ Simplify.

$= 2x - 5x + 10$ Group like terms.

$= -3x + 10$ Combine like terms.

or $10 - 3x$

(b) $\quad -(2y - 3) + (3y - 5) = -2y + 3 + 3y - 5 \qquad$ Clear parentheses.

$$= -2y + 3y + 3 - 5 \qquad \text{Group like terms.}$$

$$= 1y + (-2) \qquad \text{Combine like terms.}$$

$$= y - 2 \qquad \text{Simplify.}$$

(c) $\quad 3(2a - b + 5a) - 8b = 3(2a + 5a - b) - 8b \qquad$ Combine like terms inside the parentheses.

$$= 3(7a - b) - 8b$$

$$= 3 \cdot 7a - 3 \cdot b - 8b \qquad \text{Clear parentheses.}$$

$$= 21a - 3b - 8b \qquad \text{Simplify.}$$

$$= 21a - 11b \qquad \text{Combine like terms.}$$

(d) $\quad 6m - 2(m + n) = 6m + (-2)(m + n) \qquad$ Rename terms.

$$= 6m + (-2)m + (-2)n \qquad \text{Clear parentheses.}$$

$$= 6m - 2m - 2n \qquad \text{Simplify.}$$

$$= 4m - 2n \qquad \text{Combine like terms.}$$

Caution: To avoid making an error in Example 4-30, part **(d)**, you first rename $6m - 2(m + n)$ as $6m + (-2)(m + n)$.

EXAMPLE 4-31: Show how it is possible to make an error when $6m - 2(m + n)$ is not first renamed as $6m + (-2)(m + n)$.

Solution

$\quad\quad\quad\quad\quad$ Wrong Method

$6m - 2(m + n) = 6m - 2 \cdot m + 2 \cdot n \qquad$ No! Multiply each term inside the parentheses by -2, not 2.

$$= 6m - 2m + 2n \qquad \text{Simplify.}$$

$$= (6 - 2)m + 2n \qquad \text{Combine like terms.}$$

$$= 4m + 2n \longleftarrow \text{wrong answer (The correct answer is } 4m - 2n.)$$

4-5. Evaluating Expressions

A. Evaluate algebraic expressions in one variable.

To **evaluate an algebraic expression in one variable** given a numerical value for that variable, you first substitute the given value for the variable in each term that the variable appears in, and then evaluate using the Order of Operations.

EXAMPLE 4-32: Evaluate $2x^2 - 5x - 6$ for **(a)** $x = 3$ $\qquad$ **(b)** $x = -2$.

Solution

(a) $\quad 2x^2 - 5x - 6 = 2(3)^2 - 5(3) - 6 \qquad\qquad$ Substitute 3 for x.

$$= 2 \cdot 9 - 5(3) - 6 \qquad\qquad \text{Evaluate using the Order of Operations}$$
$$\qquad\qquad\qquad\qquad\qquad [\text{see Example 3-32}].$$

$$= 18 - 5(3) - 6$$

$$= 18 - 15 - 6$$

$$= 3 - 6$$

$$= -3$$

(b) $2x^2 - 5x - 6 = 2(-2)^2 - 5(-2) - 6$ Substitute -2 for x.

$$= 2 \cdot 4 - 5(-2) - 6$$ Evaluate using the Order of Operations.

$$= 8 - 5(-2) - 6$$

$$= 8 + 10 - 6$$

$$= 18 - 6$$

$$= 12$$

Note 1: For $x = 3$: $2x^2 - 5x - 6 = -3$

Note 2: For $x = -2$: $2x^2 - 5x - 6 = 12$

Caution: To avoid making an error, always use parentheses when substituting a numerical value for a variable, as in Example 4-32.

B. Evaluate algebraic expressions in two or more variables.

To **evaluate an algebraic expression in two or more variables** given a numerical value for each different variable, you first substitute each given value for the associated variable and then evaluate using the Order of Operations.

EXAMPLE 4-33: Evaluate $\dfrac{-b + \sqrt{b^2 - 4ac}}{2a}$ for $a = 2$, $b = -1$, and $c = -3$.

Solution: $\dfrac{-b + \sqrt{b^2 - 4ac}}{2a} = \dfrac{-(-1) + \sqrt{(-1)^2 - 4(2)(-3)}}{2(2)}$ Substitute 2 for a, -1 for b, and -3 for c.

$$= \frac{1 + \sqrt{1 - 4(2)(-3)}}{2(2)}$$ Evaluate using the Order of Operations.

$$= \frac{1 + \sqrt{1 + 24}}{2(2)}$$

$$= \frac{1 + \sqrt{25}}{2(2)}$$

$$= \frac{1 + 5}{2(2)}$$

$$= \frac{6}{2(2)}$$

$$= \frac{2(3)}{2(2)}$$

$$= \frac{3}{2} \text{ or } 1\frac{1}{2} \text{ or } 1.5$$

Note: For $a = 2$, $b = -1$, and $c = -3$, $\dfrac{-b + \sqrt{b^2 - 4ac}}{2a} = \dfrac{3}{2}$ or $1\frac{1}{2}$ or 1.5.

4-6. Evaluating Formulas

A. Identify the parts of an equation.

An **equation** is a **mathematical sentence** that contains an **equality symbol.** Every equation has three parts: a **left member,** an equality symbol, and a **right member.**

EXAMPLE 4-34: Identify the three different parts of the following equations: **(a)** $m + 3 = 2$
(b) $2n = 6$ **(c)** $4w - 3 = 1$ **(d)** $2x + 3y = -5$ **(e)** $C = \frac{5}{9}(F - 32)$

Solution

equality symbol

(a) $m + 3 \stackrel{.}{=} 2$

(b) $2n = 6$

(c) left member ⟷ $4w - 3 = 1$ ⟷ right member

(d) $2x + 3y = -5$

(e) $C = \frac{5}{9}(F - 32)$

B. Identify the parts of a formula

A **formula** is an equation that contains two or more variables and represents a known phenomenon. A **formula is solved for a given variable** when that variable is isolated in one member of the equation. When a formula is solved for a given variable, the algebraic expression in the other member of the equation is called the **solution.**

EXAMPLE 4-35: Identify the variable that each formula is solved for and the solution in each of the following: **(a) Temperature Formula:** $C = \frac{5}{9}(F - 32)$ **(b) Distance Formula:** $d = rt$
(c) Volume Formula: $V = lwh$

Solution
(a) The formula $C = \frac{5}{9}(F - 32)$ is solved for C and the solution is $\frac{5}{9}(F - 32)$.
(b) The formula $d = rt$ is solved for d and the solution is rt.
(c) The formula $V = lwh$ is solved for V and the solution is lwh.

C. Evaluate formulas to solve problems.

To **evaluate a formula that is solved for a specific variable** given a measurement value for each of the other variables, you substitute each given value for the associated variable and then evaluate using the Order of Operations.

EXAMPLE 4-36: Water boils at 212 degrees Fahrenheit F. Find the temperature in degrees Celsius C at which water boils by evaluating the temperature formula $C = \frac{5}{9}(F - 32)$.

Solution: The question asks you to evaluate $C = \frac{5}{9}(F - 32)$ for $F = 212$.

$C = \frac{5}{9}(F - 32)$ Write the given formula.

$\quad = \frac{5}{9}(\mathbf{212} - 32)$ Substitute 212 for F.

$\quad = \frac{5}{9} \cdot 180$ Evaluate using the Order of Operations.

$\quad = 100$ *Think:* $\frac{5}{9} \cdot 180 = \frac{5}{9} \cdot \frac{180}{1} = \frac{5}{\cancel{9}} \cdot \frac{\cancel{9} \cdot 20}{1} = 100$

Note: $C = 100$ means the Celsius temperature at which water boils is **100°C.**

D. Evaluate a formula when one of the variables represents a rate.

To **evaluate a formula when one of the variables represents a rate,** you may first need to rename to get a common unit of measure that can be eliminated.

EXAMPLE 4-37: What distance d can a car travel in a time t of 20 minutes at a constant rate r of 45 mph (miles per hour) using the distance formula $d = rt$?

different time units

Solution: The question asks you to evaluate $d = rt$ for $r = 45$ miles/hour and $t = 20$ minutes.

Caution: Before evaluating $d = rt$, you must have a common time unit.

20 minutes $= \frac{20}{60}$ hour $= \frac{1}{3}$ hour	Rename to get a common time unit.
$d = rt$	Write the given formula.
$= 45 \dfrac{\text{miles}}{\text{hour}} \cdot \dfrac{1}{3} \text{hour}$	Substitute 45 mph for r and $\frac{1}{3}$ hour for t.
$= 45 \dfrac{\text{miles}}{\cancel{\text{hour}}} \cdot \dfrac{1}{3} \dfrac{\cancel{\text{hour}}}{1}$	Eliminate the common time unit.
$= (45 \cdot \frac{1}{3})$ miles	Multiply a measure by a number.
$= 15$ miles	Simplify.

Note: In 20 minutes, at a constant rate of 45 mph, a car can travel 15 miles.

RAISE YOUR GRADES

Can you . . . ?

☑ write indicated products of real numbers and/or variables
☑ write indicated quotients of real numbers and/or variables
☑ identify the commutative properties
☑ identify the associative properties
☑ identify the distributive properties
☑ identify properties involving zero or one
☑ identify properties involving negative signs
☑ identify the subtraction and division properties
☑ compute using the distributive properties
☑ compute using the Order of Operations
☑ identify algebraic expressions and terms
☑ clear parentheses using the distributive properties
☑ identify numerical coefficients and literal parts in terms
☑ identify like terms and unlike terms
☑ combine like terms
☑ clear parentheses and combine like terms
☑ evaluate algebraic expressions in one or more variables
☑ identify the parts of an equation
☑ identify the parts of a formula
☑ evaluate formulas to solve problems
☑ evaluate formulas when one of the variables represents a rate

SUMMARY

1. The following are five correct ways to write the indicated product of the expression
   (a) 3 times w (b) -8 times z (c) r times s:

   (a) 3 times w $= 3 \cdot w$ $= 3(w)$ $= (3)w$ $= (3)(w)$ $= 3w$
   (b) -8 times z $= -8 \cdot z$ $= -8(z)$ $= (-8)z$ $= (-8)(z)$ $= -8z$ ⟶ preferred form
   (c) r times s $= r \cdot s$ $= r(s)$ $= (r)s$ $= (r)(s)$ $= rs$

2. The following are five correct ways to write the indicated quotient of the following variables and real numbers: (a) y divided by 2   (b) -5 divided by w (c) r divided by s
   (d) 3 divided by $(x + 3)$

(a) y divided by 2 $\quad = y \div 2 \quad = y/2 \quad = y \cdot \dfrac{1}{2} \quad = \dfrac{1}{2} \cdot y \quad = \dfrac{y}{2}$ ⟵ preferred form

(b) -5 divided by $w = -5 \div w = -5/w = -5 \cdot \dfrac{1}{w} = \dfrac{1}{w}(-5) = \dfrac{-5}{w}$

(c) r divided by $s \quad = r \div s \quad = r/s \quad = r \cdot \dfrac{1}{s} \quad = \dfrac{1}{s} \cdot r \quad = \dfrac{r}{s}$

(d) 3 divided by $(x + 3) = 3 \div (x + 3) = 3/(x + 3) = 3 \cdot \dfrac{1}{x + 3} = \dfrac{1}{x + 3} \cdot 3 = \dfrac{3}{x + 3}$

3. **Commutative Property of Addition**
   If a and b are real numbers, then $a + b = b + a$.
4. **Commutative Property of Multiplication**
   If a and b are real numbers, then $ab = ba$.
5. Real numbers are not commutative with respect to subtraction or division.
6. **Associative Property of Addition**
   If a, b, and c are real numbers, then $(a + b) + c = a + (b + c)$.
7. **Associative Property of Multiplication**
   If a, b, and c are real numbers, then $(ab)c = a(bc)$.
8. Real numbers are not associative with respect to subtraction or division.
9. **Distributive Property of Multiplication Over Addition**
   If a, b, and c are real numbers, then $a(b + c) = ab + ac$
   $\qquad\qquad$ and $(b + c)a = ba + ca$.
10. **Distributive Property of Multiplication Over Subtraction**
    If a, b, and c are real numbers, then $a(b - c) = ab - ac$
    $\qquad\qquad$ and $(b - c)a = ba - ca$.
11. The common properties involving zero or one are

    Identity Property for Addition: $\qquad a + 0 = 0 + a = a$

    Identity Property for Multiplication: $a \cdot 1 = 1 \cdot a = a$

    Additive Inverse Property: $\qquad a + (-a) = -a + a = 0$

    Multiplicative Inverse Property: $\quad a \cdot \dfrac{1}{a} = \dfrac{1}{a} \cdot a = 1$

    Zero-Factor Property: $\qquad\qquad a \cdot 0 = 0 \cdot a = 0$

    Zero-Product Property: $\qquad\qquad ab = 0$ means $a = 0$ or $b = 0$

    Unit-Fraction Property: $\qquad\qquad \dfrac{a}{b} = 1$ means $a = b$

    Zero-Dividend Property: $\qquad\qquad 0 \div a = \dfrac{0}{a} = 0 \ (a \neq 0)$

    Zero-Divisor Property: $\qquad\qquad a \div 0$ and $\dfrac{a}{0}$ are not defined.

12. **Properties Involving Negative Signs**
    If a and b are real numbers, then
    (a) $-a = -1 \cdot a = a(-1)$
    (b) $-(-a) = a$
    (c) $ab = -a(-b)$
    (d) $-(ab) = -a \cdot b = a(-b) = -(-a)(-b)$
13. **Subtraction Property**
    If a and b are real numbers, then $a - b = a + (-b)$.
14. **Division Property**
    If a and b are real numbers, then $\dfrac{a}{b} = a \cdot \dfrac{1}{b}$ or $\dfrac{1}{b} \cdot a$.
15. To compute using the distributive properties, you must multiply each term inside the parentheses separately by the term outside the parentheses.

16. To compute without using the distributive properties, you use the Order of Operations.
17. A number, variable, or the sum, difference, product, quotient, or square root of numbers and variables is called an algebraic expression.
18. In an algebraic expression that does not contain grouping symbols, the addition symbols separate the terms.
19. To clear parentheses by evaluating a product like $(3 + 1)5$ or $3(7 - 2)$, it is almost always easier to compute without using the distributive properties. However, to clear parentheses in an algebraic expression like $-2(x + 3)$ or $(m - n)3$, you must use the distributive properties because it is otherwise impossible to add different terms like x and 3 or to subtract different terms like m and n.
20. When parentheses have a positive sign (or no sign) in front of them, you can clear parentheses by just writing the same algebraic expression that is inside the parentheses.
21. When parentheses have a negative sign in front of them, you can clear parentheses by writing the opposite of each term inside the parentheses.
22. In a term, the number that multiplies the variable(s) is called the numerical coefficient.
23. The part of a term that is not the numerical coefficient is called the literal part of the term.
24. Terms that have the same literal part are called like terms.
25. Terms that have different literal parts are called unlike terms.
26. To combine like terms in an algebraic expression, you use the distributive properties to add or subtract the numerical coefficients of like terms and then write the same like literal part on the sum or difference.
27. Only like terms can be combined.
28. To clear parentheses and combine like terms, you first combine any like terms inside the parentheses, then clear the parentheses, and then combine like terms.
29. To evaluate an algebraic expression in one variable given a numerical value for that variable, you first substitute the given value for the variable in each term that the variable appears in, and then evaluate using the Order of Operations.
30. To avoid making an error, always use parentheses when substituting a numerical value for a variable.
31. To evaluate an algebraic expression in two or more variables given a numerical value for each different variable, you first substitute each given value for the associated variable and then evaluate using the Order of Operations.
32. An equation is a mathematical sentence that contains an equality symbol.
33. Every equation has three parts; a left member, an equality symbol, and a right member.
34. A formula is an equation that contains two or more variables and represents a known phenomenon.
35. A formula is solved for a given variable when that variable is isolated on one side of the equation.
36. When a formula is solved for a given variable, the algebraic expression on the other side of the equation is called the solution for the given variable.
37. To evaluate a formula that is solved for a specific variable, given a measurement value for each of the other variables, you substitute each given value for the associated variable and then evaluate using the Order of Operations.
38. To evaluate a formula when one of the variables represents a rate, you may first need to rename to get a common unit of measure that can be eliminated.

SOLVED PROBLEMS

PROBLEM 4-1 Write each indicated product in five ways for the following expressions:
(a) 4 times m (b) -5 times n (c) x times y

Solution: Recall that the traditional "times" symbol $\times$ is never used with variables because it looks too much like the letter x [see Example 4-1]:

(a) 4 times m $=$ $4 \cdot m =$ $4(m) =$ $(4)m =$ $(4)(m) = 4m$

(b) -5 times $n = -5 \cdot n = -5(n) = (-5)n = (-5)(n) = -5n$ ⟵ preferred form

(c) x times y $=$ $x \cdot y =$ $x(y) =$ $(x)y =$ $(x)(y) = xy$

PROBLEM 4-2   Write each indicated quotient in five ways for the following expressions:
(a) m divided by 8   (b) -3 divided by n (c) x divided by y

Solution: Recall that the traditional division symbol $\div$ and slash division symbol $/$, although acceptable, are seldom used to write an indicated quotient. The preferred form of division notation is $\dfrac{a}{b}$ [see Example 4-2]:

(a) m divided by 8 $= m \div 8 = m/8 = m \cdot \dfrac{1}{8} = \dfrac{1}{8} \cdot m = \dfrac{m}{8}$

(b) -3 divided by $n = -3 \div n = -3/n = -3 \cdot \dfrac{1}{n} = \dfrac{1}{n}(-3) = \dfrac{-3}{n}$ ← preferred form

(c) x divided by y $= x \div y = x/y = x \cdot \dfrac{1}{y} = \dfrac{1}{y} \cdot x = \dfrac{x}{y}$

PROBLEM 4-3   Identify each missing number and identify the property by name when possible:
(a) $3 + ? = 2 + 3$ (b) $?(-5) = -5 \cdot 9$ (c) $(? + 3) + 6 = 8 + (3 + 6)$
(d) $(4 \cdot 6)? = 4(6 \cdot 9)$ (e) $?(5 + 3) = 2 \cdot 5 + 2 \cdot 3$ (f) $(5 - 2)3 = 5 \cdot 3 - 2 \cdot ?$
(g) $? + 0 = 4$ (h) $5 + ? = 0$ (i) $6 - ? = 6$ (j) $0 - ? = -5$ (k) $5 - ? = 0$
(l) $7 \cdot 0 = ?$ (m) $5 \cdot ? = 0$ (n) $0 \div 6 = ?$ (o) $2 \div 0 = ?$ (p) $8 \cdot ? = 1$ (q) $3 \cdot ? = 3$
(r) $7 \div ? = 7$ (s) $? \div 5 = 1$ (t) $-? = -1 \cdot 5$ (u) $-(-2) = ?$ (v) $3 \cdot 4 = -3 \cdot ?$
(w) $-(2 \cdot 5) = ?(-5)$ (x) $3 - 5 = 3 + ?$ (y) $\frac{3}{4} = 3 \cdot ?$

Solution
(a) $3 + 2 = 2 + 3$ by the Commutative Property of Addition [see Example 4-4].
(b) $9(-5) = -5 \cdot 9$ by the Commutative Property of Multiplication [see Example 4-5].
(c) $(8 + 3) + 6 = 8 + (3 + 6)$ by the Associative Property of Addition [see Example 4-8].
(d) $(4 \cdot 6)9 = 4(6 \cdot 9)$ by the Associative Property of Multiplication [see Example 4-9].
(e) $2(5 + 3) = 2 \cdot 5 + 2 \cdot 3$ by the Distributive Property of Multiplication Over Addition [see Example 4-12].
(f) $(5 - 2)3 = 5 \cdot 3 - 2 \cdot 3$ by the Distributive Property of Multiplication Over Subtraction [see Example 4-13].
(g) $4 + 0 = 4$ by the Identity Property for Addition [see Example 4-14].
(h) $5 + (-5) = 0$ by the Additive Inverse Property [see Example 4-14].
(i) $6 - 0 = 6$ [see Example 4-14].
(j) $0 - 5 = -5$ [see Example 4-14].
(k) $5 - 5 = 0$ [see Example 4-14].
(l) $7 \cdot 0 = 0$ by the Zero-Factor Property [see Example 4-14].
(m) $5 \cdot 0 = 0$ by the Zero-Product Property [see Example 4-14].
(n) $0 \div 6 = 0$ by the Zero-Dividend Property [see Example 4-14].
(o) $2 \div 0$ is not defined by the Zero-Divisor Property [see Example 4-14].
(p) $8 \cdot \frac{1}{8} = 1$ by the Multiplicative Inverse Property [see Example 4-15].
(q) $3 \cdot 1 = 3$ by the Identity Property for Multiplication [see Example 4-15].
(r) $7 \div 1 = 7$ [see Example 4-15].
(s) $5 \div 5 = 1$ by the Unit-Fraction Property [see Example 4-15].
(t) $-5 = -1 \cdot 5$ [see Example 4-16].
(u) $-(-2) = 2$ [see Example 4-16].
(v) $3 \cdot 4 = -3(-4)$ [see Example 4-16].
(w) $-(2 \cdot 5) = 2(-5)$ [see Example 4-16].
(x) $3 - 5 = 3 + (-5)$ by the Subtraction Property [see Example 4-17].
(y) $\frac{3}{4} = 3 \cdot \frac{1}{4}$ by the Division Property [see Example 4-18].

PROBLEM 17-4   Identify each term in (a) $x + 3$ (b) $y - 2$ (c) mn (d) $\dfrac{w}{2}$

(e) $2a - 3b + 2$ (f) $-4h + \dfrac{3}{h} - 2\sqrt{h}$ (g) $2(x + 5)$ (h) $3(y - 4)$ (i) $+(a^2 - a - 1)$

(j) $-(x - y)$ (k) 3 (l) w (m) $\dfrac{m + 1}{4}$ (n) $\dfrac{a - b}{c}$

Solution: Recall that in an algebraic expression that does not contain grouping symbols, the addition symbols separate the terms. Also, to identify terms when grouping symbols are present, you must first clear the grouping symbols [see Example 4-22]:

(a) In $x + 3$, the terms are x and 3.
(b) In $y - 2$, the terms are y and -2 because $y - 2 = y + (-2)$.
(c) The product of two or more numbers and variables like mn is a term.

(d) The quotient of numbers and variables like $\dfrac{w}{2}$ is a term.

(e) In $2a - 3b + 2$, the terms are $2a$, $-3b$, and 2.

(f) In $-4h + \dfrac{3}{h} - 2\sqrt{h}$, the terms are $-4h, \dfrac{3}{h}$, and $-2\sqrt{h}$.

(g) In $2(x + 5)$, the terms are $2x$ and 10 because $2(x + 5) = 2 \cdot x + 2 \cdot 5 = 2x + 10$.
(h) In $3(y - 4)$, the terms are $3y$ and -12 because $3(y - 4) = 3 \cdot y - 3 \cdot 4 = 3y - 12$.
(i) In $+(a^2 - a - 1)$, the terms are a^2, $-a$, and -1 because $+(a^2 - a - 1) = a^2 - a - 1$.
(j) In $-(x - y)$, the terms are $-x$ and y because $-(x - y) = -x + y$.
(k) A number like 3 is a constant term.
(l) A variable like w is a letter term.

(m) In $\dfrac{m + 1}{4}$, the terms are $\dfrac{m}{4}$ and $\dfrac{1}{4}$ because $\dfrac{m + 1}{4} = \dfrac{m}{4} + \dfrac{1}{4}$.

(n) In $\dfrac{a - b}{c}$, the terms are $\dfrac{a}{c}$ and $-\dfrac{b}{c}$ because $\dfrac{a - b}{c} = \dfrac{a}{c} - \dfrac{b}{c}$.

PROBLEM 4-5 Clear parentheses in (a) $5(u + 2)$ (b) $(v - 3)(-4)$ (c) $-2(x + y)$
(d) $(2m - 5n)3$ (e) $+(8h - 5k)$ (f) $-(y^2 - 2y + 5)$.

Solution: Recall that to clear parentheses in algebraic expressions like those in Problem 4-5, you must use the distributive properties because it is otherwise impossible to combine the unlike terms inside the parentheses [see Examples 4-23, 4-24, and 4-25]:

(a) $5(u + 2) = 5 \cdot u + 5 \cdot 2 = 5u + 10$ (b) $(v - 3)(-4) = v(-4) - 3(-4) = -4v + 12$
(c) $-2(x + y) = (-2)x + (-2)y = -2x - 2y$ (d) $(2m - 5n)3 = 2m \cdot 3 - 5n \cdot 3 = 6m - 15n$
(e) $+(8h - 5k) = 8h - 5k$ (f) $-(y^2 - 2y + 5) = -y^2 + 2y - 5$

PROBLEM 4-6 Combine like terms in (a) $2x + 5x$ (b) $3y - y$ (c) $w - 5w$

(d) $6ab - 5ab$ (e) $3u + 2 - 4u$ (f) $m - 5n + m$ (g) $\dfrac{c}{2} + c - \dfrac{1}{2}$

(h) $6s - 5r + 2s + 3r$.

Solution: Recall that to combine like terms in an algebraic expression, you use the distributive properties to add or subtract the numerical coefficients of like terms and then write the same literal part on the sum or difference [see Example 4-29]:

(a) $2x + 5x = (2 + 5)x = 7x$ (b) $3y - y = 3y - 1y = (3 - 1)y = 2y$
(c) $w - 5w = 1w - 5w = (1 - 5)w = -4w$ (d) $6ab - 5ab = (6 - 5)ab = 1ab = ab$
(e) $3u + 2 - 4u = 3u - 4u + 2 = (3 - 4)u + 2 = -1u + 2 = -u + 2$ or $2 - u$
(f) $m - 5n + m = m + m - 5n = 1m + 1m - 5n = (1 + 1)m - 5n = 2m - 5n$

(g) $\dfrac{c}{2} + c - \dfrac{1}{2} = \dfrac{1}{2}c + 1c - \dfrac{1}{2} = \left(\dfrac{1}{2} + 1\right)c - \dfrac{1}{2} = \dfrac{3}{2}c - \dfrac{1}{2}$ or $\dfrac{3c - 1}{2}$

(h) $6s - 5r + 2s + 3r = 6s + 2s + (-5r) + 3r = (6 + 2)s + (-5 + 3)r = 8s + (-2)r = 8s - 2r$

PROBLEM 4-7 Clear parentheses and combine like terms in (a) $3(x + 2) + 2x$
(b) $-2(3y - 4) - 5$ (c) $-(m - n) + (m + n)$ (d) $2(3a - b + 5a) - 8b$ (e) $2u - 3(u - v)$.

Solution: Recall that to clear parentheses and combine like terms, you first combine any like terms inside the parentheses, then clear the parentheses, and then combine like terms [see Examples 4-30 and 4-31]:

(a) $3(x + 2) + 2x = 3 \cdot x + 3 \cdot 2 + 2x = 3x + 6 + 2x = 3x + 2x + 6 = 5x + 6$

(b) $-2(3y - 4) - 5 = (-2)3y - (-2)4 - 5 = -6y + 8 - 5 = -6y + 3$

(c) $-(m - n) + (m + n) = -m + n + m + n = -m + m + n + n = 0 + 2n = 2n$

(d) $2(3a - b + 5a) - 8b = 2(8a - b) - 8b = 2 \cdot 8a - 2 \cdot b - 8b = 16a - 2b - 8b = 16a - 10b$

(e) $2u - 3(u - v) = 2u + (-3)(u - v) = 2u + (-3)u - (-3)v = -u + 3v$ or $3v - u$

PROBLEM 4-8 Evaluate the following algebraic expressions: (a) $5x$ for $x = 2$
(b) $-2y$ for $y = -4$ (c) $w + 3$ for $w = -6$ (d) $u - 2$ for $u = 1$ (e) $2v + 3$ for $v = 8$
(f) $3a - 5$ for $a = -2$ (g) $x^2 + 5x - 6$ for $x = 2$ (h) $x^2 + 5x - 6$ for $x = -2$

(i) $\dfrac{-b - \sqrt{b^2 - 4ac}}{2a}$ for $a = 6, b = 1,$ and $c = -12$

Solution: Recall that to evaluate an algebraic expression given a numerical value for each different variable, you first substitute each given value for the associated variable and then evaluate using the Order of Operations [see Examples 4-32 and 4-33]:

(a) $5x = 5(2) = 10$ (b) $-2y = -2(-4) = 8$ (c) $w + 3 = (-6) + 3 = -3$

(d) $u - 2 = (1) - 2 = -1$ (e) $2v + 3 = 2(8) + 3 = 16 + 3 = 19$

(f) $3a - 5 = 3(-2) - 5 = -6 - 5 = -11$

(g) $x^2 + 5x - 6 = (2)^2 + 5(2) - 6 = 4 + 10 - 6 = 8$

(h) $x^2 + 5x - 6 = (-2)^2 + 5(-2) - 6 = 4 - 10 - 6 = -12$

(i) $\dfrac{-b - \sqrt{b^2 - 4ac}}{2a} = \dfrac{-(1) - \sqrt{(1)^2 - 4(6)(-12)}}{2(6)} = \dfrac{-1 - \sqrt{289}}{12} = \dfrac{-1 - 17}{12} = \dfrac{-18}{12} = -\dfrac{3}{2}$

PROBLEM 4-9 Solve each problem by evaluating the given formula:

(a) Water freezes at 0 degrees Celsius C. Find the temperature in degrees Fahrenheit F at which water freezes by evaluating the **temperature formula** $F = \frac{9}{5}C + 32$.

(b) Find the ideal weight w for a man whose height h is 72 inches using the **male ideal weight formula** $w = 5\frac{1}{2}h - 231$.

(c) Find the ideal weight w for a woman whose height h is 60 inches using the **female ideal weight formula** $w = 5\frac{1}{4}h - 216$.

(d) What distance d can a car travel in a time t of 40 minutes at a constant rate r of 60 mph using the **distance formula** $d = rt$?

(e) What is the altitude a in kilometers when the average annual temperature t is 0 degrees Celsius using the **altitude/temperature formula** $a = 0.16(15 - t)$?

(f) What is the amount A at the end of a 9 month time period t if \$500 principal P is invested at a simple interest rate r of 6% per year using the **amount formula** $A = P(1 + rt)$?

Solution: Recall that to evaluate a formula that is solved for a specific variable given a measurement value for each of the other variables, you substitute each given value for the associated variable and then evaluate using the Order of Operations. To evaluate a formula when one of the variables represents a rate, you may first need to rename to get a common unit of measure that can be eliminated [see Examples 4-36 and 4-37]:

(a) The question asks you to evaluate $F = \frac{9}{5}C + 32$ for $C = 0$:

$$F = \tfrac{9}{5}C + 32 = \tfrac{9}{5}(0) + 32 = 0 + 32 = 32 \text{ degrees Fahrenheit}$$

(b) The question asks you to evaluate $w = 5\frac{1}{2}h - 231$ for $h = 72$:

$$w = 5\tfrac{1}{2}h - 231 = 5\tfrac{1}{2}(72) - 231 = 396 - 231 = 165 \text{ pounds}$$

(c) The question asks you to evaluate $w = 5\frac{1}{4}h - 216$ for $h = 60$:

$$w = 5\tfrac{1}{4}h - 216 = 5\tfrac{1}{4}(60) - 216 = 315 - 216 = 99 \text{ pounds}$$

(d) The question asks you to evaluate $d = rt$ for $r = 60\,\dfrac{\text{miles}}{\text{hour}}$ and $t = .40$ minutes:

$$40 \text{ minutes} = \tfrac{40}{60} \text{ hour} = \tfrac{2}{3} \text{ hour}$$

$$d = rt = 60\,\frac{\text{miles}}{\cancel{\text{hour}}} \cdot \frac{2\,\cancel{\text{hour}}}{3}\,\frac{}{1} = \left(60 \cdot \frac{2}{3}\right) \text{ miles} = 40 \text{ miles}$$

(e) The question asks you to evaluate $a = 0.16(15 - t)$ for $t = 0$:

$$a = 0.16(15 - t) = 0.16(15 - \mathbf{0}) = 0.16(15) = 2.4 \text{ kilometers}$$

(f) The question asks you to evaluate $A = P(1 + rt)$ for $P = \$500$, $r = 6\%$, and $t = 9$ months:

$$9 \text{ months} = \frac{9}{12} \text{ year} = \frac{3}{4} \text{ year}$$

$$A = P(1 + rt) = \mathbf{\$500}\left(1 + \frac{6\%}{\cancel{\text{year}}} \cdot \frac{3\,\cancel{\text{year}}}{4}\,\frac{}{1}\right) = \$500\left(1 + 0.06 \cdot \frac{3}{4}\right) = \$500(1 + 0.045)$$

$$= \$500(1.045) = \$522.50$$

Supplementary Exercises

PROBLEM 4-10   Write five correct ways to indicate the following expressions: **(a)** 7 times u
(b) -3 times v **(c)** x times y **(d)** w divided by -2 **(e)** 9 divided by z **(f)** a divided by b

PROBLEM 4-11   Identify each missing number and identify the property by name when possible:

(a) $? + 3 = 3 + (-5)$ **(b)** $2 \cdot ? = -3 \cdot 2$ **(c)** $(4 + 5) + 2 = 4 + (? + 2)$
(d) $(2 \cdot 3)5 = 2(? \cdot 5)$ **(e)** $8 + ? = 8$ **(f)** $? + (-2) = 0$ **(g)** $? - 0 = 5$ **(h)** $? - 4 = -4$
(i) $? - 3 = 0$ **(j)** $? \cdot 5 = 0$ **(k)** $0 \cdot ? = 0$ **(l)** $? \div (-2) = 0$ **(m)** $2 \div ?$ is not defined
(n) $? \cdot 1 = 6$ **(o)** $? \cdot \frac{2}{3} = 1$ **(p)** $? \cdot 7 = -7$ **(q)** $? \div 1 = -6$ **(r)** $-2 \div ? = 1$
(s) $3(8 + 9) = ? \cdot 8 + ? \cdot 9$ **(t)** $(6 + 2)5 = 6 \cdot ? + 2 \cdot ?$ **(u)** $-(-5) = ?$ **(v)** $-2 \cdot 9 = 2 \cdot ?$
(w) $3 \cdot 4 = -3 \cdot ?$ **(x)** $2 - 8 = 2 + ?$ **(y)** $\frac{2}{3} = 2 \cdot ?$ **(z)** $?(-1) = -7$

PROBLEM 4-12   Identify each term in the following: **(a)** $u + 7$ **(b)** $v - 5$
(c) $3x^2 - 5x - 4$ **(d)** $2\sqrt{y} - y\sqrt{2} + y - 2$ **(e)** $-(m + n)$ **(f)** $(a - 6)$ **(g)** $-3(b - 5)$

(h) $2(9 - c)$ **(i)** 2 **(j)** p **(k)** $-3hk$ **(l)** $\dfrac{xyz}{5}$ **(m)** $\dfrac{x + 3}{x}$ **(n)** $\dfrac{m - n}{n}$

(o) $w\sqrt{5^2 - 4^2}$ **(p)** $\dfrac{a + b - c}{2}$ **(q)** $-5(2x^2 + 3x - 2)$

PROBLEM 4-13   Clear parentheses and/or combine like terms in the following: **(a)** $-3(x + 2)$
(b) $(2 - y)8$ **(c)** $5(m + n)$ **(d)** $(x - y)(-5)$ **(e)** $3h^2 - 2h + 5$ **(f)** $-(5 - 8k)$
(g) $6u + u$ **(h)** $3v - 7v$ **(i)** $9p - p$ **(j)** $3mn - 8mn$ **(k)** $2x - 5x + x$

(l) $5y - 8 + y - 5$ **(m)** $\dfrac{w}{4} - \dfrac{3w}{4}$ **(n)** $5b - 6a + b + 3a - 2b$ **(o)** $2(h - 2) + 2h$

(p) $m - 4(m - n) - n$ **(q)** $-(a - b + c) + 2a - 4c$ **(r)** $3 + 2(v - 5)$ **(s)** $u - (u + v)$
(t) $2x - 3(4x - 5) + 6$ **(u)** $a + (a - b)$ **(v)** $x^2 + 2x + 1 - (x^2 - 3x + 5)$
(w) $2y^2 - 6y + 5 + (3y^2 + 5y - 4)$

PROBLEM 4-14   Evaluate the following algebraic expressions: **(a)** $3v$ for $v = 8$
(b) $-5u$ for $u = 2$ **(c)** $x + 5$ for $x = 2$ **(d)** $y - 7$ for $y = -8$ **(e)** $5w + 6$ for $w = -1$

(f) $2z - 9$ for $z = 6$ **(g)** $a^2 + 3a - 9$ for $a = 3$ **(h)** $a^2 + 3a - 9$ for $a = -3$

(i) $\dfrac{-b + \sqrt{b^2 - 4ac}}{2a}$ for $a = 6, b = -13$ and $c = 6$

(j) $\dfrac{-b - \sqrt{b^2 - 4ac}}{2a}$ for $a = 6, b = -13$ and $c = 6$ **(k)** $2u - 8v$ for $u = \dfrac{1}{2}$ and $v = -\dfrac{1}{4}$

(l) $-6a - 9b$ for $a = -\dfrac{1}{3}$ and $b = \dfrac{2}{3}$

PROBLEM 4-15   Solve each problem by evaluating each given formula:

(a) Find the shoe-size number n for a man who has feet each measuring 12 inches in length l when standing using the **male shoe-size formula** $n = 3l - 25$.

(b) Find the shoe-size number n for a woman who has feet each measuring $9\frac{1}{3}$ inches in length l when standing using the **female shoe-size formula** $n = 3l - 22$.

(c) Find the average temperature t in degrees Celsius at an altitude a of 12 kilometers using the **temperature/altitude formula** $t = 15 - 6.25a$.

(d) Find the Fahrenheit F temperature for the answer to problem **c** using the **temperature formula** $F = \dfrac{9}{5}C + 32$.

(e) What distance d can a car travel in a time t of 15 minutes at a constant rate r of 15 mph using the **distance formula** $d = rt$?

(f) What is the amount A at the end of an 18-month time period t if \$800 principal P is invested at a simple interest rate r of 8% per year using the **amount formula** $A = P(1 + rt)$?

(g) How much time t will it take to travel a distance d of 300 miles at a constant rate r of 45 mph using the **time formula** $t = d \div r$?

(h) What is the constant rate r that is necessary to travel a distance d of 100 miles in a time t of $1\frac{1}{2}$ hours using the **rate formula** $r = d \div t$?

Answers to Supplementary Exercises

(4-10)   **(a)** $7 \text{ times } u = 7 \cdot u = 7(u) = (7)u = (7)(u) = 7u$

(b) $-3 \text{ times } v = -3 \cdot v = -3(v) = (-3)v = (-3)(v) = -3v$

(c) $x \text{ times } y = x \cdot y = x(y) = (x)y = (x)(y) = xy$

(d) $w \text{ divided by } -2 = w \div (-2) = w/(-2) = w \cdot \dfrac{1}{-2} = \dfrac{1}{-2} \cdot w = \dfrac{w}{-2}$

(e) $9 \text{ divided by } z = 9 \div z = 9/z = 9 \cdot \dfrac{1}{z} = \dfrac{1}{z} \cdot 9 = \dfrac{9}{z}$

(f) $a \text{ divided by } b = a \div b = a/b = a \cdot \dfrac{1}{b} = \dfrac{1}{b} \cdot a = \dfrac{a}{b}$

(4-11)   **(a)** -5; Commutative Property of Addition
(b) -3; Commutative Property of Multiplication   **(c)** 5; Associative Property of Addition
(d) 3; Associative Property of Multiplication   **(e)** 0; Identity Property for Addition
(f) 2; Additive Inverse Property   **(g)** 5   **(h)** 0   **(i)** 3   **(j)** 0; Zero- Product Property
(k) any real number; Zero- Factor Property   **(l)** 0; Zero-Dividend Property
(m) 0; Zero-Divisor Property   **(n)** 6; Identity Property for Multiplication
(o) $\frac{3}{2}$; Multiplicative Inverse Property   **(p)** -1 **(q)** -6 **(r)** -2; Unit-Fraction Property

(s) 3; Distributive Property of Multiplication Over Addition
(t) 5; Distributive Property of Multiplication Over Addition   **(u)** 5   **(v)** -9 **(w)** -4
(x) -8; Subtraction Property   **(y)** $\frac{1}{3}$; Division Property   **(z)** 7

(4-12)   **(a)** $u, 7$ **(b)** $v, -5$ **(c)** $3x^2, -5x, -4$ **(d)** $2\sqrt{y}, -y\sqrt{2}, y, -2$ **(e)** $-m, -n$

(f) $a, -6$ **(g)** $-3b, 15$ **(h)** $18, -2c$ **(i)** 2   **(j)** p **(k)** $-3hk$ **(l)** $\dfrac{xyz}{5}$ **(m)** $1, \dfrac{3}{x}$

(n) $\dfrac{m}{n}, -1$ **(o)** $3w$ **(p)** $\dfrac{a}{2}, \dfrac{b}{2}, -\dfrac{c}{2}$ **(q)** $-10x^2, -15x, 10$

(4-13)   **(a)** $-3x - 6$ **(b)** $16 - 8y$ **(c)** $5m + 5n$ **(d)** $5y - 5x$ **(e)** $3h^2 - 2h + 5$
(f) $8k - 5$ **(g)** $7u$ **(h)** $-4v$ **(i)** $8p$ **(j)** $-5mn$ **(k)** $-2x$ **(l)** $6y - 13$
(m) $-\dfrac{w}{2}$ or $-\dfrac{1}{2}w$ **(n)** $4b - 3a$ or $-3a + 4b$ **(o)** $4h - 4$ **(p)** $-3m + 3n$ or $3n - 3m$
(q) $a + b - 5c$ **(r)** $2v - 7$ **(s)** $-v$ **(t)** $-10x + 21$ or $21 - 10x$ **(u)** $2a - b$
(v) $5x - 4$ **(w)** $5y^2 - y + 1$

(4-14)   **(a)** 24   **(b)** -10 **(c)** 7   **(d)** -15 **(e)** 1   **(f)** 3   **(g)** 9   **(h)** -9 **(i)** $\dfrac{3}{2}$

(j) $\dfrac{2}{3}$ **(k)** 3   **(l)** -4

(4-15)   **(a)** size 11   **(b)** size 6   **(c)** $-60°$ **(d)** $-76°F$ **(e)** $3\frac{3}{4}$ miles   **(f)** \$896
(g) $6\frac{2}{3}$ hours   **(h)** $66\frac{2}{3}$ mph

5 LINEAR EQUATIONS IN ONE VARIABLE

THIS CHAPTER IS ABOUT

☑ **Identifying Linear Equations in One Variable**
☑ **Solving Linear Equations Using Rules**
☑ **Solving Linear Equations Containing Like Terms and/or Parentheses**
☑ **Solving Linear Equations Containing Fractions, Decimals, or Percents**
☑ **Solving Literal Equations and Formulas for a Given Variable**

5-1. Identifying Linear Equations in One Variable

A. Identify linear equations in one variable.

An equation in one variable that can be written in standard form as $Ax + B = C$, where A, B, and C are real numbers ($A \neq 0$) and x is any variable is called a **linear equation in one variable.**

EXAMPLE 5-1: Which of the following are linear equations in one variable? **(a)** $y + 5 = -2$

(b) $w - 4 = 5$ **(c)** $2z = 6$ **(d)** $\dfrac{u}{5} = -2$ **(e)** $-3v - 6 = -2$ **(f)** $x\sqrt{2} + 3 = 5$

(g) $\dfrac{a}{2} + 0.3 = 1.8$ **(h)** $\dfrac{b}{-3} - 4 = 5$ **(i)** $2m + 3n = 6$ **(j)** $2 + 3 = 5$

(k) $2x^2 + 4 = 3$ **(l)** $2\sqrt{y} - 6 = 4$ **(m)** $\dfrac{2}{w} + 5 = 7$

Solution
(a) $y + 5 = -2$ or $1y + 5 = -2$ is a linear equation in one variable.
(b) $w - 4 = 5$ or $1w + (-4) = 5$ is a linear equation in one variable.
(c) $2z = 6$ or $2z + 0 = 6$ is a linear equation in one variable.

(d) $\dfrac{u}{5} = -2$ or $\dfrac{1}{5}u + 0 = -2$ is a linear equation in one variable.

(e) $-3v - 6 = -2$ or $-3v + (-6) = -2$ is a linear equation in one variable.
(f) $x\sqrt{2} + 3 = 5$ or $(\sqrt{2})x + 3 = 5$ is a linear equation in one variable.

(g) $\dfrac{a}{2} + 0.3 = 1.8$ or $\dfrac{1}{2}a + 0.3 = 1.8$ is a linear equation in one variable.

(h) $\dfrac{b}{-3} - 4 = 5$ or $-\dfrac{1}{3}b + (-4) = 5$ is a linear equation in one variable.

(i) $2m + 3n = 6$ is not a linear equation in one variable because it has two variables.
(j) $2 + 3 = 5$ is not a linear equation in one variable because it has no variables.
(k) $2x^2 + 4 = 3$ is not a linear equation in one variable because the variable has an exponent greater than 1.

96

(**l**) $2\sqrt{y} - 6 = 4$ is not a linear equation in one variable because the variable is in the radicand.

(**m**) $\dfrac{2}{w} + 5 = 7$ is not a linear equation in one variable because the variable is in the denominator.

B. Identify the eight special types of linear equations in one variable.

Every linear equation in one variable that does not contain like terms or parentheses can be easily categorized as one of **eight special types of linear equations in one variable.**

EXAMPLE 5-2: List the eight special types of linear equations in one variable.

Solution: If A, B, and C are real numbers ($A \neq 0$) and x is any variable, then
1. $x + B = C$ is called an **addition equation** [see Example 5-5].
2. $x - B = C$ is called a **subtraction equation** [see Example 5-6].
3. $\quad Ax = C$ is called a **multiplication equation** [see Example 5-7].

4. $\quad \dfrac{x}{A} = C$ is called a **division equation** [see Example 5-8].

5. $Ax + B = C$ is called a **multiplication-addition equation** [see Example 5-9].
6. $Ax - B = C$ is called a **multiplication-subtraction equation** [see Example 5-10].

7. $\dfrac{x}{A} + B = C$ is called a **division-addition equation** [see Example 5-11].

8. $\dfrac{x}{A} - B = C$ is called a **division-subtraction equation** [see Example 5-12].

Given a linear equation in one variable that does not contain like terms or parentheses, you may need to first write the equation in standard form before it can be categorized as one of the eight special types of linear equations in one variable.

EXAMPLE 5-3: Which special type of linear equation in one variable is given by each of the following equations?

(**a**) $y + 5 = -2$ (**b**) $w - 4 = 5$ (**c**) $2z = 6$ (**d**) $\dfrac{u}{5} = -2$ (**e**) $2x + 3 = 9$

(**f**) $-3v - 6 = -2$ (**g**) $\dfrac{a}{2} + 3 = -1$ (**h**) $\dfrac{b}{-3} - 4 = 5$ (**i**) $-c + 2 = -3$

(**j**) $-5 - 2r = 0$ (**k**) $2 = \dfrac{s}{3} + 5$ (**l**) $-1 = -2 - \dfrac{t}{8}$

Solution
(**a**) $y + 5 = -2$ is an addition equation.
(**b**) $w - 4 = 5$ is a subtraction equation.
(**c**) $2z = 6$ is a multiplication equation.

(**d**) $\dfrac{u}{5} = -2$ is a division equation.

(**e**) $2x + 3 = 9$ is a multiplication-addition equation.
(**f**) $-3v - 6 = -2$ is a multiplication-subtraction equation.

(**g**) $\dfrac{a}{2} + 3 = -1$ is a division-addition equation.

(**h**) $\dfrac{b}{-3} - 4 = 5$ is a division-subtraction equation.

(**i**) $-c + 2 = -3$ or $-1c + 2 = -3$ is a multiplication-addition equation.

(j) $-5 - 2r = 0$ or $-2r - 5 = 0$ is a multiplication-subtraction equation.

(k) $2 = \dfrac{s}{3} + 5$ or $\dfrac{s}{3} + 5 = 2$ is a division-addition equation.

(l) $-1 = -2 - \dfrac{t}{8}$ or $\dfrac{t}{-8} - 2 = -1$ is a division-subtraction equation.

5-2. Solving Linear Equations Using Rules

A. Check a proposed solution of a linear equation in one variable.

Every linear equation in one variable has exactly one solution. The **solution of a linear equation in one variable** is the real number that can replace the variable to make both members of the equation equal. To **check a proposed solution of a linear equation in one variable,** you first substitute the proposed solution for the variable and then compute using the Order of Operations to get a number sentence. If you get a **true number sentence,** then the proposed solution **checks** and is the one and only solution of the original linear equation in one variable. If you get a **false number sentence,** then the proposed solution does not check and is not the solution of the original linear equation in one variable.

EXAMPLE 5-4: Which one of the following is a solution of $3x + 2 = 5$: 0 or 1?

Solution: Check $x = 0$ in $3x + 2 = 5$. | Check $x = 1$ in $3x + 2 = 5$.

$3x + 2 = 5$		given equation		$3x + 2 = 5$		given equation
$3(0) + 2$	5	Substitute 0 for x.		$3(1) + 2$	5	Substitute 1 for x.
$0 + 2$	5	Compute.		$3 + 2$	5	Compute.
2	5	false ($2 \neq 5$)		5	5	true ($5 = 5$)

$x = 0$ is not a solution because $2 \neq 5$. | $x = 1$ is a solution because $5 = 5$.

Note: Because every linear equation in one variable has exactly one solution, the one and only solution of $3x + 2 = 5$ is $x = 1$.

B. Solve an addition equation ($x + B = C$).

To **solve an addition equation** in the form $x + B = C$, you subtract using the following Subtraction Rule for Equations.

Subtraction Rule for Equations
If a, b, and c are real numbers and $a = b$, then $a - c = b - c$.

Note: The Subtraction Rule for Equations states that if you subtract the same term from both members of an equation, the solution(s) will not change.
To solve an addition equation using the Subtraction Rule for Equations, you subtract the numerical addend from both members of the addition equation to isolate the variable in one member.

EXAMPLE 5-5: Solve $y + 5 = -2$.

variable addend
numerical addend
sum

Solution:

$y + 5 = -2$ *Think:* In $y + 5 = -2$, the numerical addend is 5.

$y + 5 - 5 = -2 - 5$ Subtract the numerical addend 5 from both members to isolate the variable y in one member.

$y + 0 = -2 - 5$

y is isolated $\longrightarrow$ $y = -2 - 5$

$y = -7$ $\longleftarrow$ proposed solution

Check: $\underline{y + 5 = -2}$ ←—— original equation

$\dfrac{-7 + 5 \quad | \quad -2}{}$ Substitute the proposed solution -7 for y in the original equation to see if you get a true number sentence.

$-2 \quad | \quad -2$ ←—— $y = -7$ checks

Note: The one and only solution of $y + 5 = -2$ is $y = -7$.

C. Solve a subtraction equation ($x - B = C$).

To **solve a subtraction equation** in the form $x - B = C$, you add using the Addition Rule for Equations.

Addition Rule for Equations

If a, b, and c are real numbers and $a = b$, then $a + c = b + c$.

Note: The Addition Rule for Equations states that if you add the same term to both members of an equation, the solution(s) will not change.

To solve a subtraction equation using the Addition Rule for Equations, you add the subtrahend to both members of the subtraction equation to isolate the variable in one member.

EXAMPLE 5-6: Solve $w - 4 = 5$.

           minuend

               subtrahend

                   difference

Solution: $w - 4 = 5$ *Think:* In $w - 4 = 5$, 4 is the subtrahend.

$w - 4 + 4 = 5 + 4$ Add the subtrahend 4 to both members to isolate the variable w in one member.

$w + 0 = 5 + 4$

w is isolated ——→ $w = 5 + 4$

$w = 9$ ←—— proposed solution

Check: $w - 4 = 5$ ←—— original equation

$\dfrac{9 - 4 \quad | \quad 5}{5 \quad | \quad 5}$ Substitute the proposed solution 9 for w.

$5 \quad | \quad 5$ ←—— $w = 9$ checks

Note: The one and only solution of $w - 4 = 5$ is $w = 9$.

D. Solve a multiplication equation ($Ax = C$).

To **solve a multiplication equation** in the form $Ax = C$, you divide using the Division Rule for Equations.

Division Rule for Equations

If a, b, and c are real numbers ($c \neq 0$) and $a = b$, then $\dfrac{a}{c} = \dfrac{b}{c}$.

Note: The Division Rule for Equations states that if you divide both members of an equation by the same nonzero term, the solution(s) will not change.

To solve a multiplication equation using the Division Rule for Equations, you divide both members of the multiplication equation by the numerical factor to isolate the variable in one member.

EXAMPLE 5-7: Solve $2z = 6$.

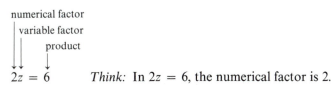

           numerical factor

             variable factor

                product

Solution: $2z = 6$ *Think:* In $2z = 6$, the numerical factor is 2.

$$\frac{2z}{2} = \frac{6}{2}$$

Divide both members by the numerical factor 2 to isolate the variable z in one member.

$$\frac{\cancel{2}z}{\cancel{2}} = \frac{6}{2}$$

z is isolated $\longrightarrow$ $z = \dfrac{6}{2}$

$z = 3 \longleftarrow$ proposed solution

Check: $\overline{2z = 6} \longleftarrow$ original equation

$\underline{2(3)} \ \big| \ 6$ Substitute the proposed solution 3 for z.

$\ \ 6 \ \big| \ 6 \longleftarrow z = 3$ checks

Note: The one and only solution of $2z = 6$ is $z = 3$.

E. Solve a division equation $\left(\dfrac{x}{A} = C\right)$.

To **solve a division equation** in the form $\dfrac{x}{A} = C$, you multiply using the Multiplication Rule for Equations.

Multiplication Rule for Equations
If a, b, and c are real numbers ($c \neq 0$) and $a = b$, then $ac = bc$.

Note: The Multiplication Rule for Equations states that if you multiply both members of an equation by the same nonzero term, the solution(s) will not change.

To solve a division equation using the Multiplication Rule for Equations, you multiply both members of the division equation by the numerical divisor to isolate the variable in one member.

EXAMPLE 5-8: Solve $\dfrac{u}{5} = -2$.

Solution

numerical divisor
variable dividend
quotient

$\dfrac{u}{5} = -2$ *Think:* In $\dfrac{u}{5} = -2$, the numerical divisor is 5.

$5 \cdot \dfrac{u}{5} = 5(-2)$ Multiply both members by the numerical divisor 5 to isolate the variable u in one member.

$\dfrac{\cancel{5}}{1} \cdot \dfrac{u}{\cancel{5}} = 5(-2)$

u is isolated $\longrightarrow$ $u = 5(-2)$

$u = -10 \longleftarrow$ proposed solution

Check: $\dfrac{u}{5} = -2 \longleftarrow$ original equation

$\dfrac{-10}{5} \ \Big| \ -2$ Substitute the proposed solution -10 for u.

$\ \ -2 \ \big| \ -2 \longleftarrow u = -10$ checks

Note: The one and only solution of $\dfrac{u}{5} = -2$ is $u = -10$.

The Addition, Subtraction, Multiplication, and Division Rules for Equations are collectively called the **rules for linear equations.**

F. Solve a multiplication-addition equation ($Ax + B = C$).

To **solve a multiplication-addition equation** in the form $Ax + B = C$, you first use the Subtraction Rule for Equations to isolate the **variable term** in one member and then use the Division Rule for Equations to isolate the variable itself.

EXAMPLE 5-9: Solve $2x + 3 = 9$.

Solution

variable term
$$\overset{\downarrow}{2x} + 3 = 9 \qquad \text{Identify the variable term.}$$

$$2x + 3 - 3 = 9 - 3 \qquad \begin{array}{l}\text{Use the Subtraction Rule for Equations to}\\ \text{isolate the variable term in one member.}\end{array}$$

$2x$ is isolated $\longrightarrow$ $2x = 6$

$$\frac{2x}{2} = \frac{6}{2} \qquad \begin{array}{l}\text{Use the Division Rule for Equations}\\ \text{to isolate the variable } x.\end{array}$$

x is isolated $\longrightarrow$ $x = 3$ $\longleftarrow$ proposed solution

Check: $2x + 3 = 9$ $\longleftarrow$ original equation

$$\begin{array}{c|c} 2(3) + 3 & 9 \\ 6 + 3 & 9 \\ 9 & 9 \end{array} \quad \begin{array}{l}\text{Substitute the proposed solution 3 for } x.\\ \text{Compute.}\\ \longleftarrow x = 3 \text{ checks}\end{array}$$

Note: The one and only solution of $2x + 3 = 9$ is $x = 3$.

G. Solve a multiplication-subtraction equation ($Ax - B = C$).

To **solve a multiplication-subtraction equation** in the form $Ax - B = C$, you first use the Addition Rule for Equations to isolate the variable term in one member and then use the Division Rule for Equations to isolate the variable itself.

EXAMPLE 5-10: Solve $-3v - 6 = -2$.

Solution

variable term
$$\overset{\downarrow}{-3v} - 6 = -2 \qquad \text{Identify the variable term.}$$

$$-3v - 6 + 6 = -2 + 6 \qquad \begin{array}{l}\text{Use the Addition Rule for Equations to isolate the}\\ \text{variable term in one member.}\end{array}$$

$-3v$ is isolated $\longrightarrow$ $-3v = 4$

$$\frac{-3v}{-3} = \frac{4}{-3} \qquad \begin{array}{l}\text{Use the Division Rule for Equations}\\ \text{to isolate the variable } v.\end{array}$$

v is isolated $\longrightarrow$ $v = \dfrac{4}{-3}$ or $\dfrac{-4}{3}$ or $-\dfrac{4}{3}$ $\longleftarrow$ proposed solution

Check: $\overline{-3v - 6 = -2}$ $\longleftarrow$ original equation

$$-3\left(-\frac{4}{3}\right) - 6 \;\Big|\; -2 \qquad \text{Substitute the proposed solution } -\frac{4}{3} \text{ for } v.$$

$$4 - 6 \;\Big|\; -2 \qquad \text{Compute.}$$

$$-2 \;\Big|\; -2 \longleftarrow v = -\frac{4}{3} \text{ checks}$$

Note: The one and only solution of $-3v - 6 = -2$ is $v = -\dfrac{4}{3}$.

H. Solve a division-addition equation $\left(\dfrac{x}{A} + B = C\right)$.

To **solve a division-addition equation** in the form $\dfrac{x}{A} + B = C$, you first use the Subtraction Rule for Equations to isolate the variable term in one member and then use the Multiplication Rule for Equations to isolate the variable itself.

EXAMPLE 5-11: Solve $\dfrac{a}{2} + 3 = -1$.

Solution

variable term
↓

$$\dfrac{a}{2} + 3 = -1 \qquad \text{Identify the variable term.}$$

$$\dfrac{a}{2} + 3 - 3 = -1 - 3 \qquad \text{Use the Subtraction Rule for Equations to isolate the variable term in one member.}$$

$\dfrac{a}{2}$ is isolated $\longrightarrow$ $\dfrac{a}{2} = -4$

$$2 \cdot \dfrac{a}{2} = 2(-4) \qquad \text{Use the Multiplication Rule for Equations to isolate the variable } a.$$

a is isolated $\longrightarrow$ $a = -8$ $\longleftarrow$ proposed solution

Check: $\dfrac{a}{2} + 3 = -1$ $\longleftarrow$ original equation

$$\dfrac{-8}{2} + 3 \ \bigg|\ -1 \qquad \text{Substitute the proposed solution } -8 \text{ for } a.$$

$$-4 + 3 \ \bigg|\ -1 \qquad \text{Compute.}$$

$$-1 \ \bigg|\ -1 \longleftarrow a = -8 \text{ checks}$$

Note: The one and only solution of $\dfrac{a}{2} + 3 = -1$ is $a = -8$.

I. Solve a division-subtraction equation $\left(\dfrac{x}{A} - B = C\right)$.

To **solve a division-subtraction equation** in the form $\dfrac{x}{A} - B = C$, you first use the Addition Rule for Equations to isolate the variable term in one member and then use the Multiplication Rule for Equations to isolate the variable itself.

EXAMPLE 5-12: Solve $\dfrac{b}{-3} - 4 = 5$.

Solution

variable term
↓

$$\dfrac{b}{-3} - 4 = 5 \qquad \text{Identify the variable term.}$$

$$\dfrac{b}{-3} - 4 + 4 = 5 + 4 \qquad \text{Use the Addition Rule for Equations to isolate the variable term in one member.}$$

$\dfrac{b}{-3}$ is isolated $\longrightarrow$ $\dfrac{b}{-3} = 9$

$$-3 \cdot \frac{b}{-3} = -3 \cdot 9 \qquad \text{Use the Multiplication Rule for Equations to isolate the variable } b.$$

b is isolated $\longrightarrow$ $b = -27$ $\longleftarrow$ proposed solution

Check: $\dfrac{b}{-3} - 4 = 5$ $\longleftarrow$ original equation

$$\begin{array}{c|c} \dfrac{-27}{-3} - 4 & 5 \qquad \text{Substitute the proposed solution } -27 \text{ for } b. \\[2ex] 9 - 4 & 5 \qquad \text{Compute.} \\[1ex] 5 & 5 \longleftarrow b = -27 \text{ checks} \end{array}$$

Note: The one and only solution of $\dfrac{b}{-3} - 4 = 5$ is $b = -27$.

J. Solve linear equations in one variable that do not contain like terms or parentheses.

Given a linear equation in one variable that does not contain like terms or parentheses and that is not in the form of one of the eight special types of linear equations, you can always rename the given linear equation as one of the special types and then solve using the rules for linear equations.

EXAMPLE 5-13: Solve **(a)** $-x + 2 = -3$ **(b)** $5 - 2y = 0$ **(c)** $2 = \dfrac{m}{3} + 5$

(d) $-1 = -3 + \dfrac{n}{-5}$

Solution

(a) $-x + 2 = -3$ Rename using $-x = -1x$.

     $-1x + 2 = -3$ $\longleftarrow$ multiplication-addition equation

   $-1x + 2 - 2 = -3 - 2$ Use the Subtraction Rule for Equations.

         $-1x = -5$

       $\dfrac{-1x}{-1} = \dfrac{-5}{-1}$ Use the Division Rule for Equations.

          $x = 5$ $\longleftarrow$ solution

(b) $5 - 2y = 0$

   $-1(5 - 2y) = -1(0)$ Multiply both members by -1.

    $-5 + 2y = 0$

     $2y - 5 = 0$ $\longleftarrow$ multiplication-subtraction equation

  $2y - 5 + 5 = 0 + 5$ Use the Addition Rule for Equations.

        $2y = 5$

       $\dfrac{2y}{2} = \dfrac{5}{2}$ Use the Division Rule for Equations.

      $y = \dfrac{5}{2}$ or $2\dfrac{1}{2}$ or 2.5 $\longleftarrow$ solution

(c) $2 = \dfrac{m}{3} + 5$ Interchange the left and right members.

    $\dfrac{m}{3} + 5 = 2$ $\longleftarrow$ division-addition equation

  $\dfrac{m}{3} + 5 - 5 = 2 - 5$ Use the Subtraction Rule for Equations.

$$\frac{m}{3} = -3$$

$$3 \cdot \frac{m}{3} = 3(-3) \qquad \text{Use the Multiplication Rule for Equations.}$$

$$m = -9 \longleftarrow \text{solution}$$

(d) $\qquad -1 = -3 + \dfrac{n}{-5}$

$$-3 + \frac{n}{-5} = -1 \qquad \text{Interchange the left and right members.}$$

$$\frac{n}{-5} + (-3) = -1 \qquad \text{Use the commutative property of addition.}$$

$$\frac{n}{-5} - 3 = -1 \longleftarrow \text{division-subtraction equation}$$

$$\frac{n}{-5} - 3 + 3 = -1 + 3 \qquad \text{Use the Addition Rule for Equations.}$$

$$\frac{n}{-5} = 2$$

$$-5 \cdot \frac{n}{-5} = -5(2) \qquad \text{Use the Multiplication Rule for Equations.}$$

$$n = -10 \longleftarrow \text{solution}$$

5-3. Solving Linear Equations Containing Like Terms and/or Parentheses

A. Solve linear equations in one variable containing like terms in only one member.

To **solve a linear equation in one variable that contains like terms in only one member** use the following steps:

1. Combine like terms.

2. Rename the equation from Step 1 when necessary to get one of the eight special types of linear equations.

3. Solve the equation using the rules for linear equations.

EXAMPLE 5-14: Solve $8 = 3x - 4 + x$.

Solution

$$\overset{\text{like terms}}{8 = 3\overset{\downarrow}{x} - 4 + \overset{\downarrow}{x}} \qquad \text{Identify like terms.}$$

$$8 = 4x - 4 \qquad \text{Combine like terms.}$$

$$4x - 4 = 8 \longleftarrow \text{multiplication-subtraction equation}$$

$$4x = 12 \qquad \text{Use the Addition Rule: } 4x - 4 + 4 = 8 + 4$$

$$x = 3 \qquad \text{Use the Division Rule: } \frac{4x}{4} = \frac{12}{4}$$

Check: $8 = 3x - 4 + x \longleftarrow$ original equation

$$\begin{array}{c|l}
8 & 3(3) - 4 + (3) \qquad \text{Substitute the proposed solution 3 for } x. \\
8 & 9 - 4 + 3 \qquad \text{Compute.} \\
8 & 5 + 3 \\
8 & 8 \longleftarrow x = 3 \text{ checks}
\end{array}$$

Note: The one and only solution of $8 = 3x - 4 + x$ is $x = 3$.

B. Solve linear equations in one variable containing like terms in both members.

To **solve a linear equation in one variable that contains like terms in both members,** use the following steps:

1. Combine like terms in each member when possible.
2. Collect and combine like terms all in one member.
3. Rename the equation to get one of the eight special types of linear equations in one variable when necessary.
4. Solve the equation using the rules for linear equations.

EXAMPLE 5-15: Solve $-y + 8 - 2y = 3y - 6 + y$.

Solution

$$-y + 8 - 2y = 3y - 6 + y \qquad \text{Identify like terms}$$

$$-3y + 8 = 4y - 6 \qquad \text{Combine like terms in each member.}$$

$$-3y + 3y + 8 = 4y + 3y - 6 \qquad \text{Collect like terms all in one member.}$$

$$8 = 4y + 3y - 6 \longleftarrow \text{ like terms are all in one member}$$

$$8 = 7y - 6 \qquad \text{Combine like terms.}$$

$$7y - 6 = 8 \longleftarrow \text{ multiplication-subtraction equation}$$

$$7y = 14 \qquad \text{Use the Addition Rule: } 7y - 6 + 6 = 8 + 6$$

$$y = 2 \qquad \text{Use the Division Rule: } \frac{7y}{7} = \frac{14}{7}$$

Check: $-y + 8 - 2y = 3y - 6 + y \longleftarrow$ original equation

$-(2) + 8 - 2(2)$	$3(2) - 6 + (2)$	
$-2 + 8 - 4$	$6 - 6 + 2$	Substitute the proposed solution 2 for y.
$6 - 4$	$0 + 2$	Compute.
2	$2 \longleftarrow y = 2$ checks	

Note: The one and only solution of $-y + 8 - 2y = 3y - 6 + y$ is $y = 2$.

C. Solve linear equations in one variable containing parentheses.

To **solve a linear equation in one variable containing parentheses,** use the following steps:

1. Combine like terms inside the parentheses when possible.
2. Clear parentheses using the distributive properties.
3. Solve the equation using the rules for linear equations.

EXAMPLE 5-16: Solve $2(7 - 5u - 4) = u - 4(5u - 6 - 3u)$.

Solution

$$2(7 - 5u - 4) = u - 4(5u - 6 - 3u) \qquad \text{Identify like terms inside parentheses.}$$

$$2(3 - 5u) = u - 4(2u - 6) \qquad \text{Combine like terms inside parentheses.}$$

$$2(3 - 5u) = u + (-4)(2u - 6)$$

$$2(3) - 2(5u) = u + (-4)(2u) - (-4)(6) \qquad \text{Use the distributive property to clear parentheses.}$$

$$6 - 10u = u - 8u + 24 \longleftarrow \text{ parentheses cleared}$$

$$6 - 10u = -7u + 24 \qquad \text{Solve as before [see Example 5-15].}$$

$$6 - 10u + 7u = -7u + 7u + 24$$

$$6 - 3u = 24$$

$$-3u + 6 = 24$$

$$-3u = 18$$

$$u = -6 \longleftarrow \text{ proposed solution}$$

Check:

$$2(7 - 5u - 4) = u - 4(5u - 6 - 3u) \longleftarrow \text{original equation}$$

$2(7 - 5(-6) - 4)$	$(-6) - 4(5(-6) - 6 - 3(-6))$	Substitute the proposed solution -6 for u.
$2(7 + 30 - 4)$	$-6 - 4(-30 - 6 + 18)$	Compute.
$2(37 - 4)$	$-6 - 4(-36 + 18)$	
$2(33)$	$-6 - 4(-18)$	
66	$-6 + 72$	
66	$66 \longleftarrow u = -6$ checks	

Note: The one and only solution of $2(7 - 5u - 4) = u - 4(5u - 6 - 3u)$ is $u = -6$.

D. Solve equations in one variable that simplify as true number sentences.

If an equation in one variable simplifies as a true number sentence then every real number is a solution of that equation.

EXAMPLE 5-17: Solve $2 - 5w = 4w + 2 - 9w$.

Solution

$$2 - 5w = 4w + 2 - 9w \qquad \text{Identify like terms.}$$

$$2 - 5w = -5w + 2 \qquad \text{Combine like terms.}$$

$$2 - 5w + 5w = -5w + 5w + 2 \qquad \text{Collect like terms.}$$

$2 = 2$ *Stop!* The true number sentence $2 = 2$ means that every real number is a solution of $2 - 5w = 4w + 2 - 9w$.

Note: The equation $2 - 5w = 4w + 2 - 9w$ is not a linear equation in one variable because it simplifies as $2 = 2$ or $0w + 2 = 2$ and therefore cannot be written in the form $Aw + B = C$ where $A \neq 0$.

E. Solve equations in one variable that simplify as false number sentences.

If an equation in one variable simplifies as a false number sentence then there are no solutions of that equation.

EXAMPLE 5-18: Solve $1 - z = 2 + 3z - 3 - 4z$.

Solution

$$1 - z = 2 + 3z - 3 - 4z \qquad \text{Identify like terms.}$$

$$1 - z = -1 - z \qquad \text{Combine like terms.}$$

$$1 - z + z = -1 - z + z \qquad \text{Collect like terms.}$$

$1 = -1$ *Stop!* The false number sentence $1 = -1$ means that there is no solution of $1 - z = 2 + 3z - 3 - 4z$.

Note: The equation $1 - z = 2 + 3z - 3 - 4z$ is not a linear equation in one variable because it simplifies as $1 = -1$ or $0z + 1 = -1$ and therefore cannot be written in the form $Az + B = C$ where $A \neq 0$.

5-4. Solving Linear Equations Containing Fractions, Decimals, or Percents

A. Solve linear equations in one variable containing fractions.

To **solve a linear equation in one variable containing fractions,** you first **clear the fractions** and then use the rules for linear equations.

To clear fractions, you
1. Find the least common denominator (LCD) of the fractions in the given equation.
2. Multiply both members of the given equation by the LCD.
3. Clear parentheses in the equation to clear all fractions.

EXAMPLE 5-19: Solve $\frac{2}{3}x - \frac{1}{2} = \frac{1}{4}$.

Solution: The LCD of $\frac{2}{3}$, $\frac{1}{2}$, and $\frac{1}{4}$ is 12 [see Section 1-7, Part D].

$$12(\tfrac{2}{3}x - \tfrac{1}{2}) = 12(\tfrac{1}{4}) \qquad \text{Multiply both members by the LCD 12.}$$

$$12(\tfrac{2}{3}x) - 12(\tfrac{1}{2}) = 12(\tfrac{1}{4}) \qquad \text{Clear parentheses to clear fractions.}$$

$$8x - 6 = 3 \longleftarrow \text{fractions cleared}$$

$$8x = 9 \qquad \text{Solve using the Addition Rule and Division Rule}$$

$$x = \tfrac{9}{8} \text{ or } 1\tfrac{1}{8} \text{ or } 1.125 \longleftarrow \text{proposed solution}$$

Check: $\quad \frac{2}{3}x - \frac{1}{2} = \frac{1}{4} \longleftarrow$ original equation

$$\begin{array}{c|c} \frac{2}{3}(\frac{9}{8}) - \frac{1}{2} & \frac{1}{4} \\ \frac{3}{4} - \frac{1}{2} & \frac{1}{4} \\ \frac{1}{4} & \frac{1}{4} \end{array}$$

$\quad$ Substitute the proposed solution $\frac{9}{8}$ for x.
$\quad$ Compute.
$\longleftarrow x = \frac{9}{8}$ checks

Note: The one and only solution of $\frac{2}{3}x - \frac{1}{2} = \frac{1}{4}$ is $x = \frac{9}{8}$ or $1\frac{1}{8}$ or 1.125.

B. Solve linear equations in one variable containing decimals.

To **solve a linear equation in one variable containing decimals,** you can use the rules for linear equations.

EXAMPLE 5-20: Solve $0.5y + 0.2 = 0.3$

Solution

$$\overbrace{0.5y}^{\text{variable term}} + 0.2 = 0.3$$

$$0.5y + 0.2 - \mathbf{0.2} = 0.3 - \mathbf{0.2} \qquad \begin{array}{l}\text{Use the Subtraction Rule for Equations to isolate} \\ \text{the variable term } 0.5y \text{ in one member.}\end{array}$$

$0.5y$ is isolated $\longrightarrow \quad 0.5y = 0.1$

$$\frac{0.5y}{\mathbf{0.5}} = \frac{0.1}{\mathbf{0.5}} \qquad \begin{array}{l}\text{Use the Division Rule for Equations to} \\ \text{isolate the variable } y.\end{array}$$

y is isolated $\longrightarrow \quad y = 0.2 \text{ or } \dfrac{1}{5} \longleftarrow$ proposed solution

Check: $\quad 0.5y + 0.2 = 0.3 \longleftarrow$ original equation

$$\begin{array}{c|c} 0.5(\mathbf{0.2}) + 0.2 & 0.3 \\ 0.1 + 0.2 & 0.3 \\ 0.3 & 0.3 \end{array}$$

$\quad$ Substitute the proposed solution 0.2 for y.
$\quad$ Compute.
$\longleftarrow y = 0.2$ or $\dfrac{1}{5}$ checks

Note: The one and only solution of $0.5y + 0.2 = 0.3$ is $y = 0.2$ or $\dfrac{1}{5}$.

An equation containing decimals can also be solved by first **clearing decimals** and then using the rules for linear equations.

To clear decimals, you
1. Find the LCD of the fraction form of the decimals from the given equation.
2. Multiply both members of the given equation by the LCD.
3. Clear parentheses in the equation to clear all decimals.

EXAMPLE 5-21: Solve $0.5y + 0.2 = 0.3$ by first clearing decimals.

Solution: The LCD of $\frac{5}{10}$ (0.5), $\frac{2}{10}$ (0.2), and $\frac{3}{10}$ (0.3) is 10 [see Section 1-7, Part D].

$$10(0.5y + 0.2) = \mathbf{10}(0.3) \qquad \text{Multiply both members by the LCD 10.}$$

$$10(0.5y) + 10(0.2) = 10(0.3) \qquad \text{Clear parentheses to clear decimals.}$$

$$5y + 2 = 3 \longleftarrow \text{decimals cleared}$$

$$5y = 1 \qquad \text{Solve using the rules for linear equations.}$$

$$y = \frac{1}{5} \text{ or } 0.2 \longleftarrow \text{same solution as found in Example 5-20}$$

Note: To solve a linear equation in one variable containing decimals, you can either use the rules for linear equations directly without clearing decimals or you can first clear decimals and then use the rules for linear equations. In practice, the best method to use is the method that is easiest for you.

C. Solve linear equations containing percents.

To **solve a linear equation in one variable containing percents,** you
1. Clear percents by renaming each percent as either a fraction or a decimal.
2. Clear fractions or decimals in the equation.
3. Solve the equation using the rules for linear equations.

EXAMPLE 5-22: Solve $x + 25\%x = 20$.

Solution

	Renaming the percent as a decimal	Renaming the percent as a fraction
	$x + 25\%x = 20$	$x + 25\%x = 20$
Clear percents.	$x + 0.25x = 20$	$x + \frac{1}{4}x = 20$
Clear decimals or fractions.	$\mathbf{100}(x + 0.25x) = \mathbf{100}(20)$	$\mathbf{4}(x + \frac{1}{4}x) = \mathbf{4}(20)$
	$100(x) + 100(0.25x) = 100(20)$	$4(x) + 4(\frac{1}{4}x) = 4(20)$
	$100x + 25x = 2000$	$4x + 1x = 80$
Solve as before	$125x = 2000$	$5x = 80$ Solve as before.
	$x = \frac{2000}{125}$	$x = \frac{80}{5}$
	$x = 16$	$x = 16$

$$\begin{array}{r|l} \textit{Check:} \quad x + 25\%x &= 20 \\ \hline 16 + 25\% \, (\mathbf{16}) & 20 \\ 16 + 0.25(16) & 20 \\ 16 + 4 & 20 \\ 20 & 20 \end{array} \qquad \begin{array}{r|l} \textit{Check:} \quad x + 25\%x &= 20 \\ \hline 16 + 25\% \, (\mathbf{16}) & 20 \\ 16 + \frac{1}{4}(16) & 20 \\ 16 + 4 & 20 \\ 20 & 20 \end{array}$$

Note: The one and only solution of $x + 25\%x = 20$ is $x = 16$.

5-5. Solving Literal Equations and Formulas for a Given Variable

A. Identify literal equations.

An equation containing more than one different variable is called a **literal equation.** Every formula is a literal equation because every formula contains two or more different variables.

EXAMPLE 5-23: Which of the following are literal equations? **(a)** $P = 2(l + w)$
(b) $2x + 3y = 5$ **(c)** $2x + 3x = 5$ **(d)** $2x + 3 = 5$ **(e)** $2 + 3 = 5$

Solution
(a) $P = 2(l + w)$ is a formula (perimeter of a rectangle) and therefore is a literal equation.
(b) $2x + 3y = 5$ is a literal equation because it contains more than one variable.
(c) $2x + 3x = 5$ is not a literal equation because it contains only one variable x.
(d) $2x + 3 = 5$ is not a literal equation because it contains only one variable x.
(e) $2 + 3 = 5$ is not a literal equation because it contains no variables.

B. Solve a literal equation for a given variable that is contained in only one term.

To **solve a literal equation for a given variable that is contained in only one term,** you first isolate the term containing the given variable in one member of the literal equation and then isolate the given variable itself. When a literal equation is solved for a given variable, the algebraic expression in the other member is called the **solution of the literal equation with respect to the given variable.**

EXAMPLE 5-24: Which variable is the literal equation $y = mx + b$ solved for and what is the solution with respect to that variable?

Solution: The literal equation $y = mx + b$ is solved for y and the solution is $mx + b$.

EXAMPLE 5-25: Solve $2x + 3y = 5$ for x.

Solution

term containing x

$$2x + 3y = 5 \qquad \text{Identify the term containing the given variable } x.$$

$$2x + 3y - 3y = 5 - 3y \qquad \begin{array}{l}\text{Use the Subtraction Rule for Equations to isolate} \\ \text{the term containing the given variable } x.\end{array}$$

$2x$ is isolated $\longrightarrow$ $2x = 5 - 3y$

$$\frac{2x}{2} = \frac{5 - 3y}{2} \qquad \begin{array}{l}\text{Use the Division Rule for Equations} \\ \text{to isolate the given variable } x.\end{array}$$

x is isolated $\longrightarrow$ $x = \dfrac{5 - 3y}{2}$ or $\dfrac{5}{2} - \dfrac{3y}{2}$ or $\dfrac{5}{2} - \dfrac{3}{2}y$ or $-\dfrac{3}{2}y + \dfrac{5}{2}$ $\longleftarrow$ proposed solution

Check: $2x + 3y = 5$ $\longleftarrow$ original equation

$$\begin{array}{c|c} 2\left(\dfrac{5 - 3y}{2}\right) + 3y & 5 \qquad \text{Substitute the proposed solution } \dfrac{5 - 3y}{2} \text{ for } x. \\[2mm] 5 - 3y + 3y & 5 \qquad \text{Simplify.} \\[1mm] 5 + 0 & 5 \\[1mm] 5 & 5 \longleftarrow \dfrac{5 - 3y}{2} \text{ checks} \end{array}$$

C. Solve a literal equation for a given variable that is contained in two or more like terms.

To **solve a literal equation for a given variable that is contained in two or more like terms,** you
1. Collect and combine any like terms.
2. Isolate the term containing the given variable in one member of the literal equation.
3. Isolate the variable itself.

EXAMPLE 5-26: Solve $-3m + 7n = 2n + 10$ for n.

Solution

like terms containing n

$$-3m + 7n = 2n + 10$$ Identify the like terms containing the given variable.

$$-3m + 7n - 2n = 2n - 2n + 10$$ Collect like terms.

$$-3m + 5n = 10$$ Combine like terms.

$$-3m + 3m + 5n = 10 + 3m$$ Use the Addition Rule for Equations to isolate the term containing the given variable n.

$5n$ is isolated $\longrightarrow$ $5n = 10 + 3m$

$$\frac{5n}{5} = \frac{10 + 3m}{5}$$ Use the Division Rule for Equations to isolate the given variable n.

n is isolated $\longrightarrow$ $n = \dfrac{10 + 3m}{5}$ or $\dfrac{3m + 10}{5}$ or $\dfrac{3m}{5} + 2$ or $\dfrac{3}{5}m + 2$ $\longleftarrow$ solution (Check as before.)

D. Solve a literal equation for a given variable that is contained in parentheses.

To **solve a literal equation for a given variable that is contained in parentheses,** you:
1. Combine any like terms inside the parentheses.
2. Clear parentheses if the given variable is contained inside the parentheses.
3. Collect and combine any like terms.
4. Isolate the term containing the given variable in one member of the literal equation.
5. Isolate the given variable itself.

EXAMPLE 5-27: Solve $P = 2(l + w)$ for w.

Solution

term containing w

$$P = 2(l + w)$$ Identify the given variable w as being contained inside the parentheses.

$$P = 2(l) + 2(w)$$ Use a distributive property to clear parentheses.

$$P = 2l + 2w$$ $\longleftarrow$ parentheses cleared

$2w$ is in the left member $\longrightarrow$ $2l + 2w = P$ Rename to get the term containing the given variable w in the left member.

$$2l - 2l + 2w = P - 2l$$ Use the Subtraction Rule for Equations to isolate the term containing the given variable w.

$2w$ is isolated $\longrightarrow$ $2w = P - 2l$

$$\frac{2w}{2} = \frac{P - 2l}{2}$$ Use the Division Rule for Equations to isolate the given variable w.

w is isolated $\longrightarrow$ $w = \dfrac{P - 2l}{2}$ or $\dfrac{P}{2} - l$ or $\dfrac{1}{2}P - l$ $\longleftarrow$ solution (Check as before.)

Note: The formula $w = \dfrac{1}{2}P - l$ means that the width w of a rectangle always equals one-half the perimeter P minus the length l.

RAISE YOUR GRADES

Can you . . . ?

☑ identify linear equations in one variable
☑ identify the eight special types of linear equations in one variable
☑ check a proposed solution of a linear equation in one variable
☑ identify the rules for linear equations
☑ solve linear equations using the rules for linear equations
☑ solve linear equations in one variable containing like terms and/or parentheses
☑ solve an equation in one variable that simplifies as a true or false number sentence
☑ solve linear equations in one variable containing fractions, decimals, or percents
☑ solve literal equations for a given variable

SUMMARY

1. An equation in one variable that can be written in standard form as $Ax + B = C$, where A, B, and C are real numbers ($A \neq 0$) and x is any variable is called a linear equation in one variable.
2. If A, B, and C are real numbers ($A \neq 0$) and x is any variable, then the eight special types of linear equations in one variable are
   (a) $x + B = C$ (addition equation)
   (b) $x - B = C$ (subtraction equation)
   (c) $Ax = C$ (multiplication equation)

   (d) $\dfrac{x}{A} = C$ (division equation)

   (e) $Ax + B = C$ (multiplication-addition equation)
   (f) $Ax - B = C$ (multiplication-subtraction equation)

   (g) $\dfrac{x}{A} + B = C$ (division-addition equation)

   (h) $\dfrac{x}{A} - B = C$ (division-subtraction equation)

3. Every linear equation in one variable that does not contain like terms or parentheses can be categorized as one of the eight special types of linear equations in one variable.
4. Every linear equation in one variable has exactly one solution.
5. The solution of a linear equation in one variable is the real number that can replace the variable to make both members of the equation equal.
6. To check a proposed solution of a linear equation in one variable, you first substitute the proposed solution for the variable and then compute using the Order of Operations to get a number sentence. If you get a true number sentence then the proposed solution checks and is the one and only solution to the original linear equation in one variable. If you get a false number sentence, then the proposed solution does not check and is not a solution of the original linear equation in one variable.
7. To solve a linear equation in one variable that does not contain like terms or parentheses, you use the following rules for linear equations:
   (a) **Addition Rule for Equations**
       If a, b, and c are real numbers and $a = b$, then $a + c = b + c$.
   (b) **Subtraction Rule for Equations**
       If a, b, and c are real numbers and $a = b$, then $a - c = b - c$.
   (c) **Multiplication Rule for Equations**
       If a, b, and c are real numbers ($c \neq 0$) and $a = b$, then $ac = bc$.
   (d) **Division Rule for Equations**
       If a, b, and c are real numbers ($c \neq 0$) and $a = b$, then $\dfrac{a}{c} = \dfrac{b}{c}$.

8. To solve an addition equation in the form $x + B = C$, you use the Subtraction Rule for Equations to subtract the numerical addend from both members of the addition equation to isolate the variable in one member.

9. To solve a subtraction equation in the form $x - B = C$, you use the Addition Rule for Equations to add the subtrahend to both members of the subtraction equation to isolate the variable in one member.

10. To solve a multiplication equation in the form $Ax = C$, you use the Division Rule for Equations to divide both members of the multiplication equation by the numerical factor to isolate the variable in one member.

11. To solve a division equation in the form $\dfrac{x}{A} = C$, you use the Multiplication Rule for Equations to multiply both members of the division equation by the numerical divisor to isolate the variable in one member.

12. To solve a multiplication-addition equation in the form $Ax + B = C$, you first use the Subtraction Rule for Equations to isolate the variable term in one member and then use the Division Rule for Equations to isolate the variable itself.

13. To solve a multiplication-subtraction equation in the form $Ax - B = C$, you first use the Addition Rule for Equations to isolate the variable term in one member and then use the Division Rule for Equations to isolate the variable itself.

14. To solve a division-addition equation in the form $\dfrac{x}{A} + B = C$, you first use the Subtraction Rule for Equations to isolate the variable term in one member and then use the Multiplication Rule for Equations to isolate the variable itself.

15. To solve a division-subtraction equation in the form $\dfrac{x}{A} - B = C$, you first use the Addition Rule for Equations to isolate the variable term in one member and then use the Multiplication Rule for Equations to isolate the variable itself.

16. To solve a linear equation in one variable containing like terms and/or parentheses, you
   (a) Combine any like terms that are inside the parentheses.
   (b) Clear any parentheses using the distributive properties.
   (c) Combine any like terms in each member.
   (d) Collect all like terms in one member.
   (e) Combine any like terms from Step (d).
   (f) Rename the equation (when necessary) to get one of the eight special types of linear equations.
   (g) Solve the equation using the rules for linear equations.

17. If an equation in one variable simplifies as a true number sentence then every real number is a solution of that equation.

18. If an equation in one variable simplifies as a false number sentence then there are no solutions of that equation.

19. A given equation that simplifies as a true or false number sentence is not a linear equation in one variable because it cannot be written as $Ax + B = C$ where $A \neq 0$.

20. To solve a linear equation in one variable containing fractions, you first clear the fractions and then use the rules for linear equations.

21. To clear fractions, you:
   (a) Find the LCD of the fractions in the given equation.
   (b) Multiply both members of the given equation by the LCD.
   (c) Clear parentheses in the equation to clear all fractions.

22. To solve a linear equation in one variable containing decimals, you can either use the rules for linear equations directly without clearing decimals or you can first clear decimals and then use the rules for linear equations. In practice, the best method to use is the method that is easiest for you.

23. To solve a linear equation in one variable containing percents, you
   (a) Clear percents by renaming each percent as either a fraction or a decimal.
   (b) Clear fractions or decimals in the equation.
   (c) Solve the equation using the rules for linear equations.

24. An equation containing more than one different variable is called a literal equation.

25. Every formula is a literal equation because every formula contains two or more different variables.
26. To solve a literal equation for a given variable that is contained in only one term, you first isolate the term containing the given variable in one member of the literal equation and then isolate the given variable itself.
27. When a literal equation is solved for a given variable, the algebraic expression in the other member is called the solution of the literal equation with respect to the given variable.
28. To solve a literal equation for a given variable, you:
    (a) Combine any like terms inside parentheses.
    (b) Clear parentheses if the given variable is contained inside the parentheses.
    (c) Collect and combine any like terms.
    (d) Isolate the term containing the given variable in one member of the literal equation.
    (e) Isolate the given variable itself.

SOLVED PROBLEMS

PROBLEM 5-1   Which of the following are linear equations in one variable? (a) $2x = 3$

(b) $y - 8 = 4$ (c) $\frac{z}{5} = -8$ (d) $\frac{w}{2} - 3 = 7$ (e) $m + 3 = 3 + m$ (f) $2 + n = -5$

(g) $3x + 2y = 6$ (h) $0.3x + 0.2 = 0.6$ (i) $3 + 2 = 5$ (j) $1 - 2r = 2 + 2r$

(k) $\frac{a}{2} + 5 = 4$ (l) $\frac{2}{a} + 5 = 4$ (m) $b\sqrt{5} - 2 = 3$ (n) $5\sqrt{b} + 2 = 3$ (o) $x^3 = 1$

(p) $2m + 3n + 6 = 8 + 3n$

Solution: Recall that a linear equation in one variable is any equation in one variable that can be written in standard form as $Ax + B = C$, where A, B, and C are real numbers ($A \neq 0$) and x is any variable [see Example 5-1]:

$$\overbrace{\text{standard form}}$$

(a) $2x = 3$ or $2x + 0 = 3$ is a linear equation in one variable.
(b) $y - 8 = 4$ or $1y + (-8) = 4$ is a linear equation in one variable.

(c) $\frac{z}{5} = -8$ or $\frac{1}{5}z + 0 = -8$ is a linear equation in one variable.

(d) $\frac{w}{2} - 3 = 7$ or $\frac{1}{2}w + (-3) = 7$ is a linear equation in one variable.

(e) $m + 3 = 3 + m$ or $3 = 3$ is not a linear equation in one variable because $3 = 3$ cannot be written as $Ax + B = C$ where $A \neq 0$.
(f) $2 + n = -5$ or $1n + 2 = -5$ is a linear equation in one variable.
(g) $3x + 2y = 6$ is not a linear equation in one variable because there are two different variables.
(h) $0.3x + 0.2 = 0.6$ is a linear equation in one variable in standard form.
(i) $3 + 2 = 5$ is not a linear equation in one variable because there are no variables.
(j) $1 - 2r = 2 + 2r$ or $-4r + 1 = 2$ is a linear equation in one variable.

(k) $\frac{a}{2} + 5 = 4$ or $\frac{1}{2}a + 5 = 4$ is a linear equation in one variable.

(l) $\frac{2}{a} + 5 = 4$ is not a linear equation because there is a variable in the denominator.

(m) $b\sqrt{5} - 2 = 3$ or $(\sqrt{5})b + (-2) = 3$ is a linear equation in one variable.
(n) $5\sqrt{b} + 2 = 3$ is not a linear equation in one variable because there is a variable under the radical symbol.

(o) $x^3 = 1$ is not a linear equation in one variable because the variable has an exponent greater than one.

(p) $2m + 3n + 6 = 8 + 3n$ or $2m + 6 = 8$ is a linear equation in one variable.

PROBLEM 5-2 List and name the eight special types of linear equations in one variable.

Solution: Recall that if A, B, and C are real numbers ($A \neq 0$) and x is any variable, then

(a) $x + B = C$ is called an addition equation [see Example 5-5].
(b) $x - B = C$ is called a subtraction equation [see Example 5-6].
(c) $Ax = C$ is called a multiplication equation [see Example 5-7].

(d) $\dfrac{x}{A} = C$ is called a division equation [see Example 5-8].

(e) $Ax + B = C$ is called a multiplication-addition equation [see Example 5-9].
(f) $Ax - B = C$ is called a multiplication-subtraction equation [see Example 5-10].

(g) $\dfrac{x}{A} + B = C$ is called a division-addition equation [see Example 5-11].

(h) $\dfrac{x}{A} - B = C$ is called a division-subtraction equation [see Example 5-12].

PROBLEM 5-3 Identify the special type of linear equation in one variable given by each of the following:

(a) $2 - x = 0$ **(b)** $-2 - x = 0$ **(c)** $5y = 1$ **(d)** $m + 2 = -5$ **(e)** $4 = n - 2$

(f) $-6 = \dfrac{u}{4}$ **(g)** $-3 + \dfrac{v}{5} = 6$ **(h)** $-15 = 2 + \dfrac{w}{3}$

Solution: Recall that a given linear equation in one variable that does not contain like terms or parentheses may need to be written in standard form before it can be categorized as one of the eight special types of linear equations in one variable [see Example 5-3]:

$$\overbrace{}^{\text{standard form}}$$

(a) $2 - x = 0$ or $-1x + 2 = 0$ is a multiplication-addition equation.
(b) $-2 - x = 0$ or $-1x - 2 = 0$ is a multiplication-subtraction equation.
(c) $5y = 1$ is a multiplication equation.
(d) $m + 2 = -5$ is an addition equation.
(e) $4 = n - 2$ or $n - 2 = 4$ is a subtraction equation.

(f) $-6 = \dfrac{u}{4}$ or $\dfrac{u}{4} = -6$ is a division equation.

(g) $-3 + \dfrac{v}{5} = 6$ or $\dfrac{v}{5} - 3 = 6$ is a division-subtraction equation.

(h) $-15 = 2 + \dfrac{w}{3}$ or $\dfrac{w}{3} + 2 = -15$ is a division-addition equation.

PROBLEM 5-4 Determine if the given real number is a solution of the given linear equation in one variable in each of the following: **(a)** $\dfrac{3}{2}$; $2x = 3$ **(b)** -3; $y + 5 = 2$ **(c)** 6; $\dfrac{w}{-2} = 3$

(d) 0; $z - 4 = -5$ **(e)** $-\dfrac{9}{2}$; $2m + 3 = -6$ **(f)** 3; $3n - 5 = 4$ **(g)** -4; $\dfrac{u}{4} + 3 = 2$

(h) -10, $\dfrac{v}{-2} - 5 = -1$ **(i)** 9; $2a + 5 = 3a - 4$ **(j)** 0; $2(b - 3) = -5 - (2 - b)$

(k) -2; $\dfrac{1}{2}x + \dfrac{3}{4} = -\dfrac{1}{4}$ **(l)** -2.4; $-0.5y - 0.2 = 1$ **(m)** 6; $w - 50\%w = 2.5$

Solution: Recall that to check a proposed solution of a given linear equation in one variable, you first substitute the proposed solution for the variable and then compute using the Order of Operations to get a number sentence. If you get a true number sentence, then the proposed solution checks and it is the one and only solution of the given equation. If you get a false number sentence, then the proposed solution does not check and it is not the solution of the given equation [see Example 5-4]:

(a) $2x = 3$

$2\left(\dfrac{3}{2}\right) \mid 3$

$\quad\quad 3 \mid 3$

$\quad\quad\quad$ true $(3 = 3)$

$\dfrac{3}{2}$ is the solution of $2x = 3$.

(b) $y + 5 = 2$

$-3 + 5 \mid 2$

$\quad\quad 2 \mid 2$

$\quad\quad\quad$ true $(2 = 2)$

-3 is the solution of $y + 5 = 2$.

(c) $\dfrac{w}{-2} = 3$

$\dfrac{6}{-2} \mid 3$

$\;\; -3 \mid 3$

$\quad\quad\quad$ false $(-3 \neq 3)$

6 is not the solution of $\dfrac{w}{-2} = 3$.

(d) $z - 4 = -5$

$\quad 0 - 4 \mid -5$

$\quad\;\; -4 \mid -5$

$\quad\quad\quad$ false $(-4 \neq -5)$

0 is not the solution of $z - 4 = -5$.

(e) $2m + 3 = -6$

$2\left(-\dfrac{9}{2}\right) + 3 \mid -6$

$\quad\quad -9 + 3 \mid -6$

$\quad\quad\quad\;\; -6 \mid -6$

$\quad\quad\quad\quad$ true $(-6 = -6)$

$-\dfrac{9}{2}$ is the solution of $2m + 3 = -6$.

(f) $3n - 5 = 4$

$3(3) - 5 \mid 4$

$\;\; 9 - 5 \mid 4$

$\quad\quad\; 4 \mid 4$

$\quad\quad\quad$ true $(4 = 4)$

3 is the solution of $3n - 5 = 4$.

(g) $\dfrac{u}{4} + 3 = 2$

$\dfrac{-4}{4} + 3 \mid 2$

$\quad -1 + 3 \mid 2$

$\quad\quad\quad 2 \mid 2$

$\quad\quad\quad$ true $(2 = 2)$

-4 is the solution of $\dfrac{u}{4} + 3 = 2$.

(h) $\dfrac{v}{-2} - 5 = -1$

$\dfrac{-10}{-2} - 5 \mid -1$

$\quad\;\; 5 - 5 \mid -1$

$\quad\quad\quad 0 \mid -1$

$\quad\quad\quad$ false $(0 \neq -1)$

-10 is not the solution of $\dfrac{v}{-2} - 5 = -1$.

(i) $2a + 5 = 3a - 4$

$2(9) + 5 \mid 3(9) - 4$

$\;\; 18 + 5 \mid 27 - 4$

$\quad\quad 23 \mid 23$

$\quad\quad\quad$ true $(23 = 23)$

9 is the solution of $2a + 5 = 3a - 4$.

(j) $2(b - 3) = -5 - (2 - b)$

$2(0 - 3) \mid -5 - (2 - 0)$

$\;\; 2(-3) \mid -5 - 2$

$\quad\; -6 \mid -7 \leftarrow$ false $(-6 \neq -7)$

0 is not the solution of $2(b - 3) = -5 - (2 - b)$.

(k) $\dfrac{1}{2}x + \dfrac{3}{4} = -\dfrac{1}{4}$

$\dfrac{1}{2}(-2) + \dfrac{3}{4} \mid -\dfrac{1}{4}$

$\quad -1 + \dfrac{3}{4} \mid -\dfrac{1}{4}$

$\quad\quad\;\; -\dfrac{1}{4} \mid -\dfrac{1}{4} \leftarrow$ true $\left(-\dfrac{1}{4} = -\dfrac{1}{4}\right)$

-2 is the solution of $\dfrac{1}{2}x + \dfrac{3}{4} = -\dfrac{1}{4}$.

(l)

$$-0.5y - 0.2 = 1$$

$-0.5(-2.4) - 0.2$	1
$1.2 - 0.2$	1
1	1 $\longleftarrow$ true $(1 = 1)$

(m) $w - 50\%w = 2.5$

$6 - 50\%(6)$	2.5
$6 - \dfrac{1}{2}(6)$	2.5
$6 - 3$	2.5
3	2.5 $\longleftarrow$ false $(3 \neq 2.5)$

-2.4 is the solution of $-0.5y - 0.2 = 1.$　　6 is not the solution of $w - 50\%w = 2.5.$

PROBLEM 5-5　Solve the following equations using the rules for linear equations: **(a)** $x + 2 = 5$

(b) $y - 2 = -7$　　**(c)** $3c = 12$　　**(d)** $\dfrac{d}{2} = -4$　　**(e)** $4r + 2 = 14$　　**(f)** $3s - 1 = -10$

(g) $\dfrac{t}{2} + 3 = 5$　　**(h)** $\dfrac{u}{-5} - 3 = 1$　　**(i)** $-5 = 5 - z$　　**(j)** $3 = 2 - 4g$　　**(k)** $1 = \dfrac{m}{-3} + 1$

(l) $-3 = -5 + \dfrac{y}{-2}$

Solution:　Recall that to solve a linear equation in one variable that does not contain like terms or parentheses, you

1. Rename when necessary as one of the eight special types of linear equations in one variable.
2. Use the rules for linear equations to isolate the variable in one member of the linear equation.

(a)　　$x + 2 = 5$

$$x + 2 - 2 = 5 - 2$$

$$x = 3 \ [\text{See Example 5-5.}]$$

(b)　　$y - 2 = -7$

$$y - 2 + 2 = -7 + 2$$

$$y = -5 \ [\text{See Example 5-6.}]$$

(c) $3c = 12$

$$\frac{3c}{3} = \frac{12}{3}$$

$$c = 4 \ [\text{See Example 5-7.}]$$

(d) $\dfrac{d}{2} = -4$

$$2 \cdot \frac{d}{2} = 2(-4)$$

$$d = -8 \ [\text{See Example 5-8.}]$$

(e)　　$4r + 2 = 14$

$$4r + 2 - 2 = 14 - 2$$

$$4r = 12$$

$$\frac{4r}{4} = \frac{12}{4}$$

$$r = 3 \ [\text{See Example 5-9.}]$$

(f)　　$3s - 1 = -10$

$$3s - 1 + 1 = -10 + 1$$

$$3s = -9$$

$$\frac{3s}{3} = \frac{-9}{3}$$

$$s = -3 \ [\text{See Example 5-10.}]$$

(g)　　$\dfrac{t}{2} + 3 = 5$

$$\frac{t}{2} + 3 - 3 = 5 - 3$$

$$\frac{t}{2} = 2$$

$$2 \cdot \frac{t}{2} = 2 \cdot 2$$

$$t = 4 \ [\text{See Example 5-11.}]$$

(h)　　$\dfrac{u}{-5} - 3 = 1$

$$\frac{u}{-5} - 3 + 3 = 1 + 3$$

$$\frac{u}{-5} = 4$$

$$-5 \cdot \frac{u}{-5} = -5 \cdot 4$$

$$u = -20 \ [\text{See Example 5-12.}]$$

(i)
$$-5 = 5 - z$$
$$5 - z = -5$$
$$-z + 5 = -5$$
$$-1z + 5 = -5$$
$$-1z + 5 - 5 = -5 - 5$$
$$-1z = -10$$
$$\frac{-1z}{-1} = \frac{-10}{-1}$$
$$z = 10 \text{ [See Example 5-13,}$$
$$\text{part (a).]}$$

(j)
$$3 = 2 - 4g$$
$$2 - 4g = 3$$
$$-4g + 2 = 3$$
$$-4g + 2 - 2 = 3 - 2$$
$$-4g = 1$$
$$\frac{-4g}{-4} = \frac{1}{-4}$$
$$g = -\frac{1}{4} \text{ [See Example 5-13,}$$
$$\text{part (b).]}$$

(k)
$$1 = \frac{m}{-3} + 1$$
$$\frac{m}{-3} + 1 = 1$$
$$\frac{m}{-3} + 1 - 1 = 1 - 1$$
$$\frac{m}{-3} = 0$$
$$-3 \cdot \frac{m}{-3} = -3 \cdot 0$$
$$m = 0 \text{ [See Example 5-13,}$$
$$\text{part (c).]}$$

(l)
$$-3 = -5 + \frac{y}{-2}$$
$$-5 + \frac{y}{-2} = -3$$
$$\frac{y}{-2} - 5 = -3$$
$$\frac{y}{-2} - 5 + 5 = -3 + 5$$
$$\frac{y}{-2} = 2$$
$$-2 \cdot \frac{y}{-2} = -2(2)$$
$$y = -4 \text{ [See Example 5-13,}$$
$$\text{part (d).]}$$

PROBLEM 5-6 Solve by first combining like terms: **(a)** $x - 5 - 6x = -10$
(b) $4y + 8 = 2y - 6$ **(c)** $-w + 8 - 2w = 3w - 6 + w$ **(d)** $2 - 5m = 4m + 2 - 9m$
(e) $1 - n = 2 + 3n - 3 - 4n$ **(f)** $5(3 - 2u) = -5$ **(g)** $3(v + 2) = 2(3 + v - 2v)$

Solution: Recall that to solve a linear equation in one variable that contains like terms and/or parentheses, you:
1. Combine any like terms inside the parentheses.
2. Clear any parentheses using the distributive properties.
3. Combine any like terms in each member.
4. Collect and combine the like terms in one member.
5. Rename the equation when necessary to get one of the eight special types of linear equations in one variable.
6. Solve the equation in one variable using the rules for linear equations.

(a)
$$x - 5 - 6x = -10$$
$$1x - 6x - 5 = -10$$
$$-5x - 5 = -10$$
$$-5x = -5$$
$$x = 1$$
[See Example 5-14.]

(b)
$$4y + 8 = 2y - 6$$
$$4y - 2y + 8 = 2y - 2y - 6$$
$$2y + 8 = -6$$
$$2y = -14$$
$$y = -7$$
[See Example 5-15.]

(c)
$$-w + 8 - 2w = 3w - 6 + w$$
$$-1w - 2w + 8 = 3w + 1w - 6$$
$$-3w + 8 = 4w - 6$$
$$-3w - 4w + 8 = 4w - 4w - 6$$
$$-7w = -14$$
$$w = 2$$
[See Example 5-15.]

(d)
$$2 - 5m = 4m + 2 - 9m$$
$$2 - 5m = 4m - 9m + 2$$
$$2 - 5m = -5m + 2$$
$$2 - 5m + 5m = -5m + 5m + 2$$
$$2 = 2 \longleftarrow \text{true number sentence}$$

Every real number is a solution of
$2 - 5m = 4m + 2 - 9m$
[See Example 5-17.]

(e)
$$1 - n = 2 + 3n - 3 - 4n$$
$$1 - n = 2 - 3 + 3n - 4n$$
$$1 - n = -1 - n$$
$$1 - n + n = -1 - n + n$$
$$1 = -1 \longleftarrow \text{false number sentence}$$

There are no solutions of
$1 - n = 2 + 3n - 3 - 4n$
[See Example 5-18.]

(f)
$$5(3 - 2u) = -5$$
$$5(3) - 5(2u) = -5$$
$$15 - 10u = -5$$
$$-10u + 15 = -5$$
$$-10u = -20$$
$$u = 2$$

(g)
$$3(v + 2) = 2(3 + v - 2v)$$
$$3(v + 2) = 2(3 - v)$$
$$3 \cdot v + 3 \cdot 2 = 2 \cdot 3 - 2 \cdot v$$
$$3v + 6 = 6 - 2v$$
$$3v + 2v + 6 = 6 - 2v + 2v$$
$$5v + 6 = 6$$
$$5v = 0$$
$$v = 0$$

[See Example 5-16.]

PROBLEM 5-7 Solve by first clearing fractions, decimals, or percents:

(a) $\frac{1}{2}x - \frac{1}{2} = \frac{3}{4}x + \frac{1}{2}$ **(b)** $\dfrac{y - 2}{6} = \dfrac{y + 3}{4} - 1$ **(c)** $\frac{1}{2}(2 - w) + 1 = \frac{2}{3}(3w - 1) + \frac{1}{6}$

(d) $0.03m - 1.2 = 0.06$ **(e)** $0.5(2 - n) = 0.3$ **(f)** $u - 25\%u = 1.5$

Solution: Recall that to solve linear equations containing fractions, decimals, or percents, you
1. Clear all fractions, decimals, and percents.
2. Combine like terms and/or clear parentheses.
3. Rename as one of the eight special types of linear equations in one variable.
4. Use the rules for linear equations. [See Examples 5-19, 5-21, and 5-22.]

(a) $\frac{1}{2}x - \frac{1}{2} = \frac{3}{4}x + \frac{1}{2}$

The LCD for $\frac{1}{2}$ and $\frac{3}{4}$ is 4.
$$4(\tfrac{1}{2}x - \tfrac{1}{2}) = 4(\tfrac{3}{4}x + \tfrac{1}{2})$$
$$4(\tfrac{1}{2}x) - 4(\tfrac{1}{2}) = 4(\tfrac{3}{4}x) + 4(\tfrac{1}{2})$$
$$2x - 2 = 3x + 2$$
$$2x - 3x - 2 = 3x - 3x + 2$$
$$-1x - 2 = 2$$
$$-1x = 4$$
$$x = -4$$

(b) $\dfrac{y - 2}{6} = \dfrac{y + 3}{4} - 1$

The LCD for $\dfrac{y - 2}{6}$ and $\dfrac{y + 3}{4}$ is 12.
$$12 \cdot \frac{y - 2}{6} = 12\left(\frac{y + 3}{4} - 1\right)$$
$$12 \cdot \frac{y - 2}{6} = 12 \cdot \frac{y + 3}{4} - 12(1)$$
$$\tfrac{12}{6}(y - 2) = \tfrac{12}{4}(y + 3) - 12$$
$$2(y - 2) = 3(y + 3) - 12$$
$$2y - 4 = 3y + 9 - 12$$
$$2y - 4 = 3y - 3$$
$$2y - 3y - 4 = 3y - 3y - 3$$
$$-1y - 4 = -3$$
$$-1y = 1$$
$$y = -1$$

(c) $\frac{1}{2}(2 - w) + 1 = \frac{2}{3}(3w - 1) + \frac{1}{6}$

The LCD for $\frac{1}{2}, \frac{2}{3},$ and $\frac{1}{6}$ is 6.

$6[\frac{1}{2}(2 - w) + 1] = 6[\frac{2}{3}(3w - 1) + \frac{1}{6}]$

$6 \cdot \frac{1}{2}(2 - w) + 6(1) = 6 \cdot \frac{2}{3}(3w - 1) + 6 \cdot \frac{1}{6}$

$3(2 - w) + 6 = 4(3w - 1) + 1$

$6 - 3w + 6 = 12w - 4 + 1$

$-3w + 12 = 12w - 3$

$-3w - 12w + 12 = 12w - 12w - 3$

$-15w + 12 = -3$

$-15w = -15$

$w = 1$

(d) $0.03m - 1.2 = 0.06$

The LCD for $\frac{3}{100}(0.03), \frac{12}{10}(1.2),$ and $\frac{6}{100}(0.06)$ is 100.

$100(0.03m - 1.2) = 100(0.06)$

$100(0.03m) - 100(1.2) = 100(0.06)$

$3m - 120 = 6$

$3m = 126$

$m = 42$

(e) $0.5(2 - n) = 0.3$

The LCD for $\frac{5}{10}(0.5)$ and $\frac{3}{10}(0.3)$ is 10.

$10[0.5(2 - n)] = 10(0.3)$

$10(0.5)(2 - n) = 10(0.3)$

$5(2 - n) = 3$

$10 - 5n = 3$

$-5n + 10 = 3$

$-5n = -7$

$n = \frac{7}{5}$ or 1.4

(f) $u - 25\%u = 1.5$

$1u - 0.25u = 1.5$

$0.75u = 1.5$

$u = 2$

PROBLEM 5-8 Solve the following literal equations and formulas:
(a) Solve for c: $s = c + m$ [retail sales formula]
(b) Solve for x: $x - 4y = -2x + 5$ [literal equation]
(c) Solve for r: $d = rt$ [distance formula] (d) Solve for l: $n = 3l - 25$ [shoe-size formula]
(e) Solve for a: $P = 2(a + b)$ [perimeter formula] (f) Solve for b: $A = \frac{1}{2}bh$ [area formula]

Solution: Recall that to solve a literal equation for a given variable, you
1. Clear fractions, decimals, and percents.
2. Combine like terms and/or clear parentheses.
3. Isolate the term containing the given variable.
4. Isolate the given variable itself.

(a)

$s = c + m$

$c + m = s$

$c + m - m = s - m$

$c = s - m$

[See Example 5-25.]

(b)

$x - 4y = -2x + 5$

$x + 2x - 4y = -2x + 2x + 5$

$3x - 4y = 5$

$3x - 4y + 4y = 4y + 5$

$3x = 4y + 5$

$x = \frac{4y + 5}{3}$

or

$x = \frac{4y}{3} + \frac{5}{3}$

or

$x = \frac{4}{3}y + \frac{5}{3}$

[See Example 5-26.]

(c) $d = rt$

$rt = d$

$\frac{rt}{t} = \frac{d}{t}$

$r = \frac{d}{t}$

[See Example 5-25.]

(d)
$$n = 3l - 25$$
$$3l - 25 = n$$
$$3l - 25 + 25 = n + 25$$
$$3l = n + 25$$
$$l = \frac{n + 25}{3}$$

or $\quad l = \frac{n}{3} + \frac{25}{3}$

or $\quad l = \frac{1}{3}n + \frac{25}{3}$

[See Example 5-25.]

(e)
$$P = 2(a + b)$$
$$2(a + b) = P$$
$$2a + 2b = P$$
$$2a + 2b - 2b = P - 2b$$
$$2a = P - 2b$$
$$a = \frac{P - 2b}{2}$$

or $\quad a = \frac{P}{2} - b$

or $\quad a = \frac{1}{2}P - b$

[See Example 5-27.]

(f)
$$A = \frac{1}{2}bh$$
$$\frac{1}{2}bh = A$$
$$2\left(\frac{1}{2}bh\right) = 2(A)$$
$$2 \cdot \frac{1}{2}bh = 2A$$
$$bh = 2A$$
$$\frac{b\cancel{h}}{\cancel{h}} = \frac{2A}{h}$$
$$b = \frac{2A}{h}$$

[See Example 5-25.]

Supplementary Exercises

PROBLEM 5-9   Determine if the given real number is a solution of the given linear equation in one variable: **(a)** $-8; x + 6 = -2$ **(b)** $9; y - 5 = 4$ **(c)** $-\frac{4}{3}; -3w = 4$

(d) $-12; \frac{z}{2} = -5$ **(e)** $-1; -3u + 2 = 5$ **(f)** $\frac{3}{2}; 2v - 4 = -1$ **(g)** $-2; \frac{a}{-2} + 1 = 3$

(h) $15; \frac{b}{5} - 2 = 1$ **(i)** $\frac{9}{4}; 2 - 3m = m - 2 + 5m$ **(j)** $0; -2(5 - n) + 5 = 3(n - 2)$

(k) $\frac{1}{3}; \frac{3}{4}h - \frac{1}{2} = \frac{1}{4}$ **(l)** $1; 0.2k + 0.3 = 0.5$ **(m)** $-8; r - 75\%r = -2$

PROBLEM 5-10   Solve the following linear equations: **(a)** $x + 5 = -2$ **(b)** $y - 4 = 2$

(c) $8 = w + 5$ **(d)** $7 = z - 2$ **(e)** $3m = 24$ **(f)** $\frac{n}{4} = 5$ **(g)** $-12 = 6a$

(h) $-9 = \frac{b}{-2}$ **(i)** $3n + 7 = -5$ **(j)** $-6 = 3v + 3$ **(k)** $2h - 10 = 6$ **(l)** $5 = 2k - 5$

(m) $\frac{r}{5} + 6 = 2$ **(n)** $3 = \frac{s}{2} + 5$ **(o)** $\frac{c}{-5} - 3 = -5$ **(p)** $-5 = \frac{d}{2} - 3$

(q) $5p - 12 - 3p = -7$ **(r)** $-q + 4q + 6 = -3q - 4 + q$ **(s)** $-2(3w - 5) = -8$

(t) $3(1 + x) + 2 = 5(x + 1) - 2x$ **(u)** $5y - 2 = 8y + 3(1 - y)$ **(v)** $\frac{1}{2}z + \frac{5}{6} = -\frac{2}{3}$

(w) $\frac{2m + 3}{2} + 1 = \frac{5 - m}{5}$ **(x)** $\frac{3}{4}(n - 4) = 2 - \frac{1}{2}(2n + 3)$ **(y)** $2.5a + 6 = 0.25$

(z) $b + 40\%b = 0.7$

PROBLEM 5-11   Solve the following literal equations and formulas:

(a) Solve for r: $s = r - d$ [retail sales formula]   **(b)** Solve for d: $s = r - d$
(c) Solve for x: $x - y = 5$ [literal equation]   **(d)** Solve for y: $x - y = 5$
(e) Solve for m: $4m - 5n = 2$ [literal equation]   **(f)** Solve for n: $4m - 5n = 2$
(g) Solve for x: $Ax + By = C$ [literal equation]   **(h)** Solve for y: $Ax + By = C$
(i) Solve for b: $y = mx + b$ [slope-intercept formula]   **(j)** Solve for m: $y = mx + b$
(k) Solve for P: $PB = A$ [percent formula]   **(l)** Solve for t: $d = rt$ [distance formula]
(m) Solve for l: $V = lwh$ [volume formula]   **(n)** Solve for r: $I = Prt$ [interest formula]
(o) Solve for h: $A = \frac{1}{2}bh$ [area formula]   **(p)** Solve for B: $V = \frac{1}{3}Bh$ [volume formula]
(q) Solve for P: $A = P(1 + rt)$ [amount formula]   **(r)** Solve for r: $A = P(1 + rt)$
(s) Solve for l: $n = 3l - 22$ [shoe-size formula]   **(t)** Solve for h: $w = 5\frac{1}{2}(h - 40)$ [weight formula]
(u) Solve for h: $A = \frac{1}{2}h(b_1 + b_2)$ [area formula]   **(v)** Solve for b_1: $A = \frac{1}{2}h(b_1 + b_2)$
(w) Solve for F: $C = \frac{5}{9}(F - 32)$ [temperature formula]
(x) Solve for t: $a = 0.16(15 - t)$ [altitude formula]
(y) Solve for v_0: $v = v_0 - 32t$ [velocity formula]   **(z)** Solve for t: $v = v_0 - 32t$

Answers to Supplementary Exercises

(5-9)   **(a)** yes   **(b)** yes   **(c)** yes   **(d)** no   **(e)** yes   **(f)** yes   **(g)** no   **(h)** yes
(i) no   **(j)** no   **(k)** no   **(l)** yes   **(m)** yes

(5-10)   **(a)** -7 **(b)** 6   **(c)** 3   **(d)** 9   **(e)** 8   **(f)** 20   **(g)** -2 **(h)** 18   **(i)** -4
(j) -3 **(k)** 8   **(l)** 5   **(m)** -20 **(n)** -4 **(o)** 10   **(p)** -4 **(q)** $\frac{5}{2}$ or $2\frac{1}{2}$ or 2.5
(r) -2 **(s)** 3   **(t)** every real number   **(u)** no solutions   **(v)** -3 **(w)** $-\frac{5}{4}$ or $-1\frac{1}{4}$ or -1.25 **(x)** 2   **(y)** $-\frac{23}{10}$ or $-2\frac{3}{10}$ or -2.3 **(z)** $\frac{1}{2}$ or 0.5

(5-11)   **(a)** $r = s + d$ **(b)** $d = r - s$ **(c)** $x = y + 5$ **(d)** $y = x - 5$

(e) $m = \dfrac{5n + 2}{4}$ or $\dfrac{5n}{4} + \dfrac{1}{2}$ or $\dfrac{5}{4}n + \dfrac{1}{2}$ **(f)** $n = \dfrac{2 - 4m}{-5}$ or $-\dfrac{2}{5} + \dfrac{4m}{5}$ or $\dfrac{4}{5}m - \dfrac{2}{5}$

(g) $x = \dfrac{C - By}{A}$ or $\dfrac{C}{A} - \dfrac{By}{A}$ or $\dfrac{C}{A} - \dfrac{B}{A}y$ or $-\dfrac{B}{A}y + \dfrac{C}{A}$

(h) $y = \dfrac{C - Ax}{B}$ or $\dfrac{C}{B} - \dfrac{Ax}{B}$ or $\dfrac{C}{B} - \dfrac{A}{B}x$ or $-\dfrac{A}{B}x + \dfrac{C}{B}$ **(i)** $b = y - mx$

(j) $m = \dfrac{y - b}{x}$ or $\dfrac{y}{x} - \dfrac{b}{x}$ **(k)** $P = \dfrac{A}{B}$ **(l)** $t = \dfrac{d}{r}$ **(m)** $l = \dfrac{V}{wh}$ **(n)** $r = \dfrac{I}{Pt}$ **(o)** $h = \dfrac{2A}{b}$

(p) $B = \dfrac{3V}{h}$ **(q)** $P = \dfrac{A}{1 + rt}$ **(r)** $r = \dfrac{A - P}{Pt}$ or $\dfrac{A}{Pt} - \dfrac{1}{t}$

(s) $l = \dfrac{n + 22}{3}$ or $\dfrac{n}{3} + \dfrac{22}{3}$ or $\dfrac{1}{3}n + \dfrac{22}{3}$ **(t)** $h = \dfrac{w + 220}{5\frac{1}{2}}$ or $\dfrac{w}{5\frac{1}{2}} + 40$ or $\dfrac{2}{11}w + 40$

(u) $h = \dfrac{2A}{b_1 + b_2}$ **(v)** $b_1 = \dfrac{2A - hb_2}{h}$ or $\dfrac{2A}{h} - b_2$ **(w)** $F = \dfrac{9C + 160}{5}$ or $\dfrac{9C}{5} + 32$ or $\dfrac{9}{5}C + 32$

(x) $t = \dfrac{2.4 - a}{0.16}$ or $15 - \dfrac{a}{0.16}$ or $15 - 6.25a$ **(y)** $v_0 = v + 32t$

(z) $t = \dfrac{v - v_0}{-32}$ or $\dfrac{v}{-32} + \dfrac{v_0}{32}$ or $-\dfrac{1}{32}v + \dfrac{1}{32}v_0$ or $\dfrac{1}{32}v_0 - \dfrac{1}{32}v$ or $\dfrac{v_0 - v}{32}$

6 APPLICATIONS OF LINEAR EQUATIONS

THIS CHAPTER IS ABOUT

☑ **Translating Words to Symbols**
☑ **Solving Number Problems Using Linear Equations**
☑ **Solving Geometry Problems Using Linear Equations**
☑ **Solving Currency Value Problems Using Linear Equations**
☑ **Solving Investment Problems Using Linear Equations [$I = Prt$]**
☑ **Solving Uniform Motion Problems Using Linear Equations [$d = rt$]**

Agreement: In this chapter the words "linear equation" will mean *a linear equation in one variable.*

6-1. Translating Words to Symbols

To help **translate words to symbols,** you can substitute the appropriate symbols given in Table 6-1 for the corresponding **key words.**

Table 6-1

Key Words	Math Symbol
the sum of / plus / added to / joined with / increased by / more than / more / and / combined with	$+$
the difference between / minus / subtracted from / take away / decreased by / less than / less / reduced by / diminished by / exceeds	$-$
the product of / times / multiplied by / equal amounts of / of / times as much	$\cdot$
the quotient of / divides / divided by / ratio / separated into equal amounts of / goes into / over	$\div$
is the same as / is equal to / equals / is / was / earns are / makes / gives / the result is / leaves / will be	$=$
twice / double / two times / twice as much as	$2 \cdot$
half / one-half of / one-half times / half as much as	$\frac{1}{2} \cdot$
what number / what part / a number / the number / what amount / what percent / what price	n (any letter can be used)
twice a (the) number / double a (the) number	$2n$
half a (the) number / one-half a (the) number	$\frac{1}{2}n$

EXAMPLE 6-1: Translate each of the following to a linear equation using key words:

(a) Three times a number is 24.
(b) Four less than a number is the same as -1.
(c) Six subtracted from two-thirds of a number gives 2.
(d) Seven times the quantity, x plus 2, is equal to 35.

Solution:

(a) *Identify:*   Three times a number is 24.   Identify key words [see Table 6-1].

   Translate: 3 $\cdot$ n $=$ 24   Translate to symbols.

   $$3n = 24 \longleftarrow \text{equation}$$

(b) *Identify:*   Four less than a number is the same as -1.

   Translate: $n - 4$ $=$ -1 *Think:* "4 less than n" means $n - 4$.

   $$n - 4 = -1 \longleftarrow \text{equation}$$

(c) *Identify:*   Six subtracted from two-thirds of a number gives 2.

   Translate: $\dfrac{2}{3} \cdot n - 6$ $= 2$

   $$\frac{2}{3}n - 6 = 2 \longleftarrow \text{equation}$$

(d) *Identify:*   Seven times the quantity, x plus 2, is equal to 35.

   Translate: 7 $\cdot$ $(x + 2)$ $=$ 35

   $$7(x + 2) = 35 \longleftarrow \text{equation}$$

6-2. Solving Number Problems Using Linear Equations

To **solve a number problem using a linear equation,** you

1. *Read* the problem very carefully several times.
2. *Identify* the unknown numbers.
3. *Decide* how to represent the unknown numbers using one variable.
4. *Translate* the problem to a linear equation using key words.
5. *Solve* the linear equation.
6. *Interpret* the solution of the linear equation with respect to each represented unknown number to find the proposed solutions of the original problem.
7. *Check* to see if the proposed solutions satisfy all the conditions of the original problem.

EXAMPLE 6-2: Solve the following number problem using a linear equation.

Solution:

1. *Read:*   Two numbers have a sum of 65. Four times the smaller number is equal to the larger number. Find the two numbers.

2. *Identify:*   The unknown numbers are $\begin{cases} \text{the smaller number} \\ \text{the larger number} \end{cases}$.

3. *Decide:*   Let $n =$ the smaller number
   then $65 - n =$ the larger number [see the following *Note*].

4. *Translate:* Four times the smaller number is equal to the larger number.   Identify key words.

   4   $\cdot$ n $=$ $65 - n$ Translate words to symbols.

$$4n = 65 - n \quad \longleftarrow \text{linear equation}$$

5. *Solve:* $\quad 4n + n = 65 - n + n \quad$ [See Example 5-15.]

$$5n = 65$$

$$n = 13$$

6. *Interpret:* $\quad n = 13$ means the smaller number is 13. $\longleftarrow$

proposed solutions

$$65 - n = 65 - 13 = 52 \text{ means the larger number is 52.}$$

7. *Check:* Did you find two numbers?   Yes: 13 and 52

Do the two numbers have a sum of 65?   Yes: $13 + 52 = 65$

Is four times the smaller number equal to the larger number?   Yes: $4 \cdot 13 = 52$

Therefore, the two numbers are 13 and 52.

Note: To represent two unknown numbers that have a sum of 65, you can

let $n =$ the smaller number and $65 - n =$ the larger number

or

let $n =$ the larger number and $65 - n =$ the smaller number

because in both cases, $n + (65 - n) = n + 65 - n = n - n + 65 = 0 + 65 = 65$.

EXAMPLE 6-3: Let $n =$ the larger number and $65 - n =$ the smaller number for the problem in Example 6-2.

Solution:

Translate: Four times the smaller number is equal to the larger number.

$$4 \quad \cdot \quad (65 - n) \quad = \quad n$$

$$4(65 - n) = n \quad \longleftarrow \text{linear equation}$$

Solve:

$$4(65) - 4(n) = n$$

$$260 - 4n = n$$

$$260 - 4n + 4n = n + 4n$$

$$260 = 5n$$

larger number $\longrightarrow n = 52 \longleftarrow$ same solutions as found

smaller number $\longrightarrow 65 - n = 65 - 52 = 13 \longleftarrow$ in Example 6-2

Integers that differ by one are called **consecutive integers.**

EXAMPLE 6-4: Write three consecutive integers starting with 5.

Solution: $\quad\quad\quad\quad 5 \longleftarrow$ first given consecutive integer

$$5 + 1 = 6 \longleftarrow \text{second consecutive integer}$$

$$6 + 1 = 7 \longleftarrow \text{third consecutive integer}$$

Even integers that differ by two are called **consecutive even integers.**

EXAMPLE 6-5: Write four consecutive even integers starting with 12.

Solution: $\quad\quad\quad\quad 12 \longleftarrow$ first given consecutive even integer

$$12 + 2 = 14 \longleftarrow \text{second consecutive even integer}$$

$$14 + 2 = 16 \longleftarrow \text{third consecutive even integer}$$

$$16 + 2 = 18 \longleftarrow \text{fourth consecutive even integer}$$

Odd integers that differ by two are called **consecutive odd integers.**

EXAMPLE 6-6: Write five consecutive odd integers starting with -5.

Solution:

$$-5 \longleftarrow \text{first given consecutive odd integer}$$

$$-5 + 2 = -3 \longleftarrow \text{second consecutive odd integer}$$

$$-3 + 2 = -1 \longleftarrow \text{third consecutive odd integer}$$

$$-1 + 2 = 1 \longleftarrow \text{fourth consecutive odd integer}$$

$$1 + 2 = 3 \longleftarrow \text{fifth consecutive odd integer}$$

A special type of number problem is the **consecutive integer problem.**

EXAMPLE 6-7: Solve the following consecutive integer problem using a linear equation.

Solution:

1. *Read:* The sum of three consecutive even integers is 78. What are the integers?

2. *Identify:* The unknown numbers are $\begin{cases} \text{the first consecutive even integer} \\ \text{the second consecutive even integer} \\ \text{the third consecutive even integer} \end{cases}$.

3. *Decide:* Let $n = $ the first consecutive even integer
then $n + 2 = $ the second consecutive even integer
and $n + 4 = $ the third consecutive even integer [see the following *Note 1*].

4. *Translate:* The sum of three consecutive even integers is 78.

$$n + (n + 2) + (n + 4) = 78$$

$$n + n + 2 + n + 4 = 78 \longleftarrow \text{linear equation}$$

5. *Solve:*

$$3n + 6 = 78$$

$$3n = 72$$

$$n = 24$$

6. *Interpret:* $n = 24$ means the first consecutive even integer is 24.
$n + 2 = 24 + 2 = 26$ means the second consecutive even integer is 26.
$n + 4 = 24 + 4 = 28$ means the third consecutive even integer is 28.

7. *Check:* Did you find three consecutive even integers? Yes: 24, 26, and 28
Do the three consecutive even integers have a sum of 78? Yes: $24 + 26 + 28 = 78$

Note 1: To represent three consecutive even [or odd] integers, you

let $n = $ the first consecutive even [or odd] integer
then $n + 2 = $ the second consecutive even [or odd] integer
and $n + 4 = $ the third consecutive even [or odd] integer

because consecutive even [or odd] integers differ by two and $(n + 2) + 2 = n + 4$.

Note 2: To represent three consecutive integers, you

let $n = $ the first consecutive integer
then $n + 1 = $ the second consecutive integer
and $n + 2 = $ the third consecutive integer

because consecutive integers differ by one and $(n + 1) + 1 = n + 2$.

6-3. Solving Geometry Problems Using Linear Equations

To **solve a geometry problem using a linear equation,** you

1. *Read* the problem very carefully several times.
2. *Draw a picture* to help visualize the problem.
3. *Identify* the unknown measures.

4. *Decide* how to represent the unknown measures using one variable.
5. *Translate* the problem to a linear equation using the correct geometry formula from Appendix Table 7.
6. *Solve* the linear equation.
7. *Interpret* the solution of the linear equation with respect to each represented unknown measure to find the proposed solutions of the original problem.
8. *Check* to see if the proposed solutions satisfy all the conditions of the original problem.

EXAMPLE 6-8: Solve the following geometry problem using a linear equation.

Solution:

1. *Read:*   The perimeter of a rectangle is 64 feet. The length of the rectangle is 10 feet longer than the width. Find the area of the rectangle.

2. *Draw a picture:*

width

length

3. *Identify:*   The unknown measures are $\begin{cases} \text{the length of the rectangle} \\ \text{the width of the rectangle} \end{cases}$.

4. *Decide:*   Let l = the length of the rectangle
                then $l - 10$ = the width of the rectangle [see the following *Note*].

5. *Translate:* $P = 2(l + w)$ ⟵ perimeter formula for a rectangle from Appendix Table 7

   $64 = 2[l + (l - 10)]$ ⟵ linear equation   *Think:* $w = l - 10$

6. *Solve:* $64 = 2(l + l - 10)$ [see Example 5-16.]

   $64 = 2(2l - 10)$

   $64 = 2(2l) - 2(10)$

   $64 = 4l - 20$

   $84 = 4l$

   $21 = l$

7. *Interpret:* $l = 21$ means the length of the rectangle is 21 feet.
                $l - 10 = 21 - 10 = 11$ means the width of the rectangle is 11 feet.
                $l = 21$ ft and $w = 11$ ft means the area of the rectangle is

   $A = lw = (21 \text{ ft})(11 \text{ ft}) = 231 \text{ ft}^2$ ⟵ proposed solution

8. *Check:*   Did you find the length and the width?   Yes: $l = 21$ feet and $w = 11$ feet
                Is the length of the rectangle 10 feet more than the width?   Yes: $21 - 10 = 11$ [feet]
                Is the perimeter of the rectangle 64 feet?   Yes: $2(21 + 11) = 2(32) = 64$ [feet]

The area of the rectangle is 231 square feet, or 231 ft².

Note: To represent the unknown length of a rectangle that is 10 units longer than the width, you can

let l = the length and $l - 10$ = the width

                     or

let w = the width and $w + 10$ = the length

because l is 10 more than $l - 10$ and $w + 10$ is 10 more than w.

EXAMPLE 6-9: Let w = the width of the rectangle and $w + 10$ = the length of the rectangle for the problem in Example 6-8.

Solution:

Translate:

$$P = 2(l + w)$$

$$64 = 2[(w + 10) + w] \longleftarrow \text{linear equation} \qquad \textit{Think: } l = w + 10$$

Solve:

$$64 = 2(w + 10 + w)$$

$$64 = 2(2w + 10)$$

$$64 = 2(2w) + 2(10)$$

$$64 = 4w + 20$$

$$44 = 4w$$

$$w = 11 \text{ [feet]} \longleftarrow \text{width}$$

$$w + 10 = 11 + 10 = 21 \text{ [feet]} \longleftarrow \text{length}$$

area of rectangle $\longrightarrow A = lw = (21 \text{ ft})(11 \text{ ft}) = 231 \text{ ft}^2 \longleftarrow$ same solution as found in Example 6-8

6-4. Solving Currency Value Problems Using Linear Equations

To **solve a currency value problem using a linear equation,** you

1. *Read* the problem very carefully several times.
2. *Identify* the unknown currency amounts.
3. *Decide* how to represent the unknown number of each currency using one variable.
4. *Make a table* to help represent the unknown value of each currency using

(denomination of currency)(number of currency) = (value of currency)

5. *Translate* the problem to a linear equation.
6. *Solve* the linear equation.
7. *Interpret* the solution of the linear equation with respect to each represented unknown number and the value of each currency to find the proposed solutions of the original problem.
8. *Check* to see if the proposed solutions satisfy all the conditions of the original problem.

EXAMPLE 6-10: Solve the following currency value problem using a linear equation:

Solution:

1. *Read:* There are 37 dimes and quarters in all. The combined value of the dimes and quarters is \$4.60. How many dimes are there? What is the value of the quarters?

2. *Identify:* The unknown currency amounts are $\left\{ \begin{array}{l} \text{the number of dimes} \\ \text{the number of quarters} \\ \text{the value of the dimes} \\ \text{the value of the quarters} \end{array} \right\}$.

3. *Decide:* Let d = the number of dimes
then $37 - d$ = the number of quarters [see the following *Note*].

4. *Make a table:*

	denomination d [in cents]	number n [of coins]	value $(v = dn)$ [in cents]
dimes	10	d	$10d$
quarters	25	$37 - d$	$25(37 - d)$
combined	——	37	460

5. *Translate:* The value of the dimes combined with the value of the quarters is $4.60.

$$10d \qquad\qquad + \qquad\qquad 25(37-d) \qquad = \quad 460 \text{ [cents]}$$

$$10d + 25(37-d) = \quad 460 \longleftarrow \text{linear equation}$$

6. *Solve:*

$$10d + 25(37) - 25(d) = \quad 460$$
$$10d + 925 - 25d = \quad 460$$
$$925 - 15d = \quad 460$$
$$-15d = \quad -465$$
$$d = \quad 31$$

7. *Interpret:*

$d = 31$ means there are 31 dimes.
$37 - d = 37 - 31 = 6$ means there are 6 quarters.
$10d = 10(31) = 310$ means the value of the dimes is 310 cents or $3.10.
$25(37 - d) = 25(6) = 150$ means the value of the quarters is 150 cents or $1.50.

8. *Check:*

Are there 37 dimes and quarters in all? Yes: $31 + 6 = 37$
Is the combined value of the dimes and quarters $4.60? Yes: $3.10 + $1.50 = $4.60

There are 31 dimes. The value of the quarters is $1.50.

Note: To represent 37 dimes and quarters, you can

let d = the number of dimes and $37 - d$ = the number of quarters

or

let q = the number of quarters and $37 - q$ = the number of dimes
because $d + (37 - d) = 37$ and $q + (37 - q) = 37$.

EXAMPLE 6-11: Let q = the number of quarters and $37 - q$ = the number of dimes for the problem in Example 6-10.

Solution:

Make a table:

	denomination d [in cents]	number n [of coins]	value $v = dn$ [in cents]
quarters	25	q	$25q$
dimes	10	$37 - q$	$10(37 - q)$
combined	——	37	460

Translate: The value of the dimes combined with the value of the quarters is $4.60.

$$10(37-q) \qquad\qquad + \qquad\qquad 25q \qquad = \quad 460 \text{ [cents]}$$

$$10(37-q) + 25q = 460 \longleftarrow \text{linear equation}$$

Solve:

$$10(37) - 10(q) + 25q = 460$$
$$370 - 10q + 25q = 460$$
$$370 + 15q = 460$$
$$15q = 90$$
$$q = 6 \text{ [quarters]}$$

number of dimes $\longrightarrow 37 - q = 37 - 6 = 31$ [dimes]
value of quarters $\longrightarrow 25q = 25(6) = 150$ [cents]

same solutions as found in Example 6-10

6-5. Solving Investment Problems Using Linear Equations [$I = Prt$]

To **solve an investment problem using a linear equation,** you

1. *Read* the problem very carefully several times.
2. *Identify* the unknown investment amounts.
3. *Decide* how to represent the unknown amounts of principal using one variable.
4. *Make a table* to help represent the unknown amounts of interest using

$$\text{Interest} = (\text{Principal})(\text{rate})(\text{time}), \text{ or } I = Prt$$

5. *Translate* the problem to a linear equation.
6. *Solve* the linear equation.
7. *Interpret* the solution of the linear equation with respect to each represented unknown amount to find the proposed solutions of the original problem.
8. *Check* to see if the proposed solutions satisfy all the conditions of the original problem.

EXAMPLE 6-12: Solve the following investment problem using a linear equation:

Solution:

1. *Read:* Two amounts of principal are invested totaling $20,000. One amount is invested at 9% per year and the other at 12%. Together they earn $1920 in one year. How much is invested at 9%? How much does the 12% investment earn?

2. *Identify:* The unknown investment amounts are $\left\{\begin{array}{l}\text{the principal invested at 9\%} \\ \text{the principal invested at 12\%} \\ \text{the interest earned at 9\%} \\ \text{the interest earned at 12\%}\end{array}\right\}$.

3. *Decide:* Let P = the principal invested at 9%
then $20{,}000 - P$ = the principal invested at 12% [see the following *Note*].

4. *Make a table:*

	Principal P [in dollars]	rate r [as a percent]	time t [in years]	Interest [$I = Prt$] [in dollars]
principal at 9%	P	9%	1	$P(9\%)1$ or $9\%P$
principal at 12%	$20{,}000 - P$	12%	1	$(20{,}000 - P)(12\%)1$ or $12\%(20{,}000 - P)$
combined	$20{,}000$	——	1	1920

5. *Translate:* The principal at 9% plus the principal at 12% earns $1920.

$$9\%P \quad + \quad 12\%(20{,}000 - P) \quad = \quad 1920 \longleftarrow \text{linear equation}$$

6. *Solve:*
$$0.09P + 0.12(20{,}000 - P) = 1920$$
$$0.09P + 0.12(20{,}000) - 0.12P = 1920$$
$$0.09P + 2400 - 0.12P = 1920$$
$$2400 - 0.03P = 1920$$
$$-0.03P = -480$$
$$P = 16{,}000$$

7. *Interpret:* $P = 16{,}000$ means the principal invested at 9% is $16,000.
$20{,}000 - P = 20{,}000 - 16{,}000 = 4000$ means the principal invested at 12% is $4000.
$9\%P = (0.09)(16{,}000) = 1440$ means the interest earned at 9% is $1440.
$12\%(20{,}000 - P) = (0.12)(4000) = 480$ means the interest earned at 12% is $480.

8. *Check:*   Do the two amounts of principal total $20,000?   Yes: 16,000 + 4000 = 20,000
Do the two amounts of interest earned total $1920?   Yes: 1440 + 480 = 1920

The principal invested at 9% is $16,000. The amount of interest earned by the 12% investment is $480.

Note: To represent two unknown amounts of principal totaling $20,000, you can

let P = the principal invested at 9% and 20,000 − P = the principal invested at 12%

or

let P = the principal invested at 12% and 20,000 − P = the principal invested at 9%
because in both cases, $P + (20,000 − P) = 20,000$.

EXAMPLE 6-13: Let P = the principal invested at 12% and 20,000 − P = the principal invested at 9% for the problem in Example 6-12.

Solution:

Make a table:

	Principal P [in dollars]	rate r [as a percent]	time t [in years]	Interest $[I = Prt]$ [in dollars]
principal at 12%	P	12%	1	$P(12\%)1$ or $12\%P$
principal at 9%	20,000 − P	9%	1	$(20,000 − P)(9\%)(1)$ or $9\%(20,000 − P)$
combined	20,000	——	1	1920

Translate: The principal at 12% plus the principal at 9% earns $1920.

$$12\%P \quad + \quad 9\%(20,000 − P) \quad = \quad 1920 \longleftarrow \text{linear equation}$$

Solve:

$$0.12P + 0.09(20,000 − P) = 1920$$

$$0.12P + 0.09(20,000) − 0.09(P) = 1920$$

$$0.12P + 1800 − 0.09P = 1920$$

$$1800 + 0.03P = 1920$$

$$0.03P = 1920 − 1800$$

$$0.03P = 120$$

principal invested at 12% $\longrightarrow P = 4000$ [dollars]

principal invested at 9% $\longrightarrow 20,000 − P = 20,000 − 4000$

$$= \$16,000 \longleftarrow$$

same solutions as found in Example 6-12

interest earned by the 12% investment $\longrightarrow 12\%P = (0.12)(4000)$

$$= \$480 \longleftarrow$$

6-6. Solving Uniform Motion Problems Using Linear Equations [$d = rt$]

To **solve a uniform motion problem using a linear equation,** you

1. *Read* the problem very carefully several times.
2. *Draw a picture* to help visualize the problem.
3. *Identify* the unknown distances, rates, or times.
4. *Decide* how to represent the unknown rates or times using one variable.
5. *Make a table* to help represent the unknown distances using

$$\text{distance} = (\text{rate})(\text{time}), \text{ or } d = rt$$

6. *Translate* the problem to a linear equation.

7. *Solve* the linear equation.

8. *Interpret* the solution of the linear equation with respect to each represented unknown (distance, rate, or time) to find the proposed solutions of the original problem.

9. *Check* to see if the proposed solutions satisfy all the conditions of the original problem.

EXAMPLE 6-14: Solve the following uniform motion problem using a linear equation:

Solution:

1. *Read:*   Starting at the same place, two planes flew in opposite directions. The uniform rate of one plane was 100 miles per hour [mph] faster than the uniform rate of the other plane. In 6 hours the two planes were 4800 miles apart. What was the rate of the faster plane? How far did the slower plane travel?

2. *Draw a picture:*

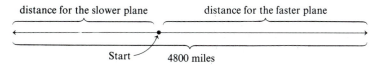

3. *Identify:*   The unknown rates and distances are $\begin{cases} \text{the rate of the slower plane} \\ \text{the rate of the faster plane} \\ \text{the distance for the slower plane} \\ \text{the distance for the faster plane} \end{cases}$.

4. *Decide:*   Let $\quad r =$ the rate of the slower plane in mph
then $r + 100 =$ the rate of the faster plane in mph [see the following *Note*].

5. *Make a table:*

	rate r [in mph]	time t [in hours]	distance $[d = rt]$ [in miles]
slower plane	r	6	$6r$
faster plane	$r + 100$	6	$6(r + 100)$
combined	—	6	4800

6. *Translate:*

The distance for the slower plane plus the distance for the faster plane is 4800 miles.

$$6r \quad + \quad 6(r + 100) \quad = \quad 4800$$

$$6r + 6(r + 100) = 4800 \quad \longleftarrow \text{ linear equation}$$

7. *Solve:*

$$6r + 6(r) + 6(100) = 4800$$

$$12r + 600 = 4800$$

$$12r = 4200$$

$$r = 350$$

8. *Interpret:* $r = 350$ means the uniform rate of the slower plane is 350 mph.
$r + 100 = 350 + 100 = 450$ means the uniform rate of the faster plane is 450 mph.
$6r = 6(350) = 2100$ means the slower plane traveled 2100 miles.
$6(r + 100) = 6(450) = 2700$ means the faster plane traveled 2700 miles.

9. *Check:*   Is the rate of one plane 100 mph faster than the other plane?
Yes: $450 - 100 = 350$
Is the total distance traveled by the two planes 4800 miles?
Yes: $2100 + 2700 = 4800$

The rate of the faster plane is 450 mph. The distance traveled by the slower plane is 2100 miles.

Note: To represent an unknown rate that is 100 mph faster [or slower] than another unknown rate, you can

let r = the slower rate and $r + 100$ = the faster rate

or

let r = the faster rate and $r - 100$ = the slower rate

because $r + 100$ is 100 more than r, and r is 100 more than $r - 100$.

EXAMPLE 6-15: Let r = the rate of the faster plane and $r - 100$ = the rate of the slower plane for the problem in Example 6-14.

Solution:

	rate r [in mph]	time t [in hours]	distance $[d = rt]$ [in miles]
Make a table: faster plane	r	6	$6r$
slower plane	$r - 100$	6	$6(r - 100)$
combined	——	6	4800

Translate: The distance for the faster plane plus distance for the slower plane is 4800 miles.

$$6r \quad\quad + \quad\quad 6(r - 100) \quad\quad = 4800$$

$$6r + 6(r - 100) = 4800 \longleftarrow \text{linear equation}$$

Solve:

$$6r + 6(r) - 6(100) = 4800$$

$$12r - 600 = 4800$$

$$12r = 5400$$

rate of faster plane $\longrightarrow r = 450 \,[\text{mph}] \longleftarrow$ same solutions as found in Example 6-14

distance for slower plane $\longrightarrow 6(r - 100) = 6(350) = 2100 \,[\text{miles}]$

RAISE YOUR GRADES

Can you ... ?

☑ translate words to symbols using key words
☑ solve a number problem using a linear equation
☑ solve a geometry problem using a linear equation
☑ solve a currency value problem using a linear equation
☑ solve an investment problem using a linear equation
☑ solve a uniform motion problem using a linear equation

SUMMARY

1. To help translate words to symbols, you substitute the appropriate symbols given in Table 6-1 for the corresponding key words.
2. Both "x less than y" and "x subtracted from y" mean $y - x$.
3. To solve a number problem, a geometry problem, a currency value problem, an investment problem, or a uniform motion problem using a linear equation, you
   (a) *Read* the problem carefully several times.
   (b) *Draw a picture* to help visualize the problem, when appropriate.

(c) *Identify* the unknowns.

(d) *Decide* how to represent the unknowns using one variable.

(e) *Make a table* to help represent the unknowns when appropriate.

(f) *Translate* the problem to a linear equation.

(g) *Solve* the linear equation.

(h) *Interpret* the solution of the linear equation with respect to each represented unknown to find the proposed solutions of the original problem.

(i) *Check* to see if the proposed solutions satisfy all the conditions of the original problem.

4. To represent two unknown numbers that have a given sum like 10 you can
   let n = the smaller number and $10 - n$ = the larger number

or

   let n = the larger number and $10 - n$ = the smaller number
   because in both cases, $n + (10 - n) = 10$.

5. To represent two unknown numbers so that one number is 10 more [or 10 less] than another number, you can
   let n = the larger number and $n - 10$ = the smaller number

or

   let n = the smaller number and $n + 10$ = the larger number
   because n is 10 more than $n - 10$, and $n + 10$ is 10 more than n.

6. To represent two unknown numbers so that the difference between the numbers is 10, you can
   let n = the larger number, then $n - 10$ = the smaller number

or

   let n = the smaller number, then $n + 10$ = the larger number
   because $n - (n - 10) = n - n + 10 = 10$ and $(n + 10) - n = n + 10 - n = 10$.

7. To represent four unknown consecutive integers, you

   let $\quad$ n = the first consecutive integer,
   then $n + 1$ = the second consecutive integer,
   and $n + 2$ = the third consecutive integer,
   and $n + 3$ = the fourth consecutive integer
   because $(n + 1) + 1 = n + 2$ and $(n + 2) + 1 = n + 3$.

8. To represent four unknown consecutive even [odd] integers, you

   let $\quad$ n = the first consecutive even [odd] integer,
   then $n + 2$ = the second consecutive even [odd] integer,
   and $n + 4$ = the third consecutive even [odd] integer,
   and $n + 6$ = the fourth consecutive even [odd] integer
   because $(n + 2) + 2 = n + 4$ and $(n + 4) + 2 = n + 6$.

SOLVED PROBLEMS

PROBLEM 6-1   Translate the following words to symbols using key words:

(a) The sum of 5 and 2.   (b) The difference between 5 and 2.

(c) The product of 5 and 2.   (d) The quotient of 5 and 2.

(e) Six more than the sum of 5 and 2.   (f) Six less than the difference between 5 and 2.

(g) Six times the product of 5 and 2.   (h) Six divided by the quotient of 5 and 2.

(i) 5 less 2   (j) 5 less than 2   (k) 5 subtracted from 2

(l) The product of 2 and another number, decreased by 5, equals 9.

(m) Five times the quantity, $x + 2$, is equal to 25.

(n) The product of 3 and another number, less twice the same number, is 20.

(o) One-half of a number, increased by 4, is the same as 8.

Solution: Recall that to help translate words to symbols, you can substitute the appropriate symbols given in Table 6-1 for the corresponding key words [see Example 6-1]:

(a) The sum of 5 and 2.

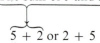

$5 + 2$ or $2 + 5$

(b) The difference between 5 and 2.

larger number ⟶ $5 - 2$ ⟵ smaller number

(c) The product of 5 and 2

$5 \cdot 2$ or $2 \cdot 5$

(d) The quotient of 5 and 2.

first number given ⟶ $5 \div 2$ ⟵ second number given

(e) Six more than the sum of 5 and 2

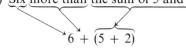

$6 + (5 + 2)$

(f) Six less than the difference between 5 and 2

$(5 - 2) - 6$

(g) Six times the product of 5 and 2

$6 \cdot (5 \cdot 2)$

(h) Six divided by the quotient of 5 and 2

$6 \div (5 \div 2)$

(i) 5 less 2

$5 - 2$

(j) 5 less than 2

$2 - 5$

(k) 5 subtracted from 2

$2 - 5$

(l) The product of 2 and another number, decreased by 5, equals 9.

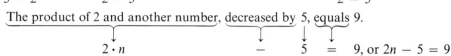

$2 \cdot n$ $-$ 5 $=$ 9, or $2n - 5 = 9$

(m) Five times the quantity, $x + 2$, is equal to 25.

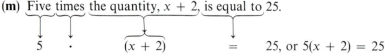

5 $\cdot$ $(x + 2)$ $=$ 25, or $5(x + 2) = 25$

(n) The product of 3 and another number, less twice the same number, is 20.

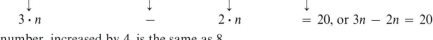

$3 \cdot n$ $-$ $2 \cdot n$ $= 20$, or $3n - 2n = 20$

(o) One-half of a number, increased by 4, is the same as 8.

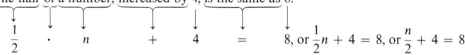

$\frac{1}{2}$ $\cdot$ n $+$ 4 $=$ 8, or $\frac{1}{2}n + 4 = 8$, or $\frac{n}{2} + 4 = 8$

PROBLEM 6-2 Write five of each of the following: **(a)** consecutive even integers beginning with 26 **(b)** consecutive odd integers beginning with -11.

Solution: Recall that consecutive integers differ by one, and consecutive even [or odd] integers differ by two [see Examples 6-4, 6-5, and 6-6]:

(a)   26 ⟵ given first consecutive even integer

$26 + 2 = 28$ ⟵ second consecutive even integer

$28 + 2 = 30$ ⟵ third consecutive even integer

$30 + 2 = 32$ ⟵ fourth consecutive even integer

$32 + 2 = 34$ ⟵ fifth consecutive even integer

(b) -11 ⟵ given first consecutive odd integer

$-11 + 2 = -9$ ⟵ second consecutive odd integer

$-9 + 2 = -7$ ⟵ third consecutive odd integer

$-7 + 2 = -5$ ⟵ fourth consecutive odd integer

$-5 + 2 = -3$ ⟵ fifth consecutive odd integer

PROBLEM 6-3 Solve each of the following number problems using a linear equation:

(a) The difference between two numbers is 16. The larger number is 25 more than one-half of the smaller number. What are the numbers?

(b) The sum of three numbers is 47. The second number is twice the first number. The third number is 2 more than the second number. What are the numbers?

(c) The sum of two consecutive integers is 111. What are the integers?

(d) The sum of three consecutive odd integers is 135. What are the integers?

Solution: Recall that to solve a number problem using a linear equation, you

1. *Read* the problem very carefully several times.
2. *Identify* the unknown numbers.
3. *Decide* how to represent the unknown numbers using one variable.
4. *Translate* the problem to a linear equation using key words.
5. *Solve* the linear equation.
6. *Interpret* the solution of the linear equation with respect to each represented unknown number to find the proposed solutions of the original problem.
7. *Check* to see if the proposed solutions satisfy all of the conditions of the original problem.
   [See Examples 6-2, 6-3, and 6-7.]

(a) *Identify:* The unknown numbers are $\begin{cases} \text{the larger number} \\ \text{the smaller number} \end{cases}$.

	Case 1		**Case 2**

Decide: **Case 1:** Let $n = $ the larger number   or   **Case 2:** let $n = $ the smaller number
then $n - 16 = $ the smaller number   then $n + 16 = $ the larger number

because $n - (n - 16) = n - n + 16 = 16$ and $(n + 16) - n = n + 16 - n = 16$.

Translate: **Case 1:** The larger number is 25 more than one-half of the smaller number.

$$n = 25 + \tfrac{1}{2} \cdot (n - 16)$$

$$n = 25 + \tfrac{1}{2}(n - 16) \longleftarrow \text{linear equation}$$

Solve:
$$2 \cdot n = 2 \cdot 25 + 2 \cdot \tfrac{1}{2}(n - 16) \qquad \textit{Think: The LCD is 2.}$$

$$2n = 50 + 1(n - 16)$$

$$2n = 50 + n - 16$$

$$2n - n = 50 + n - n - 16$$

$$n = 50 - 16$$

Interpret:
$$n = 34$$
$$n - 16 = 18$$
$\Big\rangle$ solution [Check as before.]

Translate: **Case 2:** The larger number is 25 more than one-half of the smaller number.

$$n + 16 = 25 + \tfrac{1}{2} \cdot n$$

$$n + 16 = 25 + \tfrac{1}{2}n \longleftarrow \text{linear equation}$$

Solve:
$$2(n + 16) = 2 \cdot 25 + 2 \cdot \tfrac{1}{2}n \qquad \textit{Think: The LCD is 2.}$$

$$2n + 32 = 50 + n$$

$$2n - n + 32 = 50 + n - n$$

$$n + 32 = 50$$

Interpret:
$$n = 18$$
$$n + 16 = 34$$
$\Big\rangle$ same solutions as found in Case 1

(b) *Identify:*   The unknown numbers are $\begin{cases} \text{the first number} \\ \text{the second number} \\ \text{the third number} \end{cases}$.

Decide: Let $n =$ the first number

then $2n =$ the second number, because $2n$ is twice n

and $2n + 2 =$ the third number, because $2n + 2$ is 2 more than $2n$.

Translate: The sum of the three numbers is 47.

$$n + 2n + (2n + 2) \qquad = 47$$

$$n + 2n + 2n + 2 = 47 \quad \longleftarrow \quad \text{linear equation}$$

Solve: $$5n + 2 = 47$$

$$5n = 45$$

Interpret: $$n = \;\; 9$$

$$2n = 18 \quad \longleftrightarrow \quad \text{solutions [Check as before]}$$

$$2n + 2 = 20$$

(c) *Identify:*   The unknown numbers are $\begin{cases} \text{the first consecutive integer} \\ \text{the second consecutive integer} \end{cases}$.

Decide: Let $n =$ the first consecutive integer

then $n + 1 =$ the second consecutive integer.

Translate: The sum of two consecutive integers is 111.

$$n + (n + 1) \qquad = 111$$

$$n + n + 1 = 111 \quad \longleftarrow \quad \text{linear equation}$$

Solve: $$2n + 1 = 111$$

$$2n = 110$$

Interpret: $$n = \;\; 55$$

$$n + 1 = \;\; 56 \quad \longleftrightarrow \quad \text{solutions [Check as before.]}$$

(d) *Identify:*   The unknown numbers are $\begin{cases} \text{the first consecutive odd integer.} \\ \text{the second consecutive odd integer.} \\ \text{the third consecutive odd integer.} \end{cases}$.

Decide: Let $n =$ the first consecutive odd integer

then $n + 2 =$ the second consecutive odd integer

and $n + 4 =$ the third consecutive odd integer because $(n + 2) + 2 = n + 4$.

Translate: The sum of three consecutive odd integers is 135.

$$n + (n + 2) + (n + 4) \qquad = 135$$

$$n + n + 2 + n + 4 = 135 \quad \longleftarrow \quad \text{linear equation}$$

Solve: $$3n + 6 = 135$$

$$3n = 129$$

Interpret: $$n = \;\; 43$$

$$n + 2 = \;\; 45 \quad \longleftrightarrow \quad \text{solutions [Check as before.]}$$

$$n + 4 = \;\; 47$$

PROBLEM 6-4 Solve each geometry problem using a linear equation:

(a) The perimeter of a triangle is 41 meters. The longest side is 3 times the shortest side. The third side is 8 meters shorter than the longest side. How long is each side?

(b) The area of a rectangle is 192 in.^{2} [square inches]. Find the perimeter of the rectangle if its length is 16 inches.

Solution: Recall that to solve a geometry problem using a linear equation, you

1. *Read* the problem very carefully several times.
2. *Draw a picture* to help visualize the problem.
3. *Identify* the unknown measures.
4. *Decide* how to represent the unknown measures using one variable.
5. *Translate* the problem to a linear equation using the correct geometry formula from Appendix Table 7.
6. *Solve* the linear equation.
7. *Interpret* the solution of the linear equation with respect to each represented unknown measure to find the proposed solutions of the original problem.
8. *Check* to see if the proposed solutions satisfy all the conditions of the original problem.
   [See Examples 6-8 and 6-9.]

(a) *Identify:*   The unknown measures are $\begin{cases} \text{the shortest side} \\ \text{the longest side} \\ \text{the third side} \end{cases}$.

Draw a picture:

Decide: Let s = the shortest side
then $3s$ = the longest side because $3s$ is 3 times s
and $3s - 8$ = the third side because $3s - 8$ is 8 less than $3s$.

Translate: $P = a + b + c$ ⟵ perimeter formula for a triangle from Appendix Table 7

$$41 = s + 3s + (3s - 8)$$

$$41 = s + 3s + 3s - 8 \quad \longleftarrow \text{ linear equation}$$

Solve: $41 = 7s - 8$

$$49 = 7s$$

Interpret: $s = 7$ [shortest side]

$3s = 21$ [longest side]   solutions [Check as before.]

$3s - 8 = 13$ [third side]

(b) *Identify:*   The unknown measures are $\begin{cases} \text{the width} \\ \text{the perimeter} \end{cases}$.

Draw a picture:

Decide: Let l = the length
and w = the width
then $2(l + w)$ = the perimeter (from Appendix Table 7).

Translate: $A = lw \longleftarrow$ area formula for a rectangle

$$192 = (16)w$$

$$16w = 192 \longleftarrow \text{linear equation}$$

Solve: $w = 12$ [inches]

Interpret: $2(l + w) = 56$ [inches] $\longleftarrow$ solution [Check as before.]

PROBLEM 6-5 Solve each currency value problem using a linear equation:

(a) A person changed $2.50 into dimes and nickels at a local bank. There were twice as many dimes as nickels in the change. How many nickels were there? What was the monetary value of the dimes?

(b) A theater sold a total of 621 children's and adult's movie tickets for $2725. A children's ticket sold for $3 and an adult's ticket sold for $5. How many children's tickets were sold? What was the monetary value of the adult's tickets sold?

Solution: Recall that to solve a currency value problem using a linear equation, you

1. *Read* the problem very carefully several times.
2. *Identify* the unknown currency amounts.
3. *Decide* how to represent the unknown number of each currency using one variable.
4. *Make a table* to help represent the unknown value of each currency using:

(denomination of currency)(number of currency) = value of currency

5. *Translate* the problem to a linear equation.
6. *Solve* the linear equation.
7. *Interpret* the solution of the linear equation with respect to each represented unknown number and value of each currency to find the proposed solutions of the original problem.
8. *Check* to see if the proposed solutions satisfy all the conditions of the original problem.
   [See Examples 6-10 and 6-11.]

(a) *Identify:* The unknown currency amounts are $\left\{ \begin{array}{l} \text{the number of nickels} \\ \text{the number of dimes} \\ \text{the value of the nickels} \\ \text{the value of the dimes} \end{array} \right\}$.

Decide: Let $n =$ the number of nickels   or   let $d =$ the number of dimes
then $2n =$ the number of dimes   then $\frac{1}{2}d =$ the number of nickels.

Make a table:

	denomination d [in cents]	number n [of coins]	value ($v = dn$) [in cents]
nickels	5	n	$5n$
dimes	10	$2n$	$10(2n)$
combined	——	——	250

Translate: The value of the nickels combined with the value of the dimes is $2.50.

$$5n \qquad + \qquad 10(2n) \qquad = \qquad 250 \text{ [cents]}$$

$$5n + 10(2n) = 250 \longleftarrow \text{linear equation}$$

Solve: $5n + 20n = 250$

$$25n = 250$$

Interpret: number of nickels $\longrightarrow n = 10$ [coins]

value of the dimes $\longrightarrow 10(2n) = 200$ [cents] solutions [Check as before.]

(b) *Identify:* The unknown ticket amounts are $\begin{cases} \text{the number of child tickets} \\ \text{the number of adult tickets} \\ \text{the value of the child tickets} \\ \text{the value of the adult tickets} \end{cases}$.

Decide: Let n = the number of child tickets

then $621 - n$ = the number of adult tickets

or

let n = the number of adult tickets

then $621 - n$ = the number of child tickets.

Make a table:

	denomination d [in dollars]	number n [of tickets]	value ($v = dn$) [in dollars]
child	3	n	$3n$
adult	5	$621 - n$	$5(621 - n)$
combined	——	621	2725

Translate: The value of the child tickets plus the value of the adult tickets is \$2725.

$$3n \qquad + \qquad 5(621 - n) \qquad = \qquad 2725 \text{ [dollars]}$$

$$3n + 5(621 - n) = 2725 \quad \longleftarrow \text{ linear equation}$$

Solve:
$$3n + 5(621) - 5(n) = 2725$$
$$3n + 3105 - 5n = 2725$$
$$3105 - 2n = 2725$$

Interpret:
$$-2n = -380$$

number of child tickets $\longrightarrow n = 190$ [tickets]

value of adult tickets $\longrightarrow 5(621 - n) = 2155$ [dollars]

solutions [Check as before.]

PROBLEM 6-6 Solve each investment problem using a linear equation:

(a) Manuel invested in two different bonds, each of which paid 8% interest. He invested \$100 more in Bond A than in Bond B. The total annual interest from the bonds is \$48. How much is invested in each type of bond?

(b) After one year, part of a \$6000 investment made a 12% profit. The other part made an 8% loss. The net gain was \$400. How much was invested at 12%? What was the amount of the 8% loss?

Solution: Recall that to solve an investment problem using a linear equation, you

1. *Read* the problem very carefully several times.
2. *Identify* the unknown investment amounts.
3. *Decide* how to represent the unknown amounts of principal using one variable.
4. *Make a table* to help represent the unknown amounts of interest using

$$\text{Interest} = (\text{Principal})(\text{rate})(\text{time}), \text{ or } I = Prt$$

5. *Translate* the problem to a linear equation.
6. *Solve* the linear equation.
7. *Interpret* the solution of the linear equation with respect to each represented unknown amount to find the proposed solutions of the original problem.
8. *Check* to see if the proposed solutions satisfy all the conditions of the original problem. [See Examples 6-12 and 6-13.]

(a) *Identify:* The unknown investment amounts are $\left\{\begin{array}{l}\text{the amount invested in Bond A}\\\text{the amount invested in Bond B}\end{array}\right\}$.

Decide: Let P = the amount invested in Bond A
then $P - 100$ = the amount invested in Bond B.

Make a table:

	Principal P [in dollars]	rate r [as a percent]	time t [in years]	Interest ($I = Prt$) [in dollars]
Bond A	P	8%	1	$8\%P$
Bond B	$P - 100$	8%	1	$8\%(P - 100)$
combined	——	8%	1	48

Translate: The amount of interest for Bond A plus amount of interest for Bond B equals $48.

$$8\%P \quad + \quad 8\%(P - 100) \quad = \quad 48$$

$$8\%P + 8\%(P - 100) = 48 \longleftarrow \text{linear equation}$$

Solve:

$$0.08P + 0.08(P - 100) = 48$$
$$0.08P + 0.08P - 8 = 48$$
$$0.16P - 8 = 48$$
$$0.16P = 56$$

Interpret: amount invested in Bond A $\longrightarrow P = 350$ [dollars]
amount invested in Bond B $\longrightarrow P - 100 = 250$ [dollars] $\left.\right\rangle$ solutions [Check as before.]

(b) *Identify:* The unknown investment amounts are $\left\{\begin{array}{l}\text{the principal that earned 12\% profit}\\\text{the principal that earned 8\% loss}\\\text{the amount of the 12\% profit}\\\text{the amount of the 8\% loss}\end{array}\right\}$.

Decide: Let P = the principal that earned a 12% profit
then $6000 - P$ = the principal that earned an 8% loss.

Make a table:

	Principal P [in dollars]	rate r [as a percent]	time t [in years]	Interest ($I = Prt$) [in dollars]
12% profit	P	12%	1	$12\%P$
8% loss	$6000 - P$	8%	1	$8\%(6000 - P)$
combined	6000	——	1	400

Translate: The amount of 12% profit minus the amount of 8% loss equals $400.

$$12\%P \quad - \quad 8\%(6000 - P) \quad = \quad 400 \text{ [dollars]}$$

$$12\%P - 8\%(6000 - P) = 400 \longleftarrow \text{linear equation}$$

Solve:

$$0.12P - 0.08(6000 - P) = 400$$
$$0.12P - 0.08(6000) + 0.08P = 400$$
$$0.12P - 480 + 0.08P = 400$$
$$-480 + 0.2P = 400$$
$$0.2P = 880$$
$$P = 4400$$

Interpret: principal that earned a 12% profit $\longrightarrow P = 4400$ [dollars]

amount of the 8% loss $\longrightarrow 8\%(6000 - P) = 128$ [dollars] $\Big\}$ solutions [Check as before.]

PROBLEM 6-7 Solve each uniform motion problem using a linear equation:

(a) Two people walking toward each other are $3\frac{1}{2}$ miles apart. Their uniform walking rates differ by 1 mph. They will meet in $\frac{1}{2}$ hour. What is the uniform rate of each walker? What distance will each walker travel?

(b) One train travels from New York to Chicago at an average rate of 40 mph. A second train leaves New York for Chicago on a parallel track 3 hours later, traveling at an average rate of 70 mph. How long will it take the second train to catch up to the first train? How far from New York will both trains be then?

(c) A person traveled 1040 miles in 12 hours— 8 hours by train and 4 hours by plane. The train's uniform rate was only one-third of the plane's uniform rate. What were the uniform rates for the train and plane? How far did the person travel by train? How far did the person travel by plane?

(d) A certain woman drives from her home to her place of work in 45 minutes at a constant rate. The same woman drives from her place of work to her home at a constant rate that is 20 mph slower. The total commuting time is 2 hours. How far is her home from her work place? What is the constant rate for each part of her commute?

Solution: Recall that to solve a uniform motion problem using a linear equation, you

1. *Read* the problem very carefully several times.
2. *Draw a picture* to help visualize the problem.
3. *Identify* the unknown distances, rates, or times.
4. *Decide* how to represent the unknown rates or times using one variable.
5. *Make a table* to help represent the unknown distances using

$$\text{distance} = (\text{rate})(\text{time}), \text{ or } d = rt$$

6. *Translate* the problem to a linear equation.
7. *Solve* the linear equation.
8. *Interpret* the solution of the linear equation with respect to each represented unknown distance, rate, or time to find the proposed solutions of the original problem.
9. *Check* to see if the proposed solutions satisfy all the conditions of the original problem. [See Examples 6-14 and 6-15.]

(a) *Draw a picture:*

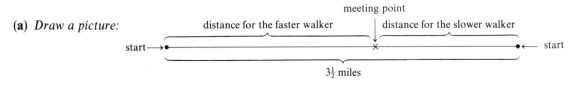

Identify: The unknown rates and distances are $\begin{cases} \text{the uniform rate of the faster walker} \\ \text{the uniform rate of the slower walker} \\ \text{the distance for the faster walker} \\ \text{the distance for the slower walker} \end{cases}$.

Decide: Let r = the uniform rate of the faster walker in mph
then $r - 1$ = the uniform rate of the slower walker in mph.

Make a table:

	rate r [in mph]	time t [in hours]	distance $(d = rt)$ [in miles]
faster walker	r	$\frac{1}{2}$	$\frac{1}{2}r$
slower walker	$r - 1$	$\frac{1}{2}$	$\frac{1}{2}(r - 1)$
combined	——	$\frac{1}{2}$	$3\frac{1}{2}$

Translate: 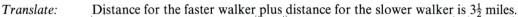 Distance for the faster walker plus distance for the slower walker is $3\frac{1}{2}$ miles.

$$\frac{1}{2}r \qquad + \qquad \frac{1}{2}(r-1) \qquad = 3\frac{1}{2}$$

$$\frac{1}{2}r + \frac{1}{2}(r-1) = \frac{7}{2} \longleftarrow \text{linear equation}$$

Solve:
$$2 \cdot \tfrac{1}{2}r + 2 \cdot \tfrac{1}{2}(r-1) = 2 \cdot \tfrac{7}{2} \qquad \textit{Think: The LCD is 2.}$$

$$r + (r-1) = 7$$

$$2r - 1 = 7$$

$$2r = 8$$

Interpret:

rate of the faster walker $\longrightarrow r = 4\,[\text{mph}]$

rate of the slower walker $\longrightarrow r - 1 = 3\,[\text{mph}]$

distance for faster walker $\longrightarrow \frac{1}{2}r = 2\,[\text{miles}]$

distance for slower walker $\longrightarrow \frac{1}{2}(r-1) = 1\frac{1}{2}\,[\text{miles}]$

solutions [Check as before.]

(b) *Draw a picture:*

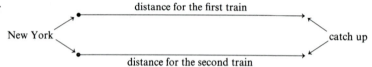

distance for the first train

New York   catch up

distance for the second train

Identify: The unknown times and distances are $\begin{cases} \text{the time for the first train} \\ \text{the time for the second train} \\ \text{the distance for the first train} \\ \text{the distance for the second train} \end{cases}$.

Decide: Let $\quad t =$ the time for the first train in hours,
then $t - 3 =$ the time for the second train in hours.

Make a table:

	rate r [in mph]	time t [in hours]	distance ($d = rt$) [in miles]
first train	40	t	$40t$
second train	70	$t - 3$	$70(t - 3)$

Translate: The distance for the first train equals the distance for the second train.

$$40t \qquad\qquad = \qquad\qquad 70(t-3)$$

$$40t = 70(t - 3) \longleftarrow \text{linear equation}$$

Solve:
$$40t = 70(t) - 70(3)$$

$$40t = 70t - 210$$

$$-30t = -210$$

Interpret:

time for the first train $\longrightarrow t = 7\,[\text{hours}]$

time for the second train $\longrightarrow t - 3 = 4[\text{hours}]$

distance for the first train $\longrightarrow 40t = 280\,[\text{miles}]$

distance for the second train $\longrightarrow 70(t - 3) = 280\,[\text{miles}]$

solutions [Check as before.]

Therefore the second train will catch up to the first train in 4 hours and both trains will then be 280 miles from New York.

(c) *Draw a picture:*

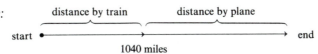

distance by train   distance by plane

start   1040 miles   end

Identify: The unknown rates and distances are $\begin{cases}\text{the uniform rate for the train}\\\text{the uniform rate for the plane}\\\text{the distance for the train}\\\text{the distance for the plane}\end{cases}$.

Decide: Let r = the uniform rate of the train in mph
then $3r$ = the uniform rate of the plane in mph.

Make a table:

	rate r [in mph]	time t [in hours]	distance ($d = rt$) [in miles]
train	r	8	$8r$
plane	$3r$	4	$4(3r)$ or $12r$
combined	——	12	1040

Translate: The distance by train plus the distance by plane is 1040 miles.

$$8r \qquad + \qquad 12r \qquad = 1040$$

$$8r + 12r = 1040 \longleftarrow \text{ linear equation}$$

Solve: $$20r = 1040$$

Interpret:

uniform rate for the train $\longrightarrow r = 52\,[\text{mph}]$

uniform rate for the plane $\longrightarrow 3r = 156\,[\text{mph}]$

distance for the train $\longrightarrow 8r = 416\,[\text{miles}]$

distance for the plane $\longrightarrow 12r = 624\,[\text{miles}]$

solutions [Check as before.]

(d) *Draw a picture:*

distance from home to work

home   work

distance from work to home

Identify: The unknown rates and distances are $\begin{cases}\text{the constant rate from home to work}\\\text{the constant rate from work to home}\\\text{the distance from home to work}\\\text{the distance from work to home}\end{cases}$.

Decide: Let r = the constant rate from home to work in mph
then $r - 20$ = the constant rate from work to home in mph.

Make a table:

	rate r [in mph]	time t [in hours]	distance ($d = rt$) [in miles]
from home	r	$\frac{3}{4}$	$\frac{3}{4}r$
from work	$r - 20$	$1\frac{1}{4}$	$1\frac{1}{4}(r - 20)$

Translate: $\underbrace{\text{The distance from home to work}}$ $\underbrace{\text{equals}}$ $\underbrace{\text{the distance from work to home.}}$

$$\tfrac{3}{4}r \qquad\qquad = \qquad\qquad 1\tfrac{1}{4}(r - 20)$$

$$\tfrac{3}{4}r = \tfrac{5}{4}(r - 20) \longleftarrow \text{linear equation}$$

Solve:

$$4 \cdot \tfrac{3}{4}r = 4 \cdot \tfrac{5}{4}(r - 20) \qquad \textit{Think: } \text{The LCD is 4.}$$

$$3r = 5(r - 20)$$

$$3r = 5(r) - 5(20)$$

$$3r = 5r - 100$$

$$-2r = -100$$

Interpret:

constant rate from home to work $\longrightarrow r = 50 \,[\text{mph}]$

constant rate from work to home $\longrightarrow r - 20 = 30 \,[\text{mph}]$

distance from home to work $\longrightarrow \tfrac{3}{4}r = 37\tfrac{1}{2} \,[\text{miles}]$

distance from work to home $\longrightarrow 1\tfrac{1}{4}(r - 20) = 37\tfrac{1}{2} \,[\text{miles}]$

solutions [Check as before.]

Supplementary Exercises

PROBLEM 6-8 Translate each word sentence to a linear equation and then solve the linear equation:

(a) Twelve more than three times a number is thirty.
(b) A number added to twenty equals eleven.
(c) Eight less than a number is thirty-six.
(d) One-fourth of a number is eighteen.
(e) A number increased by sixteen is the same as nineteen.
(f) One-third of a number increased by twice the same number is thirty-five.
(g) Five times a number diminished by ten leaves seven.
(h) If six is subtracted from four times a number, then the result is three.
(i) The product of a number and nine, less one, is forty-four.
(j) One-half a number added to two times the same number is eighty.
(k) Thirteen more than one-eighth of a number equals two and one-half.
(l) The quotient of a number divided by four is ten.
(m) Six times the sum of two and another number gives one hundred sixty-eight.
(n) A number is multiplied by five and then increased by twenty to get forty-two.
(o) Three-fourths of a number is decreased by six to get twenty-one.
(p) Two-thirds of a number, less thirteen, is equal to two.
(q) A number is 8 more than twice the difference of the same number and 2.
(r) Three-eighths of a number, less one-half the same number, gives negative five.
(s) A number increased by one-third of itself is the same as eight.
(t) One-half of a number equals one-fourth of the same number plus six.
(u) Nine is added to one-third of a number to leave nine.
(v) Three-fifths of a number joined with forty-seven will be eighty.
(w) Six subtracted from six times a number is seventy-two.
(x) Thirty subtracted from eight times a number is equal to five.
(y) One and one-half is added to one-third of a number to get one-fourth.
(z) A number, when increased by four times its half, equals fifteen.

PROBLEM 6-9   Solve each number problem using a linear equation:

(a) The difference between two numbers is 14. The larger number is equal to 3 times the smaller number, less 4. Find the numbers.

(b) The sum of three numbers is 38. The second number is 2 more than three times the first number. The third number equals one-half the difference between the first two numbers. What are the numbers?

(c) The sum of three consecutive odd integers is 171. Find the integers.

(d) If the first consecutive integer is divided by 4 and the second consecutive integer is divided by 2, the sum of the quotients would be 5. What are the integers?

PROBLEM 6-10   Solve each geometry problem using a linear equation:

(a) The sum of three angles of a triangle is 180 degrees [180°]. One angle of a triangle is 4 times the second angle. The third angle of the triangle is 30° more than the second angle. Find the measures of all the angles.

(b) Two angles of a triangle are equal. The third angle is equal to the sum of the other two angles. How many degrees are in each angle?

(c) The perimeter of a rectangle is 110 feet. The length is 5 feet shorter than twice the width. Find the area of the rectangle.

(d) The area of a rectangle is 240 square meters [240m²]. The width of the rectangle is 15m. Find the perimeter of the rectangle.

PROBLEM 6-11   Solve each currency value problem using a linear equation:

(a) There are $4.65 in dimes and quarters. There are 8 more dimes than quarters. How many dimes are there? What is the value of the quarters?

(b) A person changed $200 into five- and ten-dollar bills. Then the person had 32 bills in all. How many ten-dollar bills did the person have? What is the total value of the five-dollar bills?

(c) A stationery store sold 320 of two different types of pens for $190.55. One type of pen cost 25¢, while the other type of pen cost 80¢. How many 80-cent pens were sold? What is the value of the 25-cent pens that were sold?

(d) The daily payroll for 100 skilled and unskilled laborers is $8320. Each unskilled laborer earns $64 per day, while each skilled laborer earns $112 per day. How many unskilled laborers are there? What is the value of the daily payroll for skilled laborers?

PROBLEM 6-12   Solve each investment problem using a linear equation:

(a) What amount of money, invested at 12%, will return $600 annually?

(b) After 1 year, an 18% simple interest loan is paid off with $7080. What amount of money was originally borrowed?

(c) A partial amount of an $8000 investment is at 5% per year and the other part at 10%. The part invested at 10% earns $50 more than the part invested at 5%. What amount is invested at 10%? How much interest does the 5% investment earn per year?

(d) A partial amount of a $6000 investment made a 12% profit. The other part of the investment made an 8% loss. The net loss was $200. How much was invested at 12%? What was the amount of the 8% loss?

PROBLEM 6-13   Solve each uniform motion problem using a linear equation:

(a) Two cars traveling in opposite directions pass each other. One car's average speed is 60 mph, while the other car's average speed is 45 mph. How long will it take until the two cars are 420 miles apart? How far will the faster car have traveled by that point?

(b) Two planes traveling toward each other are 900 miles apart at noon. The constant rate of one plane is 166 mph and the constant speed of the other plane is 194 mph. At what time will the planes pass each other? How far will the slower plane have traveled by then?

(c) Two planes leave the same airport 10 minutes apart, traveling in the same direction. The uniform rate of the first plane is 258 km/h and the uniform rate of the second plane is 301 km/h. How long before the second plane overtakes the first plane? How far will they both have traveled by that point?

(d) A car traveled at a constant rate for 2 hours. Because of road construction, the car was forced to travel at 30 km/h less for the next 30 minutes. The total distance driven was 185 km. What was the rate for each part of the trip? How far did the car travel during each part of the trip?

(e) A person leaves for a visit to another city at 9 A.M. The average rate traveling to the city is 50 mph, and the average rate returning home is 45 mph and 20 minutes longer over the same route. How far is the drive to the city? How long does the return trip take?

(f) Two trains leave the same station 48 minutes apart, traveling on parallel tracks in the same direction. Because the uniform rate of the second train is 10 mph faster than the first train, it takes 4 hours for the second train to catch up to the first train. Find the uniform rate of each train. How far did each train travel until they met?

(g) Two people start from the same point and walk in opposite directions. One person walks $\frac{2}{3}$ mph faster than the other person. In 45 minutes, they are 5 miles apart. How fast is each person walking? How much of the distance does each person cover?

(h) Two cars are 275 miles apart and traveling toward each other. Their average speeds differ by 10 mph. They will meet in $2\frac{1}{2}$ hours. What is the average speed for each car? How much of the distance will each car travel?

Answers to Supplementary Exercises

(6-8) **(a)** $12 + 3n = 30$; $n = 6$ **(b)** $n + 20 = 11$; $n = -9$ **(c)** $n - 8 = 36$; $n = 44$
(d) $\frac{1}{4}n = 18$; $n = 72$ **(e)** $n + 16 = 19$; $n = 3$ **(f)** $\frac{1}{3}n + 2n = 35$; $n = 15$
(g) $5n - 10 = 7$; $n = \frac{17}{5}$ or $3\frac{2}{5}$ **(h)** $4n - 6 = 3$; $n = \frac{9}{4}$ or $2\frac{1}{4}$ **(i)** $9n - 1 = 44$; $n = 5$
(j) $\frac{1}{2}n + 2n = 80$; $n = 32$ **(k)** $13 + \frac{1}{8}n = 2\frac{1}{2}$; $n = -84$ **(l)** $\frac{n}{4} = 10$; $n = 40$
(m) $6(2 + n) = 168$; $n = 26$ **(n)** $5n + 20 = 42$; $n = \frac{22}{5}$ or $4\frac{2}{5}$ **(o)** $\frac{3}{4}n - 6 = 21$; $n = 36$
(p) $\frac{2}{3}n - 13 = 2$; $n = \frac{45}{2}$ or $22\frac{1}{2}$ **(q)** $n = 8 + 2(n - 2)$; $n = -4$
(r) $\frac{3}{8}n - \frac{1}{2}n = -5$; $n = 40$ **(s)** $n + \frac{1}{3}n = 8$; $n = 6$ **(t)** $\frac{1}{2}n = \frac{1}{4}n + 6$; $n = 24$
(u) $\frac{1}{3}n + 9 = 9$; $n = 0$ **(v)** $\frac{3}{5}n + 47 = 80$; $n = 55$ **(w)** $6n - 6 = 72$; $n = 13$
(x) $8n - 30 = 5$; $n = \frac{35}{8}$ or $4\frac{3}{8}$ **(y)** $\frac{1}{3}n + 1\frac{1}{2} = \frac{1}{4}$; $n = -\frac{15}{4}$ or $-3\frac{3}{4}$
(z) $n + 4 \cdot \frac{n}{2} = 15$; $n = 5$

(6-9) **(a)** 23, 9   **(b)** 7, 23, 8   **(c)** 55, 57, 59   **(d)** 6, 7

(6-10) **(a)** 100°, 25°, 55°   **(b)** 45°, 45°, 90°   **(c)** 700 ft^2 **(d)** 62 m

(6-11) **(a)** 19, $2.75   **(b)** 8, $120   **(c)** 201, $29.75   **(d)** 60, $4480

(6-12) **(a)** $5000   **(b)** $6000   **(c)** $3000, $250   **(d)** $1400, $368

(6-13) **(a)** 4 h, 240 mi   **(b)** 2:30 P.M., 415 mi   **(c)** 1 h, 301 km
(d) 80 km/h, 50 km/h, 160 km, 25 km   **(e)** 150 mi, 3 h 20 min
(f) 50 mph, 60 mph, 240 mi   **(g)** 3 mph, $3\frac{2}{3}$ mph, $2\frac{1}{4}$ mi, $2\frac{3}{4}$ mi
(h) 50 mph, 60 mph, 125 mi, 150 mi

7 EXPONENTS

THIS CHAPTER IS ABOUT

☑ **Using Whole Numbers as Exponents**
☑ **Multiplying with Exponents**
☑ **Dividing with Exponents**
☑ **Finding Powers of Powers**
☑ **Using Integers as Exponents**

7-1. Using Whole Numbers as Exponents

A. Identify the Fundamental Rules for Whole-Number Exponents.

Recall: A short way to write the repeated factors is exponential notation. In the exponential notation 2^4, 2 is called the base, 4 is called the exponent or power, and 2^4 means 2 is repeated as a factor 4 times, or $2^4 = 2 \cdot 2 \cdot 2 \cdot 2$.

It is possible to evaluate base-2 terms with whole-number exponents like 2^4, 2^3, 2^2, 2^1, and 2^0, using a "halfing" pattern.

EXAMPLE 7-1: Evaluate the following base-2 terms using a "halfing" pattern: **(a)** 2^4 **(b)** 2^3 **(c)** 2^2 **(d)** 2^1 **(e)** 2^0

Solution: **(a)** $2^4 = 2 \cdot 2 \cdot 2 \cdot 2 = 16$
  (b) $2^3 = 2 \cdot 2 \cdot 2 = 8 \longleftarrow$ half of 16
  (c) $2^2 = 2 \cdot 2 = 4 \longleftarrow$ half of 8
  (d) $2^1 = 2 \longleftarrow$ half of 4   **halfing pattern**
  (e) $2^0 = 1 \longleftarrow$ half of 2

Note 1: The exponential notation 2^1 is just another way to write the number 2: $2^1 = 2$

Note 2: The exponential notation 2^0 is just another way to write the number 1: $2^0 = 1$

The previous base-2 examples are generalized in the following rules:

Fundamental Rules for Whole-Number Exponents

r repeated as a factor n times

If n is a whole number, then **(a)** $r^n = \overbrace{r \cdot r \cdot r \cdots r}$ $[n > 1]$
  (b) $r^1 = r$ always
  (c) $r^0 = 1$ $[r \neq 0]$
  (d) 0^0 is not defined.

B. Rename terms without exponents as exponential notation.

To **rename terms without exponents as exponential notation**, you can use parts **(a)** and **(b)** of the Fundamental Rules for Whole-Number Exponents, as shown in Example 7-2.

EXAMPLE 7-2: Rename each term using exponential notation: **(a)** 3   **(b)** y **(c)** ww

(d) $-xxx$ **(e)** $\dfrac{m}{2} \cdot \dfrac{m}{2} \cdot \dfrac{m}{2} \cdot \dfrac{m}{2}$ **(f)** $(3n)(3n)(3n)(3n)(3n)(3n)$ **(g)** $uuuvv$

repeated factor
| number of repeated factors
↓↓

Solution: **(a)** $3 = 3^1$ (because $r^1 = r$)   **(b)** $y = y^1$ **(c)** $ww = w^2$

(d) $-xxx = -(xxx) = -x^3$ **(e)** $\dfrac{m}{2} \cdot \dfrac{m}{2} \cdot \dfrac{m}{2} \cdot \dfrac{m}{2} = \left(\dfrac{m}{2}\right)^4$ **(f)** $(3n)(3n)(3n)(3n)(3n)(3n) = (3n)^6$

(g) $uuuvv = u^3v^2$

C. Read exponential notation.
It is important to be able to read exponential notation correctly.

EXAMPLE 7-3: Read each exponential notation in Example 7-2.

Solution: **(a)** 3^1 is read as "three to the first power."
    (b) y^1 is read as "y to the first power."
    (c) w^2 is read as "w to the second power" or "w squared."
    (d) $-x^3$ is read as "the opposite of x to the third power" or "the opposite of x cubed."
    (e) $\left(\dfrac{m}{2}\right)^4$ is read as "the quantity m divided by two to the fourth power."
    (f) $(3n)^6$ is read as "the quantity three times n to the sixth power."
    (g) u^3v^2 is read as "u cubed times v squared" or just "u cubed v squared."

D. Rename exponential notation as terms without exponents.
To **rename exponential notation as a term without exponents,** you use the Fundamental Rule for Whole-Number Exponents, as shown in Example 7-4.

EXAMPLE 7-4: Rename each exponential notation as a term without exponents: **(a)** 5^0 **(b)** u^0

(c) 4^1 **(d)** z^1 **(e)** x^2 **(f)** $-y^3$ **(g)** $\left(\dfrac{3}{v}\right)^2$ **(h)** $(2w)^3$ **(i)** m^3n^2

Solution: **(a)** $5^0 = 1$ because $r^0 = 1$ if $r \neq 0$ **(b)** $u^0 = 1$ $[u \neq 0]$ **(c)** $4^1 = 4$ because $r^1 = r$

(d) $z^1 = z$ **(e)** $x^2 = xx$ **(f)** $-y^3 = -(yyy) = -yyy$ **(g)** $\left(\dfrac{3}{v}\right)^2 = \dfrac{3}{v} \cdot \dfrac{3}{v}$ or $\dfrac{9}{vv}$

(h) $(2w)^3 = (2w)(2w)(2w)$ or $8www$ **(i)** $m^3n^2 = mmmnn$

Caution: $-r^n$ does not mean $(-r)^n$

EXAMPLE 7-5: Show that $-3^2 \neq (-3)^2$.

Solution: $-3^2 \neq (-3)^2$ because $-3^2 = -(3 \cdot 3) = -9$
                but $(-3)^2 = (-3)(-3) = +9$
                and $-9 \neq +9$

E. Special rules involving exponents of 0 and 1.
There are several **special rules involving exponents of 0 and 1.**

EXAMPLE 7-6: List the special rules involving exponents of 0 and 1.

Solution: If n is any whole number, then
(a) $r^0 = 1$ $[r \neq 0]$ **(b)** $0^n = 0$ $[n \neq 0]$ **(c)** 0^0 is not defined   **(d)** $r^1 = r$

(e) $1^n = 1$ **(f)** $-1^n = -1$ **(g)** $(-1)^n = \begin{cases} 1 \text{ if } n \text{ is even} \\ -1 \text{ if } n \text{ is odd} \end{cases}$

It will save you time and effort if you become familiar with these special rules involving exponents of 0 and 1.

EXAMPLE 7-7: Evaluate: **(a)** 4^0 **(b)** 0^4 **(c)** 4^1 **(d)** 1^4 **(e)** -1^4 **(f)** -1^3 **(g)** $(-1)^4$ **(h)** $(-1)^3$ **(i)** 0^0

Solution: **(a)** $4^0 = 1$ by definition   **(b)** $0^4 = 0$ because $0^4 = 0 \cdot 0 \cdot 0 \cdot 0 = 0$
   (c) $4^1 = 4$ by definition   **(d)** $1^4 = 1$ because $1^4 = 1 \cdot 1 \cdot 1 \cdot 1 = 1$
   (e) $-1^4 = -1$ because $-1^4 = -(1 \cdot 1 \cdot 1 \cdot 1) = -1$
   (f) $-1^3 = -1$ because $-1^3 = -(1 \cdot 1 \cdot 1) = -1$
   (g) $(-1)^4 = 1$ (4 is even) because $(-1)^4 = (-1)(-1)(-1)(-1) = +1$ or 1
   (h) $(-1)^3 = -1$ (3 is odd) because $(-1)^3 = (-1)(-1)(-1) = -1$
   (i) 0^0 is not defined.

7-2. Multiplying with Exponents

A. Multiply with like bases.
Exponential notation like $x^5 x^2$ is called a **product of powers with like bases.** A product of powers with like bases can always be simplified.

EXAMPLE 7-8: Simplify $x^5 x^2$.

Solution: $x^5 x^2 = (xxxxx)(xx)$ *Think:* $r^n = \overbrace{rrr \cdots r}^{n \text{ times}}$

$\qquad\qquad = xxxxxxx$

$\qquad\qquad = x^7 \longleftarrow$ simplest form

Note: $x^5 x^2 = x^7$ and $5 + 2 = 7$. That is, $x^5 x^2 = x^{5+2} = x^7$.

The previous note is generalized in the following rule:

Multiply with Like Bases Rule
If m and n are whole numbers, then $r^m r^n = r^{m+n}$.

Note: To multiply with like bases, you add the exponents of the like bases.

EXAMPLE 7-9: Multiply with like bases: **(a)** $xx^3 x^2$ **(b)** $m^2 m^4 mn^3 n^5$

Solution: **(a)** $xx^3 x^2 = x^1 x^3 x^2$ *Think:* $x = x^1$

$\qquad\qquad = x^{1+3+2}$ Multiply by adding exponents of like bases.

$\qquad\qquad = x^6 \longleftarrow$ simplest form

   (b) $m^2 m^4 mn^3 n^5 = m^2 m^4 m^1 n^3 n^5$ *Think:* $m = m^1$

$\qquad\qquad\qquad = m^{2+4+1} n^{3+5}$ Multiply by adding exponents of like bases.

$\qquad\qquad\qquad = m^7 n^8 \longleftarrow$ simplest form

Caution: When the bases are different, the exponents cannot be added.

EXAMPLE 7-10: Can $m^7 n^8$ be simplified?

Solution: No. $m^7 n^8$ cannot be simplified because m and n are different bases.

Caution: $r^m r^n$ does not mean r^{mn}.

EXAMPLE 7-11: Show that $x^3x^2 \neq x^{3 \cdot 2}$.

Solution: $x^3x^2 \neq x^{3 \cdot 2}$ because $x^3x^2 = x^{3+2} = x^5$

and $x^{3 \cdot 2} = x^6$

and $x^5 \neq x^6$.

B. Raise a product to a power.

Exponential notation, like $(5x)^2$, is called a **product to a power**. A product to a power can always be simplified.

EXAMPLE 7-12: Simplify $(5x)^2$.

Solution: $(5x)^2 = (5x)(5x)$ *Think:* $r^2 = rr$ where $r = 5x$

$= (5 \cdot 5)(xx)$

$= 25x^2 \longleftarrow$ simplest form

Note: $(5x)^2 = 5^2x^2 = 25x^2$.

The previous *Note* is generalized in the following rule:

Raise a Product to a Power Rule

If n is a whole number, then $(rs)^n = r^n s^n$.

Note: To raise a product to a power, you raise each factor of the product to the power.

EXAMPLE 7-13: Raise each product to the given power: **(a)** $(3x)^2$ **(b)** $(-2y)^3$
(c) $(-3mn)^2$ **(d)** $(\frac{3}{4}w)^2$ **(e)** $(-a)^4$ **(f)** $(-b)^3$

Solution:

(a) $(3x)^2 = 3^2x^2$ **(b)** $(-2y)^3 = (-2)^3y^3$ **(c)** $(-3mn)^2 = (-3)^2m^2n^2$ Raise each factor to

$= 9x^2$ $= -8y^3$ $= 9m^2n^2$ the power.

(d) $(\frac{3}{4}w)^2 = (\frac{3}{4})^2w^2$ **(e)** $(-a)^4 = (-1a)^4$ **(f)** $(-b)^3 = (-1b)^3$

$= \frac{9}{16}w^2$ $= (-1)^4a^4$ $= (-1)^3b^3$

$= 1a^4$ $= -1b^3$

$= a^4$ $= -b^3$ [See the following *Shortcut.*]

To calculate parts **(e)** and **(f)** of Example 7-13 more quickly you can use *Shortcut 7-1:*

Shortcut 7-1: $(-r)^n = \begin{cases} r^n \text{ if } n \text{ is even} \\ -r^n \text{ if } n \text{ is odd} \end{cases}$

EXAMPLE 7-14: Calculate the following using *Shortcut 7-1:* **(a)** $(-a)^4$ **(b)** $(-b)^3$

Solution: **(a)** $(-a)^4 = a^4$ because 4 is even.   **(b)** $(-b)^3 = -b^3$ because 3 is odd.

Caution: rs^n does not mean $(rs)^n$.

EXAMPLE 7-15: Show that $2 \cdot 5^2 \neq (2 \cdot 5)^2$.

Solution: $2 \cdot 5^2 \neq (2 \cdot 5)^2$ because $2 \cdot 5^2 = 2(5 \cdot 5) = 2(25) = 50$

and $(2 \cdot 5)^2 = (10)^2 = 10 \cdot 10 = 100$

and $50 \neq 100$.

7-3. Dividing with Exponents

A. Divide with like bases.

Exponential notation like $\dfrac{x^5}{x^2}$ [$x \neq 0$] is called a **quotient of powers with like bases.** A quotient of powers with like bases can always be simplified.

EXAMPLE 7-16: Simplify $\dfrac{x^5}{x^2}$ [$x \neq 0$].

Solution:
$$\frac{x^5}{x^2} = \frac{xxxxx}{xx} \qquad \textit{Think: } r^n = \overbrace{rrr \cdots r}^{n \text{ times}}$$

$$= \frac{xxx}{1} \cdot \frac{xx}{xx}$$

$$= xxx \cdot 1$$

$$= xxx$$

$$= x^3 \longleftarrow \text{ simplest form}$$

Note: $\dfrac{x^5}{x^2} = x^3$ and $5 - 2 = 3$. That is, $\dfrac{x^5}{x^2} = x^{5-2} = x^3$.

The previous *Note* is generalized in the following rule:

Divide with Like Bases Rule

If m and n are whole numbers, then $\dfrac{r^m}{r^n} = r^{m-n}$ [$r \neq 0$].

Note: To divide with like bases, you subtract the exponents of the like bases.

EXAMPLE 7-17: Divide with like bases. (Assume all variables are nonzero.)

(a) $\dfrac{w^7}{w^3}$ **(b)** $\dfrac{x^2}{x}$ **(c)** $\dfrac{y}{y}$ **(d)** $\dfrac{m^3 m^5}{m^2}$

Solution:

(a) $\dfrac{w^7}{w^3} = w^{7-3}$ **(b)** $\dfrac{x^2}{x} = \dfrac{x^2}{x^1}$ Divide by subtracting exponents with like bases.

$\qquad\qquad = w^4 \qquad\qquad\qquad = x^{2-1}$

$\qquad\qquad\qquad\qquad\qquad\qquad\quad = x^1$

$\qquad\qquad\qquad\qquad\qquad\qquad\quad = x$

(c) $\dfrac{y}{y} = \dfrac{y^1}{y^1}$ **(d)** $\dfrac{m^3 m^5}{m^2} = \dfrac{m^{3+5}}{m^2}$ First multiply by adding exponents of like bases.

$\qquad = y^{1-1} \qquad\qquad\qquad = \dfrac{m^8}{m^2}$

$\qquad = y^0 \qquad\qquad\qquad\quad\; = m^{8-2}$ Then divide by subtracting exponents with like bases.

$\qquad = 1 \qquad\qquad\qquad\qquad = m^6$

Caution: When the bases are different, the exponents cannot be subtracted.

EXAMPLE 7-18: Can $\dfrac{x^3}{y^2}$ be simplified?

Solution: No. $\dfrac{x^3}{y^2}$ cannot be simplified because x and y are different bases.

Caution: $\dfrac{r^m}{r^n}$ does not mean $r^{m \div n}$.

EXAMPLE 7-19: Show that $\dfrac{x^6}{x^2} \neq x^{6 \div 2}$.

Solution: $\dfrac{x^6}{x^2} \neq x^{6 \div 2}$ because $\dfrac{x^6}{x^2} = x^{6-2} = x^4$

$$\text{and } x^{6 \div 2} = x^3$$
$$\text{and } x^4 \neq x^3.$$

B. Raise a quotient to a power.

Exponential notation like $\left(\dfrac{x}{5}\right)^2$ is called a **quotient to a power.** A quotient to a power can always be simplified.

EXAMPLE 7-20: Simplify $\left(\dfrac{x}{5}\right)^2$.

Solution: $\left(\dfrac{x}{5}\right)^2 = \dfrac{x}{5} \cdot \dfrac{x}{5} \qquad$ *Think:* $r^2 = rr$ where $r = \dfrac{x}{5}$.

$$= \dfrac{xx}{5 \cdot 5}$$

$$= \dfrac{x^2}{25} \longleftarrow \text{ simplest form}$$

Note: $\left(\dfrac{x}{5}\right)^2 = \dfrac{x^2}{5^2} = \dfrac{x^2}{25}.$

The previous *Note* is generalized in the following rule:

Raise a Quotient to a Power Rule

If n is a whole number, then $\left(\dfrac{r}{s}\right)^n = \dfrac{r^n}{s^n} \; [s \neq 0]$.

Note: To raise a quotient to a power, you raise both the numerator and denominator to the power.

EXAMPLE 7-21: Raise each quotient to the given power. (Assume all variables are nonzero.)

(a) $\left(\dfrac{4}{5}\right)^2$ **(b)** $\left(\dfrac{m}{2}\right)^4$ **(c)** $\left(-\dfrac{3}{n}\right)^2$ **(d)** $\left(-\dfrac{u}{v}\right)^5$

Solution:

(a) $\left(\dfrac{4}{5}\right)^2 = \dfrac{4^2}{5^2}$ **(b)** $\left(\dfrac{m}{2}\right)^4 = \dfrac{m^4}{2^4}$ **(c)** $\left(-\dfrac{3}{n}\right)^2 \overset{\text{even} \atop \downarrow}{=} \overbrace{\left(\dfrac{3}{n}\right)^2}^{\text{Shortcut 7-1}}$ **(d)** $\left(-\dfrac{u}{v}\right)^5 \overset{\text{odd} \atop \downarrow}{=} -\overbrace{\left(\dfrac{u}{v}\right)^5}^{\text{Shortcut 7-1}}$

$$= \dfrac{16}{25} \qquad\qquad = \dfrac{m^4}{16} \qquad\qquad = \dfrac{3^2}{n^2} \qquad\qquad = -\dfrac{u^5}{v^5}$$

$$\qquad\qquad\qquad\qquad\qquad\qquad\qquad\qquad\quad = \dfrac{9}{n^2}$$

Caution: $\dfrac{r^n}{s}$ does not mean $\left(\dfrac{r}{s}\right)^n$.

EXAMPLE 7-22: Show that $\dfrac{3^2}{4} \neq \left(\dfrac{3}{4}\right)^2$.

Solution: $\dfrac{3^2}{4} \neq \left(\dfrac{3}{4}\right)^2$ because $\dfrac{3^2}{4} = \dfrac{3 \cdot 3}{4} = \dfrac{9}{4}$

and $\left(\dfrac{3}{4}\right)^2 = \dfrac{3^2}{4^2} = \dfrac{9}{16}$

and $\dfrac{9}{4} \neq \dfrac{9}{16}$.

7-4. Finding Powers of Powers

A. Raise a power to a power.

Exponential notation like $(x^5)^2$ is called a **power to a power.** A power to a power can always be simplified.

EXAMPLE 7-23: Simplify $(x^5)^2$.

Solution: $(x^5)^2 = x^5 x^5$ *Think:* $r^2 = rr$ when $r = x^5$.

$= x^{5+5}$

$= x^{10}$ ⟵ simplest form

Note: $(x^5)^2 = x^{5 \cdot 2} = x^{10}$.

The previous *Note* is generalized in the following rule:

Raise a Power to a Power Rule

If m and n are whole numbers, then $(r^m)^n = r^{mn}$.

Note: To raise a power to a power, you multiply the exponents.

EXAMPLE 7-24: Raise each power to the given power: **(a)** $(x^3)^4$ **(b)** $(-x^3)^4$ **(c)** $-(x^3)^4$
(d) $(-x^4)^3$ **(e)** $(y^2)^1$ **(f)** $(y^1)^2$ **(g)** $(w^3)^0$

Solution: **(a)** $(x^3)^4 = x^{3 \cdot 4}$ **(b)** $(-x^3)^4 = \overset{\text{even}}{\underset{\text{Shortcut 7-1}}{(x^3)^4}}$ **(c)** $-(x^3)^4 = -x^{3 \cdot 4}$

$= x^{12}$ $= x^{12}$ $= -x^{12}$

(d) $(-x^4)^3 = \overset{\text{odd}}{\underset{\text{Shortcut 7-1}}{-(x^4)^3}}$ **(e)** $(y^2)^1 = \overset{\text{by definition}}{y^2}$ or $(y^2)^1 = y^{2 \cdot 1}$

$= -x^{12}$ $= y^2$

(f) $(y^1)^2 = y^{1 \cdot 2}$ **(g)** $(w^3)^0 = \overset{\text{by definition}}{1}$ or $(w^3)^0 = w^{3 \cdot 0}$

$= y^2$ $= w^0$

$= 1 \quad [w \neq 0]$

Caution: $(r^m)^n$ does not mean r^{m+n}.

EXAMPLE 7-25: Show that $(x^5)^2 \neq x^{5+2}$.

Solution: $(x^5)^2 \neq x^{5+2}$ because $(x^5)^2 = x^{5 \cdot 2} = x^{10}$
and $x^{5+2} = x^7$
and $x^{10} \neq x^7$.

B. Raise a product and/or quotient containing powers to a power.

To raise a product and/or quotient containing powers to a power, you raise each part of the product and/or quotient to the power.

EXAMPLE 7-26: Raise each product and/or quotient containing powers to the given power. (Assume all variables are nonzero.):

(a) $(4x^3)^2$ (b) $(-2y^5)^3$ (c) $\left(\dfrac{w^4}{5}\right)^2$ (d) $\left(-\dfrac{3m^5}{2n}\right)^2$ (e) $\dfrac{(4w^5)^2}{w^8}$ (f) $\dfrac{(x^2)^3(x^4)^2}{x(x^3)^4}$

Solution:

(a) $(4x^3)^2 = 4^2(x^3)^2$ (b) $(-2y^5)^3 = (-2)^3(y^5)^3$ (c) $\left(\dfrac{w^4}{5}\right)^2 = \dfrac{(w^4)^2}{5^2}$

$\qquad = 16x^6$ $\qquad\qquad\qquad = -8y^{15}$

$\qquad\qquad\qquad\qquad\qquad\qquad\qquad\qquad\qquad = \dfrac{w^8}{25}$

(d) $\left(-\dfrac{3m^5}{2n}\right)^2 = \left(\dfrac{3m^5}{2n}\right)^2$ (e) $\dfrac{(4w^5)^2}{w^8} = \dfrac{4^2(w^5)^2}{w^8}$ (f) $\dfrac{(x^2)^3(x^4)^2}{x(x^3)^4} = \dfrac{x^6 x^8}{x^1 x^{12}}$

$\qquad = \dfrac{3^2(m^5)^2}{2^2 n^2}$ $\qquad\qquad\quad = \dfrac{16w^{10}}{w^8}$ $\qquad\qquad\qquad = \dfrac{x^{14}}{x^{13}}$

$\qquad = \dfrac{9m^{10}}{4n^2}$ $\qquad\qquad\qquad = 16w^{10-8}$ $\qquad\qquad\qquad = x^1$

$\qquad\qquad\qquad\qquad\qquad\qquad = 16w^2$ $\qquad\qquad\qquad\qquad = x$

7-5. Using Integers as Exponents

A. Identify the Fundamental Rules for Integer Exponents.

The base-2 "halfing" pattern shown in the beginning of Section 7-1 to evaluate 2^4, 2^3, 2^2, 2^1, and 2^0 can be extended using a "halfing" pattern for base-2 numbers with negative exponents. This is shown in the following example.

EXAMPLE 7-27: Evaluate the following base-2 terms using a halfing pattern: (a) 2^{-1} (b) 2^{-2}
(c) 2^{-3} (d) 2^{-4}

Solution:

$2^4 = 2 \cdot 2 \cdot 2 \cdot 2 = 16$

$2^3 = 2 \cdot 2 \cdot 2 = 8 \longleftarrow$ half of 16

$2^2 = 2 \cdot 2 = 4 \longleftarrow$ half of 8

$2^1 = 2 \longleftarrow$ half of 4   halfing pattern from the beginning of Section 7-1

$2^0 = 1 \longleftarrow$ half of 2

(a) $2^{-1} = \dfrac{1}{2}\left[\text{or } \dfrac{1}{2^1}\right] \longleftarrow$ half of 1

(b) $2^{-2} = \dfrac{1}{4}\left[\text{or } \dfrac{1}{2^2}\right] \longleftarrow$ half of $\frac{1}{2}$

(c) $2^{-3} = \dfrac{1}{8}\left[\text{or } \dfrac{1}{2^3}\right] \longleftarrow$ half of $\frac{1}{4}$ continued halfing pattern

(d) $2^{-4} = \dfrac{1}{16}\left[\text{or } \dfrac{1}{2^4}\right] \longleftarrow$ half of $\frac{1}{8}$

Note 1: The exponential notation 2^{-1} is just another way to write $\dfrac{1}{2^1}$: $2^{-1} = \dfrac{1}{2^1}$

Note 2: The exponential notation 2^{-2} is just another way to write $\dfrac{1}{2^2}$: $2^{-2} = \dfrac{1}{2^2}$.

Note 3: The exponential notation 2^{-3} is just another way to write $\dfrac{1}{2^3}$: $2^{-3} = \dfrac{1}{2^3}$.

Note 4: The exponential notation 2^{-4} is just another way to write $\dfrac{1}{2^4}$: $2^{-4} = \dfrac{1}{2^4}$.

B. Rename terms containing negative exponents using only positive exponents.

The Fundamental Rule for Integer Exponents states that to rename a nonzero term containing a negative exponent in the numerator (denominator), you move the factor with the negative exponent to the denominator (numerator) and make the exponent positive by writing the opposite of the negative exponent.

The Fundamental Rules for Integer Exponents

If n is an integer, then **(a)** $r^{-n} = \dfrac{1}{r^n}\ [r \neq 0]$;

$$\textbf{(b)}\ \frac{1}{r^{-n}} = r^n\ [r \neq 0].$$

EXAMPLE 7-28: Rename each term using only positive exponents. (Assume all variables are nonzero.):

(a) 3^{-2} **(b)** $\dfrac{1}{2^{-3}}$ **(c)** x^{-5} **(d)** $\dfrac{1}{y^{-4}}$ **(e)** $3x^{-1}$ **(f)** $\dfrac{2}{5y^{-3}}$ **(g)** $\dfrac{3}{4}w^{-6}$ **(h)** $\dfrac{2x^{-3}}{5w^{-2}}$

Solution: **(a)** $3^{-2} = \dfrac{1}{3^2}$ **(b)** $\dfrac{1}{2^{-3}} = 2^3$ **(c)** $x^{-5} = \dfrac{1}{x^5}$ **(d)** $\dfrac{1}{y^{-4}} = y^4$

(e) $3x^{-1} = 3 \cdot \dfrac{1}{x^1}$ **(f)** $\dfrac{2}{5y^{-3}} = \dfrac{2}{5} \cdot \dfrac{1}{y^{-3}}$ **(g)** $\dfrac{3}{4}w^{-6} = \dfrac{3}{4} \cdot \dfrac{1}{w^6}$ **(h)** $\dfrac{2x^{-3}}{5w^{-2}} = \dfrac{2w^2}{5x^3}$

$\qquad\qquad = \dfrac{3}{x^1} \qquad\qquad\quad = \dfrac{2}{5} \cdot y^3 \qquad\qquad = \dfrac{3}{4w^6}$

$\qquad\qquad = \dfrac{3}{x} \qquad\qquad\quad = \dfrac{2y^3}{5}$

C. Simplify terms containing negative exponents.

All the **Simplification Rules for Whole-Number Exponents** found in Sections 7-2, 7-3, and 7-4 can be extended to include integer exponents, as stated in the following rule:

Simplification Rules for Integer Exponents

If m and n are integers, then

(a) $r^m r^n = r^{m+n}$ **(b)** $(rs)^n = r^n s^n$ **(c)** $(r^m)^n = r^{mn}$ **(d)** $(-r)^n = \begin{cases} r^n \text{ if } n \text{ is even} \\ -r^n \text{ if } n \text{ is odd} \end{cases}$

(e) $\dfrac{r^m}{r^n} = r^{m-n}\ [r \neq 0]$ **(f)** $\left(\dfrac{r}{s}\right)^n = \dfrac{r^n}{s^n}\ [s \neq 0]$ **(g)** $r^{-n} = \dfrac{1}{r^n}\ [r \neq 0]$ **(h)** $\dfrac{1}{r^{-n}} = r^n\ [r \neq 0]$

Note: The Simplification Rules for Integer Exponents include all of the Simplification Rules for Whole-Number Exponents and the Fundamental Rules for Integer Exponents.

EXAMPLE 7-29: Simplify the following and rename the answer using only positive exponents:

(a) $x^4 x^{-5}$ **(b)** $(5y)^{-2}$ **(c)** $(w^3)^{-2}$ **(d)** $\dfrac{m^{-3}}{m^{-2}}$ **(e)** $\left(\dfrac{2}{n}\right)^{-3}$ **(f)** $\left(\dfrac{a}{b}\right)^{-2}$

Solution: **(a)** $x^4 x^{-5} = x^{4+(-5)}$ or $x^4 x^{-5} = \dfrac{x^4}{x^5}$ **(b)** $(5y)^{-2} = 5^{-2} y^{-2}$ or $(5y)^{-2} = \dfrac{1}{(5y)^2}$

$\qquad\qquad = x^{-1} \qquad\qquad = x^{4-5} \qquad\qquad\qquad = \dfrac{1}{5^2} \cdot \dfrac{1}{y^2} \qquad\qquad = \dfrac{1}{5^2 y^2}$

$\qquad\qquad = \dfrac{1}{x} \qquad\qquad\quad = x^{-1} \qquad\qquad\qquad\quad = \dfrac{1}{25y^2} \qquad\qquad = \dfrac{1}{25y^2}$

$\qquad\qquad\qquad\qquad\qquad = \dfrac{1}{x}$

(c) $(w^3)^{-2} = w^{3(-2)}$ or $(w^3)^{-2} = \dfrac{1}{(w^3)^2}$ **(d)** $\dfrac{m^{-3}}{m^{-2}} = m^{-3-(-2)}$ or $\dfrac{m^{-3}}{m^{-2}} = \dfrac{m^2}{m^3}$

$\qquad\qquad = w^{-6} \qquad\qquad\qquad = \dfrac{1}{w^6} \qquad\qquad\qquad = m^{-1} \qquad\qquad\quad = m^{2-3}$

$\qquad\qquad = \dfrac{1}{w^6} \qquad\qquad\qquad\qquad\qquad\qquad\qquad = \dfrac{1}{m} \qquad\qquad\quad = m^{-1}$

$\qquad\qquad\qquad\qquad\qquad\qquad\qquad\qquad\qquad\qquad\qquad\qquad\qquad\qquad = \dfrac{1}{m}$

(e) $\left(\dfrac{2}{n}\right)^{-3} = \dfrac{2^{-3}}{n^{-3}}$ **(f)** $\left(\dfrac{a}{b}\right)^{-2} = \dfrac{a^{-2}}{b^{-2}}$

$\qquad\qquad = \dfrac{n^3}{2^3} \qquad\qquad\qquad = \dfrac{b^2}{a^2}$

$\qquad\qquad = \dfrac{n^3}{8}$

Shortcut 7-2: To **raise a nonzero fraction to a given negative power,** you write the reciprocal fraction and make the exponent positive by writing the opposite of the given negative power:

$$\left(\frac{r}{s}\right)^{-n} = \left(\frac{s}{r}\right)^n \qquad [r \neq 0 \text{ and } s \neq 0]$$

EXAMPLE 7-30: Simplify the following using *Shortcut 7-2.* (Assume all variables are nonzero.):

(a) $\left(\dfrac{2}{n}\right)^{-3}$ **(b)** $\left(\dfrac{a}{b}\right)^{-2}$

Solution: **(a)** $\left(\dfrac{2}{n}\right)^{-3} = \left(\dfrac{n}{2}\right)^3$ **(b)** $\left(\dfrac{a}{b}\right)^{-2} = \left(\dfrac{b}{a}\right)^2 \longrightarrow$ use *Shortcut 7-2*

$\qquad\qquad\qquad = \dfrac{n^3}{2^3} \qquad\qquad\qquad = \dfrac{b^2}{a^2} \nwarrow$

$\qquad\qquad\qquad\qquad\qquad\qquad\qquad\qquad\qquad\qquad\searrow$ same results as found in Example 7-29

$\qquad\qquad\qquad = \dfrac{n^3}{8} \longleftarrow$

RAISE YOUR GRADES
Can you . . . ?

☑ rename terms without exponents as exponential notation
☑ rename exponential notation as terms without exponents
☑ use the special rules for exponents involving 0 and 1
☑ multiply with like bases
☑ raise a product to a power

☑ divide with like bases
☑ raise a quotient to a power
☑ raise a power to a power
☑ raise a product and/or quotient containing powers to a power
☑ rename terms cotaining negative exponents using only positive exponents
☑ simplify terms containing negative exponents

SUMMARY

1. To rename terms without exponents as exponential notation or to rename exponential notation as terms without exponents, you use the **Fundamental Rules for Whole-Number Exponents:**

$$\overbrace{r \text{ repeated as a factor } n \text{ times}}$$

If n is a whole number, then (a) $r^n = \overbrace{rrr \cdots r}$ $[n > 1]$
  (b) $r^1 = r$ always
  (c) $r^0 = 1$ $[r \neq 0]$
  (d) 0^0 is not defined.

2. The special exponents of rules involving 0 and 1 are:
   (a) $r^0 = 1 \, [r \neq 0]$ (b) $0^n = 0 \, [n \neq 0]$ (c) 0^0 is not defined.

   (d) $r^1 = r$ (e) $1^n = 1$ (f) $-1^n = -1$ (g) $(-1)^n = \begin{cases} 1 \text{ if } n \text{ is even} \\ -1 \text{ if } n \text{ is odd} \end{cases}$

3. **Simplification Rules for Whole-Number and Integer Exponents**
   If m and n are whole numbers or integers:

   (a) $r^m r^n = r^{m+n}$ (b) $(rs)^n = r^n s^n$ (c) $(-r)^n = \begin{cases} r^n \text{ if } n \text{ is even} \\ -r^n \text{ if } n \text{ is odd} \end{cases}$

   (d) $\dfrac{r^m}{r^n} = r^{m-n} \, [r \neq 0]$ (e) $\left(\dfrac{r}{s}\right)^n = \dfrac{r^n}{s^n} \, [s \neq 0]$ (f) $(r^m)^n = r^{mn}$

4. **Fundamental Rules for Integer Exponents**

   If n is an integer, then $r^{-n} = \dfrac{1}{r^n}$ $[r \neq 0]$;

   $$\dfrac{1}{r^{-n}} = r^n \quad [r \neq 0].$$

5. To raise a nonzero fraction to a given negative power, you write the reciprocal fraction and make the exponent positive by writing the opposite of the given negative power:

   $$\left(\frac{r}{s}\right)^{-n} = \left(\frac{s}{r}\right)^n \quad [r \neq 0 \text{ and } s \neq 0]$$

6. In working with exponents, there are some general rules that you should always keep in mind:
   (a) $-r^n$ does not mean $(-r)^n$.
   (b) rs^n does not mean $(rs)^n$.
   (c) When the bases are different, exponents cannot be added or subtracted.
   (d) $r^m r^n$ does not mean r^{mn}.

   (e) $\dfrac{r^m}{r^n}$ does not mean $r^{m \div n}$.

   (f) $\dfrac{r^n}{s}$ does not mean $\left(\dfrac{r}{s}\right)^n$.

   (g) $(r^m)^n$ does not mean r^{m+n}.

SOLVED PROBLEMS

PROBLEM 7-1   Rename each term using exponential notation: **(a)** 5   **(b)** w **(c)** yy

(d) xxx **(e)** $(-m)(-m)(-m)(-m)$ **(f)** $-mmmm$ **(g)** $\dfrac{w}{3} \cdot \dfrac{w}{3} \cdot \dfrac{w}{3} \cdot \dfrac{w}{3} \cdot \dfrac{w}{3}$ **(h)** $\dfrac{wwww}{3}$

(i) $(2n)(2n)(2n)(2n)(2n)(2n)$ **(j)** $2nnnnnn$ **(k)** $xxyyy$ **(l)** $mmmm(-n)(-n)$

Solution: Recall that to rename terms without exponents as exponential notation, you can use parts **(a)** and **(b)** of the Fundamental Rule for Whole-Number Exponents [see Example 7-2]:

(a) $5 = 5^1$ **(b)** $w = w^1$ **(c)** $yy = y^2$ **(d)** $xxx = x^3$ **(e)** $(-m)(-m)(-m)(-m) = (-m)^4$

(f) $-mmmm = -m^4$ **(g)** $\dfrac{w}{3} \cdot \dfrac{w}{3} \cdot \dfrac{w}{3} \cdot \dfrac{w}{3} \cdot \dfrac{w}{3} = \left(\dfrac{w}{3}\right)^5$ **(h)** $\dfrac{wwww}{3} = \dfrac{w^4}{3}$

(i) $(2n)(2n)(2n)(2n)(2n)(2n) = (2n)^6$ **(j)** $2nnnnnn = 2n^6$ **(k)** $xxyyy = x^2y^3$
(l) $mmmm(-n)(-n) = m^4(-n)^2.$

PROBLEM 7-2   Rename each exponential notation as a term without exponents: **(a)** 3^0 **(b)** 10^1

(c) x^1 **(d)** w^0 **(e)** y^3 **(f)** $(-h)^2$ **(g)** $-h^2$ **(h)** $\left(\dfrac{k}{2}\right)^3$ **(i)** $\dfrac{k^3}{2}$ **(j)** $(5a)^4$

(k) $5a^4$ **(l)** b^3c^2 **(m)** $(-r)^5s^4$ **(n)** $-2m^3$ **(o)** $-(2m)^3$ **(p)** $(-2m)^3$ **(q)** $5m^0$

(r) $(5m)^0$ **(s)** $\dfrac{x^2y}{z^3}$ **(t)** 0^0 **(u)** 0^5 **(v)** 1^5 **(w)** -1^5 **(x)** -1^4 **(y)** $(-1)^5$

(z) $(-1)^6$

Solution: Recall that to rename exponential notation as a term without exponents, you can use the Fundamental Rule for Whole-Number Exponents and the special rules for exponents involving 0 and 1 [see Examples 7-4, 7-6, and 7-7]:

(a) $3^0 = 1$ **(b)** $10^1 = 10$ **(c)** $x^1 = x$ **(d)** $w^0 = 1 \ (w \neq 0)$ **(e)** $y^3 = yyy$

(f) $(-h)^2 = (-h)(-h)$ **(g)** $-h^2 = -(hh) = -hh$ **(h)** $\left(\dfrac{k}{2}\right)^3 = \dfrac{k}{2} \cdot \dfrac{k}{2} \cdot \dfrac{k}{2}$ **(i)** $\dfrac{k^3}{2} = \dfrac{kkk}{2}$

(j) $(5a)^4 = (5a)(5a)(5a)(5a)$ **(k)** $5a^4 = 5(aaaa) = 5aaaa$ **(l)** $b^3c^2 = (bbb)(cc) = bbbcc$
(m) $(-r)^5s^4 = (-r)(-r)(-r)(-r)(-r)ssss$ **(n)** $-2m^3 = -2mmm$
(o) $-(2m)^3 = -(2m)(2m)(2m)$ **(p)** $(-2m)^3 = (-2m)(-2m)(-2m)$ **(q)** $5m^0 = 5(1) = 5[m \neq 0]$

(r) $(5m)^0 = 1[m \neq 0]$ **(s)** $\dfrac{x^2y}{z^3} = \dfrac{xxy}{zzz}$ **(t)** 0^0 is not defined.   **(u)** $0^5 = 0 \ [0^n = 0 \text{ if } n \neq 0]$

(v) $1^5 = 1[1^n = 1]$ **(w)** $-1^5 = -1[-1^n = -1]$ **(x)** $-1^4 = -1[-1^n = -1]$
(y) $(-1)^5 = -1[(-1)^n = -1 \text{ if } n \text{ is odd}]$ **(z)** $(-1)^6 = 1[(-1)^n = 1 \text{ if } n \text{ is even}]$

PROBLEM 7-3   Multiply with like bases: **(a)** xx^2 **(b)** y^3y **(c)** w^2ww^5 **(d)** $4m^2m^3$

(e) $\dfrac{nn}{5}$ **(f)** $u^3u^2vv^5$ **(g)** $\dfrac{a^2a^3}{bb^2c}$ **(h)** $5r^3st^2$

Solution: Recall that to multiply with like bases, you add the exponents of the like bases [see Examples 7-9 and 7-10]:

(a) $xx^2 = x^1x^2$ **(b)** $y^3y = y^3y^1$ **(c)** $w^2ww^5 = w^2w^1w^5$ **(d)** $4m^2m^3 = 4m^{2+3}$

$\qquad\quad = x^{1+2}$ $\qquad\qquad = y^{3+1}$ $\qquad\qquad\quad = w^{2+1+5}$ $\qquad\qquad = 4m^5$

$\qquad\quad = x^3$ $\qquad\qquad\quad = y^4$ $\qquad\qquad\qquad = w^8$

(e) $\dfrac{nn}{5} = \dfrac{n^1 n^1}{5}$ **(f)** $u^3 u^2 vv^5 = u^3 u^2 v^1 v^5$ **(g)** $\dfrac{a^2 a^3}{bb^2 c} = \dfrac{a^2 a^3}{b^1 b^2 c}$

$$= \dfrac{n^{1+1}}{5} \qquad\qquad = u^{3+2} v^{1+5} \qquad\qquad = \dfrac{a^{2+3}}{b^{1+2} c}$$

$$= \dfrac{n^2}{5} \qquad\qquad = u^5 v^6 \qquad\qquad = \dfrac{a^5}{b^3 c}$$

(h) $5r^3 st^2$ cannot be simplified because r, s, and t are different bases.

PROBLEM 7-4 Raise each product to the given power: **(a)** $(5m)^2$ **(b)** $(-2n)^3$ **(c)** $(-4w)^2$
(d) $(2hk)^3$ **(e)** $(\tfrac{2}{3}n)^2$ **(f)** $(-a)^5$ **(g)** $(-b)^6$ **(h)** $(\tfrac{3}{4}xyz)^2$

Solution: Recall that to raise a product to a given power, you raise each factor of the product to the given power [see Examples 7-13 and 7-14]:

(a) $(5m)^2 = 5^2 m^2$ **(b)** $(-2n)^3 = -(2n)^3$ **(c)** $(-4w)^2 = (4w)^2$ **(d)** $(2hk)^3 = 2^3 h^3 k^3$

$$\qquad = 25m^2 \qquad\qquad = -2^3 n^3 \qquad\qquad = 4^2 w^2 \qquad\qquad = 8h^3 k^3$$

$$\qquad\qquad\qquad\qquad = -8n^3 \qquad\qquad = 16w^2$$

(e) $(\tfrac{2}{3}n)^2 = (\tfrac{2}{3})^2 n^2$ **(f)** $(-a)^5 = -a^5$ **(g)** $(-b)^6 = b^6$ **(h)** $(\tfrac{3}{4}xyz)^2 = (\tfrac{3}{4})^2 x^2 y^2 z^2$

$$\quad = \tfrac{4}{9} n^2 \qquad\qquad\qquad\qquad\qquad\qquad\qquad\qquad\qquad = \tfrac{9}{16} x^2 y^2 z^2$$

PROBLEM 7-5 Divide with like bases. (Assume all variables are nonzero.): **(a)** $\dfrac{x^9}{x^2}$ **(b)** $\dfrac{y^5}{y^4}$

(c) $\dfrac{w^8}{w^2 w^3}$ **(d)** $\dfrac{m^6 m^7}{m^4}$ **(e)** $\dfrac{2n^3}{n}$ **(f)** $\dfrac{a^5}{4a^2}$ **(g)** $\dfrac{-3bb^5 b^2}{5b}$ **(h)** $\dfrac{5u^2}{8v^3}$

Solution: Recall that to divide with like bases, you subtract the exponents of the like bases [see Examples 7-17 and 7-18]:

(a) $\dfrac{x^9}{x^2} = x^{9-2}$ **(b)** $\dfrac{y^5}{y^4} = y^{5-4}$ **(c)** $\dfrac{w^8}{w^2 w^3} = \dfrac{w^8}{w^{2+3}}$ **(d)** $\dfrac{m^6 m^7}{m^4} = \dfrac{m^{6+7}}{m^4}$ **(e)** $\dfrac{2n^3}{n} = 2 \cdot \dfrac{n^3}{n^1}$

$$\qquad = x^7 \qquad\qquad = y^1 \qquad\qquad = \dfrac{w^8}{w^5} \qquad\qquad = \dfrac{m^{13}}{m^4} \qquad\qquad = 2n^{3-1}$$

$$\qquad\qquad\qquad = y \qquad\qquad = w^{8-5} \qquad\qquad = m^{13-4} \qquad\qquad = 2n^2$$

$$\qquad\qquad\qquad\qquad\qquad = w^3 \qquad\qquad = m^9$$

(f) $\dfrac{a^5}{4a^2} = \dfrac{1}{4} \cdot \dfrac{a^5}{a^2}$ **(g)** $\dfrac{-3bb^5 b^2}{5b} = -\dfrac{3}{5} \dfrac{b^{1+5+2}}{b}$

$$= \dfrac{1}{4} a^{5-2} \qquad\qquad\qquad = -\dfrac{3}{5} \dfrac{b^8}{b^1}$$

$$= \dfrac{1}{4} a^3 \text{ or } \dfrac{a^3}{4} \qquad\qquad = -\dfrac{3}{5} b^{8-1}$$

$$\qquad\qquad\qquad\qquad\qquad = -\dfrac{3}{5} b^7 \text{ or } -\dfrac{3b^7}{5}$$

(h) $\dfrac{5u^2}{8v^3}$ cannot be simplified because u and v are different bases.

PROBLEM 7-6 Raise each quotient to the given power. (Assume all variables are nonzero.):

(a) $(\frac{3}{4})^2$ **(b)** $(-\frac{2}{3})^2$ **(c)** $(-\frac{1}{2})^3$ **(d)** $\left(\frac{x}{2}\right)^3$ **(e)** $\left(\frac{5}{y}\right)^2$ **(f)** $\left(\frac{m}{n}\right)^4$ **(g)** $\left(-\frac{w}{3}\right)^2$

(h) $\left(-\frac{2}{z}\right)^3$

Solution: Recall that to raise a quotient to a given power, you raise both the numerator and the denominator to the given power [see Example 7-21]:

(a) $\left(\frac{3}{4}\right)^2 = \frac{3^2}{4^2}$ **(b)** $\left(-\frac{2}{3}\right)^2 = \left(\frac{2}{3}\right)^2$ **(c)** $\left(-\frac{1}{2}\right)^3 = -\left(\frac{1}{2}\right)^3$ **(d)** $\left(\frac{x}{2}\right)^3 = \frac{x^3}{2^3}$

$= \frac{9}{16}$ $= \frac{2^2}{3^2}$ $= -\frac{1^3}{2^3}$ $= \frac{x^3}{8}$

$= \frac{4}{9}$ $= -\frac{1}{8}$

(e) $\left(\frac{5}{y}\right)^2 = \frac{5^2}{y^2}$ **(f)** $\left(\frac{m}{n}\right)^4 = \frac{m^4}{n^4}$ **(g)** $\left(-\frac{w}{3}\right)^2 = \left(\frac{w}{3}\right)^2$ **(h)** $\left(-\frac{2}{z}\right)^3 = -\left(\frac{2}{z}\right)^3$

$= \frac{25}{y^2}$ $= \frac{w^2}{3^2}$ $= -\frac{2^3}{z^3}$

$= \frac{w^2}{9}$ $= -\frac{8}{z^3}$

PROBLEM 7-7 Raise each power to the given power. (Assume all variables are nonzero.):

(a) $(x^3)^2$ **(b)** $(x^2)^3$ **(c)** $(-y^3)^4$ **(d)** $(-y^4)^3$ **(e)** $(w^3)^1$ **(f)** $(w^1)^3$ **(g)** $-(z^5)^0$
(h) $(z^0)^5$

Solution: Recall that to raise a power to a given power, you multiply the exponents [see Example 7-24]:

(a) $(x^3)^2 = x^{3 \cdot 2}$ **(b)** $(x^2)^3 = x^{2 \cdot 3}$ **(c)** $(-y^3)^4 = (y^3)^4$ **(d)** $(-y^4)^3 = -(y^4)^3$

$= x^6$ $= x^6$ $= y^{3 \cdot 4}$ $= -y^{4 \cdot 3}$

$= y^{12}$ $= -y^{12}$

(e) $(w^3)^1 = w^{3 \cdot 1}$ **(f)** $(w^1)^3 = w^{1 \cdot 3}$ **(g)** $-(z^5)^0 = -[(z^5)^0]$ **(h)** $(z^0)^5 = (1)^5$ or $(z^0)^5 = z^{0 \cdot 5}$

$= w^3$ $= w^3$ $= -1$ $= 1$ $= z^0$

$= 1$

PROBLEM 7-8 Raise each product and/or quotient containing powers to the given power. (Assume all variables are nonzero.): **(a)** $(2w^4)^3$ **(b)** $(-5z^3)^2$ **(c)** $(-2y^2)^3$ **(d)** $\left(\frac{x^3}{8}\right)^2$ **(e)** $\left(\frac{2u^5}{-3}\right)^3$

(f) $\left(\frac{-v^4}{6}\right)^2$ **(g)** $\left(\frac{4uv}{3w}\right)^2$ **(h)** $\left(-\frac{abc}{de}\right)^5$ **(i)** $\frac{3x^4(2x^2)^3}{(4x^5)^2}$

Solution: To raise a product and/or quotient containing powers to a given power, you raise each part of the product and/or quotient to the given power [see Example 7-26]:

(a) $(2w^4)^3 = 2^3(w^4)^3$ **(b)** $(-5z^3)^2 = (5z^3)^2$ **(c)** $(-2y^2)^3 = -(2y^2)^3$

$= 8w^{12}$ $= 5^2(z^3)^2$ $= -2^3(y^2)^3$

$= 25z^6$ $= -8y^6$

(d) $\left(\dfrac{x^3}{8}\right)^2 = \dfrac{(x^3)^2}{8^2}$ **(e)** $\left(\dfrac{2u^5}{-3}\right)^3 = -\left(\dfrac{2u^5}{3}\right)^3$ **(f)** $\left(\dfrac{-v^4}{6}\right)^2 = \left(\dfrac{v^4}{6}\right)^2$

$\qquad\qquad = \dfrac{x^6}{64}$ $\qquad\qquad\qquad = -\dfrac{2^3(u^5)^3}{3^3}$ $\qquad\qquad\qquad = \dfrac{(v^4)^2}{6^2}$

$\qquad\qquad\qquad\qquad\qquad = -\dfrac{8u^{15}}{27}$ $\qquad\qquad\qquad\qquad = \dfrac{v^8}{36}$

(g) $\left(\dfrac{4uv}{3w}\right)^2 = \dfrac{4^2u^2v^2}{3^2w^2}$ **(h)** $\left(-\dfrac{abc}{de}\right)^5 = -\left(\dfrac{abc}{de}\right)^5$ **(i)** $\dfrac{3x^4(2x^2)^3}{(4x^5)^2} = \dfrac{3x^4(8x^6)}{16x^{10}}$

$\qquad\qquad = \dfrac{16u^2v^2}{9w^2}$ $\qquad\qquad\qquad = -\dfrac{a^5b^5c^5}{d^5e^5}$ $\qquad\qquad\qquad\qquad = \dfrac{24x^{10}}{16x^{10}}$

$\qquad\qquad\qquad\qquad\qquad\qquad\qquad\qquad\qquad\qquad\qquad = \dfrac{3}{2}$

PROBLEM 7-9 Rename using only positive exponents. (Assume all variables are nonzero.):

(a) -3^{-2} **(b)** $\dfrac{1}{4^{-2}}$ **(c)** x^{-1} **(d)** $\dfrac{1}{y^{-1}}$ **(e)** $3w^{-2}$ **(f)** $\dfrac{5}{z^{-3}}$ **(g)** $\dfrac{2m^{-3}}{5}$ **(h)** $\dfrac{3}{4n^{-8}}$

(i) $\dfrac{7a^{-3}}{8b^{-2}}$

Solution: Recall that to rename a nonzero term containing a negative exponent in the numerator (denominator), you move the factor with the negative exponent to the denominator (numerator) and make the exponent positive by writing the opposite of the negative exponent [see Example 7-28]:

(a) $-3^{-2} = -(3^{-2})$ **(b)** $\dfrac{1}{4^{-2}} = 4^2$ **(c)** $x^{-1} = \dfrac{1}{x}$ **(d)** $\dfrac{1}{y^{-1}} = y$ **(e)** $3w^{-2} = 3 \cdot \dfrac{1}{w^2}$

$\qquad\quad = \dfrac{-1}{3^2}$ $\qquad\qquad\quad = 16$ $\qquad\qquad\qquad\qquad\qquad\qquad\qquad\qquad\qquad = \dfrac{3}{w^2}$

$\qquad\quad = -\dfrac{1}{9}$

(f) $\dfrac{5}{z^{-3}} = \dfrac{5}{1} \cdot \dfrac{1}{z^{-3}}$ **(g)** $\dfrac{2m^{-3}}{5} = \dfrac{2}{5m^3}$ **(h)** $\dfrac{3}{4n^{-8}} = \dfrac{3n^8}{4}$ **(i)** $\dfrac{7a^{-3}}{8b^{-2}} = \dfrac{7b^2}{8a^3}$

$\qquad\quad = 5z^3$

PROBLEM 7-10 Simplify. Rename the answer using only positive exponents. (Assume all variables are nonzero.): **(a)** $x^{-1}x^3x^{-2}$ **(b)** $\dfrac{w}{w^{-4}}$ **(c)** $(2y)^{-3}$ **(d)** $(-3y)^{-2}$ **(e)** $\left(\dfrac{5}{z}\right)^{-2}$

(f) $(m^{-2})^{-4}$ **(g)** $\dfrac{2}{uv^{-1}}$ **(h)** $\dfrac{(2x^{-3})^2}{3x^{-2}}$

Solution: Recall that all of the Simplification Rules for Whole-Number Exponents can be extended to include integer exponents [see Examples 7-29 and 7-30]:

(a) $x^{-1}x^3x^{-2} = x^{-1+3+(-2)}$ **(b)** $\dfrac{w}{w^{-4}} = \dfrac{w^1}{w^{-4}}$ **(c)** $(2y)^{-3} = \dfrac{1}{(2y)^3}$ **(d)** $(-3y)^{-2} = (3y)^{-2}$

$\qquad\qquad\qquad = x^0$ $\qquad\qquad\qquad = w^{1-(-4)}$ $\qquad\qquad\qquad\qquad = \dfrac{1}{2^3y^3}$ $\qquad\qquad\qquad = \dfrac{1}{(3y)^2}$

$\qquad\qquad\qquad = 1$ $\qquad\qquad\qquad\quad = w^5$ $\qquad\qquad\qquad\qquad\qquad = \dfrac{1}{8y^3}$ $\qquad\qquad\qquad\qquad = \dfrac{1}{9y^2}$

(e) $\left(\dfrac{5}{z}\right)^{-2} = \left(\dfrac{z}{5}\right)^{2}$

$\qquad = \dfrac{z^2}{5^2}$

$\qquad = \dfrac{z^2}{25}$

(f) $(m^{-2})^{-4} = m^{-2(-4)}$

$\qquad\qquad\quad = m^8$

(g) $\dfrac{2}{uv^{-1}} = \dfrac{2v^1}{u}$

$\qquad\quad = \dfrac{2v}{u}$

(h) $\dfrac{(2x^{-3})^2}{3x^{-2}} = \dfrac{4x^{-6}}{3x^{-2}}$

$\qquad\qquad = \dfrac{4}{3}x^{-6-(-2)}$

$\qquad\qquad = \dfrac{4}{3}x^{-4}$

$\qquad\qquad = \dfrac{4}{3x^4}$

Supplementary Exercises

PROBLEM 7-11　Multiply with whole-number exponents: **(a)** xxx　**(b)** $yyyy$　**(c)** $(8p)^2$
(d) $(-2q)^4$　**(e)** $(0.1r)^1$　**(f)** $(-1.5s)^2$　**(g)** w^2w^5　**(h)** z^8z　**(i)** $u^2u^3u^5$　**(j)** $v^4vv^5v^2$

(k) $(\frac{4}{5}x)^2$　**(l)** $(-\frac{1}{2}y)^4$　**(m)** $(3pq)^2$　**(n)** $-(-xy)^5$　**(o)** $5h^2h^7$　**(p)** $-2m^2mm^3$　**(q)** $\dfrac{k^2k}{-2}$

(r) $\dfrac{xx^2y}{5}$　**(s)** $(uvw)^2$　**(t)** $(0.2xyz)^0$　**(u)** $-(\frac{3}{8}mn)^2$　**(v)** $(-\frac{1}{4}xy)^2$　**(w)** $h^2hk^3k^5$

(x) $\dfrac{mm^2m}{n^2n^8}$　**(y)** $(5)10^3$　**(z)** $(1.3)10^4$

PROBLEM 7-12　Divide with whole-number exponents. (Assume all variables are nonzero.):

(a) $\dfrac{h^3}{h^3}$　**(b)** $\dfrac{k^5}{k^4}$　**(c)** $\dfrac{p^8}{p^5}$　**(d)** $\dfrac{-2q^4}{q^2}$　**(e)** $\dfrac{x^2x^8}{x^3}$　**(f)** $\dfrac{y^8}{y^2y^3}$　**(g)** $\dfrac{2mm^2}{m^3}$　**(h)** $\dfrac{-3n^9}{-4n^3n^5}$

(i) $\left(\dfrac{w}{8}\right)^2$　**(j)** $\left(\dfrac{3}{z}\right)^3$　**(k)** $\left(\dfrac{m}{n}\right)^5$　**(l)** $\left(-\dfrac{a}{b}\right)^4$　**(m)** $\left(\dfrac{3p}{2}\right)^3$　**(n)** $\left(\dfrac{5}{8q}\right)^2$　**(o)** $\left(\dfrac{2xy}{w}\right)^4$

(p) $-\left(\dfrac{3mn}{2pq}\right)^2$　**(q)** $\dfrac{m^2m^5m^{10}}{m^3mm^4m^4}$　**(r)** $-\dfrac{-y^5y}{-y}$　**(s)** $\left(-\dfrac{2xy}{wz}\right)^5$　**(t)** $\left(-\dfrac{8u}{vw}\right)^2$　**(u)** $\left(\dfrac{15h^5h}{k^2k^3}\right)^1$

(v) $-\left(\dfrac{12m^8m}{n^2n^5}\right)^1$　**(w)** $\left(\dfrac{25p^8q^9}{3pq}\right)^0$　**(x)** $-\left(-\dfrac{16x^8y}{15w^3z^9}\right)^0$　**(y)** $\dfrac{10^5}{10^3}$　**(z)** $\dfrac{(3.5)10^8}{(2.5)10^5}$

PROBLEM 7-13　Find whole-number powers of powers. (Assume all variables are nonzero.):
(a) $(x^2)^3$　**(b)** $(y^3)^2$　**(c)** $(-w^4)^5$　**(d)** $(-z^5)^4$　**(e)** $(5x^4)^2$　**(f)** $(2y^5)^3$　**(g)** $(-3w^3)^2$

(h) $(-2z^6)^3$　**(i)** $(-u^5)^3$　**(j)** $(-v^2)^4$　**(k)** $(x^2y^3)^5$　**(l)** $(6m^4n^2)^2$　**(m)** $\left(\dfrac{m^2}{2}\right)^3$

(n) $\left(\dfrac{n^3}{-8}\right)^2$　**(o)** $\left(\dfrac{-2}{h^5}\right)^3$　**(p)** $-\left(\dfrac{8}{k^8}\right)^2$　**(q)** $\left(\dfrac{2p^8}{3}\right)^3$　**(r)** $\left(\dfrac{8q^3}{-7}\right)^2$

(s) $\left(\dfrac{-h^3}{k^2}\right)^6$　**(t)** $\left(\dfrac{u^3}{-v^4}\right)^3$　**(u)** $\dfrac{(x^2)^3(x^4)^2}{(x^3)^4}$　**(v)** $[2x^3(3x^2)^3]^2$

(w) $(5\cdot10^4)^2$　**(x)** $(2\cdot10^8)^4$　**(y)** $\left[\dfrac{(3)10^6}{(1.5)10^2}\right]^3$　**(z)** $\dfrac{(1.2)10^{29}}{(4.8)10^{17}}$

PROBLEM 7-14 Simplify with integers as exponents. Write answers using only positive exponents. (Assume all variables are nonzero.): **(a)** x^5x^{-3} **(b)** $y^{-7}y^4y$ **(c)** $(2w)^{-3}$ **(d)** $(-5z)^{-1}$

(e) $\dfrac{u^6}{u^{-2}}$ **(f)** $\dfrac{v^{-3}}{v^2}$ **(g)** $\left(\dfrac{h}{4}\right)^{-2}$ **(h)** $\left(\dfrac{-2}{k}\right)^{-3}$ **(i)** $(m^2)^{-3}$ **(j)** $(n^{-3})^{-3}$ **(k)** $(2p^3)^{-1}$

(l) $(h^2k^{-3})^{-4}$ **(m)** $\dfrac{1}{a^{-2}}$ **(n)** $\dfrac{3}{b^{-1}}$ **(o)** $\dfrac{x}{yz^{-3}}$ **(p)** $\dfrac{uv^{-2}}{w^{-1}}$ **(q)** $\left(\dfrac{x^{-1}y^2}{z}\right)^2$ **(r)** $\left(\dfrac{mn^{-3}}{p^{-2}}\right)^{-1}$

(s) $\left(\dfrac{2ab^{-1}}{a^2b^{-3}}\right)^2$ **(t)** $\left(\dfrac{3u^{-1}v^2}{u^3v^{-1}}\right)^{-2}$ **(u)** $(3abc)^{-1}$ **(v)** $\dfrac{4}{2k^{-5}}$ **(w)** $3x^2(2x^{-2})^3$ **(x)** $\dfrac{(5y^{-2})^{-2}}{4y^{-2}}$

(y) $[3w(2w^{-1})^{-2}]^3$ **(z)** $(m^{-1})^{-1}$

Answers to Supplementary Exercises

(7-11)   **(a)** x^3 **(b)** y^4 **(c)** $64p^2$ **(d)** $16q^4$ **(e)** $0.1r$ **(f)** $2.25s^2$ **(g)** w^7 **(h)** z^9 **(i)** u^{10} **(j)** v^{12} **(k)** $\frac{16}{25}x^2$ **(l)** $\frac{1}{16}y^4$ **(m)** $9p^2q^2$ **(n)** x^5y^5 **(o)** $5h^9$

(p) $-2m^6$ **(q)** $-\dfrac{k^3}{2}$ **(r)** $\dfrac{x^3y}{5}$ **(s)** $u^2v^2w^2$ **(t)** 1 **(u)** $-\frac{9}{64}m^2n^2$ **(v)** $\frac{1}{16}x^2y^2$

(w) h^3k^8 **(x)** $\dfrac{m^4}{n^{10}}$ **(y)** 5000 **(z)** $13{,}000$

(7-12)   **(a)** 1 **(b)** k **(c)** p^3 **(d)** $-2q^2$ **(e)** x^7 **(f)** y^3 **(g)** 2 **(h)** $\frac{3}{4}n$ **(i)** $\dfrac{w^2}{64}$

(j) $\dfrac{27}{z^3}$ **(k)** $\dfrac{m^5}{n^5}$ **(l)** $\dfrac{a^4}{b^4}$ **(m)** $\dfrac{27p^3}{8}$ **(n)** $\dfrac{25}{64q^2}$ **(o)** $\dfrac{16x^4y^4}{w^4}$ **(p)** $-\dfrac{9m^2n^2}{4p^2q^2}$

(q) m^5 **(r)** $-y^5$ **(s)** $-\dfrac{32x^5y^5}{w^5z^5}$ **(t)** $\dfrac{64u^2}{v^2w^2}$ **(u)** $\dfrac{15h^6}{k^5}$ **(v)** $-\dfrac{12m^9}{n^7}$

(w) 1 **(x)** -1 **(y)** 10^2 or 100 **(z)** $(1.4)10^3$ or 1400

(7-13)   **(a)** x^6 **(b)** y^6 **(c)** $-w^{20}$ **(d)** z^{20} **(e)** $25x^8$ **(f)** $8y^{15}$ **(g)** $9w^6$

(h) $-8z^{18}$ **(i)** $-u^{15}$ **(j)** v^8 **(k)** $x^{10}y^{15}$ **(l)** $36m^8n^4$ **(m)** $\dfrac{m^6}{8}$ **(n)** $\dfrac{n^6}{64}$

(o) $-\dfrac{8}{h^{15}}$ **(p)** $-\dfrac{64}{k^{16}}$ **(q)** $\dfrac{8p^{24}}{27}$ **(r)** $\dfrac{64q^6}{49}$ **(s)** $\dfrac{h^{18}}{k^{12}}$ **(t)** $-\dfrac{u^9}{v^{12}}$ **(u)** x^2

(v) $2916x^{18}$ **(w)** $25 \cdot 10^8$ or $2{,}500{,}000{,}000$
(x) $16 \cdot 10^{32}$ or $1{,}600{,}000{,}000{,}000{,}000{,}000{,}000{,}000{,}000{,}000{,}000$
(y) $8 \cdot 10^{12}$ or $8{,}000{,}000{,}000{,}000$ **(z)** $(0.25)10^{12}$ or $250{,}000{,}000{,}000$

(7-14)   **(a)** x^2 **(b)** $\dfrac{1}{y^2}$ **(c)** $\dfrac{1}{8w^3}$ **(d)** $-\dfrac{1}{5z}$ **(e)** u^8 **(f)** $\dfrac{1}{v^5}$ **(g)** $\dfrac{16}{h^2}$ **(h)** $-\dfrac{k^3}{8}$

(i) $\dfrac{1}{m^6}$ **(j)** n^9 **(k)** $\dfrac{1}{2p^3}$ **(l)** $\dfrac{k^{12}}{h^8}$ **(m)** a^2 **(n)** $3b$ **(o)** $\dfrac{xz^3}{y}$ **(p)** $\dfrac{uw}{v^2}$

(q) $\dfrac{y^4}{x^2z^2}$ **(r)** $\dfrac{n^3}{mp^2}$ **(s)** $\dfrac{4b^4}{a^2}$ **(t)** $\dfrac{u^8}{9v^6}$ **(u)** $\dfrac{1}{3abc}$ **(v)** $2k^5$ **(w)** $\dfrac{24}{x^4}$ **(x)** $\dfrac{y^6}{100}$

(y) $\dfrac{27w^9}{64}$ **(z)** m

D. Classify polynomials with respect to degree.

The **degree of a term in one variable** is the exponent of the variable. To find the degree of a term in one variable, you identify the exponent of the variable.

EXAMPLE 8-5: Find the degree of each term in one variable: (a) 3   (b) x (c) $5w^2$ (d) $4w^3$

Solution: (a) The degree of a real number like 3 [**constant term**] is 0 because $3 = 3(1) = 3x^0$ and the exponent of x^0 is 0 [see the following *Note*].

   (b) The degree of a variable like x [**1st-degree term**] is 1 because $x = x^1$ and the exponent of x^1 is 1.

   (c) The degree of $5w^2$ [**2nd-degree term**] is 2 because the exponent of w^2 is 2.

   (d) The degree of $4w^3$ [**3rd-degree term**] is 3 because the exponent of w^3 is 3.

Note: To find the degree of a real number like 3, any variable can be used to rename 3 as a term in one variable because $3 = 3(1) = 3a^0 = 3b^0 = 3c^0 = \cdots = 3x^0 = 3y^0 = 3z^0$, and the exponent of each term in one variable is 0.

The **degree of a term in more than one variable** is the sum of the exponents of the variables. To find the degree of a term in more than one variable, you add the exponents of the different variables.

EXAMPLE 8-6: Find the degree of each term in more than one variable:

(a) mn (b) $\dfrac{abc}{5}$ (c) $4u^2v^2$ (d) x^2yz^3

Solution: (a) The degree of mn is 2 because $mn = m^1n^1$ and $1 + 1 = 2$.

   (b) The degree of $\dfrac{abc}{5}$ is 3 because $\dfrac{abc}{5} = \dfrac{a^1b^1c^1}{5}$ and $1 + 1 + 1 = 3$.

   (c) The degree of $4u^2v^2$ is 4 because $2 + 2 = 4$.

   (d) The degree of x^2yz^3 is 6 because $x^2yz^3 = x^2y^1z^3$ and $2 + 1 + 3 = 6$.

The **degree of a polynomial** is the highest degree of any of its terms. To **classify a polynomial with respect to degree,** you identify the degree of the polynomial.

EXAMPLE 8-7: Find the degree of each polynomial: (a) $3 + x$ (b) $4w^3 + 5w^2$
(c) $m^2 - 2mn + n^2$ (d) $x^2y^2 - y^3$

Solution: (a) The degree of $3 + x^1 = 3x^0 + x^1$ is 1 because $1 > 0$.

   (b) The degree of $4w^3 + 5w^2$ is 3 because $3 > 2$.

   (c) The degree of $m^2 - 2mn + n^2$ is 2 because the degree of each term m^2, $-2m^1n^1$, and n^2 is 2.

   (d) The degree of $x^2y^2 - y^3$ is 4 because the degree of x^2y^2 is 4 [2 + 2] and the degree of $-y^3$ is 3 and $4 > 3$.

E. Write descending powers.

A polynomial in one variable is in descending powers if the degree of each term decreases from left to right.

EXAMPLE 8-8: Write each polynomial in descending powers: (a) $1 - m$ (b) $4w^3 + 5w^2$
(c) $5x - 7 - x^2$

Solution: (a) $1 - m$ is in descending powers as $-m + 1$ because $-m + 1 = -m^1 + 1m^0$ and $1 > 0$.

   (b) $4w^3 + 5w^2$ is in descending powers already because $3 > 2$.

   (c) $5x - 7 - x^2$ is in descending powers as $-x^2 + 5x - 7$ because
   $-x^2 + 5x - 7 = -x^2 + 5x^1 - 7x^0$.

A polynomial in one variable is in **ascending powers** if the degree of each term increases from left to right.

EXAMPLE 8-9: Write each polynomial in ascending powers: (a) $1 - m$ (b) $4w^3 + 5w^2$
(c) $5x - 7 - x^2$

Solution: (a) $1 - m$ is already in ascending powers.
        (b) $4w^3 + 5w^2$ is in ascending powers as $5w^2 + 4w^3$.
        (c) $5x - 7 - x^2$ is in ascending powers as $-7 + 5x - x^2$.

8-2. Adding and Subtracting Polynomials

A. Add and subtract polynomials in horizontal form.

To **add or subtract polynomials in horizontal form,** you

1. Write the polynomials in horizontal form using parentheses when necessary.
2. Clear parentheses [see Section 4.3].
3. Collect like terms [see Section 4.4].
4. Combine like terms [see Section 4.5].

EXAMPLE 8-10: (a) Add $3x^2 + 4 - 2x$ and $5x - 7 - x^2$.
                      (b) Subtract $-m^2 + 3n^2 - 5mn$ from $m^2 - 2mn + n^2$.

Solution: (a) Add $3x^2 + 4 - 2x$ and $5x - 7 - x^2$.

$$\overbrace{(3x^2 + 4 - 2x) + (5x - 7 - x^2)}^{\text{horizontal addition form}}$$ [See the following *Note 1.*]

$$= 3x^2 + 4 - 2x + 5x - 7 - x^2$$ Clear parentheses.

$$= (3x^2 - x^2) + (-2x + 5x) + (4 - 7)$$ Collect like terms.

$$= 2x^2 + 3x - 3$$ Combine like terms.

  (b) Subtract $-m^2 + 3n^2 - 5mn$ from $m^2 - 2mn + n^2$.

$$\overbrace{(m^2 - 2mn + n^2) - (-m^2 + 3n^2 - 5mn)}^{\text{horizontal subtraction form}}$$ [See the following *Note 2.*]

$$= m^2 - 2mn + n^2 + m^2 - 3n^2 + 5mn$$ Clear parentheses.

$$= (m^2 + m^2) + (-2mn + 5mn) + (n^2 - 3n^2)$$ Collect like terms.

$$= 2m^2 + 3mn - 2n^2$$ Combine like terms.

Note 1: "Add a to b" means $a + b$ or $b + a$.

Note 2: "Subtract a from b" means $b - a$ only.

B. Add and subtract polynomials in vertical form.

To **add or subtract polynomials in vertical form,** you

1. Write the polynomials in vertical form by writing descending powers while aligning like terms in columns.
2. Change any subtractions to additions by writing the opposite of each term in the subtrahend [bottom polynomial].
3. Combine like terms.

EXAMPLE 8-11: (a) Add $8x - 9x^3 - 1$, $2x^3 + 7x - 4x^2$, and $8x^2 + 6$.
                      (b) Subtract $4m^2n - 5mn^2$ from $3m^2n + 5mn - 2mn^2$.

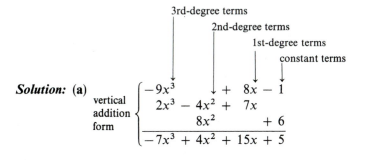

Solution: (a)   vertical addition form
$$\begin{cases} -9x^3 & + 8x - 1 \\ 2x^3 - 4x^2 + 7x \\ 8x^2 + 6 \\ \hline -7x^3 + 4x^2 + 15x + 5 \end{cases}$$

Write descending powers and align like terms in columns while leaving space for any missing terms. Combine like terms.

m^2n terms

mn terms

mn^2 terms

(b) vertical subtraction form

$$3m^2n + 5mn - 2mn^2 \longleftarrow \text{minuend}$$
$$- \quad\quad\quad + $$
$$\cancel{+}\ 4m^2n \quad\quad \cancel{\not/}\ 5mn^2 \longleftarrow \text{subtrahend}$$
$$\overline{-m^2n + 5mn + 3mn^2}$$

Collect like terms in columns [see the following *Note 1*]. Change subtraction to addition by writing the opposite of each term in the subtrahend [see the following *Note 2*]. Combine like terms.

Note 1: "Subtract a from b" means $b - a$ or $b + (-a)$.

Note 2: "Subtract $4m^2n - 5mn^2$ from $3m^2n + 5mn - 2mn^2$" means

change subtraction to addition using $a - b = a + (-b)$

$$\underbrace{(3m^2n + 5mn - 2mn^2) - (4m^2n - 5mn^2)}_{} = \underbrace{(3m^2n + 5mn - 2mn^2) + (-4m^2n + 5mn^2)}_{}$$

or

$$3m^2n + 5mn - 2mn^2$$
$$-(+4m^2n \quad\quad - 5mn^2)$$

change subtraction to addition $\longrightarrow$

$$3m^2n + 5mn - 2mn^2$$
$$- \quad\quad\quad +$$
$$\cancel{+}4m^2n \quad\quad \cancel{\not/}\ 5mn^2$$

8-3. Multiplying Polynomials

A. Multiply monomials.

To **multiply monomials,** you can use the following rule:

Multiplication Rule for Monomials

numerical coefficients

literal parts

If ar^n and bs^m are monomials, then $(ar^n)(bs^m) = \overbrace{(ab)}\overbrace{(r^ns^m)}$.

Note: To multiply monomials, you multiply the numerical coefficients and the literal parts separately.

EXAMPLE 8-12: Multiply using the Multiplication Rule for Monomials:

(a) $3x(4x^2)$ **(b)** $(2y^4) \cdot \dfrac{y^3}{5}$ **(c)** $3w(w)$ **(d)** $3z(2)$ **(e)** $(-5m^2n)(2mn)$

Solution: **(a)** $3x(4x^2) = (3 \cdot 4)xx^2$ Multiply numerical coefficients. Multiply literal parts.

$\quad\quad\quad = 12x^3 \longleftarrow$ product

(b) $(2y^4) \cdot \dfrac{y^3}{5} = (2y^4)\left(\dfrac{1}{5}y^3\right)$ Rename in ar^n form.

$\quad\quad\quad = \left(2 \cdot \dfrac{1}{5}\right)y^4y^3$ Multiply monomials.

$\quad\quad\quad = \dfrac{2}{5}y^7 \text{ or } \dfrac{2y^7}{5}$

(c) $3w(w) = (3w)(1w)$ or $3w(w) = 3ww$

$\quad\quad\quad = (3 \cdot 1)ww \quad\quad\quad = 3w^2$

$\quad\quad\quad = 3w^2$

(d) $3z(2) = (3 \cdot 2)z$ **(e)** $(-5m^2n)(2mn) = (-5 \cdot 2)(m^2nmn)$

$= 6z$ $= -10m^3n^2$

Note: To **multiply monomials with like variables,** you multiply the numerical coefficients and then add the exponents of the like variables.

Caution: When the variables are different, you cannot add the exponents of the different variables.

EXAMPLE 8-13: Multiply $(-3a^2)(-4b^3)$.

Solution: $(-3a^2)(-4b^3) = [(-3)(-4)]a^2b^3$ ⟵ different variables

$= 12a^2b^3$ ⟵ do not add exponents

B. Multiply a polynomial by a monomial.

To **multiply a polynomial by a monomial,** you can use the distributive properties.

EXAMPLE 8-14: Multiply the following using the distributive properties: $3x(4x^2 + x + 2)$

Solution: $3x(4x^2 + x + 2) = 3x(4x^2) + 3x(x) + 3x(2)$ Use a distributive property.

$= (3 \cdot 4)xx^2 + 3xx + (3 \cdot 2)x$ Multiply monomials.

$= 12x^3 + 3x^2 + 6x$ ⟵ product

Note: To multiply a polynomial by a monomial, you multiply each term of the polynomial by the monomial.

C. Multiply two binomials using the distributive properties.

To **multiply two binomials,** you can use the distributive properties.

EXAMPLE 8-15: Multiply using the distributive properties: $(3x + 2)(4x + 5)$

Solution: $(3x + 2)(4x + 5) = (3x + 2)4x + (3x + 2)5$ Distribute $3x + 2$.

$= (3x)4x + (2)4x + (3x)5 + (2)5$ Distribute $4x$ and 5.

$= 12x^2 + 8x + 15x + 10$ Multiply monomials.

$= 12x^2 + 23x + 10$ Combine like terms.

Note: To multiply two binomials, you multiply each term of one binomial by each term of the other binomial.

Caution: To multiply two binomials using the distributive properties when the second binomial is a difference, like $(a + b)(c - d)$ or $(a - b)(c - d)$, you should use brackets [ ] when distributing the second term to avoid making a sign error.

EXAMPLE 8-16: Multiply using the distributive properties: $(x + 2)(x - 3)$

Solution:

Correct Method

$(x + 2)(x - 3) = (x + 2)x - (x + 2)3$

$= (x)x + (2)x - [(x)3 + (2)3]$ Use brackets [ ] to avoid making a sign error

$= x^2 + 2x - [3x + 6]$

$= x^2 + 2x - 3x - 6$ Clear brackets.

$= x^2 - x - 6$ ⟵ correct answer

Incorrect Method

$(x + 2)(x - 3) = (x + 2)x - (x + 2)3$

wrong sign ↓   correct sign ↓

$= (x)x + (2)x - (x)3 + (2)3$ No! $-(x + 2)3 = -[(x)3 + (2)3] = -(x)3 - (2)3$

$= x^2 + 2x - 3x + 6$

$= x^2 - x + 6$ ⟵ wrong answer [The correct answer is $x^2 - x - 6$.]

D. Multiply two binomials using the FOIL method.

To **multiply two binomials,** you can use the following method:

FOIL Method for Multiplying Two Binomials

To multiply $(a + b)(c + d)$:

1. Multiply the **First** terms: $(a + b)(c + d) = ac + ? + ? + ?$

2. Multiply the **Outside** terms: $(a + b)(c + d) = ac + ad + ? + ?$

3. Multiply the **Inside** terms: $(a + b)(c + d) = ac + ad + bc + ?$

4. Multiply the **Last** terms: $(a + b)(c + d) = ac + ad + bc + bd$

5. Combine like terms when possible.

Note: The word "FOIL" can help you remember to multiply **F**irst terms, **O**utside terms, **I**nside terms, **L**ast terms.

EXAMPLE 8-17: Multiply using the FOIL method: $(3x + 2)(4x + 5)$

Solution: $(3x + 2)(4x + 5) = 12x^2 + ? + ? + ?$ *Think:* $(3x)(4x) = 12x^2$

$(3x + 2)(4x + 5) = 12x^2 + 15x + ? + ?$ *Think:* $(3x)5 = 15x$

$(3x + 2)(4x + 5) = 12x^2 + 15x + 8x + ?$ *Think:* $2(4x) = 8x$

$(3x + 2)(4x + 5) = 12x^2 + 15x + 8x + 10$ *Think:* $2(5) = 10$

$(3x + 2)(4x + 5) = 12x^2 + 23x + 10$ *Think:* $15x + 8x = 23x$

Note: To multiply two binomials, you can use either the distributive property method shown in Example 8-15, or the FOIL method shown in Example 8-17. However, you will find it much easier and quicker to use the FOIL method when multiplying two binomials.

When the two middle terms can be combined using the FOIL method, you should try to multiply and combine the middle terms in your mind.

EXAMPLE 8-18: Use the FOIL method to multiply and combine like terms mentally: $(3x + 2)(4x + 5)$

Solution: $(3x + 2)(4x + 5) = \mathbf{12x^2} + ? + ?$ 　　　*Think:* $(3x)(4x) = 12x^2$

$(3x + 2)(4x + 5) = 12x^2 + \mathbf{23x} + ?$ 　　*Think:* $(3x)5 = 15x$
　　　　　　　　　　　　　　　　　　　　　　$2(4x) = \underline{8x}$ } multiply and combine mentally
　　　　　　　　　　　　　　　　　　　　　　　　　　$23x$

$(3x + 2)(4x + 5) = 12x^2 + 23x + \mathbf{10}$ 　　*Think:* $2(5) = 10$

Caution: Middle terms that are not like terms cannot be combined.

EXAMPLE 8-19: Multiply using the FOIL method when the middle terms cannot be combined: $(x^2 - 2)(3x + 5)$

Solution: $(x^2 - 2)(3x + 5) = \mathbf{3x^3} + ? + ? + ?$ 　　*Think:* $(x^2)(3x) = 3x^3$

$(x^2 - 2)(3x + 5) = 3x^3 + \mathbf{5x^2} + ? + ?$ 　　*Think:* $(x^2)5 = 5x^2$

$(x^2 - 2)(3x + 5) = 3x^3 + 5x^2 \mathbf{-6x} + ?$ 　　*Think:* $-2(3x) = -6x$

$(x^2 - 2)(3x + 5) = 3x^3 + 5x^2 - 6x \mathbf{-10}$ 　　*Think:* $-2(5) = -10$

Note: $(x^2 - 2)(3x + 5) = 3x^3 + 5x^2 - 6x - 10$ is in simplest form because there are no like terms in $3x^3 + 5x^2 - 6x - 10$ that can be combined.

E. Multiply with trinomials.

To **multiply with trinomials,** it is usually easier to multiply in vertical form.

EXAMPLE 8-20: Multiply in vertical form: (a) $(2x + 3)(x^2 + 5x + 6)$
　　　　　　　　　　　　　　　　　　　　 (b) $(2y - y^2 + 3)(5 - 2y + y^2)$

Solution: (a)
$$\begin{array}{r} x^2 + 5x + 6 \\ 2x + 3 \end{array} \Big\}\text{ vertical multiplication form}$$

$$\begin{array}{r} 3x^2 + 15x + 18 \longleftarrow 3(x^2 + 5x + 6)\ [\textbf{first partial product}] \\ 2x^3 + 10x^2 + 12x \longleftarrow 2x(x^2 + 5x + 6)\ [\textbf{second partial product}] \\ \hline 2x^3 + 13x^2 + 27x + 18 \longleftarrow \text{combine like terms in the partial products to get the product} \end{array}$$

(b)

$$\begin{array}{r} y^2 - 2y + 5 \\ -y^2 + 2y + 3 \end{array}\Big\} \begin{array}{l}\text{write descending powers before multiplying}\\ \text{in vertical form}\end{array}$$

$$\begin{array}{rl} 3y^2 - 6y + 15 & \longleftarrow\quad 3(y^2 - 2y + 5) \quad [\textbf{first partial product}]\\ 2y^3 - 4y^2 + 10y & \longleftarrow\quad 2y(y^2 - 2y + 5) \quad [\textbf{second partial product}]\\ -y^4 + 2y^3 - 5y^2 & \longleftarrow\quad -y^2(y^2 - 2y + 5) \; [\text{third partial product}]\\ \hline -y^4 + 4y^3 - 6y^2 + 4y + 15 & \longleftarrow\quad \text{product} \end{array}$$

Note: To multiply two polynomials, you always multiply each term of one polynomial by each term of the other polynomial.

8-4. Finding Special Products

A. Multiply to get the difference of two squares.

Recall: If $a = b^2$, then a is called the square of b [see section 2-7, part C].

EXAMPLE 8-21: Find the square of **(a)** 1; **(b)** -2; **(c)** x; **(d)** $-3y$.

Solution: **(a)** 1 is the square of 1 [and -1] because $1 = 1^2$ [and $1 = (-1)^2$].
(b) 4 is the square of -2 [and 2] because $4 = (-2)^2$ [and $4 = 2^2$].
(c) x^2 is the square of x [and $-x$] because $x^2 = (x)^2$ [and $x^2 = (-x)^2$].
(d) $9y^2$ is the square of $-3y$ [and $3y$] because $9y^2 = (-3y)^2$ [and $9y^2 = (3y)^2$].

Note 1: The square of a monomial is never negative and has only even integers for exponents.

Note 2: Terms that are opposites always have the same square.

Any binomial that can be written in the polynomial form $a^2 - b^2$ is called the **difference of two squares.**

EXAMPLE 8-22: Identify which binomials are the difference of two squares:
(a) $x^2 - y^2$ **(b)** $w^2 - 49$ **(c)** $4z^2 - 9$ **(d)** $r^4 - 1$ **(e)** $16u^6 - 25$ **(f)** $a^2 + b^2$
(g) $m - n^2$

Solution: **(a)** $x^2 - y^2$ is the difference of two squares because $x^2 - y^2 = (x)^2 - (y)^2$.
(b) $w^2 - 49$ is the difference of two squares because $w^2 - 49 = (w)^2 - (7)^2$.
(c) $4z^2 - 9$ is the difference of two squares because $4z^2 - 9 = (2z)^2 - (3)^2$.
(d) $r^4 - 1$ is the difference of two squares because $r^4 - 1 = (r^2)^2 - (1)^2$.
(e) $16u^6 - 25$ is the difference of two squares because $16u^6 - 25 = (4u^3)^2 - (5)^2$.
(f) $a^2 + b^2$ is not the difference of two squares because $a^2 + b^2$ is a sum.
(g) $m - n^2$ is not the difference of two squares because m cannot be written as the square of a polynomial. That is, $m - n^2 = (\sqrt{m})^2 - (n)^2$ but $\sqrt{m}$ is not a polynomial.

Caution: Binomials with one or both terms containing a variable with an odd exponent are not the difference of two squares because a variable with an odd exponent is not the square of a polynomial.

The product of two binomials with form $(a + b)(a - b)$ or $(a - b)(a + b)$ is always the difference of two squares.

EXAMPLE 8-23: Multiply to get the difference of two squares using the FOIL method:
(a) $(a + b)(a - b)$ **(b)** $(a - b)(a + b)$

$$\overset{\textbf{F}\quad\textbf{O}\quad\textbf{I}\quad\textbf{L}}{}$$

Solution: **(a)** $(a + b)(a - b) = a^2 - ab + ab - b^2$ Multiply using the FOIL method.

$$= a^2 + 0 - b^2 \qquad \textit{Think: } -ab + ab = 0$$

$$= a^2 - b^2 \longleftarrow \text{difference of two squares}$$

$$
\begin{array}{cccc}
\text{F} & \text{O} & \text{I} & \text{L} \\
\end{array}
$$

(b) $(a - b)(a + b) = a^2 + ab - ab - b^2$ Multiply using the FOIL method.

$$= a^2 + 0 - b^2 \qquad \textit{Think: } ab - ab = 0$$

$$= a^2 - b^2 \longleftarrow \text{difference of two squares}$$

Example 8-23 is summarized in the following rule:

Product Rule for the Difference of Two Squares

For two binomials with form $a + b$ and $a - b$:

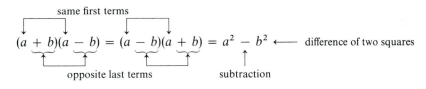

$$(a + b)(a - b) = (a - b)(a + b) = a^2 - b^2 \longleftarrow \text{difference of two squares}$$

Note: To find the product $(a + b)(a - b)$ or $(a - b)(a + b)$ using the Product Rule for the Difference of Two Squares, you

1. Write the square of a: $a^2 \; ? \; ?$
2. Write a subtraction symbol: $a^2 - \; ?$
3. Write a square of b: $a^2 - b^2 \longleftarrow \text{difference of two squares}$

EXAMPLE 8-24: Multiply to get the difference of two squares using the Product Rule for the Difference of Two Squares: **(a)** $(x + 2)(x - 2)$ **(b)** $(2m - 3n)(2m + 3n)$ **(c)** $(u + v)(v - u)$ **(d)** $(1 - y)(y + 1)$

Solution: (a) $(x + 2)(x - 2) = (x)^2 - (2)^2$ *Think:* $(a + b)(a - b) = a^2 - b^2$

$$= x^2 - 4 \longleftarrow \text{difference of two squares}$$

(b) $(2m - 3n)(2m + 3n) = (2m)^2 - (3n)^2$ *Think:* $(a - b)(a + b) = a^2 - b^2$

$$= 4m^2 - 9n^2 \longleftarrow \text{difference of two squares}$$

(c) $(u + v)(v - u) = (v + u)(v - u)$ *Think:* $u + v = v + u$

$$= (v)^2 - (u)^2 \qquad \textit{Think: } (a + b)(a - b) = a^2 - b^2$$

$$= v^2 - u^2 \longleftarrow \text{difference of two squares}$$

(d) $(1 - y)(y + 1) = (1 - y)(1 + y)$ *Think:* $y + 1 = 1 + y$

$$= (1)^2 - (y)^2 \qquad \textit{Think: } (a - b)(a + b) = a^2 - b^2$$

$$= 1 - y^2 \longleftarrow \text{difference of two squares}$$

B. Multiply to get a perfect square trinomial.

Any trinomial that can be written in the polynomial form $a^2 + 2ab + b^2$ or $a^2 - 2ab + b^2$ is called a **perfect square trinomial**.

EXAMPLE 8-25: Identify which of the following trinomials are perfect square trinomials:
(a) $m^2 + 2mn + n^2$ **(b)** $w^2 + 6w + 9$ **(c)** $4x^2 - 12xy + 9y^2$ **(d)** $u^2 - 2u + 1$
(e) $d^4 + 2d^2 + 1$ **(f)** $v^2 + 3v + 1$ **(g)** $z^2 - 5z + 4$ **(h)** $s^2 - 8s - 16$
(i) $r^2 + 4r - 4$ **(j)** $p^4 - 2p^3 + 1$

Solution: **(a)** $m^2 + 2mn + n^2$ is a perfect square trinomial because

$$m^2 + 2mn + n^2 = (m)^2 \overbrace{+ 2(m)(n)}^{2mn} + (n)^2$$

(b) $w^2 + 6w + 9$ is a perfect square trinomial because

$$w^2 + 6w + 9 = (w)^2 \overbrace{+ 2(w)(3)}^{6w} + (3)^2$$

(c) $4x^2 - 12xy + 9y^2$ is a perfect square trinomial because

$$4x^2 - 12xy + 9y^2 = (2x)^2 \overbrace{- 2(2x)(3y)}^{-12xy} + (3y)^2$$

(d) $u^2 - 2u + 1$ is a perfect square trinomial because

$$u^2 - 2u + 1 = (u)^2 \overbrace{- 2(u)(1)}^{-2u} + (1)^2$$

(e) $d^4 + 2d^2 + 1$ is a perfect square trinomial because

$$d^4 + 2d^2 + 1 = (d^2)^2 \overbrace{+ 2(d^2)(1)}^{2d^2} + (1)^2$$

(f) $v^2 + 3v + 1$ is not a perfect square trinomial because

$$v^2 + 3v + 1 \neq (v)^2 \overbrace{+ 2(v)(1)}^{2v} + (1)^2$$

(g) $z^2 - 5z + 4$ is not a perfect square trinomial because

$$z^2 - 5z + 4 \neq (z)^2 \overbrace{- 2(z)(2)}^{-4z} + (2)^2$$

(h) $s^2 - 8s - 16$ is not a perfect square trinomial because $s^2 - 8s \overset{\underset{\text{must be addition sign here}}{\downarrow}}{-} 16$.

(i) $r^2 + 4r - 4$ is not a perfect square trinomial because $r^2 + 4r \overset{\underset{\text{must be addition sign here}}{\downarrow}}{-} 4$.

(j) $p^4 - 2p^3 + 1$ is not a perfect square trinomial because

$$p^4 - 2p^3 + 1 \neq (p^2)^2 \overbrace{- 2(p^2)(1)}^{-2p^2} + (1)^2$$

The square of a binomial like $(a + b)^2$ or $(a - b)^2$ is always a perfect square trinomial.

EXAMPLE 8-26: Multiply to get a perfect square trinomial using the FOIL method:
(a) $(a + b)^2$ **(b)** $(a - b)^2$

Solution: (a) $(a + b)^2 = (a + b)(a + b)$ *Think:* $r^2 = rr$ where $r = a + b$

$$\begin{array}{cccc} \text{F} & \text{O} & \text{I} & \text{L} \end{array}$$
$$= a^2 + ab + ab + b^2 \qquad \text{Multiply using the FOIL method.}$$
$$= a^2 + 2ab + b^2 \longleftarrow \text{perfect square trinomial}$$

(b) $(a - b)^2 = (a - b)(a - b)$ *Think:* $r^2 = rr$ where $r = a - b$

$$\begin{array}{cccc} \text{F} & \text{O} & \text{I} & \text{L} \end{array}$$
$$= a^2 - ab - ab + b^2 \qquad \text{Multiply using the FOIL method.}$$
$$= a^2 - 2ab + b^2 \longleftarrow \text{perfect square trinomial}$$

Example 8-26 is summarized in the following rule:

Product Rule for Perfect Square Trinomials

For two binomials each with form $a + b$ or $a - b$

1. $(a + b)^2 = (a + b)(a + b) = a^2 + 2ab + b^2$

2. $(a - b)^2 = (a - b)(a - b) = a^2 - 2ab + b^2$

Note: To find the square of $(a + b)^2$ [or $(a - b)^2$] using the Product Rule for Perfect Square Trinomials, you

1. Write the square of a: $a^2 \quad ? \quad ? \quad ? \quad [a^2 \quad ? \quad ? \quad ?]$
2. For $(a + b)^2$ write $+2ab$. [For $(a - b)^2$ write $-2ab$]: $a^2 + 2ab \quad ? \quad ? \quad [a^2 - 2ab \quad ? \quad ?]$
3. Write an addition sign: $a^2 + 2ab + \quad ? \quad [a^2 - 2ab + \quad ?]$
4. Write the square of b: $a^2 + 2ab + b^2 [a^2 - 2ab + b^2]$

EXAMPLE 8-27: Multiply to get a perfect square trinomial using the Product Rule for Perfect Square Trinomials: (a) $(x + 1)^2$ (b) $(2m - 3n)^2$

Solution: (a) $(x + 1)^2 = (x)^2 + 2(x)(1) + (1)^2$ *Think:* $(a + b)^2 = a^2 + 2ab + b^2$

$$= x^2 + 2x + 1 \longleftarrow \text{perfect square trinomial}$$

(b) $(2m - 3n)^2 = (2m)^2 - 2(2m)(3n) + (3n)^2$ *Think:* $(a - b)^2 = a^2 - 2ab + b^2$

$$= 4m^2 - 12mn + 9n^2 \longleftarrow \text{perfect square trinomial}$$

Caution: $(r + s)^2 \neq r^2 + s^2$

EXAMPLE 8-28: Show that $(3 + 2)^2 \neq 3^2 + 2^2$.

Solution: $(3 + 2)^2 \neq 3^2 + 2^2$ because $(3 + 2)^2 = (5)^2 = 25$
and $3^2 + 2^2 = 9 + 4 = 13$
and $25 \neq 13$.

Note: $(r + s)^2 = r^2 + 2rs + s^2$ [not $r^2 + s^2$]

Caution: $(r - s)^2 \neq r^2 - s^2$

EXAMPLE 8-29: Show that $(3 - 2)^2 \neq 3^2 - 2^2$.

Solution: $(3 - 2)^2 \neq 3^2 - 2^2$ because $(3 - 2)^2 = (1)^2 = 1$

$$\text{and } 3^2 - 2^2 = 9 - 4 = 5$$
$$\text{and } 1 \neq 5.$$

Note: $(r - s)^2 = r^2 - 2rs + s^2$ and $(r + s)(r - s) = r^2 - s^2$

8-5. Dividing Polynomials

A. Divide monomials.

To **divide monomials,** you use the following rule:

Division Rule for Monomials

numerical coefficients

literal parts

If ar^n and bs^m are monomials $[b \neq 0 \text{ and } s \neq 0]$, then $\dfrac{ar^n}{bs^m} = \dfrac{a}{b} \cdot \dfrac{r^n}{s^m}$.

Note: To divide monomials, you divide the numerical coefficients and the literal parts separately.

EXAMPLE 8-30: Divide using the Division Rule for Monomials:

(a) $\dfrac{24x^5}{-8x}$ **(b)** $\dfrac{20m^2n^5}{12mn^3}$ **(c)** $18y^3 \div 12$ **(d)** $-15 \div (5w^2)$

Solution:

(a) $\dfrac{24x^5}{-8x} = \dfrac{24}{-8} \cdot \dfrac{x^5}{x}$ Divide numerical coefficients. Divide literal parts.

$\qquad = -3x^4 \longleftarrow$ quotient   *Think:* $\dfrac{x^5}{x} = \dfrac{x^5}{x^1} = x^{5-1} = x^4$

(b) $\dfrac{20m^2n^5}{12mn^3} = \dfrac{20}{12} \cdot \dfrac{m^2n^5}{mn^3}$

$\qquad = \dfrac{5}{3}mn^2 \text{ or } \dfrac{5mn^2}{3}$ *Think:* $\dfrac{m^2}{m} = m$ and $\dfrac{n^5}{n^3} = n^2$

(c) $18y^3 \div 12 = \dfrac{18y^3}{12}$ **(d)** $-15 \div (5w^2) = \dfrac{-15}{5w^2}$

$\qquad = \dfrac{18}{12} \cdot \dfrac{y^3}{1}$ $\qquad = \dfrac{-15}{5} \cdot \dfrac{1}{w^2}$

$\qquad = \dfrac{3}{2}y^3 \text{ or } \dfrac{3y^3}{2}$ $\qquad = -3 \cdot \dfrac{1}{w^2}$

$\qquad\qquad\qquad\qquad\qquad = -\dfrac{3}{w^2}$

Caution: $-15 \div 5w^2 \neq -15 \div (5w^2)$

EXAMPLE 8-31: Show that $-15 \div 5w^2 \neq -15 \div (5w^2)$.

Solution: $-15 \div 5w^2 \neq -15 \div (5w^2)$ because $-15 \div 5w^2 = \dfrac{-15}{5} \cdot w^2 = -3w^2$,

$$\text{and } -15 \div (5w^2) = \dfrac{-15}{5w^2} = -\dfrac{3}{w^2},$$

$$\text{and } -3w^2 \neq -\dfrac{3}{w^2}.$$

Note: To **divide monomials with like variables,** you divide the numerical coefficients and subtract the exponents of the like variables.

Caution: When the variables are different, you cannot subtract exponents.

EXAMPLE 8-32: Divide $-16a^5$ by $-12b^3$.

Solution: $\dfrac{-16a^5}{-12b^3} = \dfrac{-16}{-12} \cdot \dfrac{a^5}{b^3}$ ⟩ different variables

$$= \frac{4}{3} \cdot \frac{a^5}{b^3} \longleftarrow \text{ do not subtract exponents}$$

$$= \frac{4a^5}{3b^3}$$

B. Divide a polynomial by a monomial.

Recall: To add or subtract like fractions, you add or subtract the numerator and then write the same denominator.

EXAMPLE 8-33: Add and subtract the following like fractions: $\dfrac{a}{d} - \dfrac{b}{d} + \dfrac{c}{d}$

Solution: $\dfrac{a}{d} - \dfrac{b}{d} + \dfrac{c}{d} = \dfrac{a - b + c}{d}$ $\longleftarrow$ add and subtract numerators
$\phantom{Solution:\ \dfrac{a}{d} - \dfrac{b}{d} + \dfrac{c}{d} = \dfrac{a - b + c}{d}}$ $\longleftarrow$ write the same denominator

To **divide a polynomial by a monomial,** you reverse the steps used to add and subtract like fractions.

EXAMPLE 8-34: Divide the following polynomial by a monomial: $\dfrac{a - b + c}{d}$

Solution: $\dfrac{a - b + c}{d} = \dfrac{a}{d} - \dfrac{b}{d} + \dfrac{c}{d}$ $\longleftarrow$ write each term in the polynomial as a separate numerator
$\phantom{Solution:\ \dfrac{a - b + c}{d} = \dfrac{a}{d} - \dfrac{b}{d} + \dfrac{c}{d}}$ $\longleftarrow$ write the same denominator

Note: To divide a polynomial by a monomial, you divide each term of the polynomial by the monomial.

EXAMPLE 8-35: Divide $2x^2 + 3x - 4$ by $12x$.

Solution: $\dfrac{2x^2 + 3x - 4}{12x} = \dfrac{2x^2}{12x} + \dfrac{3x}{12x} - \dfrac{4}{12x}$ Divide each term of the polynomial by the monomial.

$$= \frac{x}{6} + \frac{1}{4} - \frac{1}{3x} \longleftarrow \text{ quotient}$$

C. Divide a polynomial by a binomial.
To **divide a polynomial by a binomial,** you use the following rules:

Long Division Rules for Polynomials
1. *Estimate* a term in the quotient by dividing the first term in the dividend by the first term in the divisor. Write this result in the quotient above the like term in the dividend.
2. *Multiply* the result from Step 1 by the divisor and write the product below the like terms in the dividend.
3. *Subtract* the like terms.
4. *Bring down* the next term from the dividend to form a new dividend.
5. *Repeat Steps 1 through 4* for the new dividend from Step 4.
6. *Write the remainder* when there are no more terms in the original dividend to bring down.
7. *Check* the proposed quotient and remainder by multiplying the proposed quotient by the divisor and then adding the proposed remainder to see if you get the original dividend.

EXAMPLE 8-36: Divide using the Long Division Rules for Polynomials: $\dfrac{6x^2 + 4x - 1}{2x - 1}$

Solution:

1st-degree terms
↓

1. *Estimate:* $2x - 1 \overline{\smash{)}6x^2 + 4x - 1}$ with $3x$ above   *Think:* $2x\,\overline{\smash{)}6x^2}$ means $2x - 1\,\overline{\smash{)}6x^2 + 4x}$ with $3x$ above, about $3x$

divisor
↓

2. *Multiply:* $2x - 1 \overline{\smash{)}6x^2 + 4x - 1}$ with $3x$ above
$\qquad\qquad\quad 6x^2 - 3x$

$3x \longrightarrow 3x(2x - 1) = \underbrace{6x^2 - 3x}$

1st-degree terms
2nd-degree terms

3. *Subtract:* $2x - 1 \overline{\smash{)}6x^2 + 4x - 1}$ with $3x$ above
$\qquad\qquad\quad \underline{\underset{+}{\overset{-}{\cancel{+}6x^2 \cancel{-} 3x}}}$ [See Example 8-11.]
$\qquad\qquad\qquad\quad 7x$ ← remaining 1st-degree term

4. *Bring down:* $2x - 1 \overline{\smash{)}6x^2 + 4x - 1}$ with $3x$ above ← next term in dividend
$\qquad\qquad\qquad \underline{\underset{+}{\overset{-}{\cancel{+}6x^2 \cancel{-} 3x}}}\;\downarrow$
$\qquad\qquad\qquad\qquad 7x - 1$ ← new dividend

5. *Repeat Steps 1 through 4:* $2x - 1 \overline{\smash{)}6x^2 + 4x - 1}$ with $3x + 3.5$ above
$\qquad\qquad\qquad\quad \underline{\underset{+}{\overset{-}{\cancel{+}6x^2 \cancel{-} 3x}}}$
$\qquad\qquad\qquad\qquad 7x - 1$
$\qquad\qquad\qquad\quad \underline{\underset{+}{\overset{-}{\cancel{+}7x \cancel{-} 3.5}}}$
$\qquad\qquad\qquad\qquad\quad 2.5$ ← remainder

Estimate: $2x\,\overline{\smash{)}7x}$ means $2x - 1\,\overline{\smash{)}7x - 1}$ with 3.5 above, about 3.5

Multiply: $3.5(2x - 1) = 7x - 3.5$

Subtract: $(7x - 1) - (7x - 3.5) = 2.5$

6. *Write the remainder:* $2x - 1 \overline{\smash{)}6x^2 + 4x - 1}$ with $3x + 3.5$ above $+ \dfrac{2.5}{2x - 1}$ ← divisor

Write the remainder when there are no more terms to bring down.

7. *Check:*
$\qquad\qquad 3x + 3.5$ ← proposed quotient
$\qquad\qquad \underline{2x - 1}$ ← divisor
$\qquad\quad -3x - 3.5$
$\qquad \underline{6x^2 + 7x}$
$\qquad 6x^2 + 4x - 3.5$
$\quad +\qquad\quad \underline{2.5}$ ← proposed remainder
$\qquad 6x^2 + 4x - 1$ ← original dividend $\left[3x + 3.5 + \dfrac{2.5}{2x - 1}\ \text{checks}\right]$

Multiply the proposed quotient by the divisor and then add the proposed remainder to see if you get the original dividend.

Note: $2x - 1$ does not divide $6x^2 + 4x - 1$ evenly: $\dfrac{6x^2 + 4x - 1}{2x - 1} = 3x + 3.5 + \dfrac{2.5}{2x - 1}$

Caution: Always leave room for missing terms in the dividend.

EXAMPLE 8-37: Divide $x^3 + 1$ by $x + 1$ using the Long Division Rules for Polynomials.

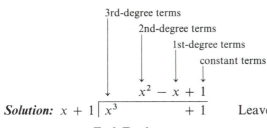

Solution:

$$x + 1 \overline{\smash{\big)}\ x^3 \qquad\quad + 1} \qquad\text{Leave room in the dividend for missing terms.}$$

quotient: $x^2 - x + 1$

3rd-degree terms / 2nd-degree terms / 1st-degree terms / constant terms

$\overline{\mp x^3 \mp x^2} \longleftarrow x^2(x+1)$

$-x^2$

$\overline{{}^+\!\!\!\diagup x^2 {}^+\!\!\!\diagup x} \longleftarrow -x(x+1)$

$x + 1$

$\overline{\mp x \mp 1} \longleftarrow 1(x+1)$

$0 \longleftarrow$ a zero remainder need not be written in the division answer

Check:

$$x^2 - x + 1 \longleftarrow \text{proposed quotient}$$
$$\underline{x + 1} \longleftarrow \text{divisor}$$
$$x^2 - x + 1$$
$$\underline{x^3 - x^2 + x}$$
$$x^3 + 0\ + 0 + 1 = x^3 + 1 \longleftarrow \text{original dividend } [x^2 - x + 1 \text{ checks}]$$

Note: $x + 1$ divides $x^3 + 1$ evenly: $\dfrac{x^3 + 1}{x + 1} = x^2 - x + 1$

Caution: $(x^3 + 1) \div (x + 1) \neq x^2 + 1 \text{ or } x^2 - 1$

RAISE YOUR GRADES

Can you . . . ?

- ☑ identify polynomials
- ☑ classify polynomials with respect to variables, terms, and degree
- ☑ write a polynomial in one variable in descending or ascending powers
- ☑ add or subtract polynomials in horizontal form
- ☑ add or subtract polynomials in vertical form
- ☑ multiply monomials using the Multiplication Rule for Monomials
- ☑ multiply a polynomial by a monomial using the distributive properties
- ☑ multiply two binomials using the distributive properties
- ☑ multiply two binomials using the FOIL method
- ☑ multiply with trinomials in vertical form
- ☑ multiply to get the difference of two squares
- ☑ multiply to get a perfect square trinomial
- ☑ divide monomials using the Division Rule for Monomials
- ☑ divide a polynomial by a monomial by reversing the steps used to add and subtract like fractions
- ☑ divide a polynomial by a binomial using the Long Division Rules for Polynomials

SUMMARY

1. A real number, variable, or the sum, difference, or product of real numbers and variables is called a polynomial.
2. Polynomials are algebraic expressions that do not contain variables in denominators or under radical symbols.
3. To classify a polynomial with respect to variables, you count the number of different variables that appear in the polynomial.
4. A polynomial that contains n different variables is called a polynomial in n variables.
5. To classify a polynomial with respect to terms, you count the number of terms that the polynomial contains.
6. A polynomial with one term is called a monomial.
7. A polynomial with two terms is called a binomial.
8. A polynomial with three terms is called a trinomial.
9. The degree of a term in one variable is the exponent of the variable.
10. The degree of a term in more than one variable is the sum of the exponents of the variables.
11. The degree of a polynomial is the highest degree of any of its terms.
12. To classify a polynomial with respect to degree, you identify the degree of the polynomial.
13. A polynomial in one variable is in descending [ascending] powers if the degree of each term decreases [increases] from left to right.
14. To add or subtract polynomials in horizontal form, you
    (a) Write the polynomials in horizontal form using parentheses when necessary.
    (b) Clear parentheses.
    (c) Collect like terms.
    (d) Combine like terms.
15. To add or subtract polynomials in vertical form, you
    (a) Write the polynomials in vertical form by writing descending powers while aligning like terms in columns.
    (b) Change any subtractions to additions by writing the opposite of each term in the subtrahend.
    (c) Combine like terms.
16. Multiplication Rule for Monomials
    If ar^n and bs^m are monomials, then $(ar^n)(bs^m) = (ab)(r^n s^m)$.
17. To multiply monomials with like variables, you multiply the numerical coefficients and add the exponents of like variables.
18. To multiply a polynomial by a monomial, you use the distributive properties to multiply each term of the polynomial by the monomial.
19. To multiply two binomials, you can use the distributive properties to multiply each term of one binomial by each term of the other binomial.
20. FOIL method for Multiplying Two Binomials

    (a) Multiply the First terms: $(a + b)(c + d) = \overset{\mathbf{F}}{\mathbf{ac}} + ? + ? + ?$

    (b) Multiply the Outside terms: $(a + b)(c + d) = ac + \overset{\mathbf{O}}{\mathbf{ad}} + ? + ?$

    (c) Multiply the Inside terms: $(a + b)(c + d) = ac + ad + \overset{\mathbf{I}}{\mathbf{bc}} + ?$

    (d) Multiply the Last terms: $(a + b)(c + d) = ac + ad + bc + \overset{\mathbf{L}}{\mathbf{bd}}$

    (e) Combine like terms when possible.
21. To multiply with trinomials, it is usually easier to multiply in vertical form.
22. To multiply two polynomials, you always multiply each term of one polynomial by each term of the other polynomial.
23. The square of a monomial is never negative and has only even integers for exponents.
24. Terms that are opposites always have the same square.

25. Any polynomial that can be written in the polynomial form $a^2 - b^2$ is called the difference of two squares.

26. Product Rule for the Difference of Two Squares for any binomials $a + b$ and $a - b$:

$$(a + b)(a - b) = (a - b)(a + b) = a^2 - b^2 \longleftarrow \text{ difference of two squares}$$

27. Any trinomial that can be written in the polynomial form $a^2 + 2ab + b^2$ or $a^2 - 2ab + b^2$ is called a perfect square trinomial.

28. Product Rule for Perfect Square Trinomials for any binomials $a + b$ or $a - b$:
    (a) $(a + b)^2 = (a + b)(a + b) = a^2 + 2ab + b^2 \longleftarrow$ ┐ perfect square trinomials
    (b) $(a - b)^2 = (a - b)(a - b) = a^2 - 2ab + b^2 \longleftarrow$ ┘

29. *Caution:* (a) $(r + s)^2 \neq r^2 + s^2$
    (b) $(r - s)^2 \neq r^2 - s^2$

30. Division Rule for Monomials

    If ar^n and bs^m are monomials $[b \neq 0$ and $s \neq 0]$, then $\dfrac{ar^n}{bs^m} = \dfrac{a}{b} \cdot \dfrac{r^n}{s^m}$.

31. To divide monomials with like variables, you divide the numerical coefficients and subtract the exponents of the like variables.

32. To divide a polynomial by a monomial, you reverse the steps used to add and subtract like fractions by dividing each term of the polynomial by the monomial.

33. Long Division Method for Polynomials
    (a) *Estimate* a term in the quotient by dividing the first term in the dividend by the first term in the divisor. Write this result in the quotient above the like term in the dividend.
    (b) *Multiply* the result from Step (a) by the divisor and write the product below the like terms in the dividend.
    (c) *Subtract* like terms.
    (d) *Bring down* the next term from the dividend to form a new dividend.
    (e) *Repeat Steps* (a) *through* (d) for the new dividend from Step (d).
    (f) *Write the remainder* when there are no more terms in the original dividend to bring down.
    (g) *Check* the proposed quotient and remainder by multiplying the proposed quotient by the divisor and then adding the proposed remainder to see if you get the original dividend.

SOLVED PROBLEMS

PROBLEM 8-1   Identify which algebraic expressions are polynomials: **(a)** $-\dfrac{5}{7}$ **(b)** w

(c) $3a$ **(d)** b^2 **(e)** $\dfrac{c}{2}$ **(f)** $\dfrac{4}{d}$ **(g)** $\dfrac{x}{y}$ **(h)** $m\sqrt{8}$ **(i)** $9\sqrt{n}$ **(j)** $z^4 + 2z^2 + 1$

(k) $h^2 - k^2$ **(l)** $6rst$ **(m)** $\dfrac{4x^2 - 8x + 5}{2x + 3}$ **(n)** $3w^2 + 2w - 5 + \dfrac{3}{w - 2}$

(o) $18y^2 \div 3y$ **(p)** $-24z^3 \div (6z^5)$

Solution: Recall that polynomials are algebraic expressions that do not have variables in denominators or under radical symbols [see Example 8-1]:

(a) $-\dfrac{5}{7}$ is a polynomial because it is a real number.

(b) w is a polynomial because it is a variable.

(c) $3a$ is a polynomial because it is the product of a real number and a variable.

(d) b^2 is a polynomial because it is the product of variables: $b^2 = bb$

(e) $\dfrac{c}{2}$ is a polynomial because it is the product of a real number and a variable: $\dfrac{c}{2} = \dfrac{1}{2} \cdot c$

(f) $\dfrac{4}{d}$ is not a polynomial because it has a variable in the denominator.

(g) $\dfrac{x}{y}$ is not a polynomial because it has a variable in the denominator.

(h) $m\sqrt{8}$ is a polynomial because it is the product of a variable and a real number.

(i) $9\sqrt{n}$ is not a polynomial because it has a variable under the radical symbol.

(j) $z^4 + 2z^2 + 1$ is a polynomial because it is a combined sum and product of real numbers and variables: $z^4 + 2z^2 + 1 = zzzz + 2zz + 1$

(k) $h^2 - k^2$ is a polynomial because it is a combined difference and product of variables: $h^2 - k^2 = hh - kk$

(l) $6rst$ is a polynomial because it is the product of a real number and variables.

(m) $\dfrac{4x^2 - 8x + 5}{2x + 3}$ is not a polynomial because it has a variable in the denominator.

(n) $3w^2 + 2w - 5 + \dfrac{3}{w - 2}$ is not a polynomial because it has a variable in the denominator.

(o) $18y^2 \div 3y$ is a polynomial because: $18y^2 \div 3y = \dfrac{18y^2}{3} \cdot y$

(p) $-24z^3 \div (6z^5)$ is not a polynomial because: $-24z^3 \div (6z^5) = \dfrac{-24z^3}{6z^5} = -\dfrac{4}{z^2}$

PROBLEM 8-2   Classify each polynomial with respect to variables: **(a)** $\sqrt{2}$ **(b)** $6u$ **(c)** $v^2 - v$ **(d)** $4abc + ac$ **(e)** $m^2 - n^2$ **(f)** $1 - 5x^2 - 3x^4$

Solution: Recall that to classify a polynomial with respect to variables, you count the number of different variables that appear in the polynomial [see Example 8-2]:

(a) $\sqrt{2}$ is a constant polynomial because no variable is shown.

(b) $6u$ is a polynomial in one variable, u.

(c) $v^2 - v$ is a polynomial in one variable, v.

(d) $4abc + ac$ is a polynomial in three variables: a, b, and c.

(e) $m^2 - n^2$ is a polynomial in two variables: m and n.

(f) $1 - 5x^2 - 3x^4$ is a polynomial in one variable, x.

PROBLEM 8-3   Classify each polynomial with respect to terms: **(a)** $\sqrt{2}$ **(b)** $6u$ **(c)** $v^2 - v$ **(d)** $4abc + ac$ **(e)** $m^2 - n^2$ **(f)** $1 - 5x^2 - 3x^4$

Solution: Recall that to classify a polynomial with respect to terms, you write a monomial, a binomial, or a trinomial for polynomials with exactly one, two, and three terms, respectively [see Example 8-4]:

(a) $\sqrt{2}$ is a monomial because it contains exactly one term: $\sqrt{2}$.

(b) $6u$ is a monomial because it contains exactly one term: $6u$.

(c) $v^2 - v$ is a binomial because it contains exactly two terms: v^2 and $-v$.

(d) $4abc + ac$ is a binomial because it contains exactly two terms: $4abc$ and ac.

(e) $m^2 - n^2$ is a binomial because it contains exactly two terms: m^2 and $-n^2$.

(f) $1 - 5x^2 - 3x^4$ is a trinomial because it contains exactly three terms: 1, $-5x^2$, and $-3x^4$.

PROBLEM 8-4   Classify each polynomial with respect to degree: **(a)** $\sqrt{2}$ **(b)** $6u$ **(c)** $v^2 - v$ **(d)** $4abc + ac$ **(e)** $m^2 - n^2$ **(f)** $1 - 5x^2 - 3x^4$

Solution: Recall that to classify a polynomial with respect to degree, you identify the highest degree of any of its terms [see Example 8-7]:

(a) The degree of $\sqrt{2}$ is 0 [constant term] because: $\sqrt{2} = (\sqrt{2})y^0$ [any variable can be used].

(b) The degree of $6u = 6u^1$ is 1.

(c) The degree of $v^2 - v = v^2 - v^1$ is 2 because $2 > 1$.

(d) The degree of $4abc + ac$ is 3 because the degree of $4abc$ is $3[1 + 1 + 1]$ and the degree of ac is $2[1 + 1]$ and $3 > 2$.

(e) The degree of $m^2 - n^2$ is 2 because the degree of both m^2 and $-n^2$ is 2.

(f) The degree of $1 - 5x^2 - 3x^4 = 1x^0 - 5x^2 - 3x^4$ is 4 because $4 > 0$ and $4 > 2$.

PROBLEM 8-5 Write each polynomial in one variable in descending powers:

(a) $2w^4 - 3$ **(b)** $4 - y$ **(c)** $x + 2 - x^2$ **(d)** $m^3 - 5m^4 + 1 - 3m^2$

Solution: Recall that to write a polynomial in one variable in descending powers, you arrange the polynomial so that the degree of each term decreases from left to right [see Example 8-8]:

(a) $2w^4 - 3$ is in descending powers because $2w^4 - 3 = 2w^4 - 3w^0$ and $4 > 0$.

(b) $4 - y = -y + 4$ is in descending powers because $-y + 4 = -y^1 + 4y^0$ and $1 > 0$.

(c) $x + 2 - x^2 = -x^2 + x + 2$ is in descending powers because $-x^2 + x + 2 = -x^2 + x^1 + 2x^0$.

(d) $m^3 - 5m^4 + 1 - 3m^2 = -5m^4 + m^3 - 3m^2 + 1$ is in descending powers.

PROBLEM 8-6 Add and subtract polynomials in horizontal form:

(a) $(3a^2 - 5a + 6) + (-7a^2 + 8a - 9)$ **(b)** Add $x^2 - 2xy + y^2$ and $-x^2 + y^2 - 5xy$

(c) $(3a^2 - 5a + 6) - (-7a^2 + 8a - 9)$ **(d)** Subtract $x^2 - 2xy + y^2$ from $-x^2 + y^2 - 5xy$.

Solution: Recall that to add or subtract polynomials in horizontal form, you

1. Write the polynomials in horizontal form using parentheses when necessary.
2. Clear parentheses.
3. Collect like terms.
4. Combine like terms.
   [See Example 8-10.]:

(a) $(3a^2 - 5a + 6) + (-7a^2 + 8a - 9) = 3a^2 - 5a + 6 - 7a^2 + 8a - 9$

$$= (3a^2 - 7a^2) + (-5a + 8a) + (6 - 9)$$

$$= -4a^2 + 3a - 3$$

(b) "Add $x^2 \quad 2xy + y^2$ and $-x^2 + y^2 - 5xy$" $= (x^2 - 2xy + y^2) + (-x^2 + y^2 - 5xy)$

$$= (x^2 - x^2) + (-2xy - 5xy) + (y^2 + y^2)$$

$$= -7xy + 2y^2$$

(c) $(3a^2 - 5a + 6) - (-7a^2 + 8a - 9) = 3a^2 - 5a + 6 + 7a^2 - 8a + 9$

$$= (3a^2 + 7a^2) + (-5a - 8a) + (6 + 9)$$

$$= 10a^2 - 13a + 15$$

(d) "Subtract $x^2 - 2xy + y^2$ from $-x^2 + y^2 - 5xy$" $= (-x^2 + y^2 - 5xy) - (x^2 - 2xy + y^2)$

$$= -x^2 + y^2 - 5xy - x^2 + 2xy - y^2$$

$$= (-x^2 - x^2) + (-5xy + 2xy) + (y^2 - y^2)$$

$$= -2x^2 - 3xy$$

PROBLEM 8-7 Add and subtract polynomials in vertical form:

(a) Add $3x^3 - 5x^2 + 7$, $-4 + 8x^2 - 2x$.

(b) Subtract $6y^3 - 1 - 9y^2 - 8y^4 - 2y$ from $3y^4 + 1 - 9y^2 + 4y^5 + 5y$.

Solution: Recall that to add or subtract polynomials in vertical form, you

1. Write the polynomials in vertical form by writing descending powers while aligning like terms in columns.
2. Change any subtractions to additions by writing the opposite of each term in the subtrahend [the bottom polynomial].

3. Combine like terms.
[See Example 8-11.]:

(a)
$$3x^3 - 5x^2 \qquad + 7$$
$$\underline{ 8x^2 - 2x - 4}$$
$$3x^3 + 3x^2 - 2x + 3$$

(b)
$$4y^5 + 3y^4 \qquad - 9y^2 + 5y + 1$$
$$+ \qquad - \qquad + \qquad + \qquad +$$
$$\underline{ \not{+} \; 8y^4 \not{+} 6y^3 \not{+} 9y^2 \not{+} 2y \not{+} 1}$$
$$4y^5 + 11y^4 - 6y^3 + 0 + 7y + 2 = 4y^5 + 11y^4 - 6y^3 + 7y + 2$$

PROBLEM 8-8   Multiply monomials: **(a)** $(2x^3)(3x^2)$ **(b)** $(4y)(-2)$ **(c)** $w(-3w^2)$

(d) $\dfrac{m}{2}(5m)$ **(e)** $(4n^3)(-3n^2)(-2n)$ **(f)** $(4ab)(5a^2b)(-2a)$ **(g)** $(-5u)(-3v)(-2w)$

Solution: Recall that to multiply monomials, you multiply the numerical coefficients and literal parts separately [see Examples 8-12 and 8-13]:

(a) $(2x^3)(3x^2) = (2 \cdot 3)x^3x^2$ **(b)** $(4y)(-2) = [4(-2)]y$ **(c)** $w(-3w^2) = -3ww^2$
$$= 6x^5 \qquad\qquad\qquad = -8y \qquad\qquad\qquad = -3w^3$$

(d) $\dfrac{m}{2}(5m) = \left(\dfrac{1}{2} \cdot 5\right)mm$ **(e)** $(4n^3)(-3n^2)(-2n) = [4(-3)(-2)]n^3n^2n$
$$= 24n^6$$
$$= \dfrac{5}{2}m^2 \text{ or } \dfrac{5m^2}{2}$$

(f) $(4ab)(5a^2b)(-2a) = [4(5)(-2)]aa^2abb$ **(g)** $(-5u)(-3v)(-2w) = [-5(-3)(-2)]uvw$
$$= -40a^4b^2 \qquad\qquad\qquad\qquad = -30uvw$$

PROBLEM 8-9   Multiply polynomials by monomials: **(a)** $5x(4x - 3)$ **(b)** $-2y^3(3y^2 - 2y + 5)$
(c) $a(a - b)$ **(d)** $-mn(m^2 - 2mn - n^2)$

Solution: Recall that to multiply a polynomial by a monomial, you use the distributive properties to multiply each term of the polynomial by the monomial [see Example 8-14]:

(a) $\mathbf{5x(4x - 3)} = \mathbf{5x}(4x) - \mathbf{5x}(3)$
$$= 20x^2 - 15x$$

(b) $\mathbf{-2y^3(3y^2 - 2y + 5)} = (\mathbf{-2y^3})(3y^2) - (\mathbf{-2y^3})(2y) + (\mathbf{-2y^3})(5)$
$$= -6y^5 + 4y^4 - 10y^3$$

(c) $\mathbf{a(a - b)} = \mathbf{aa} - \mathbf{ab}$
$$= a^2 - ab$$

(d) $\mathbf{-mn(m^2 - 2mn - n^2)} = (\mathbf{-mn})m^2 - (\mathbf{-mn})2mn - (\mathbf{-mn})n^2$
$$= -m^3n + 2m^2n^2 + mn^3$$

PROBLEM 8-10   Multiply binomials using the distributive properties: **(a)** $(x + 3)(x + 4)$
(b) $(3y + 2)(5y - 4)$ **(c)** $(3m - n)(2m + n)$ **(d)** $(w - 2)(w^2 - 3)$

Solution: Recall that to multiply two binomials, you can use the distributive properties to multiply each term of one binomial by each term of the other binomial [see Examples 8-15 and 8-16]:

(a) $\mathbf{(x + 3)(x + 4)} = \mathbf{(x + 3)}x + \mathbf{(x + 3)}4$
$$= (x)\mathbf{x} + (3)\mathbf{x} + (x)\mathbf{4} + (3)\mathbf{4}$$
$$= x^2 + 3x + 4x + 12$$
$$= x^2 + 7x + 12$$

(b) $(3y + 2)(5y - 4) = (3y + 2)5y - (3y + 2)4$

$$= (3y)5y + (2)5y - [(3y)4 + (2)4]$$

$$= 15y^2 + 10y - 12y - 8$$

$$= 15y^2 - 2y - 8$$

(c) $(3m - n)(2m + n) = (3m - n)2m + (3m - n)n$

$$= (3m)2m - (n)2m + (3m)n - (n)n$$

$$= 6m^2 - 2mn + 3mn - n^2$$

$$= 6m^2 + mn - n^2$$

(d) $(w - 2)(w^2 - 3) = (w - 2)w^2 - (w - 2)3$

$$= (w)w^2 - (2)w^2 - [(w)3 - (2)3]$$

$$= w^3 - 2w^2 - 3w - (-6)$$

$$= w^3 - 2w^2 - 3w + 6$$

PROBLEM 8-11   Multiply binomials using the FOIL method: **(a)** $(x + 4)(x + 5)$
(b) $(2y - 3)(3y + 4)$ **(c)** $(2u + v)(u - 2v)$ **(d)** $(2w^2 - 5)(3w - 4)$

Solution: Recall that to multiply two binomials using the FOIL method, you

1. Multiply the **F**irst terms: $(a + b)(c + d) = ac + ? + ? + ?$
2. Multiply the **O**utside terms: $(a + b)(c + d) = ac + ad + ? + ?$
3. Multiply the **I**nside terms: $(a + b)(c + d) = ac + ad + bc + ?$
4. Multiply the **L**ast terms: $(a + b)(c + d) = ac + ad + bc + bd$
5. Combine like terms when possible
   [See Examples 8-17, 8-18, and 8-19.]:

$$\overset{\text{F}}{\quad} \overset{\text{O} + \text{I}}{\quad} \overset{\text{L}}{\quad}$$
(a) $(x + 4)(x + 5) = x^2 + 9x + 20$ $\overset{\text{O}}{\ } \ \overset{\text{I}}{\ }$ *Think:* $5x + 4x = 9x$

$$\overset{\text{F}}{\quad} \overset{\text{O} + \text{I}}{\quad} \overset{\text{L}}{\quad}$$
(b) $(2y - 3)(3y + 4) = 6y^2 - y - 12$ $\overset{\text{O}}{\ } \ \overset{\text{I}}{\ }$ *Think:* $8y - 9y = -y$

$$\overset{\text{F}}{\quad} \overset{\text{O} + \text{I}}{\quad} \overset{\text{L}}{\quad}$$
(c) $(2u + v)(u - 2v) = 2u^2 - 3uv - 2v^2$ $\overset{\text{O}}{\ } \ \overset{\text{I}}{\ }$ *Think:* $-4uv + uv = -3uv$

$$\overset{\text{F}}{\quad} \overset{\text{O}}{\quad} \overset{\text{I}}{\quad} \overset{\text{L}}{\quad}$$
(d) $(2w^2 - 5)(3w - 4) = 6w^3 - 8w^2 - 15w + 20$ *Think:* There are no like terms to combine.

PROBLEM 8-12   Multiply with trinomials in vertical form: **(a)** $(4x - 5)(3x^2 - x + 2)$
(b) $(y^3 - 4y - 1)(2 - 3y)$ **(c)** $(m^2 + m + 1)(m^2 - m - 1)$ **(d)** $(w - 2 + w^2)(3w^2 + 1 - 2w)$

Solution: Recall that to multiply with trinomials, it is usually easier to multiply in vertical form [see Example 8-20]:

(a)
$$
\begin{array}{r}
3x^2 - \ x + 2 \\
4x - 5 \\
\hline
-15x^2 + \ 5x - 10 \quad \longleftarrow \ -5(3x^2 - x + 2) \\
12x^3 - \ 4x^2 + \ 8x \quad \longleftarrow \quad 4x(3x^2 - x + 2) \\
\hline
12x^3 - 19x^2 + 13x - 10
\end{array}
$$

(b)
$$
\begin{array}{r}
y^3 - 4y - 1 \\
-3y + 2 \quad \longleftarrow \ \text{descending powers} \\
\hline
2y^3 \qquad\qquad -8y - 2 \quad \longleftarrow \ 2(y^3 - 4y - 1) \\
-3y^4 \qquad + 12y^2 + 3y \quad \longleftarrow \qquad -3y(y^3 - 4y - 1) \\
\hline
-3y^4 + 2y^3 + 12y^2 - 5y - 2
\end{array}
$$

(e) $(2m^2n^2 - mn) \div (2mn) = \dfrac{2m^2n^2 - mn}{2mn}$

$$= \dfrac{2m^2n^2}{2mn} - \dfrac{mn}{2mn}$$

$$= mn - \dfrac{1}{2}$$

(f) $(r + s) \div r = \dfrac{r + s}{r}$

$$= \dfrac{r}{r} + \dfrac{s}{r}$$

$$= 1 + \dfrac{s}{r}$$

PROBLEM 8-17   Divide polynomials by binomials:

(a) $\dfrac{x^2 - 1}{x + 1}$

(b) Divide $y^2 - y - 12$ by $y + 3$.

(c) $(w^2 + 2 - w) \div (w + 1)$

(d) $(m^2 - n^2) \div (m - n)$

(e) $\dfrac{a^2 + ab + b^2}{a - b}$

(f) Divide $u^3 - 1$ by $u - 1$.

Solution:

(a) $\dfrac{x^2 - 1}{x + 1} = x + 1 \overline{\smash{\big)}\ \begin{array}{r} x - 1 \\ x^2 \qquad - 1 \end{array}}$

$$\underline{\mp x^2 \mp x}$$
$$-x - 1$$
$$\underline{+ \quad +}$$
$$\nleftarrow x \nleftarrow 1$$
$$0$$

(b) "Divide $y^2 - y - 12$ by $y + 3$" $= y + 3 \overline{\smash{\big)}\ \begin{array}{r} y - 4 \\ y^2 - y - 12 \end{array}}$

$$\underline{\mp y^2 \mp 3y}$$
$$-4y - 12$$
$$\underline{+ \quad +}$$
$$\nleftarrow 4y \nleftarrow 12$$
$$0$$

(c) $(w^2 + 2 - w) \div (w + 1) = w + 1 \overline{\smash{\big)}\ \begin{array}{r} w - 2 + \dfrac{4}{w + 1} \longleftarrow \text{remainder} \\[2pt] \longleftarrow \text{divisor} \\[4pt] w^2 - w + 2 \longleftarrow \text{descending powers} \end{array}}$

$$\underline{\mp w^2 \mp w}$$
$$-2w + 2$$
$$\underline{+ \quad +}$$
$$\nleftarrow 2w \nleftarrow 2$$
$$4$$

(d) $(m^2 - n^2) \div (m - n) = m - n \overline{\smash{\big)}\ \begin{array}{r} m + n \\ m^2 \qquad - n^2 \end{array}}$

$$\underline{\mp m^2 \nleftarrow mn}$$
$$mn - n^2$$
$$\underline{\nleftarrow mn \nleftarrow n^2}$$
$$0$$

(e) $\dfrac{a^2 + ab + b^2}{a - b} = a - b \overline{\smash{\big)}\ \begin{array}{r} a + 2b + \dfrac{3b^2}{a - b} \\[2pt] a^2 + ab + b^2 \end{array}}$

$$\underline{\mp a^2 \nleftarrow ab}$$
$$2ab + b^2$$
$$\underline{\mp 2ab \nleftarrow 2b^2}$$
$$3b^2$$

(f) "Divide $u^3 - 1$ by $u - 1$" $= u - 1 \;\overline{\left)\; u^3 \qquad\quad -1\right.}$ ← leave space in the dividend
for missing terms

$$u^2 + u + 1$$

$$\mp u^3 \mp u^2$$
$$\overline{u^2}$$
$$\mp u^2 \mp u$$
$$\overline{u - 1}$$
$$\mp u \mp 1$$
$$\overline{0}$$

Supplementary Exercises

PROBLEM 8-18 Classify each polynomial with respect to (1) variables, (2) terms, and (3) degree. Then write each polynomial in descending powers:

(a) 4   **(b)** $-5x$ **(c)** $3y^2$ **(d)** $-4w^3$ **(e)** $z + 2$ **(f)** $5 + 3m^2$ **(g)** $n^3 - 1$
(h) $-6 - 5a^4$ **(i)** $b^2 + b + 1$ **(j)** $u - u^3 + 5$ **(k)** $4 + v - 3v^4$ **(l)** $5c^4 - 4c^5 + 8c^3$

(m) 0   **(n)** $1 - r^2$ **(o)** $1 - 2s + s^2$ **(p)** 10^3 **(q)** $\dfrac{d}{3} + d^3\sqrt{2}$ **(r)** $x^3 + x^2 + x + 1$

(s) $2y - 3y^4 + 5 - y^2 - 4y^3$ **(t)** ab **(u)** $-3uvw$ **(v)** $x^2y - 5xy$

(w) $3mn^2 - 2m^3 + 8m^2n^2$ **(x)** $x^3 - x^2y + x^2y^2 - xy^2$ **(y)** $\dfrac{1}{x^2 - y^2}$ **(z)** $\sqrt{y^2 - 1}$

PROBLEM 8-19 Add and subtract polynomials: **(a)** $(6x^2 + 1) + (-8x^2 - 5x + 4)$
(b) $(-3y^2 + 2y - 8) + (7y^2 + 9y + 10)$ **(c)** $(z + 5z^3 - z^2) + (8 - z^2 + 10z^3)$
(d) $(-4w^2 + 3w^3 + 8 - w) + (8w^2 + 4w)$ **(e)** $(2mn^2 - 3m^2n + 5mn) + (4m^2n - 5mn)$
(f) $(5a^4 + 6 - 3a^3 + a) + (-8a^3 + 4a^2 - 5a + 9) + (-8a^4 + 2a^2 + 9)$
(g) $(-6b + 6b^5 - 2 + b^2) + (5b^4 - 8 + b - 10b^3) + (4b^4 - 8b^5 + 9b^2 - 5b^3)$
(h) $(u^2 - 2u^2v + v^2) + (-u^2 + v^2 - 3uv^2) + (8u^2v^2 + 4u^2v - 2uv^2)$
(i) Add $-3x + 7$ and $-9x + 8$.   **(j)** Add $4 - 6y$ and $-5y - 8$.
(k) Add $3w^2 - 2w + 1$ and $w^2 + 2w - 1$.   **(l)** Add $x^2 - y^2$ and $x^2 - 2xy + y^2$.
(m) Add $m^3 - n^3$, $m^3 - 3m^2n - 3mn^2$, $6mn^2 - n^3$, and $4m^2n - 2m^3$.
(n) $(6x^2 + 1) - (-8x^2 - 5x + 4)$ **(o)** $(-3y^3 + 2y - 8) - (7y^2 + 9y + 10) + (2y^3 + 5y^2)$
(p) $(z + 5z^3 - z^2) - (8 - z^2 + 10z^3)$
(q) $(-4w^2 + 3w^3 + 8 - w) + (8w^2 + 4w) - (5w^3 - 6 - w)$
(r) $(2mn^2 - 3m^2n + 5mn) - (4m^2n - 5mn)$ **(s)** $(-8a^3 + 4a^2 - 5a + 9) - (-8a^4 + 2a^2 + 9)$
(t) $(u^2 - 2u^2v + v^2) - (-u^2 + v^2 - 3uv^2)$ **(u)** Subtract $-3x + 7$ from $-9x + 8$.
(v) Subtract $4 - 6y$ from $-5y - 8$.   **(w)** Subtract $3w^2 - 2w + 1$ from $w^2 + 2w - 1$.
(x) Subtract $x^2 - y^2$ from $x^2 - 2xy + y^2$.   **(y)** Subtract $m^3 - 3m^2n - 3mn^2$ from $6mn^2 - n^3$.
(z) Subtract $abc - a^2b + b^2c - bc + ab$ from $a^2b - cb + bc^2 - abc$.

PROBLEM 8-20 Multiply polynomials: **(a)** $(5x^2)(3x^4)$ **(b)** $(-4y)(5y^3)(-3y^4)$
(c) $2w(3w + 5)$ **(d)** $-3m(2m^3 + 3m^2 - 8m + 6)$ **(e)** $(z + 3)(z + 5)$ **(f)** $(2a - 4)(a + 5)$
(g) $(4b + 3)(b - 6)$ **(h)** $(2c - 5)(3c - 4)$ **(i)** $(u + 2v)(3u + v)$ **(j)** $(m^2 - 1)(m + 1)$
(k) $(x + y)(x - y)$ **(l)** $(5s + 3r)(3r - 5s)$ **(m)** $(3 - 4t)(3 + 4t)$ **(n)** $(2w^3 - 6)(6 + 2w^3)$
(o) $(5 - 3y)(3y - 5)$ **(p)** $(-3m^2 + 2n)(5m - 4n)$ **(q)** $(z^2 - 2)(z^2 + 2z)$
(r) $(w^2 + w)(w^2 + 1)$ **(s)** $(r + \frac{1}{4})^2$ **(t)** $(2s + 3)^2$ **(u)** $(h - \frac{3}{2})^2$ **(v)** $(2k - 5)^2$
(w) $(x + 5)(x^2 + x + 3)$ **(x)** $(5 - y^2)(5y^2 + 2y + 4)$ **(y)** $(2a^2 - 3a + 5)(-5a^2 + a - 6)$
(z) $(4b - b^3 + 5b^2)(2b^3 - 4 + 5b + b^4)$

PROBLEM 8-21 Divide polynomials: **(a)** $\dfrac{-12x^2}{3}$ **(b)** $\dfrac{-10}{-5y^4}$ **(c)** Divide $8w^3$ by $-w^3$.

(d) Divide $-20z^4$ by $-5z^3$.   **(e)** $50a^6b^2 \div (-25a^5b^3)$ **(f)** $15m^3n \div 5m^5n^4$

(g) $\dfrac{-c^2 - 2c + 3}{-1}$ **(h)** $\dfrac{8d^3 - 4d^2 - 2d}{d}$ **(i)** Divide $24h^3 - 12h^2 + 15h$ by $-3h^2$.

(j) Divide $32k^5 - 16k^4 + 24k^3$ by $8k^2$.   **(k)** $(8a^2b - 4ab^2) \div (-4ab)$

(l) $(-6u^2v^3 + 18u^3v^2 - 12u^4v) \div (-6u^2v^5)$ **(m)** $(15xy - 10xz) \div 5x$ **(n)** $\dfrac{m^2 + m}{m + 1}$

(o) $\dfrac{x^3 - x^2}{x - 1}$ **(p)** $\dfrac{y^2 - 7y + 8}{y + 1}$ **(q)** Divide $w^2 + 9w + 19$ by $w + 5$.

(r) Divide $12z^2 + 7z - 18$ by $2z + 3$.   **(s)** $(35 + 3a - a^2) \div (5 - a)$

(t) $(m^3 + n^3) \div (m + n)$ **(u)** $(6b^5 - 4b^3) \div (2b^4)$ **(v)** $(-8c^4 + 12c^3 - 18c^2) \div (-6c^3)$

(w) $(r^2 + 11r + 28) \div (7 + r)$ **(x)** $(8x^3 - y^3) \div (2x - y)$ **(y)** $(p^2 - 25) \div (5 + p)$

(z) $(w^2 - 5w + 7) \div w + 3$

Answers to Supplementary Exercises

(8-18)   **(a)**   constant, monomial, degree 0 or constant term, 4
     (b)   one variable, monomial, 1st-degree, $-5x$
     (c)   one variable, monomial, 2nd-degree, $3y^2$
     (d)   one variable, monomial, 3rd-degree, $-4w^3$
     (e)   one variable, binomial, 1st-degree, $z + 2$
     (f)   one variable, binomial, 2nd-degree, $3m^2 + 5$
     (g)   one variable, binomial, 3rd-degree, $n^3 - 1$
     (h)   one variable, binomial, 4th-degree, $-5a^4 - 6$
     (i)   one variable, trinomial, 2nd-degree, $b^2 + b + 1$
     (j)   one variable, trinomial, 3rd-degree, $-u^3 + u + 5$
     (k)   one variable, trinomial, 4th-degree, $-3v^4 + v + 4$
     (l)   one variable, trinomial, 5th-degree, $-4c^5 + 5c^4 + 8c^3$
     (m) constant, monomial, degree 0 or constant term, 0
     (n)   one variable, binomial, 2nd-degree, $-r^2 + 1$
     (o)   one variable, trinomial, 2nd-degree, $s^2 - 2s + 1$
     (p)   constant, monomial, degree 0 or constant term, 10^3
     (q)   one variable, binomial, 3rd-degree, $d^3\sqrt{2} + \dfrac{d}{3}$
     (r)   one variable, 4 terms, 3rd-degree, $x^3 + x^2 + x + 1$
     (s)   one variable, 5 terms, 4th-degree, $-3y^4 - 4y^3 - y^2 + 2y + 5$
     (t)   two variables, monomial, 2nd-degree, ab
     (u)   three variables, monomial, 3rd-degree, $-3uvw$
     (v)   two variables, binomial, 3rd-degree, $x^2y - 5xy$
     (w) two variables, trinomial, 4th-degree, $8m^2n^2 + 3mn^2 - 2m^3$ or $8m^2n^2 - 2m^3 + 3mn^2$
     (x)   two variables, 4 terms, 4th-degree, $x^2y^2 - x^2y - xy^2 + x^3$ (other descending power orderings are possible as long as the first term is x^2y^2).
     (y)   not a polynomial
     (z)   not a polynomial

(8-19)   **(a)** $-2x^2 - 5x + 5$ **(b)** $4y^2 + 11y + 2$ **(c)** $15z^3 - 2z^2 + z + 8$
     (d) $3w^3 + 4w^2 + 3w + 8$ **(e)** $m^2n + 2mn^2$ **(f)** $-3a^4 - 11a^3 + 6a^2 - 4a + 24$
     (g) $-2b^5 + 9b^4 - 15b^3 + 10b^2 - 5b - 10$ **(h)** $8u^2v^2 + 2u^2v - 5uv^2 + 2v^2$
     (i) $-12x + 15$ **(j)** $-11y - 4$ **(k)** $4w^2$ **(l)** $2x^2 - 2xy$ **(m)** $m^2n + 3mn^2 - 2n^3$

(n) $14x^2 + 5x - 3$ **(o)** $-y^3 - 2y^2 - 7y - 18$ **(p)** $-5z^3 + z - 8$
(q) $-2w^3 + 4w^2 + 4w + 14$ **(r)** $-7m^2n + 10mn + 2mn^2$
(s) $8a^4 - 8a^3 + 2a^2 - 5a$ **(t)** $2u^2 - 2u^2v + 3uv^2$ **(u)** $-6x + 1$ **(v)** $y - 12$
(w) $-2w^2 + 4w - 2$ **(x)** $-2xy + 2y^2$ **(y)** $-m^3 + 3m^2n + 9mn^2 - n^3$
(z) $2a^2b - 2abc - ab - b^2c + bc^2$

(8-20)   **(a)** $15x^6$ **(b)** $60y^8$ **(c)** $6w^2 + 10w$ **(d)** $-6m^4 - 9m^3 + 24m^2 - 18m$
(e) $z^2 + 8z + 15$ **(f)** $2a^2 + 6a - 20$ **(g)** $4b^2 - 21b - 18$ **(h)** $6c^2 - 23c + 20$
(i) $3u^2 + 7uv + 2v^2$ **(j)** $m^3 + m^2 - m - 1$ **(k)** $x^2 - y^2$ **(l)** $9r^2 - 25s^2$
(m) $9 - 16t^2$ **(n)** $4w^6 - 36$ **(o)** $-9y^2 + 30y - 25$
(p) $-15m^3 + 12m^2n + 10mn - 8n^2$ **(q)** $z^4 + 2z^3 - 2z^2 - 4z$
(r) $w^4 + w^3 + w^2 + w$ **(s)** $r^2 + \frac{1}{2}r + \frac{1}{16}$ **(t)** $4s^2 + 12s + 9$ **(u)** $h^2 - 3h + \frac{9}{4}$
(v) $4k^2 - 20k + 25$ **(w)** $x^3 + 6x^2 + 8x + 15$ **(x)** $-5y^4 - 2y^3 + 21y^2 + 10y + 20$
(y) $-10a^4 + 17a^3 - 40a^2 + 23a - 30$ **(z)** $-b^7 + 3b^6 + 14b^5 + 3b^4 + 29b^3 - 16b$

(8-21)   **(a)** $-4x^2$ **(b)** $\dfrac{2}{y^4}$ **(c)** -8 **(d)** $4z$ **(e)** $\dfrac{-2a}{b}$ **(f)** $3m^8n^5$ **(g)** $c^2 + 2c - 3$

(h) $8d^2 - 4d - 2$ **(i)** $-8h + 4 - \dfrac{5}{h}$ **(j)** $4k^3 - 2k^2 + 3k$ **(k)** $-2a + b$

(l) $\dfrac{1}{v^2} - \dfrac{3u}{v^3} + \dfrac{2u^2}{v^4}$ **(m)** $3x^2y - 2x^2z$ **(n)** m **(o)** x^2 **(p)** $y - 8 + \dfrac{16}{y+1}$

(q) $w + 4 - \dfrac{1}{w+5}$ **(r)** $6z - 5.5 - \dfrac{1.5}{2z+3}$ **(s)** $a + 2 + \dfrac{25}{5-a}$ **(t)** $m^2 - mn + n^2$

(u) $3b - \dfrac{2}{b}$ **(v)** $\dfrac{4c}{3} - 2 + \dfrac{3}{c}$ **(w)** $r + 4$ **(x)** $4x^2 + 2xy + y^2$ **(y)** $p - 5$

(z) $w - 2 + \dfrac{7}{w}$

9 FACTORING

THIS CHAPTER IS ABOUT

- ☑ **Factoring with Monomials**
- ☑ **Factoring by Grouping**
- ☑ **Factoring Trinomials with Form $x^2 + bx + c$**
- ☑ **Factoring Trinomials with Form $ax^2 + bx + c$**
- ☑ **Factoring Special Products**

9-1. Factoring with Monomials

A. Factor by reversing known multiplication steps.

To find the product of two or more polynomials, you multiply those polynomials [see Section 8-3].

EXAMPLE 9-1: Multiply: **(a)** $3(4)$ **(b)** $5x(3x)$ **(c)** $2(3y - 4)$ **(d)** $(w + 2)(w + 3)$

Solution:
(a) $\overset{\text{multiply}}{\overrightarrow{3(4) = 12}}$ **(b)** $\overset{\text{multiply}}{\overrightarrow{5x(3x) = 15x^2}}$ **(c)** $\overset{\text{multiply}}{\overrightarrow{2(3y - 4) = 6y - 8}}$

(d) $\overset{\text{multiply}}{\overrightarrow{(w + 2)(w + 3) = w^2 + 5w + 6}}$

To **factor,** you reverse the multiplication steps.

EXAMPLE 9-2: Factor by reversing the multiplication steps used in Example 9-1:
(a) 12 **(b)** $15x^2$ **(c)** $6y - 8$ **(d)** $w^2 + 5w + 6$

Solution: **(a)** $\overset{\text{factor}}{\overrightarrow{12 = 3(4)}}$ Reverse the multiplication steps in Example 9-1, part **(a)**.

(b) $\overset{\text{factor}}{\overrightarrow{15x^2 = 3x(5x)}}$ Reverse the multiplication steps in Example 9-1, part **(b)**.

(c) $\overset{\text{factor}}{\overrightarrow{6y - 8 = 2(3y - 4)}}$ Reverse the multiplication steps in Example 9-1, part **(c)**.

(d) $\overset{\text{factor}}{\overrightarrow{w^2 + 5w + 6 = (w + 2)(w + 3)}}$ Reverse the multiplication steps in Example 9-1, part **(d)**.

Note: There is a special relationship between multiplication and factoring.

EXAMPLE 9-3: Show the special relationship between multiplication and factoring using
(a) $3(4) = 12$;   **(b)** $3x(5x) = 15x^2$;   **(c)** $2(3y - 4) = 6y - 8$; and
(d) $(w + 2)(w + 3) = w^2 + 5w + 6$.

Solution: **(a)** factors $\longrightarrow$ $\overline{3(4) = 12}$ $\longleftarrow$ product   To factor, you reverse the multiplication steps.

multiply

factor

(b) factors $\longrightarrow$ $\overline{3x(5x) = 15x^2}$ $\longleftarrow$ product

multiply

factor

(c) factors $\longrightarrow$ $\overline{2(3y - 4) = 6y - 8}$ $\longleftarrow$ product

multiply

factor

(d) factors $\longrightarrow$ $\overline{(w + 2)(w + 3) = w^2 + 5w + 6}$ $\longleftarrow$ product

multiply

factor

Note: A **factored form** of 12, $15x^2$, $6y - 8$, and $w^2 + 5w + 6$ is $3(4)$, $3x(5x)$, $2(3y - 4)$, and $(w + 2)(w + 3)$, respectively.

B. Factor polynomials over the integers.

There is usually more than one different factored form for a given polynomial.

EXAMPLE 9-4: Write two different factored forms for:
(a) 12   **(b)** $15x^2$ **(c)** $6y - 8$ **(d)** $w^2 + 5w + 6$

Solution:
(a) $12 = 3(4)$ because $3(4) = 12$.
     $12 = \frac{15}{2}(1.6)$ because $\frac{15}{2}(1.6) = 12$.
(b) $15x^2 = 3x(5x)$ because $3x(5x) = 15x^2$.
     $15x^2 = 3(5)xx$ because $3(5)xx = 15x^2$.
(c) $6y - 8 = 2(3y - 4)$ because $2(3y - 4) = 2(3y) - 2(4) = 6y - 8$.
     $6y - 8 = 6(y - \frac{4}{3})$ because $6(y - \frac{4}{3}) = 6(y) - 6(\frac{4}{3}) = 6y - 8$.
(d) $w^2 + 5w + 6 = (w + 2)(w + 3)$ because
     $(w + 2)(w + 3) = w^2 + 3w + 2w + 6 = w^2 + 5w + 6$.

$$w^2 + 5w + 6 = w\left(w + 5 + \frac{6}{w}\right) \text{ because}$$

$$w\left(w + 5 + \frac{6}{w}\right) = w(w) + w(5) + w\left(\frac{6}{w}\right) = w^2 + 5w + 6.$$

Note: Each factor in $3(4)$, $3x(5x)$, $2(3y - 4)$, and $(w + 2)(w + 3)$ has integers for all numerical coefficients and constants.

A polynomial that has integers for all numerical coefficients and constants is called an **integral polynomial.**

EXAMPLE 9-5: Identify which algebraic expressions are integral polynomials:
(a) 2   **(b)** $\frac{15}{2}$ **(c)** 1.6   **(d)** $\sqrt{2}$ **(e)** $3y - 4$ **(f)** $y - \frac{4}{3}$ **(g)** $w^2 + 5w + 6$

(h) $w + 5 + \dfrac{6}{w}$

Solution:
(a) 2 is an integral polynomial because 2 is an integer.
(b) $\frac{15}{2}$ is not an integral polynomial because $\frac{15}{2}$ is not an integer.
(c) 1.6 is not an integral polynomial because 1.6 is not an integer.
(d) $\sqrt{2}$ is not an integral polynomial because $\sqrt{2}$ is not an integer.
(e) $3y - 4$ is an integral polynomial because both 3 and -4 are integers.
(f) $y - \frac{4}{3}$ is not an integral polynomial because $-\frac{4}{3}$ is not an integer.
(g) $w^2 + 5w + 6$ is an integral polynomial because 1, 5, and 6 are all integers.
(h) $w + 5 + \dfrac{6}{w}$ is not an integral polynomial because it is not a polynomial.

Note: If a polynomial contains a fraction, decimal, or radical, then the polynomial is not an integral polynomial.

To **factor a polynomial over the integers,** you write the polynomial in factored form as the product of two or more integral polynomials.

Agreement: For the remainder of this text, the word "factor" will mean "factor over the integers" unless otherwise stated. That is, for the remainder of this text, a polynomial is in "factored form" if it is written as the product of two or more *integral polynomials.*

EXAMPLE 9-6: Identify which polynomials are in factored form: (a) $12 = 3(4)$ (b) $12 = \frac{15}{2}(1.6)$
(c) $15x^2 = 3x(5x)$ (d) $15x^2 = 3(5)xx$ (e) $6y - 8 = 2(3y - 4)$ (f) $6y - 8 = 6(y - \frac{4}{3})$

(g) $w^2 + 5w + 6 = (w + 2)(w + 3)$ (h) $w^2 + 5w + 6 = w\left(w + 5 + \dfrac{6}{w}\right)$

Solution:
(a) 12 is in factored form as 3(4) because 3 and 4 are both integral polynomials.
(b) 12 is not in factored form as $\frac{15}{2}(1.6)$ because $\frac{15}{2}$ and 1.6 are not integral polynomials.
(c) $15x^2$ is in factored form as $3x(5x)$ because $3x$ and $5x$ are integral polynomials.
(d) $15x^2$ is in factored form as $3(5)xx$ because 3, 5, and x are integral polynomials.
(e) $6y - 8$ is in factored form as $2(3y - 4)$ because 2 and $3y - 4$ are integral polynomials.
(f) $6y - 8$ is not in factored form as $6(y - \frac{4}{3})$ because $y - \frac{4}{3}$ is not an integral polynomial.
(g) $w^2 + 5w + 6$ is in factored form as $(w + 2)(w + 3)$ because $w + 2$ and $w + 3$ are integral polynomials.
(h) $w^2 + 5w + 6$ is not in factored form as $w\left(w + 5 + \dfrac{6}{w}\right)$ because $w + 5 + \dfrac{6}{w}$ is not a polynomial.

Note: Factoring is a very important technique in algebra. It is worth your time and effort to study and learn each factoring method presented in this chapter.

C. Factor integers.

To factor certain trinomials with form $x^2 + bx + c$ or $ax^2 + bx + c$ using the methods given in Sections 9-3 and 9-4, respectively, you must be able to **factor a given integer as the product of two integers in as many different ways as possible.**

EXAMPLE 9-7: Factor the following integers as the product of two integers in as many different ways as possible: (a) 6   (b) -6

Solution: (a) $6 = 1(6)$ —— 1 is the first positive factor of every nonzero integer

$= 2(3)$ —— 2 is the next largest positive integer that divides 6 evenly

$= \cancel{3(2)}$ ←—— equivalent to 2(3) [See the following *Note 2.*]

$= \cancel{6(1)}$ ←—— equivalent to 1(6)

$= -1(-6)$ ←—— product associated with 1(6)

$= -2(-3)$ ←—— product associated with 2(3)

Think: $a(b) = -a(-b)$

(b) $-6 = 1(-6)$

$\qquad = 2(-3)$

$\qquad = 3(-2)$ ⟵ different from $2(-3)$ [See the following *Note 3*.]

$\qquad = 6(-1)$ ⟵ different from $1(-6)$

$\qquad = $ ~~3(6)~~

$\qquad = $ ~~2(3)~~

$\qquad = $ ~~3(2)~~

$\qquad = $ ~~6(1)~~

Think: Using $a(b) = -a(-b)$ gives only equivalent factored forms. [See the following *Note 2*.]

Note 1: Both 6 and -6 can be factored as the product of two integers in exactly four different ways:

$$6 = 1(6) = 2(3) = -1(-6) = -2(-3)$$
$$-6 = 1(-6) = 2(-3) = 3(-2) = 6(-1)$$

Note 2: Products like 2(3) and 3(2) are not considered to be different factored forms because they involve the same integers [2, 3].

Note 3: Products like $2(-3)$ and $-2(3)$ are different factored forms because they involve different integers [2, -3 and -2, 3, respectively].

D. Factor monomials.

To **factor a monomial completely,** you write the monomial factors that cannot be factored further over the integers except for factors of 1 or -1.

EXAMPLE 9-8: Factor each monomial completely: **(a)** $6x^4$ **(b)** $-12m^3n^2$ **(c)** $5abc$

Solution: **(a)** $6x^4$ is factored completely as $2(3)xxxx$ [see the following *Note 1*].
(b) $-12m^3n^2$ is factored completely as $-2(2)(3)mmmnn$.
(c) $5abc$ is factored completely because 5, a, b, and c cannot be factored over the integers except for factors of 1 or -1 [see the following *Note 2*].

Note 1: A monomial is factored completely when its numerical coefficient is factored as a product of primes [see Example 1-17] and the exponent of each variable is one.

Note 2: "Cannot be factored further except for factors of 1 or -1" means that a monomial like $5abc$ is not considered to be factored further by writing it as $1(5)abc$ or $-1(-5)abc$.

To factor certain trinomials with form $x^2 + bx + c$ or $ax^2 + bx + c$ using the methods given in Sections 9-3 and 9-4, respectively, you must be able to **factor a given 2nd-degree integral monomial as the product of two 1st-degree monomials in as many different ways as possible.**

EXAMPLE 9-9: Factor as the product of two 1st-degree monomials in as many different ways as possible: **(a)** $6x^2$ **(b)** $-24y^2$

2nd-degree   1st-degree factors

Solution: **(a)** $6x^2 = x(6x)$ 　　　*Think:* $6 = 1(6)$ [See Example 9-7, part **(a)**.]

$\qquad = 2x(3x)$ 　　　　　　$= 2(3)$

$\qquad = -x(-6x)$ 　　　　　$= -1(-6)$

$\qquad = -2x(-3x)$ 　　　　$= -2(-3)$

(b) $-24y^2 = y(-24y)$ *Think:* $-24 = 1(-24)$ [See Example 9-7, part **(b)**.]

$$= 2y(-12y) \qquad\qquad = 2(-12)$$
$$= 3y(-8y) \qquad\qquad = 3(-8)$$
$$= 4y(-6y) \qquad\qquad = 4(-6)$$
$$= 6y(-4y) \qquad\qquad = 6(-4)$$
$$= 8y(-3y) \qquad\qquad = 8(-3)$$
$$= 12y(-2y) \qquad\qquad = 12(-2)$$
$$= 24y(-y) \qquad\qquad = 24(-1)$$

Note: To factor a given 2nd-degree integral monomial as the product of two 1st-degree monomials in as many different ways as possible, you write products of two 1st-degree monomials containing integral coefficients found by factoring the integral coefficient of the given 2nd-degree monomial as the product of two integers in as many different ways as possible.

E. Find the greatest common factor [GCF] of two or more monomials.

The **greatest common factor [GCF]** of two or more given monomials is the monomial with the greatest integer coefficient (or its opposite) and the greatest power of each different variable that is common to each one of the given monomials. To help find the GCF of two or more monomials, you can first factor each monomial completely.

EXAMPLE 9-10: Find the GCF of **(a)** $12x^4yz$, $-18x^3y^2$ and $24x^2y^3z$;   **(b)** $3ac$ and $-2b$.

Solution:

(a)
$$12x^4yz = \boxed{2} \cdot 2 \cdot \boxed{3} \boxed{xx} xx \boxed{y} z$$
$$-18x^3y^2 = -\boxed{2} \cdot 3 \cdot \boxed{3} \boxed{xx} xy \, y$$
$$24x^2y^3z = \boxed{2} \cdot 2 \cdot 2 \cdot \boxed{3} \boxed{xx} yy \boxed{y} z$$

Factor each monomial completely.

Circle the greatest number of each different factor that is common to each one of the given monomials.

$$\text{GCF} = 2 \cdot 3xxy$$

Write the GCF as a product.

$$= 6x^2y \text{ or } -6x^2y \longleftarrow \text{ GCF of } 12x^4yz,\ -18x^3y^2,\ \text{and } 24x^2y^3z \text{ [See the following } Note\ 1.]$$

(b) $3ac$ is factored completely
    $-2b$ is factored completely.

$$\text{GCF} = 1 \text{ or } -1 \longleftarrow \text{ GCF of } 3ac \text{ and } -2b \text{ [See the following } Note\ 2.]$$

Note 1: The GCF of $12x^4yz$, $-18x^3y^2$, and $24x^2y^3z$ $[6x^2y$ or $-6x^2y]$ does not include the variable z because z is not common to all three terms.

Note 2: The GCF of $3ac$ and $-2b$ is 1 or -1 because the largest integer coefficient (or its opposite) that is common to both 3 and -2 is 1 or -1 and ac and b have no variables in common.

F. Factor out the greatest common factor [GCF].

To factor a polynomial when each term has a common factor other than 1 or -1, you always **factor out the greatest common factor [GCF].**

EXAMPLE 9-11: Factor out the GCF: **(a)** $18x - 12$ **(b)** $-5y^2 - 10y$
(c) $12a^4b - 20a^3b^2 + 8a^2b^3$ **(d)** $w^2(w + 3) + 2(w + 3)$

Solution:
(a) The GCF of $18x$ and -12 is 6 or -6.

$$18x - 12 = \mathbf{6}(3x) - \mathbf{6}(2)$$

Factor each term using 6 as the GCF [see the following *Note 1*].

$$= \mathbf{6}(3x - 2) \longleftarrow \text{simplest form}$$

Factor out the GCF 6.

Check: Is 1 or -1 the GCF of $3x$ and -2? Yes.
Is $6(3x - 2)$ equal to the original polynomial?   Yes:

$$\mathbf{6}(3x - 2) = \mathbf{6}(3x) - \mathbf{6}(2)$$
$$= 18x - 12 \longleftarrow \text{original polynomial } [6(3x - 2) \text{ checks}]$$

(b) The GCF of $-5y^2$ and $-10y$ is $5y$ or $-5y$.

$$-5y^2 - 10y = (\mathbf{-5y})y + (\mathbf{-5y})2$$

Factor each term using $-5y$ as the GCF [see the following *Note 2*].

$$= \mathbf{-5y}(y + 2) \longleftarrow \text{simplest form}$$

Factor out the GCF $-5y$.

(c) The GCF of $12a^4b$, $-20a^3b^2$, and $8a^2b^3$ is $4a^2b$ or $-4a^2b$.

$$12a^4b - 20a^3b^2 + 8a^2b^3 = \mathbf{4a^2b}(3a^2) - \mathbf{4a^2b}(5ab) + \mathbf{4a^2b}(2b^2)$$

Factor each term using $4a^2b$ as the GCF.

$$= \mathbf{4a^2b}(3a^2 - 5ab + 2b^2) \longleftarrow \text{simplest form}$$

Factor out the GCF $4a^2b$.

(d) The GCF of $w^2(w + 3)$ and $2(w + 3)$ is $(w + 3)$ or $-(w + 3)$.

$$w^2(\mathbf{w + 3}) + 2(\mathbf{w + 3}) = (w^2 + 2)(\mathbf{w + 3}) \text{ or } (w + 3)(w^2 + 2)$$

Factor out the GCF $(w + 3)$.

Note 1: In Example 9-11, part (a), you could factor $18x - 12$ using -6 as the GCF:

$$18x - 12 = (\mathbf{-6})(-3x) + (\mathbf{-6})(2)$$
$$= \mathbf{-6}(-3x + 2)$$

Both $6(3x - 2)$ and $-6(-3x + 2)$ are considered to be correct factored forms for $18x - 12$. However, $6(3x - 2)$ is considered to be the **simplest factored form** because it has fewer negative signs than $-6(-3x + 2)$.

Note 2: In Example 9-11, part (b), you could factor $-5y^2 - 10y$ using $5y$ as the GCF:

$$-5y^2 - 10y = \mathbf{5y}(-y) + \mathbf{5y}(-2)$$
$$= \mathbf{5y}(-y - 2)$$

However, $-5y(y + 2)$ is considered to be the simplest factored form because it has fewer negative signs than $5y(-y - 2)$.

Caution: A polynomial is not factored until it is written as a product of two or more integral polynomials.

EXAMPLE 9-12: Identify which one of the following is a correct factored form of $x^2 - 6x + 8$:
(a) $x(x - 6) + 8$ (b) $x^2 - 2(3x - 4)$ (c) $(x - 2)(x - 4)$

Solution:
(a) $x^2 - 6x + 8$ is not factored as $x(x - 6) + 8$ because $x(x - 6) + 8$ is a sum.
(b) $x^2 - 6x + 8$ is not factored as $x^2 - 2(3x - 4)$ because $x^2 - 2(3x - 4)$ is a difference.
(c) $x^2 - 6x + 8$ is factored as $(x - 2)(x - 4)$ because $(x - 2)(x - 4)$ is a product and

$$\overset{\text{F} \quad \text{O} \quad \text{I} \quad \text{L}}{(x - 2)(x - 4) = x^2 - 4x - 2x + 8}$$

Multiply using the FOIL method. [See Example 8-17.]

$$= x^2 - 6x + 8 \longleftarrow \text{original polynomial } [(x - 2)(x - 4) \text{ checks}]$$

9-2. Factoring by Grouping

A. Factor a four-term polynomial by grouping.

Recall: To multiply two binomials, you can use the distributive properties [see Example 8-15].

EXAMPLE 9-13: Multiply using the distributive properties: $(x^2 + 2)(x + 3)$

Solution: $(x^2 + 2)(x + 3) = x^2(x + 3) + 2(x + 3)$ Distribute $x + 3$.

$$= (x^3 + 3x^2) + (2x + 6) \quad \text{Distribute } x^2 \text{ and } 2.$$

$$= x^3 + 3x^2 + 2x + 6$$

Recall: To factor, you reverse the multiplication steps.

EXAMPLE 9-14: Factor by reversing the multiplication steps in Example 9-13: $x^3 + 3x^2 + 2x + 6$

Solution: $x^3 + 3x^2 + 2x + 6 = (x^3 + 3x^2) + (2x + 6)$ Reverse the multiplication steps in Example 9-13.

$$= x^2(x + 3) + 2(x + 3)$$

$$= (x^2 + 2)(x + 3) \qquad \text{[See the following \textit{Note}.]}$$

Note: To factor $x^3 + 3x^2 + 2x + 6$ in Example 9-14, you first regrouped as the sum of the two binomials $(x^3 + 3x^2) + (2x + 6)$ and then factored out the GCF of each binomial to get **like-binomial factors** in $x^2(x + 3) + 2(x + 3)$.

The method for factoring certain **four-term polynomials** shown in Example 9-14 is summarized in the following method:

Factor-by-Grouping Method for Certain Four-term Polynomials

1. Write the four-term polynomial in descending powers when necessary.
2. Factor out the GCF of the four-term polynomial, other than 1 or -1, when possible.
3. Regroup the four-term polynomial as the sum of two binomials.
4. Factor out the GCF of each binomial from Step 3 to get like-binomial factors.
5. Factor out the like-binomial factors from Step 4 to get the proposed factored form.
6. Check the proposed factored form from Step 5 by multiplying the binomial factors to see if you get the original four-term polynomial.

EXAMPLE 9-15: Factor by grouping: $y - 3y^2 - 3 + y^3$

Solution:

$$y - 3y^2 - 3 + y^3 = y^3 - 3y^2 + y - 3 \qquad \text{Write in descending powers.}$$

$$= (y^3 - 3y^2) + (y - 3) \qquad \text{Regroup as the sum of two binomials.}$$

$$= y^2(y - 3) + 1(y - 3) \qquad \text{Factor out the GCF of each binomial to get like-binomial factors [see the following \textit{Note}].}$$

$$= (y^2 + 1)(y - 3) \qquad \text{Factor out the like-binomial factor.}$$

$$\overset{\text{F} \quad \text{O} \quad \text{I} \quad \text{L}}{}$$

Check: $(y^2 + 1)(y - 3) = y^3 - 3y^2 + y - 3$ Multiply using the FOIL method.

$$= y - 3y^2 - 3 + y^3 \longleftarrow \text{original polynomial } [(y^2 + 1)(y - 3) \text{ checks}]$$

Note: In Example 9-15, the binomial $y - 3$ is written as $1(y - 3)$ in preparation for factoring out the like-binomial factors in $y^2(y - 3) + 1(y - 3)$.

Caution: $y - 3y^2 - 3 + y^3$ will not factor by grouping without first writing in descending powers.

EXAMPLE 9-16: Show that $y - 3y^2 - 3 + y^3$ cannot be factored by grouping without first writing in descending powers.

Solution: $y - 3y^2 - 3 + y^3 = (y - 3y^2) + (-3 + y^3)$

$$= y(1 - 3y) - 1(3 - y^3) \qquad \text{Stop!}$$

Note: $y - 3y^2 - 3 + y^3$ is not factored as the difference $y(1 - 3y) - 1(3 - y^3)$, and $y(1 - 3y) - 1(3 - y^3)$ cannot be factored further by grouping because $(1 - 3y)$ and $(3 - y^3)$ are not like-binomial terms.

B. Factor a four-term polynomial containing like terms by grouping.

To factor trinomials with form $ax^2 + bx + c$ using one of the methods given in Section 9-4, you must be able to **factor four-term polynomials containing like terms by grouping.**

Caution: To factor a four-term polynomial containing like terms by grouping, you do not combine like terms.

EXAMPLE 9-17: Factor by grouping: $6x^2 - 4x + 9x - 6$

Solution:

$6x^2 - 4x + 9x - 6 = (6x^2 - 4x) + (9x - 6)$ — Regroup as the sum of two binomials.

$\qquad\qquad = 2x(3x - 2) + 3(3x - 2)$ — Factor out the GCF of each binomial to get like-binomial factors.

$\qquad\qquad = (2x + 3)(3x - 2) \text{ or } (3x - 2)(2x + 3)$ — Factor out the like-binomial factor.

$$\overset{\text{F} \qquad \text{O} \qquad \text{I} \qquad \text{L}}{}$$
Check: $(2x + 3)(3x - 2) = 6x^2 - 4x + 9x - 6$ ⟵ original polynomial $[(2x + 3)(3x - 2)$ checks]

Caution: Not every four-term polynomial containing like terms can be factored by grouping.

EXAMPLE 9-18: Show that $w^2 + 4w + w + 6$ cannot be factored by grouping.

Solution:

$w^2 + 4w + w + 6 = (w^2 + 4w) + (w + 6) \text{ or } w^2 + 4w + w + 6 = (w^2 + w) + (4w + 6)$

$\qquad\qquad = w(w + 4) + 1(w + 6) \qquad\qquad\qquad\qquad\qquad = w(w + 1) + 2(2w + 3)$

$w^2 + 4w + w + 6$ cannot be factored by grouping because $(w + 4)$ and $(w + 6)$ are not like-binomial factors and $(w + 1)$ and $(2w + 3)$ are not like-binomial factors.

Note: Although $w^2 + 4w + w + 6$ cannot be factored by grouping, $w^2 + 4w + w + 6$ can be factored as follows:

$w^2 + 4w + w + 6 = w^2 + 5w + 6 \qquad$ Combine like terms.

$\qquad\qquad = (w + 2)(w + 3) \qquad$ Factor using known products [see Example 9-2, part (**d**)].

C. Factor four-term polynomials completely.

To **factor a polynomial completely,** you

1. Factor out the GCF other than 1 or -1.
2. Continue factoring until each factor cannot be factored further over the integers.

EXAMPLE 9-19: Factor completely: $2y^3 + 2y^2 - 2y^2 - 2y$

Solution:

$$2y^3 + 2y^2 - 2y^2 - 2y = 2y[y^2 + y - y - 1] \qquad \text{First factor out the GCF } 2y.$$

$$= 2y[(y^2 + y) + (-y - 1)] \qquad \text{Factor the four-term polynomial by grouping.}$$

$$= 2y[y(y + 1) - 1(y + 1)] \qquad \text{Write } (-y - 1) \text{ as } -1(y + 1) \text{ to get like-binomial factors.}$$

$$= 2y[(y - 1)(y + 1)]$$

$$= 2y(y - 1)(y + 1) \text{ or } 2y(y + 1)(y - 1) \longleftarrow \text{factored completely}$$

$$\qquad\qquad\qquad\qquad\qquad\quad \text{F} \quad \text{O} \quad \text{I} \quad \text{L}$$

Check: $2y[(y - 1)(y + 1)] = 2y(y^2 + y - y - 1)$

$$= 2y^3 + 2y^2 - 2y^2 - 2y \longleftarrow \text{original polynomial } [2y(y - 1)(y + 1) \text{ checks}]$$

9-3. Factoring Trinomials with Form $x^2 + bx + c$

A. Factor trinomials with form $x^2 + bx + c$ using the *b*-and-*c* method.

Recall: To multiply two binomials, you can use the FOIL method [see Example 8-17].

EXAMPLE 9-20: Multiply using the FOIL method: $(w + 2)(w + 3)$

$$\qquad\qquad\qquad\qquad \text{F} \quad \text{O} \quad \text{I} \quad \text{L}$$

Solution: $(w + 2)(w + 3) = w^2 + 3w + 2w + 6$

$$= w^2 + (2 + 3)w + 2(3) \text{ or } w^2 + 5w + 6$$

Recall: To factor, you reverse the multiplication steps.

EXAMPLE 9-21: Factor by reversing the multiplication steps in Example 9-20: $w^2 + 5w + 6$

Solution: $w^2 + 5w + 6 = w^2 + (2 + 3)w + 2(3) \qquad \text{Reverse the multiplication steps in Example 9-20.}$

$$= (w + 2)(w + 3) \text{ or } (w + 3)(w + 2) \qquad \text{[See the following \textit{Note}.]}$$

$$\qquad\qquad\qquad\qquad\quad m \quad n \qquad m\, n \qquad\quad m \qquad n$$

Note: $w^2 + 5w + 6 = w^2 + (2 + 3)w + 2(3) = (w + 2)(w + 3)$, where $m + n = 5$ and $mn = 6$.

Caution: To factor $w^2 + 5w + 6$ using the method shown in Example 9-21, you must know how to write 5 and 6 using the integers $m = 2$ and $n = 3$ as $(2 + 3)$ and $2(3)$.

To **factor a trinomial with form $x^2 + bx + c$** using the method shown in Example 9-21, you must first find m and n so that the product of m and n equals c in $x^2 + bx + c$ and the sum of m and n equals b in $x^2 + bx + c$, as summarized in the following method:

b-and-c Method for Factoring Trinomials with the Form $x^2 + bx + c$

If $mn = c$ and $m + n = b$, then $x^2 + bx + c = (x + m)(x + n)$.

Note: The b-and-c method for factoring trinomials with the form $x^2 + bx + c$ states that if $x^2 + bx + c$ can be factored, then it can be factored as $(x + m)(x + n)$ by finding m and n so that $mn = c$ and $m + n = b$.

Caution: To factor a trinomial that does factor over the integers using the b-and-c method, the trinomial must have the form $x^2 + bx + c$. That is, the numerical coefficient of the 2nd-degree term must be 1.

EXAMPLE 9-22: Factor using the b-and-c method: **(a)** $x^2 - 3x - 18$ \qquad **(b)** $r^2 - 7rs + 12s^2$

Solution:

(a) In $x^2 - 3x - 18$, $b = -3$ and $c = -18$.   Identify b and c.

$$mn = c = -18 \qquad m + n = b = -3$$ Find m and n.

Factor -18 as the product of two integers in as many different ways as possible [see Example 9-7].

$$-18 = 1(-18) \qquad 1 + (-18) = -17$$
$$= 2(-9) \qquad 2 + (-9) = -7$$
$$= \mathbf{3(-6)} \qquad \mathbf{3 + (-6) = -3}$$ Stop! $m = 3$ and $n = -6$ will work.
$$= 6(-3)$$
$$= 9(-2) \quad \Big\} \text{ not needed}$$
$$= 18(-1)$$

$$x^2 - 3x - 18 = (x \overset{m}{+ 3})(x \overset{n}{- 6}) \text{ or } (x - 6)(x + 3) \qquad \text{Factor using } m \text{ and } n$$

$$\text{Check: } (x + 3)(x - 6) = \overset{F}{x^2} \overset{O}{- 6x} \overset{I}{+ 3x} \overset{L}{- 18}$$
$$= x^2 - 3x - 18 \longleftarrow \text{ original polynomial } [(x + 3)(x - 6) \text{ checks}]$$

(b) In $r^2 - 7rs + 12s^2$, $b = -7s$ and $c = 12s^2$.   Identify b and c in $r^2 + br + c$.

$$mn = c = 12s^2 \qquad m + n = b = -7s$$ Find m and n.

Factor $12s^2$ as the product of two 1st-degree monomials in as many different ways as possible [see Example 9-10].

$$12s^2 = 1s(12s) \qquad 1s + 12s = 13s$$
$$= 2s(6s) \qquad 2s + 6s = 8s$$
$$= 3s(4s) \qquad 3s + 4s = 7s$$
$$= -1s(-12s) \qquad -1s + (-12s) = -13s$$
$$= -2s(-6s) \qquad -2s + (-6s) = -8s$$
$$= \mathbf{-3s(-4s)} \qquad \mathbf{-3s + (-4s) = -7s}$$ Stop! $m = -3s$ and $n = -4s$ will work.

$$r^2 - 7rs + 12s^2 = (r \overset{m}{- 3s})(r \overset{n}{- 4s}) \qquad \text{Factor using } m \text{ and } n.$$

$$\text{Check: } (r - 3s)(r - 4s) = \overset{F}{r^2} \overset{O}{- 4rs} \overset{I}{- 3rs} \overset{L}{+ 12s^2}$$
$$= r^2 - 7rs + 12s^2 \longleftarrow \text{ original polynomial } [(r - 3s)(r - 4s) \text{ checks}]$$

Caution: Not all trinomials with form $x^2 + bx + c$ can be factored over the integers.

EXAMPLE 9-23: Show that $x^2 - 8x + 24$ cannot be factored over the integers.

Solution: In $x^2 - 8x + 24$, $b = -8$ and $c = 24$.

$$mn = c = 24 \qquad m + n = b = -8$$

$$24 = 1(24) \qquad 1 + 24 = 25$$
$$= 2(12) \qquad 2 + 12 = 14$$
$$= 3(8) \qquad 3 + 8 = 11$$
$$= 4(6) \qquad 4 + 6 = 10$$
$$= -1(-24) \qquad -1 + (-24) = -25$$
$$= -2(-12) \qquad -2 + (-12) = -14$$
$$= -3(-8) \qquad -3 + (-8) = -11$$
$$= -4(-6) \qquad -4 + (-6) = -10$$

There are no integers m and n such that $mn = 24$ and $m + n = -8$

$x^2 - 8x + 24$ cannot be factored over the integers because $x^2 - 8x + 24$ cannot be factored using the b-and-c method.

B. Factor trinomials with form $ax^2 + bx + c$ completely.

Recall: To factor a polynomial completely, you

1. Factor out the GCF other than 1 or -1.
2. Continue factoring until each factor cannot be factored further over the integers.

EXAMPLE 9-24: Factor completely: **(a)** $-24x^2 + 14x^3 - 2x^4$ **(b)** $5y^2 + 25y + 60$

Solution: **(a)** $-24x^2 + 14x^3 - 2x^4 = -2x^4 + 14x^3 - 24x^2$ Write in descending powers.

$$= -2x^2(x^2 - 7x + 12) \qquad \text{Factor out the GCF } -2x^2.$$

$$= -2x^2(x - 3)(x - 4) \qquad \textit{Think:} \quad -3(-4) = 12 \longleftarrow c$$
$$\text{and } -3 + (-4) = -7 \longleftarrow b$$

(b) $5y^2 + 25y + 60 = 5(y^2 + 5y + 12) \longleftarrow$ factored completely [see the following *Note*].

Note: In Example 9-24, $5y^2 + 25y + 60$ is factored completely as $5(y^2 + 5y + 12)$ because $y^2 + 5y + 12$ will not factor using the b-and-c method. That is, $y^2 + 5y + 12$ cannot be factored further over the integers.

Caution: Both $-2x^2(x^2 - 7x + 12)$ and $2x^2(-x^2 + 7x - 12)$ have the same number of negative signs. However, $-2x^2(x^2 - 7x + 12)$ is the **preferred factored form** because the b-and-c method can be used to try and factor $x^2 - 7x + 12$ further whereas the b-and-c method cannot be used to try and factor $-x^2 + 7x - 12$ further. That is, the numerical coefficient of the 2nd-degree term in $-x^2 + 7x - 12$ is -1 [not 1].

EXAMPLE 9-25: Show that $-x^2 + 7x - 12$ cannot be factored using the b-and-c method in its given form $-x^2 + bx + c$.

Solution: In $-x^2 + 7x - 12$, $b = 7$ and $c = -12$.

$$
\begin{array}{ll}
\underline{mn = c = -12} & \underline{m + n = b = 7} \\
-12 = 1(-12) & 1 + (-12) = -11 \\
 = 2(-6) & 2 + (-6) = -4 \\
 = 3(-4) & 3 + (-4) = -1 \\
 = 4(-3) & 4 + (-3) = 1 \\
 = 6(-2) & 6 + (-2) = 4 \\
 = 12(-1) & 12 + (-1) = 11
\end{array}
$$

There are no integers m and n such that $mn = -12$ and $m + n = 7$.

$-x^2 + 7x - 12$ cannot be factored using the b-and-c method in its given form $-x^2 + bx + c$.

Note: In Example 9-25, $-x^2 + 7x - 12$ can be factored over the integers as $(-x + 3)(x - 4)$ or $-(x - 3)(x - 4)$ using the method called Factoring Trinomials with Form $ax^2 + bx + c$ [see Example 9-29, part **(b)**].

To try to **factor a trinomial with form $-x^2 + bx + c$** using the b-and-c method, you first factor out the GCF -1.

EXAMPLE 9-26: Factor using the b-and-c method by first factoring out the GCF -1:
$-x^2 + 7x - 12$

Solution: $-x^2 + 7x - 12 = -(x^2 - 7x + 12)$ *Think:* $x^2 - 7x + 12$ is in the form $x^2 + bx + c$.

$$= -[(x - 3)(x - 4)]$$ *Think:* $-3(-4) = 12 \longleftarrow c$
and $-3 + (-4) = -7 \longleftarrow b$

$$= -(x - 3)(x - 4) \text{ or } -(x - 4)(x - 3) \text{ or } (3 - x)(x - 4)$$
$$\text{or } (x - 4)(3 - x) \text{ or } (x - 3)(4 - x) \text{ or } (4 - x)(x - 3)$$

9-4. Factoring Trinomials with Form $ax^2 + bx + c$

A. Factor trinomials with form $ax^2 + bx + c$ using the ac method.

Recall: To multiply two binomials, you can use the distributive properties.

EXAMPLE 9-27: Multiply using the distributive properties: $(2x + 5)(3x + 4)$

Solution: $(2x + 5)(3x + 4) = 2x(3x + 4) + 5(3x + 4)$ Distribute $(3x + 4)$.

$$= 6x^2 + 8x + 15x + 20$$ Distribute $2x$ and 5.

$$= 6x^2 + 23x + 20$$ Combine like terms.

Recall: To factor, you reverse the multiplication steps.

EXAMPLE 9-28: Factor by reversing the multiplication steps in Example 9-27: $6x^2 + 23x + 20$

Solution: $6x^2 + 23x + 20 = 6x^2 + 8x + 15x + 20$ Reverse the steps in Example 9-27.

$$= 2x(3x + 4) + 5(3x + 4)$$

$$= (2x + 5)(3x + 4)$$ [See the following *Note.*]

Note: In Example 9-28, $\overset{a}{6x^2} + \overset{\overbrace{m+n}}{23x} + \overset{c}{20} = 6x^2 + \overset{m}{8x} + \overset{n}{15x} + 20$ where $m + n = 23$, $mn = ac = 120$, and $6x^2 + 8x + 15x + 20$ can be factored by grouping as $(2x + 5)(3x + 4)$.

Caution: To **factor a trinomial with form $ax^2 + bx + c$** using the method shown in Example 9-28, you must first find integers m and n so that the product of m and n equals ac in $ax^2 + bx + c$ and the sum of m and n equals b in $ax^2 + bx + c$, as summarized in the following method.

ac Method for Factoring Trinomials with Form $ax^2 + bx + c$

If $mn = ac$ and $m + n = b$, then $ax^2 + bx + c = ax^2 \overset{\overbrace{bx}}{+ mx + nx} + c$
and $ax^2 + mx + nx + c$ can be factored by grouping.

Note: The ac method for factoring trinomials with form $ax^2 + bx + c$ states that if $ax^2 + bx + c$ can be factored as the product of two binomials, then it can be factored by grouping when you find m and n so that $mn = ac$ and $m + n = b$.

To find m and n using the ac method, you proceed in much the same way as you did for the b-and-c method shown in Section 9-3.

EXAMPLE 9-29: Factor using the ac method:
(a) $3w^2 + 13w - 10$ (b) $-x^2 + 7x - 12$ (c) $-4x^5y^3 - 10x^4y^2 - 6x^3y$

Solution:

(a) In $3w^2 + 13w - 10$, $a = 3$, $b = 13$, and $c = -10$.

$mn = ac = 3(-10) = -30$	$m + n = b = 13$
$-30 = 1(-30)$	$1 + (-30) = -29$
$= 2(-15)$	$2 + (-15) = -13$
$= 3(-10)$	$3 + (-10) = -7$
$= 5(-6)$	$5 + (-6) = -1$
$= 6(-5)$	$6 + (-5) = 1$
$= 10(-3)$	$10 + (-3) = 7$
$= \mathbf{15(-2)}$	$\mathbf{15 + (-2) = 13}$
$= 30(-1)$ ⟵ not needed	

Factor -30 as the product of two integers in as many different ways as possible [see Example 9-7].

Stop! $m = 15$ and $n = -2$ will work.

$$3w^2 + 13w - 10 = 3w^2 \overset{m}{+ 15w} \overset{n}{- 2w} - 10$$

Rename as a four-term polynomial using m and n: $13w = 15w - 2w$

$$= 3w(w + 5) - 2(w + 5)$$

Factor by grouping [see Example 9-17].

$$= (3w - 2)(w + 5) \text{ or } (w + 5)(3w - 2)$$

Check: $(3w - 2)(w + 5) = 3w^2 \overset{F}{} \overset{O}{+ 15w} \overset{I}{- 2w} \overset{L}{- 10}$

$$= 3w^2 + 13w - 10 \longleftarrow \text{original polynomial } [(3w - 2)(w + 5) \text{ checks}]$$

(b) In $-x^2 + 7x - 12$, $a = -1$, $b = 7$, and $c = -12$.

$mn = ac = -1(-12) = 12$	$m + n = b = 7$
$12 = 1(12)$	$1 + 12 = 13$
$= 2(6)$	$2 + 6 = 8$
$= \mathbf{3(4)}$	$\mathbf{3 + 4 = 7}$
$= -1(-12)$	
$= -2(-6)$	not needed
$= -3(-4)$	

Stop! $m = 3$ and $n = 4$ will work.

$$-x^2 + 7x - 12 = -x^2 \overset{m}{+ 3x} \overset{n}{+ 4x} - 12$$

Think: $7x = 3x + 4x$

$$= -x(x - 3) + 4(x - 3)$$

Factor by grouping.

$$= (-x + 4)(x - 3) \text{ or } -(x - 4)(x - 3) \text{ or } -(x - 3)(x - 4)$$

Check: $(-x + 4)(x - 3) = -x^2 \overset{F}{} \overset{O}{+ 3x} \overset{I}{+ 4x} \overset{L}{- 12}$

$$= -x^2 + 7x - 12 \longleftarrow \text{original polynomial } [(-x + 4)(x - 3) \text{ checks}]$$

(c) The GCF of $-4x^5y^3$, $-10x^4y^2$, and $-6x^3y$ is $2x^3y$ or $-2x^3y$. [See Example 9-11.]

$$-4x^5y^3 - 10x^4y^2 - 6x^3y = -2x^3y(2x^2y^2 + 5xy + 3) \qquad \text{Factor out the GCF } -2x^3y.$$

In $2x^2y^2 + 5xy + 3$, $a = 2$, $b = 5$, and $c = 3$.   *Think:* $2(xy)^2 + 5(xy) + 3$

$mn = ac = 2(3) = 6$	$m + n = b = 5$	Factor $2x^2y^2 + 5xy + 3$ using the *ac* method.
$6 = 1(6)$	$1 + 6 = 7$	
$\quad = 2(3)$	$2 + 3 = 5 \leftarrow$	Stop! $m = 2$ and
$\quad = -1(-6)$		$n = 3$ will work.
$\quad = -2(-3)$		

$-1(-6)$ and $-2(-3)$ not needed

$$
\begin{aligned}
-2x^3y(2x^2y^2 + 5xy + 3) &= -2x^3y(2x^2y^2 \overset{m}{+ 2xy} \overset{n}{+ 3xy} + 3) \\
&= -2x^3y[2xy(xy + 1) + 3(xy + 1)] \\
&= -2x^3y(2xy + 3)(xy + 1) \text{ or } -2x^3y(xy + 1)(2xy + 3) \longleftarrow \text{ factored completely}
\end{aligned}
$$

$$
\begin{aligned}
\text{\textit{Check:} } -2x^3y(2xy + 3)(xy + 1) &= -2x^3y(2x^2y^2 \overset{\text{F}}{+} \overset{\text{O}}{2xy} \overset{\text{I}}{+} \overset{\text{L}}{3xy} + 3) \\
&= -2x^3y(2x^2y^2 + 5xy + 3) \\
&= -4x^5y^3 - 10x^4y^2 - 6x^3y \longleftarrow \text{ original polynomial} \\
&\qquad\qquad\qquad\qquad\qquad\quad [-2x^3y(2xy + 3)(xy + 1) \text{ checks}]
\end{aligned}
$$

Caution: Not all trinomials with form $ax^2 + bx + c$ can be factored over the integers.

EXAMPLE 9-30: Show that $5x^2 - 10x + 8$ cannot be factored over the integers.

Solution: In $5x^2 - 10x + 8$, $a = 5$, $b = -10$, and $c = 8$.

$mn = ac = 5(8) = 40$	$m + n = b = -10$	
$40 = 1(40)$	$1 + 40 = 41$	
$\quad = 2(20)$	$2 + 20 = 22$	
$\quad = 4(10)$	$4 + 10 = 14$	
$\quad = 5(8)$	$5 + 8 = 13$	There are no integers m and n such
$\quad = -1(-40)$	$-1 + (-40) = -41$	that $mn = 40$ and $m + n = -10$.
$\quad = -2(-20)$	$-2 + (-20) = -22$	
$\quad = -4(-10)$	$-4 + (-10) = -14$	
$\quad = -5(-8)$	$-5 + (-8) = -13$	

$5x^2 - 10x + 8$ cannot be factored over the integers because $5x^2 - 10x + 8$ cannot be factored using the *ac* method.

B. Factor trinomials with form $ax^2 + bx + c$ using the trial-and-error method.

If a trinomial with form $ax^2 + bx + c$ will factor over the integers, then it will factor using the *ac* method and also using the following method:

Trial-and-Error Method for Factoring Trinomials with Form $ax^2 + bx + c$

1. Write the given trinomial in the form $ax^2 + bx + c$ where $a > 0$.
2. Identify the terms ax^2, bx, and c in the trinomial from Step 1.
3. Factor ax^2 as the product of two 1st-degree monomials in as many different ways as possible using only positive integers.

4. Factor c as the product of two integers m and n using the following sign pattern:

If b is	and c is	then m and n have
positive $(+)$	positive $(+)$	positive signs $(+, +)$
negative $(-)$	positive $(+)$	negative signs $(-, -)$
positive $(+)$	negative $(-)$	opposite signs $(+, -$ or $-, +)$
negative $(-)$	negative $(-)$	opposite signs $(+, -$ or $-, +)$

5. Multiply **trial products** with form $(dx + m)(ex + n)$ where $dx(ex) = ax^2$ and $mn = c$ until the product is the original trinomial $ax^2 + bx + c$. That is, multiply until the middle term of the product is bx in $ax^2 + bx + c$.

EXAMPLE 9-31: Factor $6x^2 - 17x + 12$ using the trial-and-error method.

Solution: In $6x^2 - 17x + 12$, $ax^2 = 6x^2$, $bx = -17x$, and $c = 12$.

$$\underline{ax^2 = 6x^2} \qquad \text{and} \qquad \underline{mn = c = 12}$$

$$6x^2 = 1x(6x) \qquad\qquad 12 = -1(-12)$$
$$= 2x(3x) \qquad\qquad = -2(-6)$$
$$= -3(-4)$$

In $bx = -17x$, $b = -17$ is negative $(-)$ and $c = 12$ is positive $(+)$ means that m and n must both be negative $(-, -)$.

$$\overbrace{(1x}^{dx} \overbrace{- 1)}^{m}\overbrace{(6x}^{ex} \overbrace{- 12)}^{n} = \overbrace{6x^2}^{ax^2} \overbrace{- 18x}^{bx} \overbrace{+ 12}^{c}$$ *Trial 1:* Error. bx should be $-17x$ [see the following *Note 1*].

$$(1x - 12)(6x - 1) = 6x^2 - 73x + 12$$ *Trial 2:* Error. [See the following *Note 2*].

$$(1x - 2)(6x - 6) = 6x^2 - 18x + 12$$ *Trial 3:* Error. [See the following *Shortcut 9-1*].

$$(1x - 6)(6x - 2) = 6x^2 - 38x + 12$$ *Trial 4:* Error.

$$(1x - 3)(6x - 4) = 6x^2 - 22x + 12$$ *Trial 5:* Error.

$$(1x - 4)(6x - 3) = 6x^2 - 27x + 12$$ *Trial 6:* Error.

$$(2x - 1)(3x - 12) = 6x^2 - 27x + 12$$ *Trial 7:* Error.

$$(2x - 12)(3x - 1) = 6x^2 - 38x + 12$$ *Trial 8:* Error.

$$(2x - 2)(3x - 6) = 6x^2 - 18x + 12$$ *Trial 9:* Error.

$$(2x - 6)(3x - 2) = 6x^2 - 22x + 12$$ *Trial 10:* Error.

$$(2x - 3)(3x - 4) = 6x^2 - \mathbf{17x} + 12$$ *Trial 11:* Correct: $bx = -17x$

$$(2x - 4)(3x - 3) = 6x^2 - 18x + 12$$ *Trial 12:* ⟵ not needed

$$6x^2 - 17x + 12 = (2x - 3)(3x - 4) \text{ or } (3x - 4)(2x - 3) \longleftarrow \text{proposed solution}$$

$$\overset{\text{F} \qquad \text{O} \qquad \text{I} \qquad \text{L}}{}$$
Check: $(2x - 3)(3x - 4) = 6x^2 - 8x - 9x + 12$

$$= 6x^2 - 17x + 12 \longleftarrow \text{original polynomial } [(2x - 3)(3x - 4) \text{ checks}]$$

Note 1: To factor $ax^2 + bx + c$ using the trial-and-error method, you concentrate on getting the middle term bx correct because the first term $ax^2 = dx(ex)$ and the last term $c = mn$ are always the same $[6x^2, 12]$ for each trial product $(dx + m)(ex + n)$.

Note 2: If the trial product $(dx + m)(ex + n)$ is in error, then switch m and n and try $(dx + n)(ex + m)$.

Shortcut 9-1: When the GCF of $ax^2 + bx + c$ is 1 or -1, you can omit all trial products where one or both binomial factors have a GCF other than 1 or -1.

EXAMPLE 9-32: Factor $6x^2 - 17x + 12$ from Example 9-31 using the trial-and-error method and Shortcut 9-1.

cross out all binomial factors that have a GCF other than 1 or −1.

Solution: $(1x - 1)(6x - 12)$ ⟵ Omit because the GCF of $6x$ and -12 is 6 or -6 [not 1 or -1].

$(1x - 12)(6x - 1) = 6x^2 - 73x + 12$ *Trial 1:* Error: bx should be $-17x$.

$(1x - 2)(6x - 6)$ ⟵ Omit because the GCF of $6x$ and -6 is 6 or -6.

$(1x - 6)(6x - 2)$ ⟵ Omit because the GCF of $6x$ and -2 is 2 or -2.

$(1x - 3)(6x - 4)$ ⟵ Omit because the GCF of $6x$ and -4 is 2 or -2.

$(1x - 4)(6x - 3)$ ⟵ Omit because the GCF of $6x$ and -3 is 3 or -3.

$(2x - 1)(3x - 12)$ ⟵ Omit because the GCF of $3x$ and -12 is 3 or -3.

$(2x - 12)(3x - 1)$ ⟵ Omit because the GCF of $2x$ and -12 is 2 or -2.

$(2x - 2)(3x - 6)$ ⟵ Omit because the GCF of $2x$ and -2 is 2 or -2.

$(2x - 6)(3x - 2)$ ⟵ Omit because the GCF of $2x$ and -6 is 2 or -2.

$(2x - 3)(3x - 4) = 6x^2 - 17x + 12$ *Trial 2:* Correct: $bx = -17x$
                                       [See the following *Note*].

$(2x - 4)(3x - 3)$ ⟵ Omit because the GCF of $2x$ and -4 is 2 or -2.

Note: To factor $6x^2 - 17x + 12$ using the trial-and-error method and *Shortcut 9-1* in Example 9-32, it was only necessary to multiply 2 of the possible 12 trial products. Without using *Shortcut 9-1* in Example 9-31, it was necessary to multiply 11 of the possible 12 trial products.

C. Factor $ax^2 + bx + c$ using the correct method.

To factor $ax^2 + bx + c$ using the correct method, you

1. Factor out the GCF of ax^2, bx, and c other than 1 or -1
2. Factor the remaining trinomial using
   (a) the *b-and-c* method if the GCF of ax^2, bx, and c is a or $-a$ [see Section 9-3];
   (b) either the *ac* method or the trial-and-error method if the GCF of ax^2, bx, and c is not a or $-a$ [see this Section 9-4].

EXAMPLE 9-33: Factor using the correct method: **(a)** $4x^2 + 10x + 4$ **(b)** $4x^2 + 8x + 4$

Solution:
(a) Because the GCF of $4x^2$, $10x$, and 4 is 2 or -2 [not $a = 4$ or -4], you first factor out the GCF and then try to factor the remaining trinomial using either the *ac* method or the trial-and-error method.

$$4x^2 + 10x + 4 = 2(2x^2 + 5x + 2)$$

ac method	trial-and-error method
In $2x^2 + 5x + 2$, $a = 2$, $b = 5$, and $c = 2$.	In $2x^2 + 5x + 2$, $ax^2 = 2x^2$, $bx = 5x$, and $c = 2$.

ac method:

$$mn = ac = 2(2) = 4 \qquad m + n = b = 5$$
$$4 = 1(4) \qquad\qquad 1 + 4 \quad = 5$$
$$= 2(2)$$
$$\left. \begin{array}{l} = -1(-4) \\ = -2(-2) \end{array} \right\} \text{not needed}$$

$2(2x^2 + 5x + 2)$

$= 2(2x^2 + 1x + 4x + 2)$

$= 2[x(2x + 1) + 2(2x + 1)]$

$= 2(x + 2)(2x + 1)$

trial-and-error method:

$$ax^2 = 2x^2 \quad \text{and} \quad mn = c = 2$$
$$2x^2 = x(2x) \qquad\qquad 2 = 1(2)$$

$2(x + 1)(2x + 2)$ ⟵ Omit [see Example 9-32].

$2(x + 2)(2x + 1) = 2(2x^2 + 5x + 2)$ *Trial 1:* Correct!

(b) Because the GCF of $4x^2$, $8x$, and 4 is $a = 4$ or -4, you first factor out the GCF and then try to factor the remaining trinomial using the *b-and-c* method.

$$4x^2 + 8x + 4 = 4(x^2 + 2x + 1)$$

b-and-c method

In $x^2 + 2x + 1$, $b = 2$ and $c = 1$.

$mn = c = 1$

$1 = 1(1)$

$= -1(-1) \longleftarrow$ not needed

$m + n = b = 2$

$1 + 1 \quad = 2$

$\overset{m}{\overbrace{\qquad}}\ \overset{n}{\overbrace{\qquad}}$

$4(x^2 + 2x + 1) = 4(x + 1)(x + 1)$ or $4(x + 1)^2 \longleftarrow$ factored completely

9-5. Factoring Special Products

A. Factor the difference of two squares.

Recall: Product Rule for the Difference of Two Squares:
For any two binomials with form $a + b$ and $a - b$
$(a + b)(a - b) = (a - b)(a + b) = a^2 - b^2 \longleftarrow$ difference of two squares

EXAMPLE 9-34: Multiply using the Product Rule for the Difference of Two Squares: $(2x + 3)(2x - 3)$

Solution: $(2x + 3)(2x - 3) = (2x)^2 - (3)^2$ *Think:* $(a + b)(a - b) = a^2 - b^2$

$$= 4x^2 - 9 \longleftarrow \text{difference of two squares}$$

Recall: To factor, you reverse the multiplication steps.

EXAMPLE 9-35: Factor by reversing the multiplication steps in Example 9-34: $4x^2 - 9$

Solution: $4x^2 - 9 = (2x)^2 - (3)^2$ Reverse the steps in Example 9-34.

$$= (2x + 3)(2x - 3)$$

Note: $4x^2 - 9 = (2x + 3)(2x - 3)$ because $4x^2 = (2x)^2$ and $9 = (3)^2$.

To **factor the difference of two squares** using the method shown in Example 9-35, you can use the following rule:

Factoring Rule for the Difference of Two Squares
For any two squares a^2 and b^2

same first terms

difference of two squares $\longrightarrow$ $a^2 - b^2 = (a + b)(a - b) = (a - b)(a + b)$ [See the following *Note.*]

$\uparrow$ subtraction   opposite last terms

Note: To factor the difference of two squares using the Factoring Rule for the Difference of Two Squares, you

1. Write the given difference of two squares in the form $a^2 - b^2$
2. Draw two pairs of parentheses: $a^2 - b^2 \quad (\quad)(\quad)$
3. Write a as the first term in each parentheses: $a^2 - b^2 \quad (a \quad)(a \quad)$
4. Write $+b$ and $-b$ as the last terms in the parentheses: $a^2 - b^2 = (a + b)(a - b)$

$$\text{or } (a - b)(a + b)$$

EXAMPLE 9-36: Factor using the Factoring Rule for the Difference of Two Squares: $9m^2 - 16n^2$

Solution: $9m^2 - 16n^2 = (3m)^2 - (4n)^2$ *Think:* $a^2 - b^2$

$$= (3m + 4n)(3m - 4n) \quad Think: \ a^2 - b^2 = (a + b)(a - b)$$
$$\text{or } (3m - 4n)(3m + 4n) \qquad\qquad \text{or } (a - b)(a + b)$$

Caution: The **sum of two squares** never factors over the integers.

EXAMPLE 9-37: Show that $4x^2 + 9$ cannot be factored over the integers.

Solution:

ac method

In $4x^2 + 9 = 4x^2 + 0x + 9$, $a = 4$, $b = 0$, and $c = 9$.

$mn = ac = 4(9) = 36$	$m + n = b = 0$
$36 = 1(36)$	$1 + 36 = 37$
$= 2(18)$	$2 + 18 = 20$
$= 3(12)$	$3 + 12 = 15$
$= 4(9)$	$4 + 9 = 13$
$= 6(6)$	$6 + 6 = 12$
$= -1(-36)$	$-1 + (-36) = -37$
$= -2(-18)$	$-2 + (-18) = -20$
$= -3(-12)$	$-3 + (-12) = -15$
$= -4(-9)$	$-4 + (-9) = -13$
$= -6(-6)$	$-6 + (-6) = -12$

There are no integers m and n such that $mn = 36$ and $m + n = 0$.

$4x^2 + 9$ cannot be factored over the integers because $4x^2 + 0x + 9$ cannot be factored using the *ac* method.

B. Factor a perfect square trinomial.

Recall: Product Rule for Perfect Square Trinomials:
For two binomials each with form $a + b$ or $a - b$:

same operation symbol

$$(a + b)(a + b) = (a + b)^2 = a^2 + 2ab + b^2$$

addition   perfect square trinomials

$$(a - b)(a - b) = (a - b)^2 = a^2 - 2ab + b^2$$

same operation symbol

EXAMPLE 9-38: Multiply using the Product Rule for Perfect Square Trinomials:
(a) $(2x + 3)^2$ **(b)** $(4m - 5n)^2$

Solution: **(a)** $(2x + 3)^2 = (2x)^2 + 2(2x)(3) + (3)^2$ *Think:* $(a + b)^2 = a^2 + 2ab + b^2$

$$= 4x^2 + 12x + 9 \longleftarrow \text{perfect square trinomial}$$

(b) $(4m - 5n)^2 = (4m)^2 - 2(4m)(5n) + (5n)^2$ *Think:* $(a - b)^2 = a^2 - 2ab + b^2$

$$= 16m^2 - 40mn + 25n^2 \longleftarrow \text{perfect square trinomial}$$

Recall: To factor, you reverse the multiplication steps.

EXAMPLE 9-39: Factor by reversing the multiplication steps in Example 9-38:
(a) $4x^2 + 12x + 9$ (b) $16m^2 - 40mn + 25n^2$

Solution: (a) $4x^2 + 12x + 9 = (2x)^2 + 2(2x)(3) + (3)^2$ Reverse the steps in Example 9-38, part (a).

$$= (2x + 3)^2$$ [See the following *Note 1*.]

(b) $16m^2 - 40mn + 25n^2 = (4m)^2 - 2(4m)(5n) + (5n)^2$ Reverse the steps in Example 9-38, part (b).

$$= (4m - 5n)^2$$ [See the following *Note 2*.]

Note 1: In Example 9-39, part (a), $4x^2 + 12x + 9 = (2x + 3)^2$ because $4x^2 = (2x)^2$, $9 = (3)^2$, and $12x = 2(2x)(3)$.

Note 2: In Example 9-39, part (b), $16m^2 - 40mn + 25n^2 = (4m - 5n)^2$ because $16m^2 = (4m)^2$, $25n^2 = (5n)^2$, and $-40mn = -2(4m)(5n)$.

To **factor a perfect square trinomial** using the method shown in Example 9-39, you can use the following rule:

Factoring Rule for Perfect Square Trinomials

For any two squares a^2 and b^2

same operation symbol

$$a^2 + 2ab + b^2 = (a + b)^2 \text{ or } (a + b)(a + b)$$ [See the following *Note*.]

perfect square trinomials   addition

$$a^2 - 2ab + b^2 = (a - b)^2 \text{ or } (a - b)(a - b)$$ [See the following *Note*.]

same operation symbol

Note: To factor a perfect square trinomial using the Factoring Rule for Perfect Square Trinomials, you

1. Write the given perfect square trinomial in the form $a^2 + 2ab + b^2$
   or $a^2 - 2ab + b^2$
2. Draw two pairs of parentheses: $a^2 + 2ab + b^2$ (   )(   )
   $a^2 - 2ab + b^2$ (   )(   )
3. Write a as the first term in each parentheses: $a^2 + 2ab + b^2$ (a)(a)
   $a^2 - 2ab + b^2$ (a)(a)
4. Write $+b$ $[-b]$ as the last term in each parentheses: $a^2 + 2ab + b^2 = (a + b)(a + b)$
   $a^2 - 2ab + b^2 = (a - b)(a - b)$

EXAMPLE 9-40: Factor using the Factoring Rule for Perfect Square Trinomials:
(a) $x^4 + 2x^2 + 1$ (b) $4m^2n^2 - 20mn + 25$

$2x^2$

Solution: (a) $x^4 + 2x^2 + 1 = (x^2)^2 + 2(x^2)(1) + (1)^2$ *Think:* $a^2 + 2ab + b^2$

$$= (x^2 + 1)^2$$ *Think:* $a^2 + 2ab + b^2 = (a + b)^2$
or $(a + b)(a + b)$

$$\text{or } (x^2 + 1)(x^2 + 1)$$

$$-20mn$$
$$\text{(b) } 4m^2n^2 - 20mn + 25 = (2mn)^2 \overbrace{- 2(2mn)(5)} + (5)^2 \qquad \textit{Think: } a^2 - 2ab + b^2$$

$$= (2mn - 5)^2 \qquad \textit{Think: } a^2 - 2ab + b^2 = (a - b)^2 \text{ or } (a - b)(a - b)$$

$$\text{or } (2mn - 5)(2mn - 5)$$

C. Factor special products completely.

Recall: To factor a polynomial completely, you

1. Factor out the GCF other than 1 or -1.
2. Continue factoring until each factor cannot be factored over the integers.

EXAMPLE 9-41: Factor $2x^7 - 4x^5 + 2x^3$ completely.

Solution:
$$\begin{aligned}
2x^7 - 4x^5 + 2x^3 &= 2x^3(x^4 - 2x^2 + 1) && \text{Factor out the GCF.} \\
&= 2x^3(x^2 - 1)^2 && \text{Factor the perfect square trinomial:} \\
& && (x^2)^2 - 2(x^2)(1) + (1)^2 \\
&= 2x^3(x^2 - 1)(x^2 - 1) && \\
&= 2x^3(x + 1)(x - 1)(x + 1)(x - 1) && \text{Factor each difference of two} \\
& && \text{squares: } (x)^2 - (1)^2 \\
&= 2x^3(x + 1)^2(x - 1)^2 \longleftarrow && \text{simplest factored form}
\end{aligned}$$

Note: $2x^7 - 4x^5 + 2x^3$ is factored completely as $2x^3(x + 1)^2(x - 1)^2$ because each factor $2x^3$, $x + 1$, and $x - 1$ cannot be factored further over the integers.

RAISE YOUR GRADES

Can you . . . ?

☑ factor polynomials using known products
☑ explain how to factor a polynomial over the integers
☑ factor an integer as the product of two integers in as many different ways as possible
☑ factor a monomial completely
☑ factor a given 2nd-degree integral monomial as a product of two 1st-degree monomials in as many different ways as possible
☑ find the greatest common factor [GCF] of two or more monomials
☑ factor out the GCF
☑ factor completely by grouping
☑ factor completely using the *b*-and-*c* method
☑ factor completely using the *ac* method
☑ factor completely using the trial-and-error method
☑ factor completely using special products

SUMMARY

1. To factor, you reverse the multiplication steps.
2. A polynomial that has integers for all numerical coefficients and constants is called an integral polynomial.

3. To factor a polynomial over the integers, you write the polynomial in factored form as the product of two or more integral polynomials.

4. By agreement, the word "factor" is understood to mean "factor over the integers" unless otherwise stated.

5. To factor a given integer as the product of two or more integers in as many different ways as possible, products like 2(3) and 3(2) are not considered to be different factored forms while products like 2(−3) and −2(3) are considered to be different factored forms.

6. To factor a monomial completely, you write the monomial as the product of monomial factors that cannot be factored further over the integers except for factors of 1 or −1.

7. To factor a given 2nd-degree integral monomial as the product of two 1st-degree monomials in as many different ways as possible, you write products of the two 1st-degree monomials containing the given variable with integral coefficients found by factoring the integral coefficient of the given 2nd-degree integral monomial in as many different ways as possible.

8. The greatest common factor [GCF] of two or more monomials is the monomial with the greatest integer coefficient [or its opposite] and the greatest power of each different variable that is common to each one of the given monomials.

9. To factor a polynomial when each term has a common factor other than 1 or −1, you always factor out the greatest common factor [GCF].

10. Factor-by-Grouping Method for Certain Four-Term Polynomials:
    (a) Write the four-term polynomial in descending powers.
    (b) Factor out the GCF of the four-term polynomial, other than 1 or −1, when possible.
    (c) Regroup the four-term polynomial as the sum of two binomials.
    (d) Factor out the GCF of each binomial from Step (c) to get like-binomial factors.
    (e) Factor out the like-binomial factors from Step (d) to get the proposed factored form.
    (f) Check the factored form from Step (e) by multiplying the polynomial factors to see if you get the original four-term polynomial.

11. To factor a four-term polynomial containing like terms by grouping, you do not combine like terms.

12. b-and-c Method for Factoring Trinomials with Form $x^2 + bx + c$:
    If $mn = c$ and $m + n = b$, then $x^2 + bx + c = (x + m)(x + n)$.

13. To factor a trinomial that will factor over the integers with form $-x^2 + bx + c$ using the b-and-c method, you first factor out the GCF -1.

14. ac Method for Factoring Trinomials with Form $ax^2 + bx + c$:

$$\text{If } mn = ac \text{ and } m + n = b, \text{ then } ax^2 + bx + c = ax^2 + \overbrace{mx + nx}^{bx} + c$$
$$\text{and } ax^2 + mx + nx + c \text{ can be factored by grouping.}$$

15. Trial-and-Error Method for Factoring Trinomials with Form $ax^2 + bx + c$:
    (a) Write the given trinomial in the form $ax^2 + bx + c$ where $a > 0$.
    (b) Identify the terms ax^2, bx, and c in the trinomial from Step (a).
    (c) Factor c as the product of two integers m and n using the following sign pattern:

If b is	and c is	then m and n have
positive (+)	positive (+)	positive signs (+, +).
negative (−)	positive (+)	negative signs (−, −).
positive (+)	negative (−)	opposite signs (+, − or −, +).
negative (−)	negative (−)	opposite signs (+, − or −, +).

    (d) Multiply trial products with form $(dx + m)(ex + n)$ where $dx(ex) = ax^2$ and $mn = c$ until the product is the original trinomial $ax^2 + bx + c$. That is, multiply until the middle term of the product is bx in $ax^2 + bx + c$.

16. When the GCF of $ax^2 + bx + c$ is 1 or −1, you can omit all trial products where one or both binomial factors has a GCF other than 1 or −1.

17. To factor $ax^2 + bx + c$, you
    (a) Factor out the GCF of ax^2, bx, and c other than 1 or −1.
    (b) Factor the remaining trinomial using
        (i) the b-and-c method if the GCF of ax^2, bx, and c is a or $-a$;

(ii) either the *ac* method or the trial-and-error method if the GCF of ax^2, bx, and c is not a or $-a$.
18. Factoring Rule for the Difference of Two Squares:
For any squares a^2 and b^2, $a^2 - b^2 = (a + b)(a - b)$ or $(a - b)(a + b)$.
19. The sum of two squares cannot be factored over the integers.
20. Factoring Rule for Perfect Square Trinomials:
For any two squares a^2 and b^2, $a^2 + 2ab + b^2 = (a + b)^2$ or $(a + b)(a + b)$
and $a^2 - 2ab + b^2 = (a - b)^2$ or $(a - b)(a - b)$.
21. To factor a polynomial containing two or more terms completely, you
(a) Factor out the GCF other than 1 or -1.
(b) Continue factoring until each factor cannot be factored further over the integers, except for factors of 1 or -1.

SOLVED PROBLEMS

PROBLEM 9-1 Identify which algebraic expressions are integral polynomials: (a) -8 (b) $\frac{3}{4}$ (c) 0.25 (d) $\sqrt{5}$ (e) $2x - 3$ (f) $\frac{1}{2}y$ (g) $3z - 1.2$ (h) $w + \sqrt{2}$ (i) $m^2 - m$

(j) $2n^2 + 3n - 7$ (k) $2\sqrt{a} + 3$ (l) $\frac{3}{b} - 6$

Solution: Recall that a polynomial that has integers for all numerical coefficients and constants is called an integral polynomial [see Example 9-5]:

(a) -8 is an integral polynomial because -8 is an integer.
(b) $\frac{3}{4}$ is not an integral polynomial because $\frac{3}{4}$ is not an integer.
(c) 0.25 is not an integral polynomial because 0.25 is not an integer.
(d) $\sqrt{5}$ is not an integral polynomial because $\sqrt{5}$ is not an integer.
(e) $2x - 3$ is an integral polynomial because both 2 and -3 are integers.
(f) $\frac{1}{2}y$ is not an integral polynomial because $\frac{1}{2}$ is not an integer.
(g) $3z - 1.2$ is not an integral polynomial because -1.2 is not an integer.
(h) $w + \sqrt{2}$ is not an integral polynomial because $\sqrt{2}$ is not an integer.
(i) $m^2 - m = 1m^2 - 1m$ is an integral polynomial because both 1 and -1 are integers.
(j) $2n^2 + 3n - 7$ is an integral polynomial because 2, 3, and -7 are all integers.
(k) $2\sqrt{a} + 3$ is not an integral polynomial because it is not a polynomial.
(l) $\frac{3}{b} - 6$ is not an integral polynomial because it is not a polynomial.

PROBLEM 9-2 Factor as the product of two integers in as many different ways as possible:
(a) 60 (b) -72

Solution: Recall that to factor a given integer as the product of two integers in as many different ways as possible, products like 2(3) and 3(2) are not considered to be different factored forms while products like 2(-3) and -2(3) are considered to be different factored forms [see Example 9-7]:

(a) $60 = 1(60) = 2(30) = 3(20) = 4(15) = 5(12) = 6(10)$ ⟵ 12 different
$60 = -1(-60) = -2(-30) = -3(-20) = -4(-15) = -5(-12) = -6(-10)$ ⟵ factored forms

(b) $-72 = 1(-72) = 2(-36) = 3(-24) = 4(-18) = 6(-12) = 8(-9)$ ⟵ 12 different
$-72 = 9(-8) = 12(-6) = 18(-4) = 24(-3) = 36(-2) = 72(-1)$ ⟵ factored forms

PROBLEM 9-3 Factor each monomial completely: (a) $24m^2$ (b) $-30w^2x^4yz^3$

Solution: Recall that to factor a monomial completely, you write the monomial as a product of monomial factors that cannot be factored further over the integers [see Example 9-8]:

(a) $24m^2 = 2(2)(2)3mm$ (b) $-30w^2x^4yz^3 = -2(3)(5)wwxxxxyzzz$

PROBLEM 9-4   Factor as the product of two 1st-degree monomials in as many different ways as possible: (a) $3x^2$ (b) $-4y^2$ (c) $12w^2$

Solution: Recall that to factor a 2nd-degree integral monomial as the product of two 1st-degree monomials in as many different ways as possible, you write products of two 1st-degree monomials containing the given variable with integral coefficients found by factoring the integral coefficient of the given 2nd-degree integral monomial in as many different ways as possible [see Example 9-9]:

(a) $3x^2 = x(3x) = -x(-3x)$ (b) $-4y^2 = y(-4y) = 2y(-2y) = 4y(-y)$

(c) $12w^2 = w(12w) = 2w(6w) = 3w(4w) = -w(-12w) = -2w(-6w) = -3w(-4w)$

PROBLEM 9-5   Find the GCF of: (a) $6w$ and 12 (b) r^2 and $2r$ (c) x^2y and xy^2
(d) m^2 and $-n^2$ (e) $6ab^4c$, $-21a^2b^3c$, and $15a^3b^2$

Solution: Recall that the greatest common factor [GCF] of two or more monomials is the monomial with the greatest integer coefficient [or its opposite] and the greatest power of each different variable that is common to each one of the given monomials [see Example 9-10]:

(a) The GCF of $6w$ and 12 is 6 or -6 because $\begin{aligned} 6w &= \boxed{2}(\boxed{3})w \\ 12 &= \boxed{2}(2)\boxed{3} \end{aligned}$ and $2 \cdot 3 = 6.$

(b) The GCF of r^2 and $2r$ is r or $-r$ because $\begin{aligned} r^2 &= r\boxed{r} \\ 2r &= 2\boxed{r} \end{aligned}.$

(c) The GCF of x^2y and xy^2 is xy or $-xy$ because $\begin{aligned} x^2y &= \boxed{x}x\boxed{y} \\ xy^2 &= \boxed{x}y\boxed{y} \end{aligned}.$

(d) The GCF of m^2 and $-n^2$ is 1 or -1 because m^2 and $-n^2$ have no variables in common and no integer factors in common except 1 or -1.

(e) The GCF of $6ab^4c$, $-21a^2b^3c$, and $15a^3b^2$ is $3ab^2$ or $-3ab^2$ because

$$6ab^4c = 2(\boxed{3})\boxed{a}(\boxed{bb})bbc$$
$$-21a^2b^3c = -7(\boxed{3})\boxed{a}a\boxed{bb}bc$$
$$15a^3b^2 = 5(\boxed{3})\boxed{a}aa\boxed{bb}$$

PROBLEM 9-6   Factor out the GCF: (a) $20w - 30$ (b) $2x^2 + x$ (c) $ab - a^2$
(d) $-8y^2 - 10y$ (e) $-6r^2 + 4r - 10$ (f) $12m^4n - 18m^3n^2 + 24m^2n^3$
(g) $2u(u - 3) - 5(u - 3)$ (h) $v^2(v + 2) + (v + 2)$ (i) $2r(3r - 2s) + s(2s - 3r)$

Solution: Recall that to factor a polynomial when each term has a common factor other than 1 or -1, you always factor out the GCF [see Example 9-11]:

(a) The GCF of $20w$ and -30 is 10 or -10: $20w - 30 = 10(2w) - 10(3) = 10(2w - 3)$
(b) The GCF of $2x^2$ and x is x or $-x$: $2x^2 + x = x(2x) + x(1) = x(2x + 1)$
(c) The GCF of ab and $-a^2$ is a or $-a$: $ab - a^2 = a(b) - a(a) = a(b - a)$
(d) The GCF of $-8y^2$ and $-10y$ is $2y$ or $-2y$: $(-2y)4y + (-2y)5 = -2y(4y + 5)$
(e) The GCF of $-6r^2$, $4r$ and -10 is 2 or -2: $2(-3r^2) + 2(2r) + 2(-5) = 2(-3r^2 + 2r - 5)$
   $\quad$ or $-2(3r^2) - 2(-2r) - 2(5) = -2(3r^2 - 2r + 5)$
(f) The GCF of $12m^4n$, $-18m^3n^2$, and $24m^2n^3$ is $6m^2n$ or $-6m^2n$:

$$12m^4n - 18m^3n^2 + 24m^2n^3 = 6m^2n(2m^2) + 6m^2n(-3mn) + 6m^2n(4n^2)$$

$$= 6m^2n(2m^2 - 3mn + 4n^2)$$

(g) $2u(u - 3) - 5(u - 3) = (2u - 5)(u - 3)$
(h) $v^2(v + 2) + (v + 2) = v^2(v + 2) + 1(v + 2) = (v^2 + 1)(v + 2)$
(i) $2r(3r - 2s) + s(2s - 3r) = 2r(3r - 2s) - s(3r - 2s) = (2r - s)(3r - 2s)$

PROBLEM 9-7   Factor completely by grouping: (**a**) $x^3 + 2x^2 + 3x + 6$
(**b**) $y^3 + 6 - 2y^2 - 3y$ (**c**) $6n^2 - 4n + 9n - 6$ (**d**) $x^2 + xy + xy - y^2$
(**e**) $12w^5 + 2w^2 - 4w^4 - 6w^3$

Solution: Recall that to factor certain four-term polynomials by grouping, you

1. Write the four-term polynomial in descending powers when necessary.
2. Factor out the GCF of the four-term polynomial, other than 1 or -1, when possible.
3. Regroup the four-term polynomial as the sum of two binomials.
4. Factor out the GCF of each binomial from Step 3 to get like-binomial factors.
5. Factor out the like-binomial factor from Step 4 to get the proposed factored form.
6. Check the proposed factored form from Step 5 by multiplying the polynomial factors to see if you get the original four-term polynomial.
   [See Section 9-2.]

(**a**) $x^3 + 2x^2 + 3x + 6 = (x^3 + 2x^2) + (3x + 6)$ Regroup as the sum of two binomials.

$\qquad\qquad\qquad\quad = x^2(x + 2) + 3(x + 2)$ Factor out the GCF of each binomial to get like-binomial factors.

$\qquad\qquad\qquad\quad = (x^2 + 3)(x + 2)$ Factor out the like-binomial factor.

$\qquad\qquad\qquad\qquad\qquad\qquad\quad$ F $\quad$ O $\quad$ I $\quad$ L
Check: $(x^2 + 3)(x + 2) = x^3 + 2x^2 + 3x + 6 \longleftarrow (x^2 + 3)(x + 2)$ $\quad$ checks [see Example 9-15].

(**b**) $y^3 + 6 - 2y^2 - 3y = y^3 - 2y^2 - 3y + 6$ First write descending powers.

$\qquad\qquad\qquad\quad = (y^3 - 2y^2) + (-3y + 6)$ Then factor by grouping [see Example 9-16].

$\qquad\qquad\qquad\quad = y^2(y - 2) - 3(y - 2)$

$\qquad\qquad\qquad\quad = (y^2 - 3)(y - 2)$ [Check as before.]

(**c**) $6n^2 - 4n + 9n - 6 = (6n^2 - 4n) + (9n - 6)$ Do not combine like terms to factor by grouping [see Example 9-17].

$\qquad\qquad\qquad\quad = 2n(3n - 2) + 3(3n - 2)$

$\qquad\qquad\qquad\quad = (2n + 3)(3n - 2)$ [Check as before.]

(**d**) $x^2 + xy + xy - y^2 = (x^2 + xy) + (xy - y^2)$ Factor by grouping.

$\qquad\qquad\qquad\quad = x(x + y) + y(x - y)$ Stop!

$x^2 + xy + xy - y^2$ cannot be factored by grouping because $(x + y)$ and $(x - y)$ are not like-binomial factors [see Example 9-18].

(**e**) $12w^5 + 2w^2 - 4w^4 - 6w^3 = 12w^5 - 4w^4 - 6w^3 + 2w^2$ Write descending powers.

$\qquad\qquad\qquad\quad = 2w^2(6w^3 - 2w^2 - 3w + 1)$ Factor out the GCF.

$\qquad\qquad\qquad\quad = 2w^2[(6w^3 - 2w^2) + (-3w + 1)]$ Factor completely [see Example 9-19].

$\qquad\qquad\qquad\quad = 2w^2[2w^2(3w - 1) - 1(3w - 1)]$

$\qquad\qquad\qquad\quad = 2w^2(2w^2 - 1)(3w - 1)$

$\qquad\qquad\qquad\qquad\qquad\qquad\qquad$ F $\quad$ O $\quad$ I $\quad$ L
Check: $2w^2(2w^2 - 1)(3w - 1) = 2w^2(6w^3 - 2w^2 - 3w + 1)$

$\qquad\qquad\qquad\quad = 12w^5 - 4w^4 - 6w^3 + 2w^2$

$\qquad\qquad\qquad\quad = 12w^5 + 2w^2 - 4w^4 - 6w^3 \longleftarrow 2w^2(2w^2 - 1)(3w - 1)$ checks.

PROBLEM 9-8   Factor completely using the *b*-and-*c* method:

(a) $x^2 - 2x - 8$ (b) $y^2 - 14y + 36$ (c) $3x^4y + 18x^3y^2 + 15x^2y^3$ (d) $-w^2 + w + 12$

Solution: Recall the *b*-and-*c* method for factoring trinomials with form $x^2 + bx + c$: If $mn = c$ and $m + n = b$, then $x^2 + bx + c = (x + m)(x + n)$.

(a) In $x^2 - 2x - 8$, $b = -2$ and $c = -8$ [see Example 9-22].

$$\underline{mn = c = -8} \qquad \underline{m + n = b = -2}$$

$$-8 = 1(-8) \qquad 1 + (-8) = -7$$

$$= 2(-4) \qquad 2 + (-4) = -2 \quad \text{Stop! } m = 2 \text{ and}$$
$$\qquad\qquad\qquad\qquad\qquad\qquad\qquad n = -4 \text{ will work}$$

$$= 4(-2) \;\Big\}\; \text{not}$$
$$= 8(-1) \;\Big\}\; \text{needed}$$

$$x^2 - 2x - 8 = (x + 2)(x - 4)$$

Check: $(x + 2)(x - 4) = x^2 - 4x + 2x - 8$

$$= x^2 - 2x - 8 \; \longleftarrow \; \text{original polynomial } [(x + 2)(x - 4) \text{ checks}]$$

(b) In $y^2 - 14y + 36$, $b = -14$ and $c = 36$ [see Example 9-23].

$$\underline{mn = c = 36} \qquad \underline{m + n = b = -14}$$

$$36 = 1(36) \qquad\qquad 1 + 36 = 37$$
$$= 2(18) \qquad\qquad 2 + 18 = 20$$
$$= 3(12) \qquad\qquad 3 + 12 = 15$$
$$= 4(9) \qquad\qquad 4 + 9 = 13$$
$$= 6(6) \qquad\qquad 6 + 6 = 12$$
$$= -1(-36) \qquad -1 + (-36) = -37$$
$$= -2(-18) \qquad -2 + (-18) = -20$$
$$= -3(-12) \qquad -3 + (-12) = -15$$
$$= -4(-9) \qquad -4 + (-9) = -13$$
$$= -6(-6) \qquad -6 + (-6) = -12$$

There are no integers m and n such that $mn = 36$ and $m + n = -14$.

$y^2 - 14y + 36$ will not factor over the integers.

(c) $3x^4y + 18x^3y^2 + 15x^2y^3 = 3x^2y(x^2 + 6xy + 5y^2)$ [See Example 9-24.]

In $x^2 + 6xy + 5y^2$, $b = 6y$ and $c = 5y^2$.   *Think:* $x^2 + bx + c$

$$\underline{mn = c = 5y^2} \qquad\qquad \underline{m + n = b = 6y}$$

$$5y^2 = y(5y) \qquad\qquad y + 5y = 6y \quad \text{Stop! } m = y \text{ and}$$

$$= -y(-5y) \; \longleftarrow \; \text{not needed} \qquad\qquad\qquad n = 5y \text{ will work.}$$

$$3x^4y + 18x^3y^2 + 15x^2y^3 = 3x^2y(x + y)(x + 5y) \; \longleftarrow \; \text{proposed factored form}$$

Check: $3x^2y(x + y)(x + 5y) = 3x^2y(x^2 + 5xy + xy + 5y^2)$

$$= 3x^2y(x^2 + 6xy + 5y^2)$$

$$= 3x^4y + 18x^3y^2 + 15x^2y^3 \; \longleftarrow \; 3x^2y(x + y)(x + 5y) \text{ checks.}$$

(d) $-w^2 + w + 12 = -(w^2 - w - 12)$ [See Example 9-26.]

$$= -[(w + 3)(w - 4)] \qquad \textit{Think: } 3(-4) = -12 \; \longleftarrow \; c$$

$$\text{and } 3 + (-4) = -1 \; \longleftarrow \; b$$

$$= -(w + 3)(w - 4) \text{ or } (w + 3)(4 - w)$$

Check: $-(w + 3)(w - 4) = -(w^2 - 4w + 3w - 12)$

$$= -(w^2 - w - 12)$$

$$= -w^2 + w + 12 \longleftarrow \quad -(w + 3)(w - 4) \text{ checks.}$$

PROBLEM 9-9   Factor completely using the *ac* method: **(a)** $2x^2 + 5x + 2$ **(b)** $-w^2 - w + 12$
(c) $6x^3y^2 - 4x^2y^3 - 2xy^4$ **(d)** $4y^2 + 11y + 5$

Solution: Recall the *ac* method for factoring trinomials with form $ax^2 + bx + c$:
If $mn = ac$ and $m + n = b$, then $ax^2 + bx + c = ax^2 + mx + nx + c$
and $ax^2 + mx + nx + c$ can be factored by grouping.
[See Examples 9-29 and 9-30.]:

(a) In $2x^2 + 5x + 2$, $a = 2$, $b = 5$, and $c = 2$.

$$\underline{mn = ac = 2(2) = 4} \qquad \underline{m + n = b = 5}$$

$4 = 1(4) \qquad\qquad\qquad 1 + 4 = 5 \longleftarrow$ Stop! $m = 1$ and
$ \qquad\qquad\qquad\qquad\qquad\qquad\qquad\qquad n = 4$ will work

$\left. \begin{array}{l} = 2(2) \\ = -1(-4) \\ = -2(-2) \end{array} \right\}$ not needed

$2x^2 + 5x + 2 = 2x^2 + 1x + 4x + 2$

$$= x(2x + 1) + 2(2x + 1)$$

$$= (x + 2)(2x + 1)$$

Check: $(x + 2)(2x + 1) = 2x^2 + x + 4x + 2$

$$= 2x^2 + 5x + 2 \longleftarrow \text{ original polynomial } [(x + 2)(2x + 1) \text{ checks}]$$

(b) In $-w^2 - w + 12$, $a = -1$, $b = -1$, and $c = 12$.

$$\underline{mn = ac = -1(12) = -12} \qquad \underline{m + n = b = -1}$$

$-12 = 1(-12) \qquad\qquad\qquad\qquad 1 + (-12) = -11$

$ = 2(-6) \qquad\qquad\qquad\qquad 2 + (-6) \ = -4$

$ = 3(-4) \qquad\qquad\qquad\qquad 3 + (-4) \ = -1 \longleftarrow$ Stop! $m = 3$ and
$ = 4(-3) \qquad\qquad\qquad\qquad\qquad\qquad\qquad\qquad\qquad n = -4$ will work.

$\left. \begin{array}{l} = 4(-3) \\ = 6(-2) \\ = 12(-1) \end{array} \right\}$ not needed

$-w^2 - w + 12 = -w^2 + 3w - 4w + 12$

$$= -w(w - 3) - 4(w - 3)$$

$$= (-w - 4)(w - 3) \text{ or } -(w + 4)(w - 3) \qquad \text{[check as before]}$$

(c) $6x^3y^2 - 4x^2y^3 - 2xy^4 = 2xy^2(3x^2 - 2xy - y^2)$

In $3x^2 - 2xy - y^2$, $a = 3$, $b = -2y$, and $c = -y^2$.   *Think: $ax^2 + bx + c$*

$$\underline{mn = ac = 3(-y^2) = -3y^2} \qquad \underline{m + n = b = -2y}$$

$-3y^2 = y(-3y) \qquad\qquad\qquad\qquad y + (-3y) = -2y \longleftarrow$ Stop! $m = y$ and
$ = 3y(-y) \longleftarrow \text{ not needed} \qquad\qquad\qquad\qquad\qquad\qquad\qquad n = -3y$ will work.

$2xy^2(3x^2 - 2xy - y^2) = 2xy^2(3x^2 + yx - 3yx - y^2)$

$$= 2xy^2[x(3x + y) - y(3x + y)]$$

$$= 2xy^2(x - y)(3x + y) \qquad \text{[check as before]}$$

(d) In $4y^2 + 11y + 5$, $a = 4$, $b = 11$, and $c = 5$.

$mn = ac = 4(5) = 20$	$m + n = b = 11$	
$20 = 1(20)$	$1 + 20 = 21$	
$ = 2(10)$	$2 + 10 = 12$	
$ = 4(5)$	$4 + 5 = 9$	There are no integers m and n such
$ = -1(-20)$	$-1 + (-20) = -21$	that $mn = 20$ and $m + n = 11$.
$ = -2(-10)$	$-2 + (-10) = -12$	
$ = -4(-5)$	$-4 + (-5) = -9$	

$4y^2 + 11y + 5$ will not factor over the integers.

PROBLEM 9-10   Factor completely using the trial-and-error method: **(a)** $6x^2 + 23x + 20$
(b) $-3y^2 + y + 2$ **(c)** $18x^3y^2 + 48x^2y^3 + 30xy^4$ **(d)** $5x^2y^2 - 10xy + 8$

Solution: Recall the trial-and-error method for factoring trinomials with form $ax^2 + bx + c$:

1. Write the given trinomial in the form $ax^2 + bx + c$ where $a > 0$ when necessary.
2. Identify the terms ax^2, bx, and c in the trinomial from Step 1.
3. Factor c as the product of two integers m and n using the sign pattern:

If b is	and c is	then m and n have
positive $(+)$	positive $(+)$	positive signs $(+, +)$.
negative $(-)$	positive $(+)$	negative signs $(-, -)$.
positive $(+)$	negative $(-)$	opposite signs $(+, -$ or $-, +)$.
negative $(-)$	negative $(-)$	opposite signs $(+, -$ or $-, +)$.

4. Multiply trial products with form $(dx + m)(ex + n)$ when $dx(ex) = ax^2$ and $mn = c$ until the product is the original trinomial $ax^2 + bx + c$. That is, multiply until the middle term of the product is bx in $ax^2 + bx + c$.
   [See Examples 9-31, 9-32, and 9-33.]

(a) In $6x^2 + 23x + 20$, $ax^2 = 6x^2$, $bx = 23x$, and $c = 20$.

$ax^2 = 6x^2$	and	$mn = c = 20$	
$6x^2 = x(6x)$		$20 = 1(20)$	In $bx = 23x$, $b = 23$ is positive $(+)$ and $c = 20$ is
$ = 2x(3x)$		$ = 2(10)$	positive $(+)$ means that m and n must both be
		$ = 4(5)$	positive $(+, +)$.

$\overset{dx \ \ m}{\overbrace{}} \ \overset{ex \ \ n}{\overbrace{}}$

$(x + 1)(6x + 20)$ ⟵ omit because the GCF of $6x$ and 20 is 2 or -2 [not 1 or -1]

$(x + 20)(6x + 1) \ = 6x^2 + 121x + 20$ *Trial 1:* Error. bx should be $23x$.
$(x + 2)(6x + 10)$ ⟵ omit

$(x + 10)(6x + 2)$ ⟵ omit

$(x + 4)(6x + 5) \ = 6x^2 + 29x + 20$ *Trial 2:* Error.

$(x + 5)(6x + 4)$ ⟵ omit

$(2x + 1)(3x + 20) = 6x^2 + 43x + 20$ *Trial 3:* Error.

$(2x + 20)(3x + 1)$ ⟵ omit

$(2x + 2)(3x + 10)$ ⟵ omit

$(2x + 10)(3x + 2)$ ⟵ omit

$(2x + 4)(3x + 5)$ ⟵ omit

$(2x + 5)(3x + 4) \ = 6x^2 + 23x + 20$ *Trial 4:* Correct: $bx = 23x$!

Check: $(2x + 5)(3x + 4) = 6x^2 + 8x + 15x + 20$

$$= 6x^2 + 23x + 20 \longleftarrow (2x + 5)(3x + 4) \quad \text{checks}$$

(b) $-3y^2 + y + 2 = -(3y^2 - y - 2)$ Make a in ay^2 positive ($a > 0$).

In $3y^2 - y - 2$, $ay^2 = 3y^2$, $by = -y$, and $c = -2$.

$\underline{ay^2 = 3y^2}$ and $\underline{mn = c = -2}$

$3y^2 = y(3y)$ $\left.\begin{array}{l} -2 = 1(-2) \\ = 2(-1) \end{array}\right\}$ In $by = -y$, $b = -1$ is negative ($-$) and $c = -2$ is negative ($-$) means m and n have opposite signs ($+, -$ or $-, +$).

$\overbrace{(y + 1)}^{m}\overbrace{(3y - 2)}^{n} = 3y^2 + y - 2$ *Trial 1:* Error. by should be $-y$.

$(y - 2)(3y + 1) = 3y^2 - 5y - 2$ *Trial 2:* Error.

$(y + 2)(3y - 1) = 3y^2 + 5y - 2$ *Trial 3:* Error.

$(y - 1)(3y + 2) = 3y^2 - y - 2$ *Trial 4:* Correct: $by = -y$!

$-3y^2 + y + 2 = -(3y^2 - y - 2)$

$$= -(y - 1)(3y + 2) \text{ or } (1 - y)(3y + 2) \quad \text{[Check as before.]}$$

(c) $18x^3y^2 + 48x^2y^3 + 30xy^4 = 6xy^2(3x^2 + 8xy + 5y^2)$

In $3x^2 + 8xy + 5y^2$, $ax^2 = 3x^2$, $bx = 8xy$, and $c = 5y^2$.

$\underline{ax^2 = 3x^2}$ and $\underline{mn = c = 5y^2}$

$3x^2 = x(3x)$ $5y^2 = y(5y)$ $\longleftarrow$ b and c are both positive means m and n are both positive ($+, +$)

$\qquad (x + y)(3x + 5y) = 3x^2 + 8xy + 5y^2$ *Trial 1:* Correct: $bx = 8xy$!

$18x^3y^2 + 48x^2y^3 + 30xy^4 = 6xy^2(x + y)(3x + 5y)$ [Check as before.]

(d) In $5x^2y^2 - 10xy + 8$, $ax^2 = 5x^2y^2$, $bx = -10xy$, and $c = 8$.

$\underline{ax^2 = 5x^2y^2}$ and $\underline{mn = c = 8}$

$5x^2y^2 = xy(5xy)$ $\left.\begin{array}{l} 8 = -1(-8) \\ = -2(-4) \end{array}\right\}$ In $bx = -10xy$, $b = -10$ is negative ($-$) and $c = 8$ is positive ($+$) means m and n are both negative ($-, -$).

$(xy - 1)(5xy - 8) = 5x^2y^2 - 13xy + 8$ *Trial 1:* Error. bx should be $-10xy$.

$(xy - 8)(5xy - 1) = 5x^2y^2 - 41xy + 8$ *Trial 2:* Error.

$(xy - 2)(5xy - 4) = 5x^2y^2 - 14xy + 8$ *Trial 3:* Error.

$(xy - 4)(5xy - 2) = 5x^2y^2 - 22xy + 8$ *Trial 4:* Error.

$5x^2y^2 - 10xy + 8$ will not factor over the integers.

PROBLEM 9-11   Factor completely using the Factoring Rule for the Difference of Two Squares:

(a) $64w^2 - 81$ **(b)** $4m^2 + 9n^2$ **(c)** $64a^2 - 8b^2$ **(d)** $2x^4 - 2y^4$

Solution: Recall the Factoring Rule for the Difference of Two Squares:
For any two squares a^2 and b^2: $a^2 - b^2 = (a + b)(a - b)$ or $(a - b)(a + b)$

(a) $64w^2 - 81 = (8w)^2 - (9)^2$

$$= (8w + 9)(8w - 9)$$
$$\text{or } (8w - 9)(8w + 9)$$

[See Example 9-36].

(b) $4m^2 + 9n^2$ cannot be factored over the integers because it is the sum of two squares [See Example 9-37.]

(c) $64a^2 - 8b^2 = 8(8a^2 - b^2)$

$8(8a^2 - b^2)$ cannot be factored further as the difference of two squares because $8a^2$ is not an integral square.

(d) $2x^4 - 2y^4 = 2(x^4 - y^4)$

$$= 2(x^2 + y^2)(x^2 - y^2)$$

$$= 2(x^2 + y^2)(x + y)(x - y)$$

[See Example 9-41.]

PROBLEM 9-12   Factor completely using the Factoring Rule for Perfect Square Trinomials:

(a) $x^2 + 8x + 16$ **(b)** $y^2 - 4y + 4$ **(c)** $9a^4 + 42a^2 + 49$
(d) $m^2 - 2mn + n^2$ **(e)** $w^2 + w + 1$ **(f)** $4b^2 - 36b - 81$
(g) $-3c^6 + 24c^4 - 48c^2$ **(h)** $6x^4y - 30x^3y^2 + 12x^2y^3$

Solution: Recall the Factoring Rule for Perfect Square Trinomials:
For any two squares a^2 and b^2: $a^2 + 2ab + b^2 = (a + b)^2$ or $(a + b)(a + b)$
                         and $a^2 - 2ab + b^2 = (a - b)^2$ or $(a - b)(a - b)$
[See Examples 9-40 and 9-41.]:

(a) $x^2 + 8x + 16 = (x)^2 + \overbrace{2(x)(4)}^{8x} + (4)^2$

$$= (x + 4)^2$$

$$\text{or } (x + 4)(x + 4)$$

(b) $y^2 - 4y + 4 = (y)^2 - \overbrace{2(y)(2)}^{-4y} + (2)^2$

$$= (y - 2)^2$$

$$\text{or } (y - 2)(y - 2)$$

(c) $9a^4 + 42a^2 + 49 = (3a^2)^2 + \overbrace{2(3a^2)(7)}^{42a^2} + (7)^2$

$$= (3a^2 + 7)^2$$

$$\text{or } (3a^2 + 7)(3a^2 + 7)$$

(d) $m^2 - 2mn + n^2 = (m)^2 - \overbrace{2(m)(n)}^{-2mn} + (n)^2$

$$= (m - n)^2$$

$$\text{or } (m - n)(m - n)$$

(e) $w^2 + w + 1$ cannot be factored using the Factoring Rule for Perfect Square Trinomials because it is not a perfect square trinomial:
$[(w)^2 + 2(w)(1) + (1)^2 = w^2 + 2w + 1]$.

(f) $4b^2 - 36b - 81$ cannot be factored using the Factoring Rule for Perfect Square Trinomials because it is not a perfect square trinomial—it has the wrong sign pattern.

(g) $-3c^6 + 24c^4 - 48c^2 = -3c^2(c^4 - 8c^2 + 16)$

$$= -3c^2[(c^2)^2 - 2(c^2)(4) + (4)^2]$$

$$= -3c^2(c^2 - 4)^2$$

$$= -3c^2(c^2 - 4)(c^2 - 4)$$

$$= -3c^2(c + 2)(c - 2)(c + 2)(c - 2) \text{ or } -3c^2(c + 2)^2(c - 2)^2$$

(h) $6x^4y - 30x^3y^2 + 12x^2y^3 = 6x^2y(x^2 - 5xy + 2y^2)$ ⟵ factored completely

Supplementary Exercises

PROBLEM 9-13   Factor out the GCF: **(a)** $5x + 10$ **(b)** $-3y + 12$ **(c)** $w^2 - w$
(d) $-z^3 - z^2$ **(e)** $2a^2 + 4a$ **(f)** $-4b^3 + 6b^2$ **(g)** $r^2s - rs^2$ **(h)** $-mn^2 - n$
(i) $2c + 3d$ **(j)** $5a^2 - 4b^2$ **(k)** $-60x^4 + 24x^3$ **(l)** $-48m^3n^2 - 30m^2n^3$
(m) $6u^3 + 3u^2 + 15u$ **(n)** $24v^6 - 12v^5 - 36v^4$ **(o)** $5x^3y - 15x^2y^2 + 10xy^2$
(p) $30a^2b - 20ab^2 - 40ab$ **(q)** $x(x - 2) + 2(x - 2)$ **(r)** $2y(3y - 5) - (3y - 5)$
(s) $w^2(w^2 - 2) + (w^2 - 2)$ **(t)** $3z^2(2z - 1) - 5(1 - 2z)$ **(u)** $3a^2 + 5a - 2a^2 - 6a$

(v) $5b^3 - b^2 - 3b^3 + 4b - 2b^3$ (w) $2x^2y + 3xy - x^2y - 2xy^2$
(x) $8hk - 5h^2 + 4hk + 10h^2k^2 - 12hk$ (y) $24d^5 - 60d^4 + 36d^3 - 48d^2$
(z) $-12a^2b - 6ab^2 - 18a - 30b$

PROBLEM 9-14   Factor completely by grouping: (a) $x^3 + 2x^2 + x + 2$
(b) $w^3 + 2w^2 + 2w + 4$ (c) $2x^3 - 6x^2 + x - 3$ (d) $15y^3 - 5y^2 + 6y - 2$
(e) $6a^4 + 15a^3 - 8a - 20$ (f) $21b^8 + 35b^5 - 6b^3 - 10$ (g) $10c^3 + 25c^2 + 15c - 20$
(h) $16d^4 - 24d^3 + 32d^2 + 64d$ (i) $Ax + By + Ay + Bx$ (j) $8Am - Bn - An + 8Bm$
(k) $12w^2x^2y^2 - 4xy + 9w^2xy - 3$ (l) $10abc^2 - 4a^2b^2c^2 + 6ab - 15$ (m) $a^2 + 2a + a + 2$
(n) $b^2 + 3b - 2b - 6$ (o) $c^2 - 3c + c - 3$ (p) $d^2 - 2d - d + 2$
(q) $4x^2 + 2x + 6x + 3$ (r) $15y^2 + 9y - 10y - 6$ (s) $12w^3 - 18w^2 + 8w^2 - 12w$
(t) $60z^4 - 45z^3 - 24z^3 + 18z^2$ (u) $u^2 + uv + uv + v^2$ (v) $m^2n^2 - mn - mn + 1$
(w) $10x^2 - 4xy + 15xy - 6y^2$ (x) $20r^2 - 12rs + 15rs - 9s^2$ (y) $a^3 + a^2b + a^2b + a$
(z) $8c^2d^2 - 4cd - 4cd + 2$

PROBLEM 9-15   Factor completely using the *b*-and-*c* method: (a) $a^2 + 2a + 1$
(b) $b^2 + 14b + 13$ (c) $c^2 + 4c + 4$ (d) $d^2 + 6d + 8$ (e) $h^2 + 9h + 18$
(f) $x^2 + 2xy + y^2$ (g) $k^2 - 4k + 3$ (h) $m^2 - 24m + 23$ (i) $n^2 - 9n + 8$
(j) $r^2 - 8r + 15$ (k) $s^2 - 11s + 8$ (l) $u^2 - 2uv + v^2$ (m) $w^2 + 6w - 7$
(n) $x^2 + 18x - 19$ (o) $y^2 + 8y - 9$ (p) $z^2 + 3z - 10$ (q) $a^2 + 11a - 12$
(r) $b^2 + bc - 2c^2$ (s) $d^2 - d - 2$ (t) $h^2 - 10h - 11$ (u) $-k^2 + 3k - 2$
(v) $3m^2 + 6m + 3$ (w) $n^3 - 7n^2 - 60n$ (x) $3x^3 + 21x^2 + 30x$ (y) $-16y + 96 - 2y^2$
(z) $r^2s - 5rs - 36s$

PROBLEM 9-16   Factor completely using the *ac* method: (a) $3a^2 + 5a + 2$ (b) $4b^2 + 9b + 2$
(c) $3c^2 + 8c - 3$ (d) $4d^2 + 4d - 3$ (e) $5h^2 - 12h + 4$ (f) $7k^2 - 29k + 4$
(g) $6m^2 - 13m - 15$ (h) $20n^2 - 17n - 24$ (i) $-10r^2 + 3r + 4$ (j) $-s^2 - s + 20$
(k) $2u^2 + 3uv + v^2$ (l) $5x^2 - xy - 4y^2$ (m) $4w^2 + 10w + 6$ (n) $9z^2 + 48z + 15$
(o) $5a^3 - 17a^2 + 6a$ (p) $7b^3 - 24b^2 + 9b$ (q) $-12c^3 - 2c^2 + 10c$
(r) $-45d^3 - 12d^2 + 9d$ (s) $24h^3 - 24h^2 - 90h$ (t) $32k^3 - 40k^2 - 100k$
(u) $72m^5 + 102m^4 - 240m^3$ (v) $80n^5 - 68n^4 - 96n^3$ (w) $-30r^4s^2 - 74r^3s^2 - 36r^2s^2$
(x) $-27x^4y^4 + 18x^3y^3 + 24x^2y^2$ (y) $10w^2 - 19w + 6$ (z) $a^4b^2 - 10a^3b + 25a^2$

PROBLEM 9-17   Factor completely using the trial-and-error method: (a) $2a^2 + 5a + 3$
(b) $2b^2 + 11b + 12$ (c) $6c^2 - 14c + 4$ (d) $3d^2 - 13d + 4$ (e) $4h^2 + 7h - 2$
(f) $12k^2 + 19k - 18$ (g) $15m^2 - 2m - 1$ (h) $12n^2 - 11n - 5$ (i) $9r^2 + 6rs + s^2$
(j) $12u^2 - 25uv + 12v^2$ (k) $2x^2y^2 + 11xy - 6$ (l) $6w^2z^2 - 5wz - 6$
(m) $-2a^2 + 3a + 9$ (n) $-9b^2 + 6b + 8$ (o) $-c^2 + 13c + 30$ (p) $-d^2 - 5d + 84$
(q) $-h^2 + h + 1$ (r) $-12k^2 + k - 18$ (s) $-9m^2n^2 + 12mn - 4$
(t) $-30r^2s^3 + 4rs^2 + 2s$ (u) $-4x^2y^2 - 12xyz - 9z^2$ (v) $-3h^2 + 17hk + 6k^2$
(w) $-5a^4 - 5a^2bc + 10b^2c^2$ (x) $-8m^3 + 8m^2n + 16mn^2$ (y) $24u^3v + 108u^2v^2 - 60uv^3$
(z) $-36x^4y^3 + 36x^3y^2 + 16x^2y$

PROBLEM 9-18   Factor special products completely: (a) $a^2 - 1$ (b) $b^2 - 64$ (c) $4 - c^2$
(d) $16 - d^2$ (e) $4h^2 - 25$ (f) $9k^2 - 64$ (g) $100 - 49m^2$ (h) $81 - 36n^2$
(i) $r^2 + s^2$ (j) $u^2 - 9v^3$ (k) $4w^2 - 16z^2$ (l) $x^2y^2 - z^2$ (m) $a^2 + 2a + 1$
(n) $b^2 + 6b + 9$ (o) $4c^2 + 20c + 25$ (p) $9d^2 + 42d + 49$ (q) $h^2 + 2hk + k^2$
(r) $16m^2 + 24mn + 9n^2$ (s) $r^2 - 4r + 4$ (t) $s^2 - 8s + 16$ (u) $16u^2 - 40u + 25$
(v) $49v^2 - 112v + 64$ (w) $5x^2 - 405y^6$ (x) $w^4 - 16$ (y) $a^4 - 13a^2 + 36$
(z) $b^4 - 5b^2 + 4$

Answers to Supplementary Exercises

(9-13)   (a) $5(x + 2)$ (b) $3(-y + 4)$ (c) $w(w - 1)$ (d) $-z^2(z + 1)$ (e) $2a(a + 2)$
(f) $2b^2(-2b + 3)$ (g) $rs(r - s)$ (h) $-n(mn + 1)$ (i) $2c + 3d$ (j) $5a^2 - 4b^2$
(k) $12x^3(-5x + 2)$ (l) $-6m^2n^2(8m + 5n)$ (m) $3u(2u^2 + u + 5)$

(n) $12v^4(2v - 3)(v + 1)$ or $-12v^4(-2v^2 + v + 3)$ (o) $5xy(x^2 - 3xy + 2y)$

(p) $10ab(3a - 2b - 4)$ or $-10ab(-3a + 2b + 4)$ (q) $(x + 2)(x - 2)$

(r) $(2y - 1)(3y - 5)$ (s) $(w^2 + 1)(w^2 - 2)$ (t) $(3z^2 + 5)(2z - 1)$ (u) $a(a - 1)$

(v) $b(-b + 4)$ (w) $xy(x - 2y + 3)$ (x) $5h^2(2k^2 - 1)$

(y) $12d^2(2d^3 - 5d^2 + 3d - 4)$ (z) $-6(2a^2b + ab^2 + 3a + 5b)$

(9-14) (a) $(x^2 + 1)(x + 2)$ (b) $(w^2 + 2)(w + 2)$ (c) $(2x^2 + 1)(x - 3)$

(d) $(5y^2 + 2)(3y - 1)$ (e) $(3a^3 - 4)(2a + 5)$ (f) $(7b^5 - 2)(3b^3 + 5)$

(g) $5(2c^3 + 5c^2 + 3c - 4)$ (h) $8d(2d^3 - 3d^2 + 4d + 8)$ (i) $(A + B)(x + y)$

(j) $(8m - n)(A + B)$ (k) $(3w^2xy - 1)(4xy + 3)$ (l) $(2abc^2 - 3)(5 - 2ab)$

(m) $(a + 1)(a + 2)$ (n) $(b - 2)(b + 3)$ (o) $(c + 1)(c - 3)$ (p) $(d - 1)(d - 2)$

(q) $(2x + 3)(2x + 1)$ (r) $(3y - 2)(5y + 3)$ (s) $2w(3w + 2)(2w - 3)$

(t) $3z^2(5z - 2)(4z - 3)$ (u) $(u + v)^2$ (v) $(mn - 1)^2$ (w) $(2x + 3y)(5x - 2y)$

(x) $(4r + 3s)(5r - 3s)$ (y) $a(a^2 + 2ab + 1)$ (z) $2(2cd - 1)^2$

(9-15) (a) $(a + 1)^2$ (b) $(b + 1)(b + 13)$ (c) $(c + 2)^2$ (d) $(d + 2)(d + 4)$

(e) $(h + 3)(h + 6)$ (f) $(x + y)^2$ (g) $(k - 1)(k - 3)$ (h) $(m - 1)(m - 23)$

(i) $(n - 1)(n - 8)$ (j) $(r - 3)(r - 5)$ (k) will not factor   (l) $(u - v)^2$

(m) $(w - 1)(w + 7)$ (n) $(x - 1)(x + 19)$ (o) $(y - 1)(y + 9)$ (p) $(z - 2)(z + 5)$

(q) $(a - 1)(a + 12)$ (r) $(b - c)(b + 2c)$ (s) $(d + 1)(d - 2)$ (t) $(h + 1)(h - 11)$

(u) $-(k - 1)(k - 2)$ (v) $3(m + 1)^2$ (w) $n(n + 5)(n - 12)$ (x) $3x(x + 2)(x + 5)$

(y) $-2(y + 12)(y - 4)$ (z) $s(r + 4)(r - 9)$

(9-16) (a) $(a + 1)(3a + 2)$ (b) $(b + 2)(4b + 1)$ (c) $(c + 3)(3c - 1)$ (d) $(2d + 3)(2d - 1)$

(e) $(h - 2)(5h - 2)$ (f) $(k - 4)(7k - 1)$ (g) $(m - 3)(6m + 5)$ (h) $(5n - 8)(4n + 3)$

(i) $(-5r + 4)(2r + 1)$ (j) $(s + 5)(-s + 4)$ (k) $(2u + v)(u + v)$

(l) $(x - y)(5x + 4y)$ (m) $2(2w + 3)(w + 1)$ (n) $3(z + 5)(3z + 1)$

(o) $a(a - 3)(5a - 2)$ (p) $b(b - 3)(7b - 3)$ (q) $-2c(c + 1)(6c - 5)$

(r) $-3d(5d + 3)(3d - 1)$ (s) $6h(2h - 5)(2h + 3)$ (t) $4k(2k - 5)(4k + 5)$

(u) $6m^3(3m + 8)(4m - 5)$ (v) $4n^3(5n - 8)(4n + 3)$ (w) $-2r^2s^2(5r + 9)(3r + 2)$

(x) $-3x^2y^2(3xy - 4)(3xy + 2)$ (y) $(2w - 3)(5w - 2)$ (z) $a^2(ab - 5)^2$

(9-17) (a) $(a + 1)(2a + 3)$ (b) $(b + 4)(2b + 3)$ (c) $2(c - 2)(3c - 1)$ (d) $(d - 4)(3d - 1)$

(e) $(h + 2)(4h - 1)$ (f) $(3k - 2)(4k + 9)$ (g) $(3m - 1)(5m + 1)$

(h) $(3n + 1)(4n - 5)$ (i) $(3r + s)^2$ (j) $(3u - 4v)(4u - 3v)$ (k) $(xy + 6)(2xy - 1)$

(l) $(2wz - 3)(3wz + 2)$ (m) $-(a - 3)(2a + 3)$ (n) $-(3b + 2)(3b - 4)$

(o) $-(c + 2)(c - 15)$ (p) $-(d - 7)(d + 12)$ (q) will not factor   (r) will not factor

(s) $-(3mn - 2)^2$ (t) $-2s(3rs - 1)(5rs + 1)$ (u) $-(2xy + 3z)^2$

(v) $-(h - 6k)(3h + k)$ (w) $-5(a^2 - bc)(a^2 + 2bc)$ (x) $-8m(m + n)(m - 2n)$

(y) $12uv(u + 5v)(2u - v)$ (z) $-4x^2y(3xy - 4)(3xy + 1)$

(9-18) (a) $(a + 1)(a - 1)$ (b) $(b + 8)(b - 8)$ (c) $(2 + c)(2 - c)$ (d) $(4 + d)(4 - d)$

(e) $(2h + 5)(2h - 5)$ (f) $(3k + 8)(3k - 8)$ (g) $(10 + 7m)(10 - 7m)$

(h) $9(3 + 2n)(3 - 2n)$ (i) will not factor   (j) will not factor   (k) $4(w - 2z)(w + 2z)$

(l) $(xy + z)(xy - z)$ (m) $(a + 1)^2$ (n) $(b + 3)^2$ (o) $(2c + 5)^2$ (p) $(3d + 7)^2$

(q) $(h + k)^2$ (r) $(4m + 3n)^2$ (s) $(r - 2)^2$ (t) $(s - 4)^2$ (u) $(4u - 5)^2$

(v) $(7v - 8)^2$ (w) $5(x + 9y^3)(x - 9y^3)$ (x) $(w^2 + 4)(w + 2)(w - 2)$

(y) $(a + 2)(a - 2)(a + 3)(a - 3)$ (z) $(b + 1)(b - 1)(b + 2)(b - 2)$

MIDTERM EXAMINATION

Chapters 1–9

Part 1: Skills and Concepts (90 questions)

1. 18 factored as a product of primes is
   (a) $2 \times 2 \times 3$ (b) 2×9 (c) $2 \times 3 \times 3$ (d) 3×6 (e) none of these.

2. The correct symbol for ? in $\frac{12}{15}$? $\frac{15}{18}$ is
   (a) $=$ (b) $<$ (c) $>$ (d) all of these   (e) none of these.

3. $\frac{18}{12}$ in simplest form is
   (a) $\frac{4}{6}$ (b) $\frac{1}{2}$ (c) $\frac{2}{3}$ (d) $\frac{8}{12}$ (e) none of these.

4. The product of $\frac{2}{3} \times \frac{3}{4}$ in simplest form is
   (a) $\frac{1}{2}$ (b) $\frac{6}{16}$ (c) $\frac{2}{4}$ (d) $\frac{3}{6}$ (e) all of these.

5. The quotient of $\frac{3}{4} \div \frac{8}{9}$ in simplest form is
   (a) $\frac{2}{3}$ (b) $\frac{3}{2}$ (c) $\frac{27}{32}$ (d) $\frac{32}{27}$ (e) none of these.

6. The sum of $\frac{2}{3} + \frac{3}{4}$ in simplest form is
   (a) $\frac{17}{12}$ (b) $\frac{3}{2}$ (c) $\frac{6}{12}$ (d) $\frac{1}{2}$ (e) none of these.

7. The difference between $\frac{7}{10} - \frac{1}{5}$ in simplest form is
   (a) $\frac{6}{5}$ (b) $\frac{1}{2}$ (c) $\frac{5}{10}$ (d) $\frac{9}{10}$ (e) none of these.

8. The correct symbol for the ? in 365.24717 ? 365.24709 is
   (a) $<$ (b) $>$ (c) $=$ (d) all of these   (e) none of these.

9. 325.963 rounded to the nearest tenth is
   (a) 325.9   (b) 325   (c) 326.0   (d) 326   (e) none of these.

10. The sum of $8.25 + 4.63 + 0.25 + 0.125 + 128$ is
    (a) 1566   (b) 141.255   (c) 141.155   (d) 140.255   (e) none of these.

11. The difference between $1.5 - 0.125$ is
    (a) 2.5   (b) 1.425   (c) 1.375   (d) 1.625   (e) none of these.

12. The product of 3.4×1.02 is
    (a) 3.468   (b) 34.68   (c) 346.8   (d) 3468   (e) none of these.

13. The quotient of $25.2 \div 0.02$ is
    (a) 126   (b) 12.6   (c) 1260   (d) 12,600   (e) none of these.

14. 0.125 renamed as an equal fraction in simplest form is
    (a) $\frac{125}{100}$ (b) $\frac{125}{1000}$ (c) $12\frac{1}{2}$ (d) $\frac{1}{8}$ (e) none of these.

15. $\frac{5}{8}$ renamed as an equal decimal is
    (a) 6.25   (b) 0.75   (c) 1.6   (d) 0.625   (e) none of these.

16. 12% renamed as an equal decimal is
    (a) 12.0   (b) 1.2   (c) 0.12   (d) 120   (e) none of these.

17. $\frac{1}{3}$ renamed as an equal percent is
    (a) $\frac{1}{3}\%$ (b) $\frac{100}{3}$ (c) $\frac{3}{100}\%$ (d) 33%   (e) none of these.

18. $\frac{2}{5}^3$ evaluated is
    (a) $\frac{8}{125}$ (b) $\frac{8}{5}$ (c) $\frac{6}{5}$ (d) all of these   (e) none of these.

19. The square root of $\frac{49}{100}$ is
    (a) $\frac{7}{10}$ (b) 0.7   (c) 70%   (d) all of these   (e) none of these.

20. The opposite of $\frac{3}{4}$ is
    (a) $\frac{3}{4}$ (b) $-\frac{3}{4}$ (c) $\frac{4}{3}$ (d) all of these   (e) none of these.

21. The reciprocal of $\frac{3}{4}$ is
    (a) $\frac{3}{4}$ (b) $-\frac{3}{4}$ (c) $\frac{4}{3}$ (d) all of these   (e) none of these.

22. The absolute value of $\frac{3}{4}$ is
    (a) $\frac{3}{4}$ (b) $-\frac{3}{4}$ (c) $\frac{4}{3}$ (d) all of these   (e) none of these.

23. Which of the following are rational numbers?
    (a) integers   (b) whole numbers   (c) natural numbers   (d) all of these
    (e) none of these.

24. Which of the following are not rational numbers?
    (a) $\frac{2}{3}$ (b) $\sqrt{4}$ (c) $-0.\overline{3}$ (d) all of these   (e) none of these.

25. The correct symbol for ? in $-5\,?\,-4$ is
    (a) $<$ (b) $>$ (c) $=$ (d) all of these   (e) none of these.

26. The sum of $-5 + 4$ is
    (a) 1   (b) -1 (c) 9   (d) -9 (e) none of these.

27. The difference of $2 - (-7)$ is
    (a) 5   (b) -5 (c) 9   (d) -9 (e) none of these.

28. The product of $-3(-2)$ is
    (a) -6 (b) -5 (c) 5   (d) 6   (e) none of these.

29. The quotient of $\dfrac{-12}{3}$ is

    (a) -4 (b) 4   (c) $\dfrac{1}{4}$ (d) $-\dfrac{1}{4}$ (e) none of these.

30. $\dfrac{3}{4}$ is equal to

    (a) $\dfrac{-3}{4}$ (b) $\dfrac{-3}{-4}$ (c) $-\dfrac{-3}{-4}$ (d) all of these   (e) none of these.

31. $-\dfrac{3}{4}$ is equal to

    (a) $\dfrac{-3}{4}$ (b) $\dfrac{3}{-4}$ (c) $-\dfrac{-3}{-4}$ (d) all of these   (e) none of these.

32. $-\dfrac{-12}{18}$ simplified is

    (a) $-\dfrac{-2}{-3}$ (b) $-\dfrac{2}{3}$ (c) $\dfrac{2}{3}$ (d) all of these   (e) none of these.

33. The sum of $\frac{1}{2} + 0.75$ is
    (a) 1.25   (b) $1\frac{1}{4}$ (c) $\frac{5}{4}$ (d) all of these   (e) none of these.

34. The difference of $2.5 - 3\frac{3}{4}$ is
 (a) 1.25   (b) 6.25   (c) -6.25 (d) all of these   (e) none of these.

35. The product of $0.5(-\frac{1}{2})(-4)(0)(-\frac{3}{4})$ is
 (a) 0   (b) $-\frac{3}{4}$ (c) $\frac{3}{4}$ (d) all of these   (e) none of these.

36. The quotient of $-\frac{1}{2} \div (-0.125)$ is
 (a) -4 (b) 4   (c) 0.0625   (d) all of these   (e) none of these.

37. $\dfrac{-1 - \sqrt{1^2 - 4(6)(-12)}}{2(6)}$ evaluated is

 (a) $\dfrac{4}{3}$ (b) $-\dfrac{3}{2}$ (c) $-\dfrac{17}{12}$ (d) all of these   (e) none of these.

38. "2 times x" can be written as
 (a) $2 \cdot x$ (b) $2(x)$ (c) $2x$ (d) all of these   (e) none of these.

39. "2 divided by x" can be written as
 (a) $x \div 2$ (b) $x \cdot \dfrac{1}{2}$ (c) $\dfrac{x}{2}$ (d) all of these   (e) none of these.

40. $2(3 \cdot 4) = (3 \cdot 4)2$ is an example of
 (a) a commutative property   (b) an associative property   (c) a distributive property
 (d) all of these   (e) none of these.

41. $2(3 \cdot 4) = (2 \cdot 3)4$ is an example of
 (a) a commutative property   (b) an associative property   (c) a distributive property
 (d) all of these   (e) none of these.

42. $2(3 - 4) = 2 \cdot 3 - 2 \cdot 4$ is an example of
 (a) a commutative property   (b) an associative property   (c) a distributive property
 (d) all of these   (e) none of these.

43. $2(3 - 4) = 2(3 + (-4))$ is an example of
 (a) a commutative property   (b) an associative property   (c) a distributive property
 (d) all of these   (e) none of these.

44. $-2(x - 3)$ equals
 (a) $-2x + 6$ (b) $6 - 2x$ (c) $6 + (-2x)$ (d) all of these   (e) none of these.

45. $2x + 3y + 5 - 3x - 2y - 4$ equals
 (a) $-x + y + 1$ (b) $-x + y - 1$ (c) $x + y + 1$ (d) all of these   (e) none of these.

46. $4m - (5m - 10n - m)$ equals
 (a) $-2m - 10n$ (b) $-10n$ (c) $10n$ (d) all of these   (e) none of these.

47. For $x = -2$; $3x^2 + 5x - 6$ equals
 (a) 16   (b) -4 (c) -28 (d) -8 (e) none of these.

48. For $a = 3$, $b = 5$, and $c = -2$; $\dfrac{-b + \sqrt{b^2 - 4ac}}{2a}$ equals

 (a) $\dfrac{1}{3}$ (b) 2   (c) 1   (d) $-\dfrac{2}{3}$ (e) none of these.

49. Which of the following are linear equations in one variable?
 (a) $3x - 5 = 2x + 4$ (b) $2x + 3 = 2x - 1$ (c) $2x + 3y = 5$ (d) all of these
 (e) none of these.

50. -3 is a solution of

(a) $2x = 6$ (b) $\dfrac{x}{-3} = -1$ (c) $2x + 6 = 12$ (d) all of these   (e) none of these.

51. The solution of $x + 5 = -2$ is
(a) 3   (b) -7 (c) -3 (d) 7   (e) none of these.

52. The solution of $y - 3 = 8$ is
(a) 11   (b) 5   (c) -5 (d) -11 (e) none of these.

53. The solution of $2w = -8$ is
(a) 4   (b) -16 (c) -4 (d) -10 (e) none of these.

54. The solution of $\dfrac{z}{-2} = -4$ is

(a) 8   (b) 2   (c) -2 (d) -8 (e) none of these.

55. The solution of $2m + 3 = 1$ is
(a) -1 (b) 2   (c) -4 (d) 8   (e) none of these.

56. The solution of $\dfrac{n}{5} - 3 = -8$ is

(a) -1 (b) -25 (c) -55 (d) $-\dfrac{11}{5}$ (e) none of these.

57. The solution of $3x - 5 = 8x + 5$ is
(a) 0   (b) 2   (c) $\frac{11}{10}$ (d) -2 (e) none of these.

58. The solution of $-2(x - 3) = -4$ is
(a) 5   (b) -1 (c) 1   (d) -5 (e) none of these.

59. The solution of $2(x) - 3 = 5x - (3 + x)$ is
(a) $-\frac{4}{3}$ (b) -3 (c) -1 (d) $-\frac{3}{4}$ (e) none of these.

60. The solution of $\frac{1}{2}y + \frac{3}{4} = \frac{1}{8}$ is
(a) $\frac{5}{4}$ (b) $-\frac{5}{4}$ (c) $\frac{5}{2}$ (d) $-\frac{5}{2}$ (e) none of these.

61. The solution of $0.2w - 0.3 = 0.1$ is
(a) 0.2   (b) -1 (c) 20   (d) 2   (e) none of these.

62. The solution of $x + 33\frac{1}{3}\%x = 16$ is
(a) 12   (b) 0   (c) 1   (d) 24   (e) none of these.

63. The solution of $2x + 3y = 5$ for y is

(a) $\dfrac{5 - 3y}{2}$ (b) $2x - 5$ (c) $\dfrac{5 - 2x}{3}$ (d) $\dfrac{2x - 5}{3}$ (e) none of these.

64. The solution of $A = 2(l + w)$ for w is

(a) $\dfrac{A}{2l}$ (b) $\dfrac{A}{2} + l$ (c) $l - \dfrac{A}{2}$ (d) $\dfrac{1}{2}A - l$ (e) none of these.

65. x^6x^2 simplified is
(a) x^8 (b) x^{12} (c) x^4 (d) x^3 (e) none of these.

66. $(-5y)^2$ simplified is
(a) $-5y^2$ (b) $-25y^2$ (c) $25y^2$ (d) $25y$ (e) none of these.

67. $\dfrac{w^6}{w^2}$ simplified is

    (a) w^3 (b) w^4 (c) w^{12} (d) w^8 (e) none of these.

68. $\left(\dfrac{a}{-2}\right)^3$ simplified is

    (a) $\dfrac{a^3}{8}$ (b) $-\dfrac{a^3}{2}$ (c) $-\dfrac{a}{8}$ (d) $-\dfrac{a^3}{8}$ (e) none of these.

69. $(z^6)^2$ simplified is

    (a) z^8 (b) z^{12} (c) z^4 (d) z^3 (e) none of these.

70. $\dfrac{(4x^3)^2}{x^5}$ simplified is

    (a) 16 (b) $4x$ (c) $16x$ (d) $\dfrac{16}{x^4}$ (e) none of these.

71. $2x^{-3}$ renamed using only positive exponents is

    (a) $\dfrac{2}{x^3}$ (b) $\dfrac{1}{2x^3}$ (c) $2x^3$ (d) $\dfrac{x^3}{2}$ (e) none of these.

72. $\left(\dfrac{b}{-4}\right)^2$ simplified is

    (a) $\dfrac{b^2}{16}$ (b) $-\dfrac{b^2}{16}$ (c) $-\dfrac{16}{b^2}$ (d) $\dfrac{1}{16b^2}$ (e) none of these.

73. The sum of $(6x^2 - 2x - 7) + (8 - 3x^2)$ is

    (a) $9x^2 - 2x + 1$ (b) $9x^2 - 2x - 15$ (c) $3x^2 + 1$ (d) $3x^2 - 2x + 1$

    (e) none of these.

74. The difference between $(8y - 7) - (5 - 2y^3)$ is

    (a) $-2y^3 + 8y - 2$ (b) $-2y^3 - 8y + 12$ (c) $2y^3 + 8y - 12$ (d) $2y^3 + 8y - 2$

    (e) none of these.

75. The product of $3w(2w^2 - w + 5)$ is

    (a) $6w^3 - 3w^2 + 15w$ (b) $6w^2 - 3w + 15$ (c) $6w^3 + 3w + 15$ (d) $6w^3 - 3w^2 + 15$

    (e) none of these.

76. The product of $(2y - 3)(3y + 4)$ is

    (a) $6y^2 + y - 12$ (b) $6y^2 - y - 12$ (c) $6y^2 - 17y - 12$ (d) $6y^2 + 17y + 12$

    (e) none of these.

77. The product of $(a - 2)(a^2 + 3a - 1)$ is

    (a) $a^3 + 5a^2 + 5a - 2$ (b) $a^3 + 5a^2 - 7a + 2$ (c) $a^3 + a^2 - 6a + 2$

    (d) $a^3 + a^2 - 7a + 2$ (e) none of these.

78. The product of $(3m - n)(3m + n)$ is

    (a) $9m^2 + n^2$ (b) $9m^2 - n^2$ (c) $9m^2 - 6mn - n^2$ (d) $9m^2 - 6mn + n^2$

    (e) none of these.

79. The product of $(r + s)^2$ is

    (a) $r^2 + 2rs + s^2$ (b) $r^2 + s^2$ (c) $r^2 - s^2$ (d) $r^2 - 2rs + s^2$ (e) none of these.

80. The product of $(2h - 3k)^2$ is

    (a) $4h^2 + 12hk + 9k^2$ (b) $4h^2 - 9k^2$ (c) $4h^2 + 9k^2$ (d) $4h^2 - 12hk + 9k^2$

    (e) none of these.

81. The quotient of $(x^2 - 12x + 3) \div (6x)$ is

    (a) $\dfrac{x^3}{6} - 2x^2 + \dfrac{x}{2}$ (b) $\dfrac{x}{6} + 2 + \dfrac{1}{2x}$ (c) $\dfrac{x}{6} - 2 + \dfrac{1}{2x}$

    (d) $\dfrac{1}{6} - 2 + \dfrac{1}{2x}$ (e) none of these.

82. The quotient of $(w^2 - 5w + 3) \div (w - 1)$ is

    (a) $w - 4 - \dfrac{1}{w - 1}$ (b) $w - 6 + \dfrac{9}{w - 1}$ (c) $w - 4$ (d) $w - 6$ (e) none of these.

83. $4x^4 - 6x^3 + 10x^2$ factored completely is

    (a) $2x^2(2x^2 - 3x + 5)$ (b) $2x^2(2x - 5)(x + 1)$ (c) $x^2(4x^2 - 6x + 10)$
    (d) will not factor over the integers   (e) none of these.

84. $y^3 + 3y^2 + 2y + 6$ factored completely is

    (a) $y(y^2 + 3y + 2) + 6$ (b) $y(y + 1)(y + 2) + 6$ (c) $(y^2 + 2)(y + 3)$
    (d) will not factor over the integers   (e) none of these.

85. $w^2 - w - 6$ factored completely is

    (a) $w(w - 1) - 6$ (b) $w^2 - (w + 6)$ (c) $(w + 2)(w - 3)$
    (d) will not factor over the integers   (e) none of these.

86. $2a^2 + 5a - 12$ factored completely is

    (a) $(2a + 3)(a - 4)$ (b) $(2a - 3)(a + 4)$ (c) $(2a - 1)(a + 12)$
    (d) will not factor over the integers   (e) none of these.

87. $16m^2 + 9n^2$ factored completely is

    (a) $(4m + 3n)^2$ (b) $(4m + 3n)(4m - 3n)$ (c) $(4m - 3n)^2$
    (d) will not factor over the integers   (e) none of these.

88. $25h^2 - 4k^2$ factored completely is

    (a) $(5h + 2k)^2$ (b) $(5h + 2k)(5h - 2k)$ (c) $(5h - 2k)^2$
    (d) will not factor over the integers   (e) none of these.

89. $a^2 + 2ab + b^2$ factored completely is

    (a) $(a + b)^2$ (b) $(a + b)(a - b)$ (c) $(a - b)^2$ (d) will not factor over the integers
    (e) none of these.

90. $4x^4 - 8x^2 + 4$ factored completely is

    (a) $4(x^2 - 1)^2$ (b) $(2x^2 - 2)^2$ (c) $4(x + 1)^2(x - 1)^2$
    (d) will not factor over the integers   (e) none of these.

Part 2: Problem Solving (10 questions)

91. One number is 4 more than 3 times another number. The difference between the two numbers is 18. What are the numbers?

92. The sum of three consecutive odd integers is 105. Find the integers.

93. The area of a rectangle is 104 m². If the width of the rectangle is 8 m, what is the perimeter?

94. The perimeter of a rectangle is 582 ft. The length of the rectangle is twice the width. What is the area of the rectangle?

95. Jerry gave Don 3 one-dollar bills in exchange for an equal value of dimes and nickels. If Done gave Jerry 50 coins in all, what was the value of the nickels?

96. A stationery store sells 3 black pens for each red pen sold. Last week the store sold $288.60 worth of pens—both black and red. If black pens cost 25¢ each and red pens cost 55¢ each, what was the value of the black pens sold last week?

97. Gail invested half her money at 6% per year and half at 8% per year to earn $378 per year. How much was Gail's total investment?

98. Richard invested $10,000, part at 8% per year and the rest at 12% per year. His annual return on investment is $1000. How much interest was earned by the 12% investment? How much was invested at 8%?

99. Starting at the same place, two planes flew in opposite directions. One plane traveled 100 mph faster than the other plane. In 6 hours the planes were 4800 miles apart. How far did the faster plane travel?

100. Two trains are 460 miles apart and traveling toward each other on parallel tracks. The constant rate of one train is 6 mph slower than the other train. After 5 hours, the two trains pass each other. How far did the slower train travel to the passing point?

Midterm Examination Answers

Part 1

1. (c)	**2.** (b)	**3.** (e)	**4.** (a)	**5.** (c)
6. (a)	**7.** (b)	**8.** (b)	**9.** (c)	**10.** (b)
11. (c)	**12.** (a)	**13.** (c)	**14.** (d)	**15.** (d)
16. (c)	**17.** (e)	**18.** (b)	**19.** (d)	**20.** (b)
21. (c)	**22.** (a)	**23.** (d)	**24.** (e)	**25.** (a)
26. (b)	**27.** (c)	**28.** (d)	**29.** (a)	**30.** (b)
31. (d)	**32.** (c)	**33.** (d)	**34.** (e)	**35.** (a)
36. (b)	**37.** (b)	**38.** (d)	**39.** (e)	**40.** (a)
41. (b)	**42.** (c)	**43.** (e)	**44.** (d)	**45.** (a)
46. (c)	**47.** (b)	**48.** (a)	**49.** (a)	**50.** (e)
51. (b)	**52.** (a)	**53.** (c)	**54.** (a)	**55.** (a)
56. (b)	**57.** (d)	**58.** (a)	**59.** (e)	**60.** (b)
61. (d)	**62.** (a)	**63.** (c)	**64.** (d)	**65.** (a)
66. (c)	**67.** (b)	**68.** (d)	**69.** (b)	**70.** (c)
71. (a)	**72.** (a)	**73.** (d)	**74.** (c)	**75.** (a)
76. (b)	**77.** (d)	**78.** (b)	**79.** (a)	**80.** (d)
81. (c)	**82.** (a)	**83.** (a)	**84.** (c)	**85.** (c)
86. (b)	**87.** (d)	**88.** (b)	**89.** (a)	**90.** (c)

Part 2

91. 7 and 25	**92.** 33, 35, and 37	**93.** 42 m	**94.** 18,818 ft^2
95. $2	**96.** $166.50	**97.** $5400	**98.** $600, $5000
99. 2700 miles	**100.** 215 miles		

10 GRAPHING

THIS CHAPTER IS ABOUT

☑ **Plotting Points and Finding Coordinates**
☑ **Finding Solutions of Linear Equations in Two Variables**
☑ **Graphing by Plotting Points**
☑ **Finding the Slope**
☑ **Graphing Using the Slope-Intercept Method**

10-1. Plotting Points and Finding Coordinates

A. Identify the parts of a rectangular coordinate system.

A **rectangular coordinate system,** or **Cartesian coordinate system,** is used to study the relationship between geometric figures and algebraic equations.

EXAMPLE 10-1: Draw and label a rectangular coordinate system.

Solution:

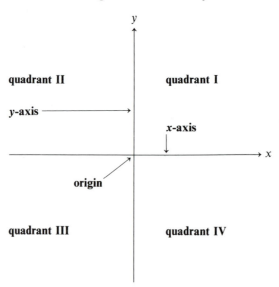

In a rectangular coordinate system:

1. The horizontal number line is called the **x-axis** and is labeled with an "*x*".
2. The vertical number line is called the **y-axis** and is labeled with a "*y*".
3. The *x*- and *y*-axes intersect at a point called the **origin.**
4. Zero is at the origin on both the *x*- and *y*-axes.
5. The positive numbers on the *x*-axis are to the right of the origin.
6. The positive numbers on the *y*-axis are above the origin.
7. The *x*- and *y*-axes are **perpendicular.** That is, the *x*- and *y*-axes intersect at right angles.
8. The *x*- and *y*-axes separate the rectangular coordinate system into four **quadrants.**
9. The four quadrants are numbered I, II, III, and IV in a counterclockwise direction starting with the upper right quadrant.

B. Identify coordinates given an ordered pair.

In the **ordered pair** (x, y), the numbers x and y are called **coordinates**. x is called the **first coordinate** or **x-coordinate,** and y is called the **second coordinate** or **y-coordinate.**

EXAMPLE 10-2: Identify the coordinates of the following ordered pairs: **(a)** (3, 4) **(b)** (4, 3)

Solution: **(a)** In (3, 4), the first coordinate or *x*-coordinate is 3 and
the second coordinate or *y*-coordinate is 4.
(b) In (4, 3), the first coordinate or *x*-coordinate is 4 and
the second coordinate or *y*-coordinate is 3.

C. Plot the point that represents a given ordered pair.

To **plot a point** on a rectangular coordinate system for a given ordered pair (x, y), you locate, draw, and label the point that represents (x, y) on the rectangular coordinate system.

To locate the point that represents a given ordered pair (x, y) on a rectangular coordinate system, you
1. start at the origin (0, 0);

2. then move $\begin{cases} \text{right } x \text{ units if } x \text{ is positive} \\ \text{left } x \text{ units if } x \text{ is negative} \\ \text{nowhere if } x \text{ is zero} \end{cases}$;

3. and then move $\begin{cases} \text{up } y \text{ units if } y \text{ is positive} \\ \text{down } y \text{ units if } y \text{ is negative} \\ \text{nowhere if } y \text{ is zero} \end{cases}$.

EXAMPLE 10-3: Plot the point that represents the ordered pair (3, −4).

Solution:

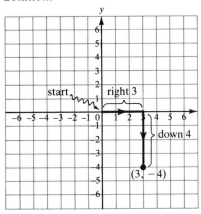

Locate: Start at the origin (0, 0),
then move right 3,
then move down 4.
Draw a bold dot to show the located point.
Label the located point with the given ordered pair.

Note: To plot a point on a rectangular coordinate system for a given ordered pair (x, y):
1. The *x*-coordinate directs left and right movement.
2. The *y*-coordinate directs up and down movement.

D. Identify the location of a point for a given ordered pair without plotting.

To identify the location of a point for a given ordered pair without plotting, you can use the following table:

If the signs of the coordinates of (x, y) are:	then the location of the point is:
(+, +)	in quadrant I
(−, +)	in quadrant II
(−, −)	in quadrant III
(+, −)	in quadrant IV
(0, +) or (0, −)	on the *y*-axis
(+, 0) or (−, 0)	on the *x*-axis
(0, 0)	at the origin

EXAMPLE 10-4: Identify the location of the following ordered pairs without plotting: **(a)** $(3, -4)$ **(b)** $(0, 2)$ **(c)** $(-5, 0)$

Solution: **(a)** In $(3, -4)$, the x-coordinate 3 is positive $(+)$,
the y-coordinate -4 is negative $(-)$, and
$(+, -)$ means that the point is in quadrant IV [see Example 10-3].
(b) In $(0, 2)$, the x-coordinate is 0 means that the point is on the y-axis.
(c) In $(-5, 0)$, the y-coordinate is 0 means that the point is on the x-axis.

Note: For the ordered pair (x, y):
1. when the x-coordinate is 0, the point is on the y-axis.
2. when the y-coordinate is 0, the point is on the x-axis.
3. when both the x- and y-coordinates are 0, the point is at the origin.

E. Write the ordered pair that is represented by a given plotted point.

To write the ordered pair that is represented by a given point on a rectangular coordinate system, you

1. Find the x-coordinate m by drawing a vertical guideline through the x-axis at $x = m$.
2. Find the y-coordinate n by drawing a horizontal guideline through the y-axis at $y = n$.
3. Use the x-coordinate m and the y-coordinate n to write the ordered pair (m, n).

EXAMPLE 10-5: Write the ordered pair that represents the point A shown on the following rectangular coordinate system:

Solution:

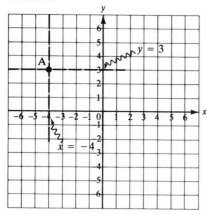

The ordered pair that represents point A is $(-4, 3)$ because the vertical guideline through A crosses the x-axis at $(-4, 0)$ and the horizontal guideline crosses the y-axis at $(0, 3)$.

Note: To write the ordered pair (x, y) for a given point on a rectangular coordinate system:

1. use a vertical guideline to find the x-coordinate.
2. use a horizontal guideline to find the y-coordinate.

10-2. Finding Solutions of Linear Equations in Two Variables

A. Identify linear equations in two variables.

Recall: An equation that can be written in standard form as $Ax + B = C$, where A, B, and C are real numbers $[A \neq 0]$ and x is any variable, is called a linear equation in one variable [see Example 5-1].

An equation that can be written in **standard form** as $Ax + By = C$ where A, B, and C are real numbers $[A$ and B are not both 0$]$ and x and y are any two different variables, is called a **linear equation in two variables.**

Note: In the equation $Ax + By = C$, the phrase "A and B are not both 0" means that the two variables are never both missing at the same time.

EXAMPLE 10-6: Identify each equation that can be written as a linear equation in two variables:

(a) $2x + 3y = 5$ (b) $3m - 4n = -2$ (c) $\frac{u}{2} = 3v + 1$ (d) $-2(5 - a) = b$

(e) $3x = 5$ (f) $2y + 3 = 0$ (g) $x^2 + 3y = 5$ (h) $\frac{2}{x} + y = 5$

(i) $2xy + 3 = 5$ (j) $2 + 3 = 5$ (k) $2x + 3y = 5z$ (l) $2\sqrt{x} + 3y = 5$

Solution:
(a) $2x + 3y = 5$ is a linear equation in two variables in standard form.
(b) $3m - 4n = -2$ or $3m + (-4)n = -2$ is a linear equation in two variables in standard form.
(c) $\frac{u}{2} = 3v + 1$ is a linear equation in two variables in standard form as $\frac{1}{2}u - 3v = 1$.
(d) $-2(5 - a) = b$ is a linear equation in two variables in standard form as $2a - b = 10$.
(e) $3x = 5$ can be written as a linear equation in two variables as $3x + 0y = 5$.
    [See the following *Note.*]
(f) $2y + 3 = 0$ can be written as a linear equation in two variables as $0x + 2y = -3$.
(g) $x^2 + 3y = 5$ is not a linear equation because x has an exponent greater than 1.
(h) $\frac{2}{x} + y = 5$ is not a linear equation because there is a variable in a denominator.
(i) $2xy + 3 = 5$ is not a linear equation because it includes a product of variables.
(j) $2 + 3 = 5$ is not a linear equation because both variables are missing at the same time.
(k) $2x + 3y = 5z$ is not a linear equation in two variables because it has more than two different
    variables.
(l) $2\sqrt{x} + 3y = 5$ is not a linear equation because it contains a variable under a radical symbol.

Note: Every linear equation in one variable can be written as a linear equation in two variables.

B. Check proposed solutions of equations in two variables.

An ordered pair (m, n) is **a solution of an equation in two variables** x and y if substituting m for x and n for y produces a true number sentence.

EXAMPLE 10-7: Check each proposed solution of $2x + 3y = 1$: (a) $(-1, 2)$ (b) $(2, -1)$.

Solution: (a) $(-1, 2)$ is not a solution of $2x + 3y = 1$ because

$$2x + 3y = 1 \longleftarrow \text{given equation}$$

$2(-1) + 3(2)$	1 *Substitute:* $(-1, 2)$ means $x = -1$ and $y = 2$.
$-2 + 6$	1 *Compute.*
4	1 *Compare:* $4 \neq 1$ means that $(-1, 2)$ does not check.

(b) $(2, -1)$ is a solution of $2x + 3y = 1$ because

$$2x + 3y = 1 \longleftarrow \text{given equation}$$

$2(2) + 3(-1)$	1 *Substitute:* $(2, -1)$ means $x = 2$ and $y = -1$.
$4 + (-3)$	1 *Compute.*
1	1 *Compare:* $1 = 1$ means $(2, -1)$ checks.

Therefore, $(2, -1)$ is a correct solution for $2x + 3y = 1$ because substituting 2 for x and -1 for y produces a true number sentence.

C. Find solutions of a linear equation in two variables.

In Example 10-7, the linear equation in two variables $2x + 3y = 1$ has more than just the one solution $(2, -1)$.

To find solutions of a linear equation in two variables like $2x + 3y = 1$, you

1. Substitute any real number m for one of the variables.
2. Solve for the remaining variable to get another real number n.
3. Write the solution as an ordered pair using the coordinates m and n from Steps 1 and 2.

EXAMPLE 10-8: Find solutions of the linear equation $2x + 3y = 1$ for: **(a)** $x = 0$ **(b)** $y = 0$
(c) $x = 1$ **(d)** $y = 1$

Solution: **(a)** Solve $2x + 3y = 1$ for $x = 0$.

$$2(0) + 3y = 1 \qquad \textit{Think: } x = 0$$

$$0 + 3y = 1 \longleftarrow \text{ linear equation in one variable}$$

$$3y = 1 \qquad \text{Solve as before [see Example 5-7].}$$

$$y = \tfrac{1}{3}$$

$x = 0$ and $y = \tfrac{1}{3}$ means $(0, \tfrac{1}{3})$ is a solution of $2x + 3y = 1$ [check as before].

(b) Solve $2x + 3y = 1$ for $y = 0$.

$$2x + 3(0) = 1 \qquad \textit{Think: } y = 0$$

$$2x + 0 = 1 \longleftarrow \text{ linear equation in one variable}$$

$$2x = 1 \qquad \text{Solve as before [see Example 5-7].}$$

$$x = \tfrac{1}{2}$$

$y = 0$ and $x = \tfrac{1}{2}$ means $(\tfrac{1}{2}, 0)$ is a solution of $2x + 3y = 1$ [check as before].

(c) Solve $2x + 3y = 1$ for $x = 1$.

$$2(1) + 3y = 1 \qquad \textit{Think: } x = 1$$

$$2 + 3y = 1 \longleftarrow \text{ linear equation in one variable}$$

$$3y = -1 \qquad \text{Solve as before [see Example 5-9].}$$

$$y = -\tfrac{1}{3}$$

$x = 1$ and $y = -\tfrac{1}{3}$ means $(1, -\tfrac{1}{3})$ is a solution of $2x + 3y = 1$ [check as before].

(d) Solve $2x + 3y = 1$ for $y = 1$.

$$2x + 3(1) = 1 \qquad \textit{Think: } y = 1$$

$$2x + 3 = 1 \longleftarrow \text{ linear equation in one variable}$$

$$2x = -2 \qquad \text{Solve as before [see Example 5-9].}$$

$$x = -1$$

$y = 1$ and $x = -1$ means $(-1, 1)$ is a solution of $2x + 3y = 1$ [check as before].

Note: For any value of x or y, there is a solution of $2x + 3y = 1$. Or, put another way, $2x + 3y = 1$ has infinitely many solutions.

Every linear equation in two variables has infinitely many solutions.

EXAMPLE 10-9: Show that each linear equation in two variables has infinitely many solutions:
(a) $2x + 3y = 6$ **(b)** $2x + 0y = 6$ or $2x = 6$

Solution: **(a)** For $x = 0, 1, 2, 3, \cdots$, the solutions of $2x + 3y = 6$ are:

$(0, 2), (1, \tfrac{4}{3}), (2, \tfrac{2}{3}), (3, 0), \cdots$, respectively. [Check as before.]

(b) For $y = 0, 1, 2, 3, \cdots$, the solutions of $2x + 0y = 6$ or $2x = 6$ are:

$(3, 0), (3, 1), (3, 2), (3, 3), \cdots$, respectively. [Check as before.]

10-3 Graphing by Plotting Points

A. Graph equations in two variables by plotting points.

The **graph of an equation in two variables** is the representation of all ordered pair solutions on a rectangular coordinate system.

To **graph an equation in two variables by plotting points,** you

1. Make a table of several solutions using values like $x = 3, 2, 1, 0, -1, -2,$ and -3.
2. Plot each solution from the table from Step 1 on a rectangular coordinate system.
3. Draw the graph of the given equation by connecting the points from Step 2.
4. Label your graph from Step 3 with the original equation.

EXAMPLE 10-10: Graph each equation in two variables by plotting points:
(a) $2x + 3y = 6$ **(b)** $y = |x|$ **(c)** $y = x^2$ **(d)** $y = 2^x$

Solution:

Make a Table	Plot Points	Draw and Label Graph

(a)

x	y	
3	0	← $2(3) + 3y = 6$
2	$\frac{2}{3}$	← $2(2) + 3y = 6$
1	$\frac{4}{3}$	← $2(1) + 3y = 6$
0	2	← $2(0) + 3y = 6$
−1	$\frac{8}{3}$	← $2(-1) + 3y = 6$
−2	$\frac{10}{3}$	← $2(-2) + 3y = 6$
−3	4	← $2(-3) + 3y = 6$

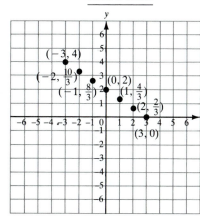
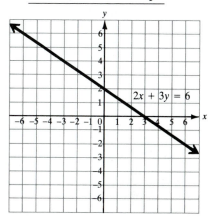

[See the following *Note. 1.*]

(b)

x	y	
3	3	← $y = |3|$
2	2	← $y = |2|$
1	1	← $y = |1|$
0	0	← $y = |0|$
−1	1	← $y = |-1|$
−2	2	← $y = |-2|$
−3	3	← $y = |-3|$

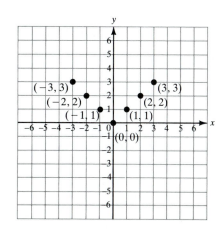
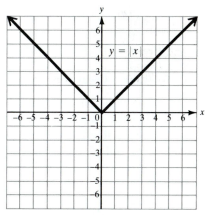

[See the following *Note 2.*]

(c)

x	y	
3	9	← $y = 3^2$
2	4	← $y = 2^2$
1	1	← $y = 1^2$
0	0	← $y = 0^2$
−1	1	← $y = (-1)^2$
−2	4	← $y = (-2)^2$
−3	9	← $y = (-3)^2$

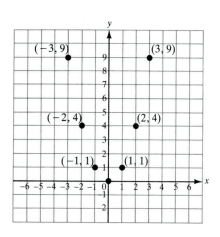
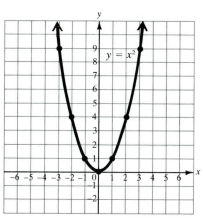

[See the following *Note 3.*]

Make a Table	Plot Points	Draw and Label Graph

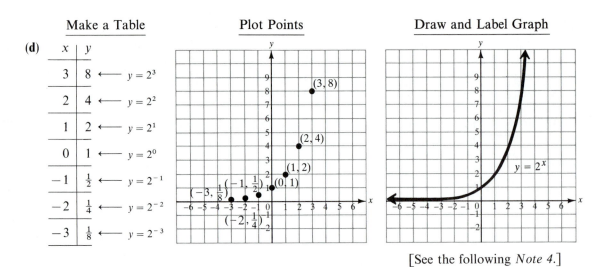

(d)

x	y	
3	8	$\leftarrow y = 2^3$
2	4	$\leftarrow y = 2^2$
1	2	$\leftarrow y = 2^1$
0	1	$\leftarrow y = 2^0$
-1	$\frac{1}{2}$	$\leftarrow y = 2^{-1}$
-2	$\frac{1}{4}$	$\leftarrow y = 2^{-2}$
-3	$\frac{1}{8}$	$\leftarrow y = 2^{-3}$

[See the following *Note 4*.]

Note 1: The graph of every linear equation in two variables is a straight line.

Note 2: The graph of an **absolute value equation in two variables** like $y = |x|$ is usually V-shaped.

Note 3: The cup-shaped graph of a **quadratic equation in two variables** like $y = x^2$ is called a **parabola.**

Note 4: The graph of an **exponential equation in two variables** like $y = 2^x$ is usually shaped like a park slide.

B. Graph linear equations in two variables using the intercept method.

A linear equation in two variables is called "linear" because its graph is always a straight line. Because the graph of every linear equation in two variables is always a straight line, you need only find and plot two ordered pair solutions to graph it. However, it is a good idea to find and plot one extra ordered pair solution to check your straight line.

To **graph a linear equation in two variables using the intercept method,** you

1. Find the **y-intercept** by substituting $x = 0$ in the original equation and then solving for y.
2. Find the **x-intercept** by substituting $y = 0$ in the original equation and then solving for x.
3. Draw a line through the x- and y-intercepts using a straightedge.
4. Find and plot a **check point** by substituting $x = 1$ [or $y = 1$] in the original equation and then solving for y [or for x].
5. Label your graph with the original equation.

EXAMPLE 10-11: Graph the following equations using the intercept method:
(a) $2x + 3y = 6$ **(b)** $x + y = 0$ **(c)** $2x = 6 \,[2x + 0y = 6]$ **(d)** $3y = 6 \,[0x + 3y = 6]$

Solution:

$$2x + 3y = 6$$

(a) For $x = 0$, $\overparen{2(0) + 3y = 6}$ or $y = 2 \longrightarrow$

x	y	
0	2	$\leftarrow y$ – intercept
3	0	$\leftarrow x$ -intercept
1	$\frac{4}{3}$	$\leftarrow$ check point

For $y = 0$, $2x + 3(0) = 6$ or $x = 3 \longrightarrow$

For $x = 1$, $2(1) + 3y = 6$ or $y = \frac{4}{3} \longrightarrow$

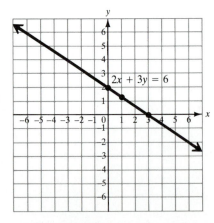

[See the following *Note 1*.]

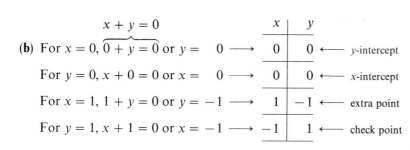
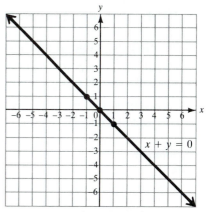

(b) For $x = 0$, $\overbrace{0 + y = 0}$ or $y = 0 \longrightarrow$

For $y = 0$, $x + 0 = 0$ or $x = 0 \longrightarrow$

For $x = 1$, $1 + y = 0$ or $y = -1 \longrightarrow$

For $y = 1$, $x + 1 = 0$ or $x = -1 \longrightarrow$

x	y	
0	0	← *y*-intercept
0	0	← *x*-intercept
1	−1	← extra point
−1	1	← check point

[See the following *Note 2*.]

$2x = 6$ or $2x + 0y = 6$

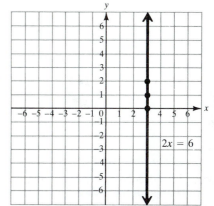

(c) For $x = 0$, $\overbrace{2(0) = 6}$
or $0 = 6$ [false]
For any value of y,
$2x + 0y = 6$ or $x = 3$

x	y	
—	—	← no *y*-intercept
3	0	← *x*-intercept
3	1	← extra point
3	2	← check point

[See the following *Note 3*.]

$3y = 6$

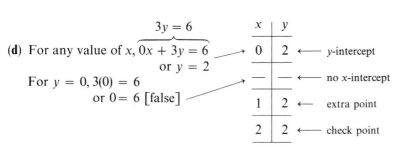
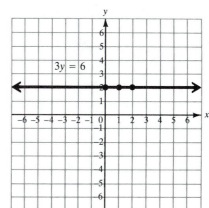

(d) For any value of x, $\overbrace{0x + 3y = 6}$
or $y = 2$
For $y = 0$, $3(0) = 6$
or $0 = 6$ [false]

x	y	
0	2	← *y*-intercept
—	—	← no *x*-intercept
1	2	← extra point
2	2	← check point

[See the following *Note 4*.]

Note 1: In $Ax + By = C$ [A, B, and $C \neq 0$], the *x*- and *y*-intercepts are always represented by different points. That is, to graph $Ax + By = C$ [A, B, and $C \neq 0$], you only need to plot the *x*- and *y*-intercepts.

Note 2: In $Ax + By = 0$ [$C = 0$], the *x*- and *y*-intercepts are always represented by the same point $(0, 0)$. That is, to graph $Ax + By = 0$, you need to find and plot an extra point because both the *x*- and *y*-intercepts are at the origin.

Note 3: For $Ax = C$ [$A \neq 0$], the graph will always be a vertical line. That is, to graph $Ax = C$ [$A \neq 0$], you need only solve $Ax = C$ for x to find the *x*-intercept and then draw a vertical line through the *x*-intercept.

Note 4: For $By = C$ [$B \neq 0$], the graph will always be a horizontal line. That is, to graph $By = C$ [$B \neq 0$], you need only solve $By = C$ for y to find the *y*-intercept and then draw a horizontal line through the *y*-intercept.

10-4. Finding the Slope

A. Find the slope of a straight line given the graph of the line.

To get from one point to another point in a rectangular coordinate system, you can either first move vertically and then horizontally, or you can first move horizontally and then vertically.

EXAMPLE 10-12: Show how to get from $(2, -1)$ to $(-4, 3)$ by **(a)** first moving vertically and then horizontally; and **(b)** first moving horizontally and then vertically.

Solution:

(a)

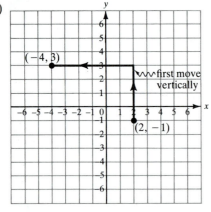

(b)

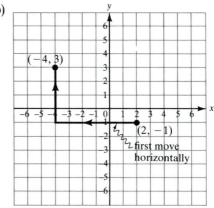

When you use only one vertical move and one horizontal move to get from one point to another point on a rectangular coordinate system, you call the direction and amount of vertical movement the **change in the y-direction** or the **change in y** and the direction and the amount of horizontal movement, the **change in the x-direction** or the **change in x.**

EXAMPLE 10-13: Find the change in y and the change in x for:

(a)

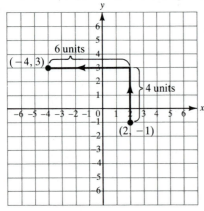

(b)

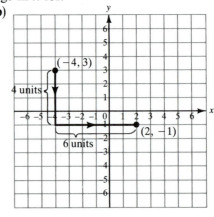

Solution:

(a) The change in y is up 4 units (or up 4) because the direction of vertical movement is up and the amount of vertical movement is 4 units.

The change in x is left 6 units (or left 6) because the direction of horizontal movement is left and the amount of horizontal movement is 6 units.

(b) The change in y is down 4 units because the direction of vertical movement is down and the amount of vertical movement is 4 units.

The change in x is right 6 units because the direction of horizontal movement is right and the amount of horizontal movement is 6 units.

To find the **slope of a straight line** given the graph of the line, you can use the following definition and formula:

Geometric Definition for Slope

For any two different points on the given graph of a straight line:

$$\text{slope} = \frac{\text{change in } y}{\text{change in } x} \qquad [\text{change in } x \neq 0]$$

where the directions up and right are positive directions (+); and the directions left and down are negative directions (−).

EXAMPLE 10-14: Find the slope of the straight line in:

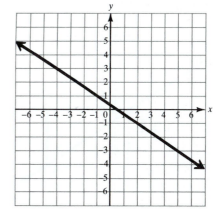

Solution:

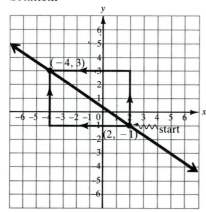

$$\text{slope} = \frac{\text{change in } y}{\text{change in } x} \quad \longleftarrow \quad \text{geometric definition for slope}$$

$$= \frac{\text{up } 4}{\text{left } 6} \quad [\text{See Example 10-13.}]$$

$$= \frac{+4}{-6} \qquad \begin{array}{l} \textit{Think: } Up \text{ is a positive direction } (+). \\ \qquad\quad Left \text{ is a negative direction } (-). \end{array}$$

$$= -\frac{2}{3} \qquad \text{Simplify when possible.}$$

Note: To find the slope of a straight line using the geometric definition, you can first move vertically or horizontally, and you can start at either chosen point on the line.

EXAMPLE 10-15: Find the slope of the straight line shown in Example 10-14 by starting at the point (−4, 3).

Solution:

$$\text{slope} = \frac{\text{change in } y}{\text{change in } x} \quad \longleftarrow \quad \text{geometric definition for slope}$$

$$= \frac{\text{down } 4}{\text{right } 6} \quad [\text{See Example 10-13.}]$$

$$= \frac{-4}{+6} \qquad \begin{array}{l} \textit{Think: } Down \text{ is a negative} \\ \qquad\quad \text{direction } (-). \\ \qquad\quad Right \text{ is a positive} \\ \qquad\quad \text{direction } (+). \end{array}$$

$$= -\frac{2}{3} \quad \longleftarrow \quad \text{same slope as found in Example 10-14}$$

To find the slope of a straight line using the geometric definition, you can choose any two different points on the line.

EXAMPLE 10-16: Find the slope of the straight line shown in Example 10-14 using two different points on the line.

Solution:

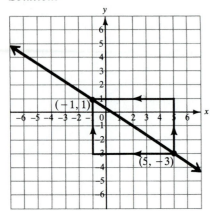

$$\text{slope} = \frac{\text{change in } y}{\text{change in } x} \quad \longleftarrow \quad \text{geometric definition for slope}$$

$$= \frac{\text{up } 4}{\text{left } 6} \quad \text{[See Example 10-13.]}$$

$$= \frac{+4}{-6} \quad \text{[See Example 10-14.]}$$

$$= -\frac{2}{3} \quad \longleftarrow \quad \text{same slope as found in Example 10-14}$$

Note: The slope of the straight line in Example 10-16 is negative (slope $= -\frac{2}{3}$).

Straight lines that fall from left to right always have negative slopes. Straight lines that rise from left to right always have positive slopes.

EXAMPLE 10-17: Find the slope of a straight line that rises from left to right on a rectangular coordinate system.

Solution:

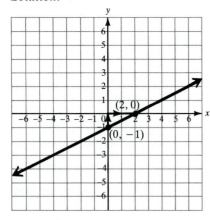

$$\text{slope} = \frac{\text{change in } y}{\text{change in } x} \quad \longleftarrow \quad \text{geometric definition for slope}$$

$$= \frac{\text{up } 1}{\text{right } 2}$$

$$= \frac{+1}{+2}$$

$$= \frac{1}{2} \quad \longleftarrow \quad \text{positive slope}$$

This line rises from left to right.

The slope of every horizontal line is zero.

EXAMPLE 10-18: Find the slope of the horizontal line in:

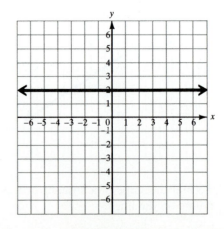

Solution:

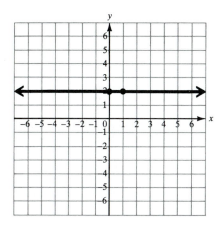

slope $= \dfrac{\text{change in } y}{\text{change in } x}$ ⟵ geometric definition for slope

$\quad = \dfrac{\text{none}}{\text{right 1}}$ *Think:* There is no vertical movement.

$\quad = \dfrac{0}{+1}$ *Think:* No movement is represented by zero.

$\quad = 0$

Note: In Example 10-18, the slope of the horizontal line is 0.

The slope of every vertical line is not defined.

EXAMPLE 10-19: Find the slope of the vertical line in:

Solution:

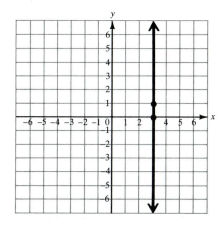

slope $= \dfrac{\text{change in } y}{\text{change in } x}$ ⟵ geometric definition for slope

$\quad = \dfrac{\text{up 1}}{\text{none}}$

$\quad = \dfrac{+1}{0}$ Stop! $\dfrac{+1}{0}$ is not defined.

Note: In Example 10-19, the slope of the vertical line is not defined.

Note: The slope of every nonvertical straight line is represented by one and only one real number. Conversely, every real number represents the slope of a nonvertical straight line.

B. Find the slope of a straight line given any two different points on the line.

Recall: To find the slope of a straight line given its graph in Example 10-14, first two different points on the graph, $(-4, 3)$ and $(2, -1)$, were identified and then the geometric definition for slope was used to find

$$\text{slope} = \frac{\text{change in } y}{\text{change in } x} = \frac{\text{up } 4}{\text{left } 6} = \frac{+4}{-6} = -\frac{2}{3}$$

Note 1: The "change in y" between $(-4, \mathbf{3})$ and $(2, -\mathbf{1})$ is the difference between the y-coordinates:

$$\text{difference between } y\text{-coordinates} \longrightarrow 3 - (-\mathbf{1}) = 3 + 1 = +4 \text{ or } 4$$

Note 2: The "change in x" between $(-\mathbf{4}, 3)$ and $(\mathbf{2}, -1)$ is the difference between the x-coordinates:

$$\text{difference between the } x\text{-coordinates} \longrightarrow -\mathbf{4} - (\mathbf{2}) = -6$$

Note 3: For the straight line containing the two points $(-4, 3)$ and $(2, -1)$:

$$\text{slope} = \frac{\text{change in } y}{\text{change in } x} = \frac{\text{difference between } y\text{-coordinates}}{\text{difference between } x\text{-coordinates}} = \frac{3 - (-1)}{-4 - (2)} = \frac{+4}{-6} = -\frac{2}{3}$$

The previous statements are generalized in the following definition:

Algebraic Definition for Slope

For any two different points (x_1, y_1) and (x_2, y_2) on a straight line:

1. The change in y is the difference between the y-coordinates: $y_1 - y_2$ or $y_2 - y_1$.
2. The change in x is the difference between the x-coordinates: $x_1 - x_2$ or $x_2 - x_1$.

3. The slope of the line is: $\text{slope} = \dfrac{\text{difference between } y\text{-coordinates}}{\text{difference between } x\text{-coordinates}} = \dfrac{y_1 - y_2}{x_1 - x_2}$ or $\dfrac{y_2 - y_1}{x_2 - x_1}$.

Note: For any two points (x_1, y_1) and (x_2, y_2): $\text{slope} = \dfrac{\text{change in } y}{\text{change in } x} = \dfrac{y_1 - y_2}{x_1 - x_2}$ or $\dfrac{y_2 - y_1}{x_2 - x_1}$.

EXAMPLE 10-20: Find the slope of the straight line through $(2, 0)$ and $(0, -1)$.

Solution: Let $(x_1, y_1) = (2, 0)$ and so $x_1 = 2$ and $y_1 = 0$
and $(x_2, y_2) = (0, -1)$ and so $x_2 = 0$ and $y_2 = -1$.

$$\text{slope} = \frac{y_1 - y_2}{x_1 - x_2} \quad \text{or} \quad \frac{y_2 - y_1}{x_2 - x_1} \quad \longleftarrow \text{ algebraic definition for slope}$$

$$= \frac{0 - (-1)}{2 - (0)} \quad \text{or} \quad \frac{-1 - (0)}{0 - (2)} \qquad \text{Substitute for } x_1, y_1, x_2, \text{ and } y_2.$$

$$= \frac{+1}{+2} \quad \text{or} \quad \frac{-1}{-2} \qquad \text{Compute.}$$

$$= \frac{1}{2} \quad \longleftarrow \quad \text{same slope as found in Example 10-17 using the geometric definition for slope}$$

Note: The slope never changes, no matter which one of the two given points you represent by (x_1, y_1) or (x_2, y_2).

EXAMPLE 10-21: Find the slope of the straight line through $(2, 0)$ and $(0, -1)$ by choosing (x_1, y_1) and (x_2, y_2) differently than in Example 10-20.

Solution: Let $(x_1, y_1) = (0, -1)$ and so $x_1 = 0$ and $y_1 = -1$
and $(x_2, y_2) = (2, 0)$ and so $x_2 = 2$ and $y_2 = 0$.

$$\text{slope} = \frac{y_1 - y_2}{x_1 - x_2} \text{ or } \frac{y_2 - y_1}{x_2 - x_1} \longleftarrow \text{algebraic definition for slope}$$

$$= \frac{-1 - (0)}{0 - (2)} \text{ or } \frac{0 - (-1)}{2 - (0)} \qquad \text{Substitute.}$$

$$= \frac{-1}{-2} \qquad \text{or } \frac{+1}{+2} \qquad \text{Compute.}$$

$$= \frac{1}{2} \longleftarrow \text{same slope as found in Example 10-20}$$

C. Find the slope of a straight line given the equation of the line.
To find the slope of a straight line given the equation of the line, you can use the following definition:

Equation Definition for Slope
Every linear equation in two variables can be written in the form $y = mx + b$ or the form $x = C$ where m, b, and C are real numbers and:

1. The graph of $y = mx + b$ is a straight line with slope m [see the following *Note*].
2. The graph of $x = C$ is a vertical line with an undefined slope.

Note: The first part of the Equation Definition for Slope states that to find the slope of a linear equation in two variables that can be written in the form $y = mx + b$, you solve the equation for y and then identify the numerical coefficient of x.

EXAMPLE 10-22: Find the slope of the straight line with equation $2x + 3y = 1$ using the Equation Definition for Slope.

Solution: $\qquad 2x + 3y = 1 \longleftarrow$ given equation

$$2x - 2x + 3y = -2x + 1 \qquad \text{Solve for } y.$$
$$3y = -2x + 1$$
$$\tfrac{1}{3}(3y) = \tfrac{1}{3}(-2x) + \tfrac{1}{3}(1)$$
$$y = -\tfrac{2}{3}x + \tfrac{1}{3} \longleftarrow y = mx + b \text{ form}$$

The graph of $y = mx + b$ is a straight line with slope m.
The slope of the straight line with equation $2x + 3y = 1$ is $-\tfrac{2}{3}$.

Note 1: The graph of $2x + 3y = 1$ is the same line shown in Example 10-14 with slope $-\tfrac{2}{3}$ because both $(-4, 3)$ and $(2, -1)$ are solutions of $2x + 3y = 1$:

$$2(-4) + 3(3) = -8 + 9 = 1 \quad \text{and} \quad 2(2) + 3(-1) = 4 + (-3) = 1$$

Note 2: Because $(-4, 3)$ and $(2, -1)$ are solutions of $2x + 3y = 1$, you can also find the slope of the straight line with equation $2x + 3y = 1$ using the algebraic definition for slope:

$$\text{slope} = \frac{y_1 - y_2}{x_1 - x_2} = \frac{3 - (-1)}{-4 - (2)} = \frac{+4}{-6} = -\frac{2}{3} \longleftarrow \text{same slope as found in Example 10-22}$$

Recall 1: For $x = C$, the graph will always be a vertical line [see Example 10-11, part (**c**)].

Recall 2: The slope of every vertical line is not defined [see Example 10-19].

The previous statements are summarized in the following statement: The graph of the equation $x = C$ is a vertical line with an undefined slope.

EXAMPLE 10-23: Find the slope of the straight line with equation $2x = 6$ using the equation definition for slope.

Solution: $2x = 6$ ⟵— given equation

$\qquad x = 3$ ⟵— $x = C$ form $\qquad$ *Think:* The y-variable is missing, so solve for x.

The graph of $x = C$ is a vertical line with an undefined slope.
The slope of the straight line with equation $2x = 6$ is not defined.

Recall 1: For $y = C$, the graph will always be a horizontal line [see Example 10-11, part **(d)**].

Recall 2: The slope of every horizontal line is 0 [see Example 10-18].

The previous statements are summarized in the following: The graph of the equation $y = C$ is a horizontal line with a slope of 0.

EXAMPLE 10-24: Find the slope of the straight line with equation $2y = 6$ using the Equation Definition for Slope.

Solution: $2y = 6$ ⟵— given equation

$\qquad y = 3$ ⟵— $y = C$ form $\qquad\qquad$ Solve for y.

$\qquad y = 0x + 3$ ⟵— $y = mx + b$ form

The graph of $y = mx + b$ is a straight line with slope m.
The slope of the straight line with equation $2y = 6$ is 0.

10-5. Graphing Using the Slope-Intercept Method

A. Write slope-intercept form for a linear equation in two variables.

Recall: The graph of $y = mx + b$ is a straight line with slope m.

The previous statement is extended in the following rule:

Slope-Intercept Form for a Linear Equation in Two Variables
The $y = mx + b$ form of a linear equation in two variables is called **slope-intercept form** because the graph of $y = mx + b$ is a straight line with slope m and y-intercept $(0, b)$:

$$\begin{array}{cc} \text{slope } m & \\ \downarrow & y\text{-intercept }(0, b) \\ & \downarrow \\ y = mx & + b \end{array}$$

EXAMPLE 10-25: Find the slope and y-intercept of the straight line with equation $3x - 4y = 12$.

Solution: $3x - 4y = 12$ ⟵— given equation

$\qquad -4y = -3x + 12 \qquad\qquad$ Solve for y.

$$\frac{-4y}{-4} = \frac{-3x}{-4} + \frac{12}{-4}$$

$$y = \frac{3}{4}x + (-3) \longleftarrow \text{slope-intercept form } [y = mx + b] \text{ for } 3x - 4y = 12$$

$\qquad\qquad$ slope $m = \frac{3}{4}$ $\qquad$ y-intercept $(0, b) = (0, -3)$

The slope and y-intercept of the straight line with equation $3x - 4y = 12$ are $m = \frac{3}{4}$ and $(0, b) = (0, -3)$, respectively.

EXAMPLE 10-26: Show that $5x + 3y - 1 = 3y + 2$ cannot be written in slope-intercept form but can be written in $x = C$ form.

Solution: $5x + 3y - 1 = 3y + 2$ ⟵ given equation

$$5x + 3y - 3y - 1 = 3y - 3y + 2$$

$$5x - 1 = 2 \qquad \textit{Think: } \text{The } y\text{-variable has been eliminated.}$$

$$5x = 3 \qquad \text{Solve for } x.$$

$$x = \tfrac{3}{5} \text{ ⟵ } x = C \text{ form for } 5x + 3y - 1 = 3y + 2$$

Note: The $x = C$ form for $5x + 3y - 1 = 3y + 2$ is $x = \tfrac{3}{5}$ means that the graph of $5x + 3y - 1 = 3y + 2$ is a vertical line with an undefined slope that goes through the x-intercept $(\tfrac{3}{5}, 0)$.

B. Graph linear equations in two variables using the slope-intercept method.

To graph a linear equation in two variables using the **slope-intercept method,** you

1. Write the given equation in slope-intercept form $y = mx + b$ by solving for y.

2. Plot the y-intercept $(0, b)$ [see Example 10-3].

3. Plot a second point using the slope m in fraction form: $m = \dfrac{r}{s}$. [See Example 10-14.]

4. Draw a line through the two plotted points using a straightedge.

5. Plot a check point using the slope in fraction form: $m = \dfrac{-r}{-s}$.

6. Label your graph with the original equation.

EXAMPLE 10-27: Graph $2x + 3y = 6$ using the slope-intercept method.

Solution: $2x + 3y = 6$ ⟵ given equation

$$3y = -2x + 6 \qquad \text{Solve for } y. \qquad\qquad \tfrac{r}{s}\text{ form of the slope}$$

$$\frac{3y}{3} = -\frac{2x}{3} + \frac{6}{3} \qquad\qquad\qquad\qquad \frac{-r}{-s}\text{ form of the slope}$$

$$y = -\frac{2}{3}x + 2 \quad \text{means the slope is } m = -\frac{2}{3} \text{ or } \frac{-2}{+3} \text{ or } \frac{+2}{-3}$$

$$\text{and the } y\text{-intercept is } (0, b) = (0, 2).$$

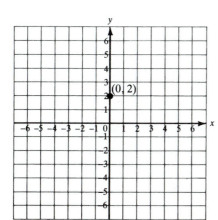

Plot the y-intercept $(0, 2)$.

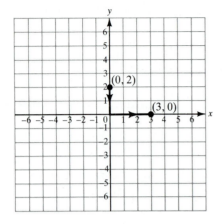

Plot a second point using the slope

$$m = \frac{-2}{+3} \begin{array}{l} \longleftarrow \text{ down 2 units} \\ \longleftarrow \text{ right 3 units} \end{array}$$

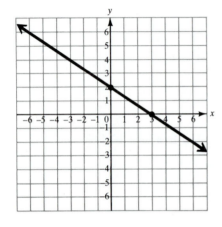

Draw the line through the two plotted points using a straightedge.

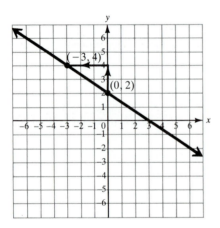

Plot a check point using the slope

$$m = \frac{+2}{-3} \begin{array}{l} \longleftarrow \text{ up 2 units} \\ \longleftarrow \text{ left 3 units} \end{array}$$

The check point $(-3, 4)$ appears to be on the straight line, which means the graph is probably correct.

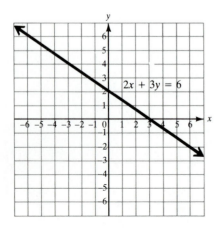

Label the graph with the original equation.

Note 1: In Example 10-27, either form of the slope $\dfrac{-2}{+3}$ or $\dfrac{+2}{-3}$ can be used to plot the second point.

Note 2: In Example 10-27, $\dfrac{+2}{-3}$ was used to check the graph because $\dfrac{-2}{+3}$ was used to plot the second point.

When the slope $m = \dfrac{r}{s}$ is used to plot a second point, use $m = \dfrac{-r}{-s}$ to plot a check point.

RAISE YOUR GRADES

Can you . . . ?

☑ identify the parts of a rectangular coordinate system
☑ identify coordinates given an ordered pair
☑ plot the point that represents a given ordered pair
☑ identify the location of a point for a given ordered pair without plotting
☑ write the ordered pair that is represented by a given plotted point
☑ identify linear equations in two variables
☑ check proposed solutions of equations in two variables
☑ find solutions of linear equations in two variables
☑ graph equations in two variables by plotting points
☑ graph linear equations in two variables using the intercept method
☑ find the slope of a straight line given the graph of the line
☑ find the slope of a straight line given any two different points on the line
☑ find the slope of a straight line given the equation of the line
☑ write slope-intercept form for a linear equation in two variables
☑ graph linear equations in two variables using the slope-intercept method

SUMMARY

1. The parts of a rectangular coordinate system are:

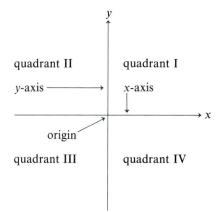

2. In the ordered pair (x, y), the numbers x and y are called coordinates.
3. In the ordered pair (x, y), x is called the first coordinate or x-coordinate and y is called the second coordinate or y-coordinate.
4. To plot a point on a rectangular coordinate system for a given ordered pair (x, y), you locate, draw, and label the point that represents (x, y) on the rectangular coordinate system.

5. To locate the point that represents a given ordered pair (x, y) on a rectangular coordinate system, you
   (a) start at the origin;

   (b) then move $\begin{cases} \text{right } x \text{ units if } x \text{ is positive} \\ \text{left } x \text{ units if } x \text{ is negative} \\ \text{nowhere if } x \text{ is zero} \end{cases}$;

   (c) and then move $\begin{cases} \text{up } y \text{ units if } y \text{ is positive} \\ \text{down } y \text{ units if } y \text{ is negative} \\ \text{nowhere if } y \text{ is zero.} \end{cases}$.

6. To identify the location of a point for a given ordered pair without plotting, you can use the following table:

If the signs of the coordinates of (x, y) are:	then the location of the point is:
$(+, +)$	in quadrant I.
$(-, +)$	in quadrant II.
$(-, -)$	in quadrant III.
$(+, -)$	in quadrant IV.
$(0, +)$ or $(0, -)$	on the y-axis.
$(+, 0)$ or $(-, 0)$	on the x-axis.
$(0, 0)$	at the origin.

7. To write the ordered pair that is represented by a given point on a rectangular coordinate system, you
   (a) Find the x-coordinate m by drawing a vertical guideline through the x-axis at $x = m$.
   (b) Find the y-coordinate n by drawing a horizontal guideline through the y-axis at $y = n$.
   (c) Use the x-coordinate m and the y-coordinate n to write the ordered pair (m, n).

8. An equation that can be written in standard form as $Ax + By = C$ where A, B, and C are real numbers [A and B are not both 0] and x and y are any two different variables is called a linear equation in two variables.

9. An ordered pair (m, n) is a solution of an equation in two variables x and y if substituting m for x and n for y produces a true number sentence.

10. To find a solution of a linear equation in two variables, you:
    (a) Substitute any real number m for one of the variables.
    (b) Solve for the remaining variable to get another real number n.
    (c) Write the solution as an ordered pair using the coordinates m and n from Steps (a) and (b).

11. Every linear equation in two variables has infinitely many solutions.

12. The graph of an equation in two variables is the representation of all ordered pair solutions on a rectangular coordinate system.

13. To graph an equation in two variables by plotting points, you
    (a) Make a table of several solutions using values like $x = 3, 2, 1, 0, -1, -2, -3$.
    (b) Plot each solution from the table in Step (a) on a rectangular coordinate system.
    (c) Draw the graph of the given equation by connecting the points from Step (b).
    (d) Label your graph from Step (c) with the original equation.

14. The graph of every linear equation in two variables is a straight line.

15. The graph of an absolute value equation in two variables like $y = |x|$ is usually V-shaped.

16. The cup-shaped graph of a quadratic equation in two variables like $y = x^2$ is called a parabola.

17. The graph of an exponential equation in two variables like $y = 2^x$ is usually shaped like a park slide.

18. To graph a linear equation in two variables using the intercept method, you
    (a) Find the y-intercept by substituting $x = 0$ in the original equation and then solving for y.
    (b) Find the x-intercept by substituting $y = 0$ in the original equation and then solving for x.
    (c) Draw a line through the x- and y-intercepts using a straightedge.

(d) Find and plot a check point by substituting $x = 1$ [or $y = 1$] in the original equation and then solving for y [or for x].

(e) Label your graph with the original equation.

19. To graph $Ax + By = C$ [A, B, and $C \neq 0$], you only need to plot the x- and y-intercepts.

20. To graph $Ax + By = C$ [$C = 0$], you need to find and plot an extra point because both the x- and y-intercepts are at the origin.

21. To graph $Ax = C$ [$A \neq 0$], you need only solve $Ax = C$ for x to find the x-intercept and then draw a vertical line through the x-intercept.

22. To graph $By = C$ [$B \neq 0$], you need only solve $By = C$ for y to find the y-intercept and then draw a horizontal line through the y-intercept.

23. When only one vertical move and one horizontal move is used to get from one point to another point on a rectangular coordinate system, the direction and amount of vertical movement is called the change in the y-direction or change in y and the direction and amount of horizontal movement is called the change in the x-direction or change in x.

24. Geometric Definition for Slope:

For any two different points on the given graph of a straight line:

$$\text{slope} = \frac{\text{change in } y}{\text{change in } x} \quad [\text{change in } x \neq 0]$$

where the directions up and right are positive directions $(+)$ and the directions down and left are negative directions $(-)$.

25. Algebraic Definition for Slope:

For any two different points (x_1, y_1) and (x_2, y_2) on a straight line:

(a) The change in y is the difference between the y-coordinates: $y_1 - y_2$ or $y_2 - y_1$.

(b) The change in x is the difference between the x-coordinates: $x_1 - x_2$ or $x_2 - x_1$.

(c) The slope of the line is slope $= \dfrac{\text{difference between } y\text{-coordinates}}{\text{difference between } x\text{-coordinates}} = \dfrac{y_1 - y_2}{x_1 - x_2}$ or $\dfrac{y_2 - y_1}{x_2 - x_1}$.

26. Equation Definition for Slope:

Every linear equation in two variables can be written in the form $y = mx + b$ or the form $x = C$ when m, b, and C are real numbers and:

(a) The graph of $y = mx + b$ is a straight line with slope m.

(b) The graph of $x = C$ is a vertical line with an undefined slope.

27. The graph of $y = C$ is a horizontal line with a slope of 0.

28. Straight lines that fall from left to right always have negative slopes.

29. Straight lines that rise from left to right always have positive slopes.

30. The slope of every nonvertical straight line is represented by one and only one real number. And, conversely, every real number represents the slope of a nonvertical straight line.

31. Slope-Intercept Form for a Linear Equation in Two Variables:

The $y = mx + b$ form of a linear equation in two variables is called slope-intercept form because the graph of $y = mx + b$ is a straight line with slope m and y-intercept $(0, b)$:

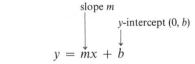

32. To graph a linear equation in two variables using the slope-intercept method, you

(a) Write the given equation in slope-intercept form $y = mx + b$ by solving for y.

(b) Plot the y-intercept $(0, b)$.

(c) Plot a second point using the slope m in fraction form: $m = \dfrac{r}{s}$.

(d) Draw a line through the two plotted points using a straightedge.

(e) Plot a check point using the slope: $m = \dfrac{-r}{-s}$.

(f) Label your graph with the original equation.

SOLVED PROBLEMS

PROBLEM 10-1   Plot each point on a rectangular coordinate system: **(a)** (2, 3)   **(b)** (−3, 4)
(c) (4, −2)   **(d)** (−1, −3)   **(e)** (0, 2)   **(f)** (−4, 0)   **(g)** (0, 0)

Solution: Recall that to locate the point that represents a given ordered pair (x, y) on a rectangular coordinate system, you
1. start at the origin;

2. then move $\begin{cases} \text{right } x \text{ units if } x \text{ is positive} \\ \text{left } x \text{ units if } x \text{ is negative} \\ \text{nowhere if } x \text{ is zero} \end{cases}$;

3. and then move $\begin{cases} \text{up } y \text{ units if } y \text{ is positive} \\ \text{down } y \text{ units if } y \text{ is negative} \\ \text{nowhere if } y \text{ is zero} \end{cases}$.

[See Example 10-3.]

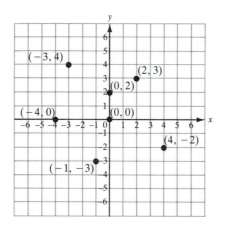

PROBLEM 10-2   Identify the location of each point without plotting: **(a)** (3, 1)   **(b)** (−2, 4)
(c) (−4, −3)   **(d)** (2, −3)   **(e)** (0, 1)   **(f)** (0, −4)   **(g)** (3, 0)   **(h)** (−2, 0)   **(i)** (0, 0)

Solution: Recall that to identify the location of a point for a given ordered pair without plotting, you can use the following table [see Example 10-4]:

If the signs of the coordinates of (x, y) are:	then the location of the point is:
(+, +)	in quadrant I.
(−, +)	in quadrant II.
(−, −)	in quadrant III.
(+, −)	in quadrant IV.
(0, +) or (0, −)	on the y-axis.
(+, 0) or (−, 0)	on the x-axis.
(0, 0)	at the origin.

(a) (3, 1) or (+3, +1) is in quadrant I because the signs of the coordinates are (+, +).
(b) (−2, 4) or (−2, +4) is in quadrant II because the signs of the coordinates are (−, +).
(c) (−4, −3) is in quadrant III because the signs of the coordinates are (−, −).
(d) (2, −3) or (+2, −3) is in quadrant IV because the signs of the coordinates are (+, −).
(e) (0, 1) or (0, +1) is on the y-axis because the signs of the coordinates are (0, +).
(f) (0, −4) is on the y-axis because the signs of the coordinates are (0, −).

(g) (3, 0) or (+3, 0) is on the *x*-axis because the signs of the coordinates are (+, 0).
(h) (−2, 0) is on the *x*-axis because the signs of the coordinates are (−, 0).
(i) (0, 0) is at the origin.

PROBLEM 10-3 Write each ordered pair that is represented by the points shown on the following rectangular coordinate system:

Solution: Recall that to write the ordered pair that is represented by a given point on a rectangular coordinate system, you:

1. Find the *x*-coordinate *m* by drawing a vertical guideline through the *x*-axis at *x* = *m*.
2. Find the *y*-coordinate *n* by drawing a horizontal guideline through the *y*-axis at *y* = *n*.
3. Use the *x*-coordinate *m* and the *y*-coordinate *n* to write the ordered pair (*m*, *n*).
   [See Example 10-5.]

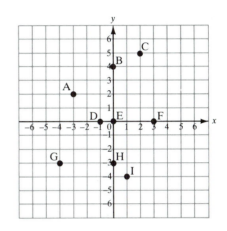

(a) The ordered pair that represents point A is (−3, 2).
(b) The ordered pair that represents point B is (0, 4).
(c) The ordered pair that represents point C is (2, 5).
(d) The ordered pair that represents point D is (−1, 0).
(e) The ordered pair that represents point E is (0, 0).
(f) The ordered pair that represents point F is (3, 0).
(g) The ordered pair that represents point G is (−4, −3).
(h) The ordered pair that represents point H is (0, −3).
(i) The ordered pair that represents point I is (1, −4).

PROBLEM 10-4 Identify each equation that can be written as a linear equation in two variables:

(a) $5r + 3s = 2$ **(b)** $5r + 3\sqrt{s} = 2$ **(c)** $5r − 3s = 2$ **(d)** $5r − 3s = 2t$ **(e)** $\frac{r}{5} + 3s = 2$
(f) $\frac{5}{r} + 3s = 2$ **(g)** $5(r − 3s) = 2$ **(h)** $5 − 3 = 2$ **(i)** $5r = 2$ **(j)** $5rs = 2$
(k) $5 − 3s = 2$ **(l)** $5r − 3s^2 = 2$

Solution: Recall that an equation that can be written in standard form as $Ax + By = C$ where *A*, *B*, and *C* are real numbers [and *A* and *B* are not both 0] and *x* and *y* are any two different variables is called a linear equation in two variables [see Example 10-6]:

(a) $5r + 3s = 2$ is a linear equation in two variables in standard form.
(b) $5r + 3\sqrt{s} = 2$ is not a linear equation in two variables because there is a variable under a radical symbol.
(c) $5r − 3s = 2$ or $5r + (−3)s = 2$ is a linear equation in two variables in standard form.
(d) $5r − 3s = 2t$ is not a linear equation in two variables because there are more than two different variables.
(e) $\frac{r}{5} + 3s = 2$ is a linear equation in two variables in standard form as $\frac{1}{5}r + 3s = 2$.
(f) $\frac{5}{r} + 3s = 2$ is not a linear equation in two variables because there is a variable in a denominator.
(g) $5(r − 3s) = 2$ is a linear equation in two variables in standard form as $5r − 15s = 2$.
(h) $5 − 3 = 2$ is not a linear equation in two variables because both variables are missing at the same time.
(i) $5r = 2$ can be written as a linear equation in two variables as $5r + 0s = 2$.
(j) $5rs = 2$ is not a linear equation in two variables because it contains a product of variables.
(k) $5 − 3s = 2$ can be written as a linear equation in two variables as $0r − 3s = −3$.
(l) $5r − 3s^2 = 2$ is not a linear equation in two variables because the variable *s* has an exponent greater than 1.

PROBLEM 10-5   Identify which one of the two given ordered pairs is a solution of the given equation in two variables:

(a) (3, 5) or (1, −3) given $2y + 4 = 3x − 5$ **(b)** (2, 0) or (0, 2) given $y = 2x^2 − 3x − 2$
(c) (0, 3) or (3, 0) given $y = |2x − 6|$

Solution: Recall that an ordered pair (m, n) is a solution of an equation in two variables x and y if substituting m for x and n for y produces a true number sentence [see Example 10-7]:

(a) (3, 5) is not a solution of:

$$2y + 4 = 3x − 5$$

$2(5) + 4$	$3(3) − 5$
$10 + 4$	$9 − 5$
14	4 ⟵ false

(1, −3) is a solution of:

$$2y + 4 = 3x − 5$$

$2(−3) + 4$	$3(1) − 5$
$−6 + 4$	$3 − 5$
$−2$	$−2$ ⟵ true

(b) (2, 0) is a solution of:

$$y = 2x^2 − 3x − 2$$

0	$2(2)^2 − 3(2) − 2$
0	$8 − 6 − 2$
0	0 ⟵ true

(0, 2) is not a solution of:

$$y = 2x^2 − 3x − 2$$

2	$2(0)^2 − 3(0) − 2$
2	$0 − 0 − 2$
2	$−2$ ⟵ false

(c) (0, 3) is not a solution of:

$$y = |2x − 6|$$

3	$|2(0) − 6|$
3	$|−6|$
3	6 ⟵ false

(3, 0) is a solution of:

$$y = |2x − 6|$$

0	$|2(3) − 6|$
0	$|0|$
0	0 ⟵ true

PROBLEM 10-6   Find solutions of each equation in two variables for $x = 1$, 0, and −1:

(a) $2y + 4 = 3x − 5$ **(b)** $y = 2x^2 − 3x − 2$ **(c)** $y = |2x − 6|$ **(d)** $y = 2^x$

Solution: Recall that to find solutions of an equation in two variables, you

1. Substitute any real number m for one of the variables.
2. Solve for the remaining variable to get another real number n.
3. Write the solution as an ordered pair using the coordinates m and n from Steps 1 and 2.
   [See Example 10-8.]

(a) For $x = 1$:

$$2y + 4 = 3x − 5$$
$$2y + 4 = 3(1) − 5$$
$$2y + 4 = −2$$
$$2y = −6$$
$$y = −3$$
$$(x, y) = (1, −3)$$

For $x = 0$:

$$2y + 4 = 3x − 5$$
$$2y + 4 = 3(0) − 5$$
$$2y + 4 = −5$$
$$2y = −9$$
$$y = −\tfrac{9}{2}$$
$$(x, y) = (0, −\tfrac{9}{2})$$

For $x = −1$:

$$2y + 4 = 3x − 5$$
$$2y + 4 = 3(−1) − 5$$
$$2y + 4 = −8$$
$$2y = −12$$
$$y = −6$$
$$(x, y) = (−1, −6)$$

(b) For $x = 1$:

$$y = 2x^2 − 3x − 2$$
$$y = 2(1)^2 − 3(1) − 2$$
$$y = 2 − 3 − 2$$
$$y = −3$$
$$(x, y) = (1, −3)$$

For $x = 0$:

$$y = 2x^2 − 3x − 2$$
$$y = 2(0)^2 − 3(0) − 2$$
$$y = 0 − 0 − 2$$
$$y = −2$$
$$(x, y) = (0, −2)$$

For $x = −1$:

$$y = 2x^2 − 3x − 2$$
$$y = 2(−1)^2 − 3(−1) − 2$$
$$y = 2 + 3 − 2$$
$$y = 3$$
$$(x, y) = (−1, 3)$$

(c) For $x = 1$:   For $x = 0$:   For $x = -1$:

$$y = |2x - 6|$$ $$y = |2x - 6|$$ $$y = |2x - 6|$$

$$y = |2(1) - 6|$$ $$y = |2(0) - 6|$$ $$y = |2(-1) - 6|$$

$$y = |-4|$$ $$y = |-6|$$ $$y = |-8|$$

$$y = 4$$ $$y = 6$$ $$y = 8$$

$$(x, y) = (1, 4)$$ $$(x, y) = (0, 6)$$ $$(x, y) = (-1, 8)$$

(d) For $x = 1$:   For $x = 0$:   For $x = -1$:

$$y = 2^x$$ $$y = 2^x$$ $$y = 2^x$$

$$y = 2^1$$ $$y = 2^0$$ $$y = 2^{-1}$$

$$y = 2$$ $$y = 1$$ $$y = \frac{1}{2}$$ *Think:* $2^{-1} = \frac{1}{2^1} = \frac{1}{2}$

$$(x, y) = (1, 2)$$ $$(x, y) = (0, 1)$$ $$(x, y) = (-1, \tfrac{1}{2})$$

PROBLEM 10-7   Graph each equation in two variables by plotting points:

(a) $2y + 4 = 3x - 5$ **(b)** $y = -x^2$ **(c)** $y = -|x|$ **(d)** $y = 2^{-x}$

Solution: Recall that to graph an equation in two variables by plotting points, you

1. Make a table of several solutions using values like $x = 3, 2, 1, 0, -1, -2, -3$.
2. Plot each solution from the table in Step 1 on a rectangular coordinate system.
3. Draw the graph of the given equation by connecting the points from Step 2.
4. Label your graph from Step 3 with the original equation [see Example 10-10]:

(a)

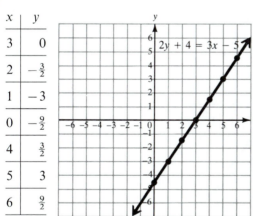

(b)

(c)

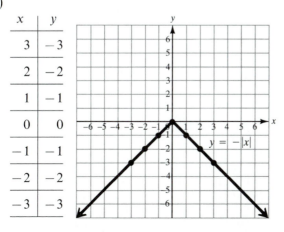

(d)

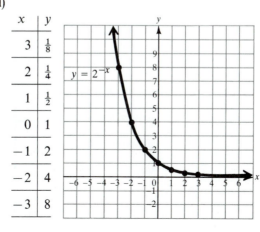

PROBLEM 10-8   Graph each linear equation in two variables using the intercept method:

(a) $2x - 3y = -6$ (b) $3x + 2y = 0$ (c) $2x = -6$ (d) $3y = -6$

Solution: Recall that to graph a linear equation in two variables using the intercept method, you

1. Find the *y*-intercept by substituting $x = 0$ in the original equation and then solving for *y*.
2. Find the *x*-intercept by substituting $y = 0$ in the original equation and then solving for *x*.
3. Draw a line through the *x*- and *y*-intercepts using a straightedge.
4. Find and plot a check point by substituting $x = 1$ [or $y = 1$] in the original equation and then solving for *y* [or for *x*].
5. Label your graph with the original equation.
[See Example 10-11.]

(a)
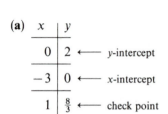

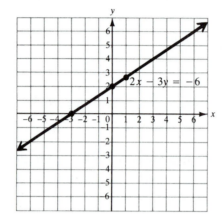

(b)
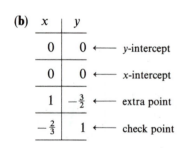

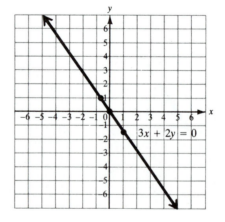

(c)
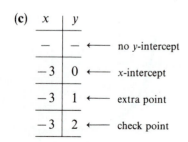

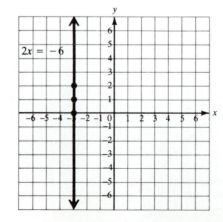

(d)

x	y	
0	-2	⟵ *y*-intercept
—	—	⟵ no *x*-intercept
1	-2	⟵ extra point
2	-2	⟵ check point

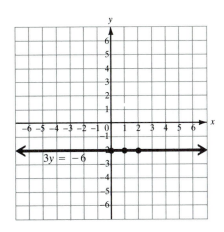

PROBLEM 10-9   Find the slope of each given straight line:

(a)

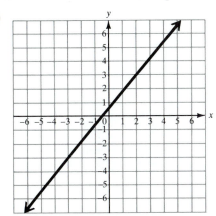

(b)

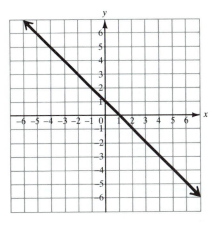

(c)

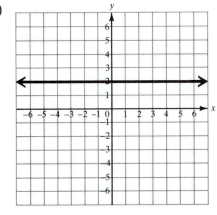

(d)

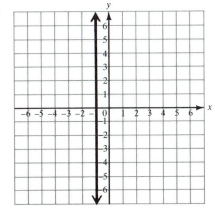

Solution: Recall that for any two points on the given graph of a straight line:

$$\text{slope} = \frac{\text{change in } y}{\text{change in } x} \qquad [\text{change in } x \neq 0]$$

where the directions up and right are positive directions $(+)$ and the directions left and down are negative directions $(-)$ [see Example 10-14]:

(a)

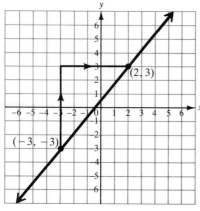

$$\text{slope} = \frac{\text{change in } y}{\text{change in } x}$$

$$= \frac{\text{up } 6}{\text{right } 5}$$

$$= \frac{+6}{+5}$$

$$= \frac{6}{5}$$

(b)

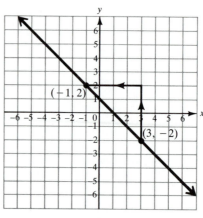

$$\text{slope} = \frac{\text{change in } y}{\text{change in } x}$$

$$= \frac{\text{up } 4}{\text{left } 4}$$

$$= \frac{+4}{-4}$$

$$= -1$$

(c)

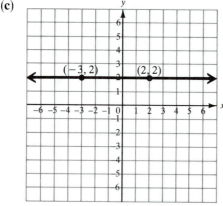

$$\text{slope} = \frac{\text{change in } y}{\text{change in } x}$$

$$= \frac{\text{none}}{\text{right } 5}$$

$$= \frac{0}{+5}$$

$$= 0$$

(d)

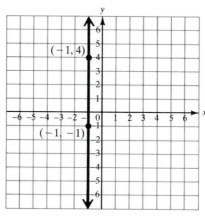

$$\text{slope} = \frac{\text{change in } y}{\text{change in } x}$$

$$= \frac{\text{up } 5}{\text{none}}$$

$$= \frac{+5}{0} \text{ Stop!}$$

The slope is not defined.

PROBLEM 10-10   Find the slope of each straight line through the two given points:

(a) $(-2, 5)$ and $(3, -4)$ (b) $(2, 3)$ and $(-5, 3)$ (c) $(4, -1)$ and $(4, 2)$

Solution: Recall that for any two points (x_1, y_1) and (x_2, y_2) on a straight line:

1. The "change in y" is the difference between the y-coordinates: $y_1 - y_2$ or $y_2 - y_1$.
2. The "change in x" is the difference between the x-coordinates: $x_1 - x_2$ or $x_2 - x_1$.

3. The slope of the line is: $\text{slope} = \dfrac{\text{difference between the } y\text{-coordinates}}{\text{difference between the } x\text{-coordinates}} = \dfrac{y_1 - y_2}{x_1 - x_2} \text{ or } \dfrac{y_2 - y_1}{x_2 - x_1}.$

  [See Example 10-20.]

(a) Let $(x_1, y_1) = (-2, 5)$ and $(x_2, y_2) = (3, -4)$.

$$\text{slope} = \frac{y_1 - y_2}{x_1 - x_2}$$

$$= \frac{5 - (-4)}{-2 - (3)}$$

$$= \frac{9}{-5}$$

$$= -\frac{9}{5}$$

(b) Let $(x_1, y_1) = (2, 3)$ and $(x_2, y_2) = (-5, 3)$

$$\text{slope} = \frac{y_1 - y_2}{x_1 - x_2}$$

$$= \frac{3 - (3)}{2 - (-5)}$$

$$= \frac{0}{7}$$

$$= 0$$

(c) Let $(x_1, y_1) = (4, -1)$ and $(x_2, y_2) = (4, 2)$.

$$\text{slope} = \frac{-1 - (2)}{4 - (4)}$$

$$= \frac{-3}{0} \quad \text{Stop!}$$

The points $(4, -1)$ and $(4, 2)$ lie on a vertical line with an undefined slope.

PROBLEM 10-11 Find the slope of the straight line for each given linear equation in two variables: **(a)** $3x - 4y = 8$ **(b)** $-4y = 8$ **(c)** $3x = 8$

Solution: Recall that every linear equation in two variables can be written in the form $y = mx + b$ or in the form $x = C$, where m, b, and C are real numbers and

1. The graph of $y = mx + b$ is a straight line with slope m.
2. The graph of $x = C$ is a vertical line with an undefined slope.
[See Example 10-22.]

(a) $3x - 4y = 8$

$$-4y = -3x + 8$$

$$\frac{-4y}{-4} = \frac{-3x}{-4} + \frac{8}{-4}$$

$$y = \frac{3}{4}x - 2$$

The slope of $3x - 4y = 8$ is $\dfrac{3}{4}$.

(b) $-4y = 8$

$$\frac{-4y}{-4} = \frac{8}{-4}$$

$$y = -2$$

$$y = 0x - 2$$

The slope of $-4y = 8$ is 0.

(c) $3x = 8$

$$x = \frac{8}{3} \quad \longleftarrow x = C \text{ form}$$

The slope of $3x = 8$ is not defined.

PROBLEM 10-12 Find the slope and y-intercept of the straight line with equation:

(a) $2x + 3y = 5$ **(b)** $3y = 5$ **(c)** $2x = 5$

Solution: Recall that the $y = mx + b$ form of a linear equation in two variables is called slope-intercept form because the graph of $y = mx + b$ is a straight line with slope m and y-intercept $(0, b)$ [see Examples 10-25 and 10-26]:

(a) $2x + 3y = 5$

$$3y = -2x + 5$$

$$y = -\tfrac{2}{3}x + \tfrac{5}{3}$$

$$\text{slope} = -\tfrac{2}{3}$$

y-intercept $= (0, \tfrac{5}{3})$

[falling line, from left to right, through $(0, \tfrac{5}{3})$]

(b) $3y = 5$

$$y = \tfrac{5}{3}$$

$$y = 0x + \tfrac{5}{3}$$

$$\text{slope} = 0$$

y-intercept $= (0, \tfrac{5}{3})$

[horizontal line through $(0, \tfrac{5}{3})$]

(c) $2x = 5$

$$x = \tfrac{5}{2} \quad \longleftarrow x = C \text{ form}$$

Slope is not defined.

no y-intercept

x-intercept $= (\tfrac{5}{2}, 0)$

[vertical line through $(\tfrac{5}{2}, 0)$]

PROBLEM 10-13 Graph each linear equation in two variables using the slope-intercept method:

(a) $3x - 4y = 8$ **(b)** $3x + 4y = 0$ **(c)** $4y = 8$ **(d)** $3x = 8$

Solution: Recall that to graph a linear equation in two variables using the slope-intercept method, you

1. Write the given equation in slope-intercept form $y = mx + b$ by solving for y.

2. Plot the y-intercept $(0, b)$.

3. Plot a second point using the slope m in fraction form: $m = \dfrac{r}{s}$.

4. Draw a line through the two plotted points using a straightedge.

5. Plot a check point using the slope $m = \dfrac{-r}{-s}$.

6. Label your graph with the original equation.
[See Example 10-27.]

(a) $3x - 4y = 8$

$$-4y = -3x + 8$$

$$y = \frac{3}{4}x - 2$$

$$y = \frac{3}{4}x + (-2) \longleftarrow y = mx + b \text{ form}$$

$$\text{slope} = \frac{3}{4} \text{ or } \frac{-3}{-4}$$

y-intercept $= (0, -2)$

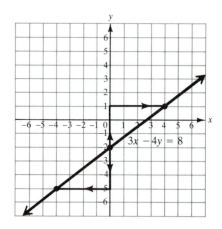

(b) $3x + 4y = 0$

$$4y = -3x$$

$$y = -\frac{3}{4}x$$

$$y = -\frac{3}{4}x + 0 \longleftarrow y = mx + b \text{ form}$$

$$\text{slope} = -\frac{3}{4} = \frac{-3}{4} = \frac{3}{-4}$$

y-intercept $= (0, 0)$

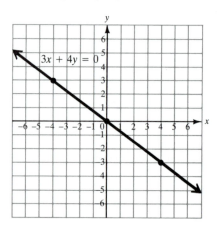

(c) $4y = 8$

$$y = 2$$

$$y = 0x + 2 \longleftarrow y = mx + b \text{ form}$$

slope $= 0$ [horizontal line]
y-intercept $= (0, 2)$

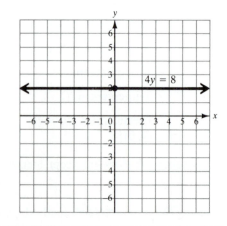

(d) $3x = 8$

$$x = \tfrac{8}{3} \longleftarrow x = C \text{ form}$$

Slope is not defined [vertical line].
x-intercept $= (\tfrac{8}{3}, 0)$

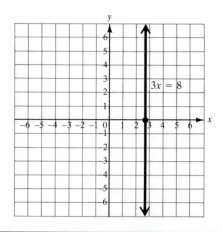

Supplementary Exercises

PROBLEM 10-14: Plot each point on a rectangular coordinate system: **(a)** $(1, 2)$ **(b)** $(-1, 2)$
(c) $(1, -2)$ **(d)** $(-1, -2)$ **(e)** $(0, 0)$ **(f)** $(2, 0)$ **(g)** $(0, 2)$ **(h)** $(-2, 0)$ **(i)** $(0, -2)$

PROBLEM 10-15: Identify the location of each point without plotting: **(a)** $(-3, 0)$ **(b)** $(2, 3)$
(c) $(0, -3)$ **(d)** $(-2, 3)$ **(e)** $(0, 3)$ **(f)** $(2, -3)$ **(g)** $(3, 0)$ **(h)** $(-2, -3)$ **(i)** $(0, 0)$

PROBLEM 10-16: Write each ordered pair represented by the points shown on the following grid:

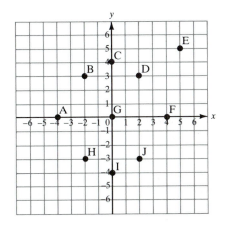

PROBLEM 10-17 Identify which of the given ordered pairs is a solution of the given equation in
two variables: **(a)** $(0, 1)$ or $(1, 0)$ given $x + y = 1$ **(b)** $(2, 0)$ or $(0, 2)$ given $y - x = 2$
(c) $(-1, -1)$ or $(1, 1)$ given $3x + y = 4$ **(d)** $(1, -1)$ or $(-1, 1)$ given $y - 3x = -4$
(e) $(2, \frac{1}{3})$ or $(\frac{1}{2}, 2)$ given $2x + 3y = 5$ **(f)** $(1, -0.4)$ or $(0.2, 1)$ given $5y - 2x = -4$
(g) $(0, 0)$ or $(1, 1)$ given $0.1x + 0.2y = 0$ **(h)** $(1, -1)$ or $(-1, 1)$ given $\frac{1}{2}y - \frac{1}{2}x = \frac{1}{4}$
(i) $(1, 0)$ or $(0, 1)$ given $3x - 5y + 2 = 2x + 3y - 6$

PROBLEM 10-18 Graph each equation in two variables by plotting points: **(a)** $2x = 3y + 6$
(b) $4x = 3y$ **(c)** $-2y = 4$ **(d)** $x = 3$ **(e)** $y = 2x^2$ **(f)** $y = 1 - x^2$ **(g)** $y = |2x|$
(h) $y = |x| + 1$ **(i)** $y = (\frac{1}{2})^x$ **(j)** $y = (\frac{1}{2})^{-x}$

PROBLEM 10-19 Graph each linear equation in two variables using the intercept method:

(a) $3x + 4y = 12$ **(b)** $3x = 4y$ **(c)** $3x = 12$ **(d)** $4y = 12$

PROBLEM 10-20 Find the slope of each straight line:

(a)

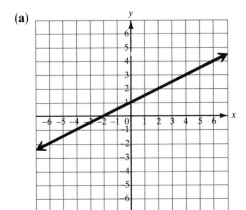

(b)

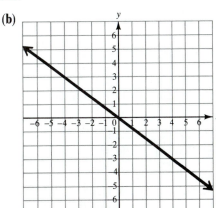

(c) 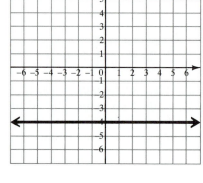   **(d)**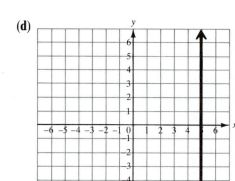

PROBLEM 10-21   Find the slope of each straight line through the two given points:

(a) $(-4, -3)$ and $(3, 0)$ **(b)** $(0, 5)$ and $(1, -2)$ **(c)** $(-1, -4)$ and $(3, -4)$
(d) $(-3, 2)$ and $(-3, -5)$

PROBLEM 10-22   Find the slope and *y*-intercept of the straight line for each given linear equation in two variables: **(a)** $2x + y = 3$ **(b)** $3y - 2x = -1$ **(c)** $5y = 2$ **(d)** $2x = -3$
(e) $5x + 2y - 3 = 0$ **(f)** $3y + 2 - 5x = 0$

PROBLEM 10-23   Graph each linear equation in two variables using the slope-intercept method:

(a) $y - 2x = 0$ **(b)** $4x + 5y = 20$ **(c)** $5y = 10$ **(d)** $-2x = 10$

Answers to Supplementary Exercises

(10-14)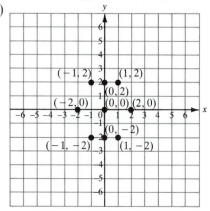

(10-15)   **(a)** on *x*-axis   **(b)** in quadrant I   **(c)** on *y*-axis   **(d)** in quadrant II   **(e)** on *y*-axis
(f) in quadrant IV   **(g)** on *x*-axis   **(h)** in quadrant III   **(i)** at origin

(10-16)   **(a)** A $(-4, 0)$ **(b)** B $(-2, 3)$ **(c)** C $(0, 4)$ **(d)** D $(2, 3)$ **(e)** E $(5, 5)$ **(f)** F $(4, 0)$
(g) G $(0, 0)$ **(h)** H $(-2, -3)$ **(i)** I $(0, -4)$ **(j)** J $(2, -3)$

(10-17)   **(a)** both $(0, 1)$ and $(1, 0)$ **(b)** $(0, 2)$ **(c)** $(1, 1)$ **(d)** $(1, -1)$ **(e)** $(2, \frac{1}{3})$
(f) $(1, -0.4)$ **(g)** $(0, 0)$ **(h)** Neither ordered pair is a solution.   **(i)** $(0, 1)$

(10-18) **(a)**

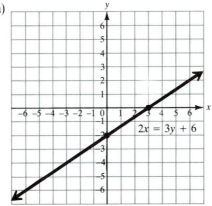

$2x = 3y + 6$

(b)

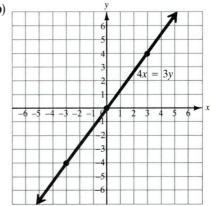

$4x = 3y$

(c)

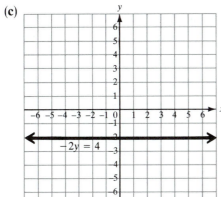

$-2y = 4$

(d)

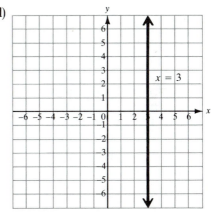

$x = 3$

(e)

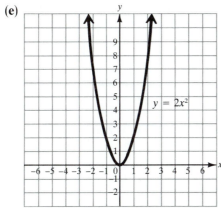

$y = 2x^2$

(f)

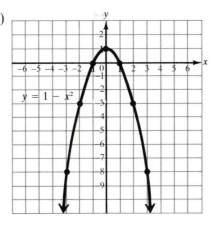

$y = 1 - x^2$

(g)

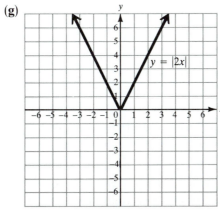

$y = |2x|$

(h)

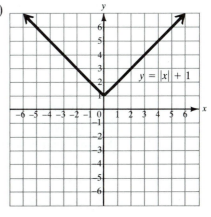

$y = |x| + 1$

(i)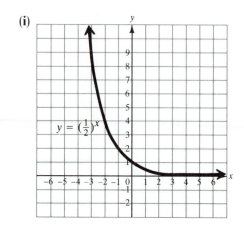

(j)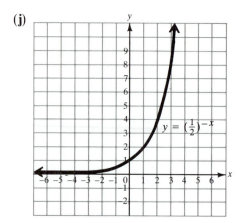

(10-19) **(a)**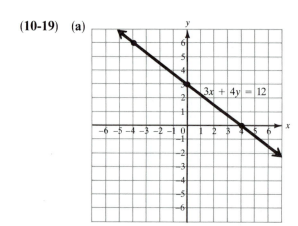

(b)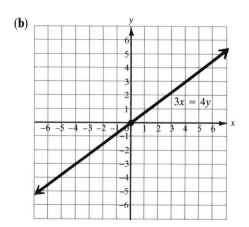

(c)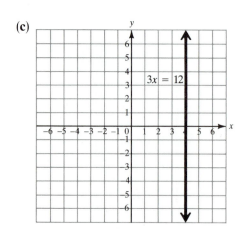

(d)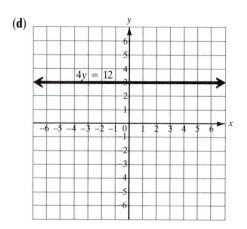

(10-20)   **(a)** $\frac{1}{2}$ **(b)** $-\frac{3}{4}$ **(c)** 0   **(d)** not defined

(10-21)   **(a)** $\frac{3}{7}$ **(b)** -7 **(c)** 0   **(d)** not defined

(10-22)   **(a)** slope $= -2$, y-intercept $= (0, 3)$ [falling line, from left to right, through $(0, 3)$]
   (b) slope $= \frac{2}{3}$, y-intercept $= (0, -\frac{1}{3})$ [rising line, from left to right, through $(0, -\frac{1}{3})$]
   (c) slope $= 0$, y-intercept $= (0, \frac{2}{5})$ [horizontal line through $(0, \frac{2}{5})$]
   (d) slope is not defined, no y-intercept, x-intercept $= (-\frac{3}{2}, 0)$ [vertical line through $(-\frac{3}{2}, 0)$]
   (e) slope $= -\frac{5}{2}$, y-intercept $= (0, \frac{3}{2})$ [falling line, from left to right, through $(0, \frac{3}{2})$]
   (f) slope $= \frac{5}{3}$, y-intercept $= (0, -\frac{2}{3})$ [rising line, from left to right, through $(0, -\frac{2}{3})$]

(10-23) **(a)**

(b)

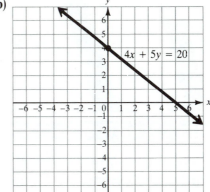

(c)

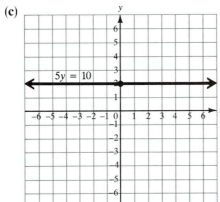

(d)

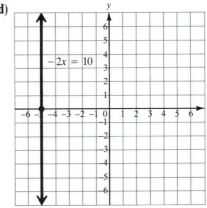

11 SYSTEMS

THIS CHAPTER IS ABOUT

☑ **Identifying System Solutions**
☑ **Solving Systems by Graphing**
☑ **Solving Systems Using the Substitution Method**
☑ **Solving Systems Using the Addition Method**

11-1. Identifying System Solutions

A. Check a proposed system solution.

Two or more equations considered at the same time are called a **system of equations,** or just a **system.** Every ordered pair that is a common solution of each system equation is called a **solution of the system** or a **system solution.**

EXAMPLE 11-1: Which ordered pair is a system solution of $\begin{cases} 3x + 4y = 8 \\ 2x + 5y = 10 \end{cases}$?

(a) $(\frac{8}{3}, 0)$ **(b)** $(0, 2)$

Solution:

(a) Check for $(x, y) = (\frac{8}{3}, 0)$ in $3x + 4y = 8$.

$3x + 4y = 8$	
$3(\frac{8}{3}) + 4(0)$	8
$8 + 0$	8
8	8 ⟵ $(\frac{8}{3}, 0)$ checks in
	$3x + 4y = 8$

Check for $(x, y) = (\frac{8}{3}, 0)$ in $2x + 5y = 10$.

$2x + 5y = 10$	
$2(\frac{8}{3}) + 5(0)$	10
$\frac{16}{3} + 0$	10
$\frac{16}{3}$	10 ⟵ $(\frac{8}{3}, 0)$ does not check in
	$2x + 5y = 10$

$(\frac{8}{3}, 0)$ is not a system solution of $\begin{cases} 3x + 4y = 8 \\ 2x + 5y = 10 \end{cases}$ because $(\frac{8}{3}, 0)$ is not a solution of both system equations.

(b) Check for $(x, y) = (0, 2)$ in $3x + 4y = 8$.

$3x + 4y = 8$	
$3(0) + 4(2)$	8
$0 + 8$	8
8	8 ⟵ $(0, 2)$ checks in
	$3x + 4y = 8$

Check for $(x, y) = (0, 2)$ in $2x + 5y = 10$.

$2x + 5y = 10$	
$2(0) + 5(2)$	10
$0 + 10$	10
10	10 ⟵ $(0, 2)$ checks in
	$2x + 5y = 10$

$(0, 2)$ is a system solution of $\begin{cases} 3x + 4y = 8 \\ 2x + 5y = 10 \end{cases}$ because $(0, 2)$ is a solution of both system equations.

Note: For an ordered pair to be a solution of a given system, the ordered pair must check in each one of the system equations. That is, if an ordered pair does not check in one of the given system equations, then that ordered pair is not a solution of the given system.

$$\begin{cases} 2x - 3y = \\ -4x + 6y = \end{cases}$$
have any points

E. Identify an

To identify an
forms.

EXAMPLE 11

independent-co

Solution:

2

In $y = -\frac{2}{3}x$

Straight lines

$$\begin{cases} 2x + 3y = \\ -4x + 6y = \end{cases}$$
lines have exa

Note: It is no
exactly one p
it should be
Example 11-
the case. Th
have differen
for finding th

11-2. So

To **solve a s**

Agreement:
equations in

To **solve a**

1. Write sl
2. Graph e
3. Identify
4. Check t

EXAMPL

B. Identify the three basic types of graphs for a system of two linear equations in two variables.

To **graph a system,** you graph each system equation on the same rectangular coordinate system. There are three basic types of graphs for a system of two linear equations in two variables.

EXAMPLE 11-2: Show and describe the three basic types of graphs for a system of two linear equations in two variables.

Solution:

(a) Coinciding lines

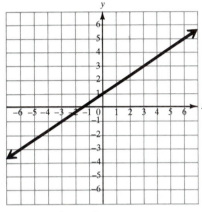

dependent system

Straight lines that lie one on top of the other are called **coinciding lines.** Coinciding lines always have the same slope and y-intercept [or x-intercept]. Because coinciding lines have a whole line of points in common, a system that graphs as coinciding lines will have infinitely many solutions [a **dependent system**]. That is, a dependent system containing two equations will always graph as two coinciding lines.

(b) Distinct parallel lines

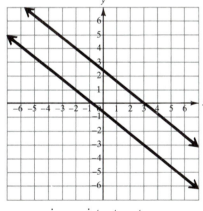

inconsistent system

Straight lines that never intersect are called **distinct parallel lines.** Distinct parallel lines always have the same slope and different y-intercepts [or x-intercepts]. Because distinct parallel lines have no points in common, a system that graphs as distinct parallel lines will have no solutions [an **inconsistent system**]. That is, an inconsistent system containing two equations will always graph as two distinct parallel lines.

(c) Intersecting lines

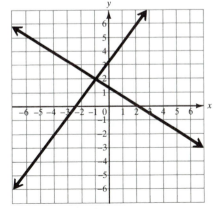

independent-consistent system

Straight lines that intersect at exactly one point are called **intersecting lines.** Intersecting lines always have different slopes. Because intersecting lines have exactly one point in common, a system that graphs as intersecting lines will have exactly one solution [an **independent-consistent system**]. That is, an independent-consistent system containing two equations will always graph as two intersecting lines.

Note: Given ar
coinciding lines
equations in tw
1. infinitely ma
2. no solutions
3. exactly one

C. Identify a
To identify a d

EXAMPLE 1

[a dependent s

Solution:

$2x - 3y = 1$

$-3y = $

$\dfrac{-3y}{-3} = $

$y = $

In $y = \frac{2}{3}x -$

Straight lines

$\begin{cases} 2x - 3y \\ -4x + 6y \end{cases}$
a whole line

D. Identify
To identify a

EXAMPLE

system].

Solution:

In $y = \frac{2}{3}x$

Straight lin

Solution:

$$x + y = 2 \qquad\qquad 3x - 2y = 1 \;\longleftarrow\; \text{given system equations}$$

$$y = -x + 2 \qquad\qquad -2y = -3x + 1 \qquad \text{Solve for } y.$$

$$y = -1x + 2 \qquad\qquad y = \frac{3}{2}x - \frac{1}{2} \;\longleftarrow\; \text{slope-intercept forms}$$

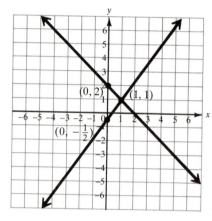

For $y = -1x + 2$: the slope $= -1$
and: the y-intercept $= (0, 2)$
[see Example 10-27].

For $y = \frac{3}{2}x - \frac{1}{2}$: the slope $= \frac{3}{2}$
and: the y-intercept $= (0, -\frac{1}{2})$.

The two graphs appear to intersect at $(1, 1)$.

Check $(1, 1)$ in $x + y = 2$.

$x + y$	$= 2$
$1 + 1$	2
2	$2 \longleftarrow$ true

Check $(1, 1)$ in $3x - 2y = 1$.

$3x - 2y$	$= 1$
$3(1) - 2(1)$	1
$3 - 2$	1
1	$1 \longleftarrow$ true

Check $(1, 1)$ in both original system equations to see if you get two true number sentences.

$(x, y) = (1, 1)$ is the one and only solution of $\begin{cases} x + y = 2. \\ 3x - 2y = 1 \end{cases}$.

Note: It was possible to solve $\begin{cases} x + y = 2 \\ 3x - 2y = 1 \end{cases}$ by graphing because the solution $(1, 1)$ has integral coordinates.

Caution: If the one and only solution of an independent-consistent system does not have integral coordinates, you will probably not be able to find the solution of the system by graphing.

Agreement: In this text, the words "solve an independent-consistent system by graphing" will mean that the system solution has integral coordinates.

11-3. Solving Systems Using the Substitution Method

A. Identify the steps to solve a system using the substitution method.

To **solve a system using the substitution method,** you

1. Solve one of the system equations for one of the variables to get a **solved equation.**
2. Substitute the solution from Step 1 into the **other system equation** to eliminate a variable.
3. Solve the equation from Step 2 for the remaining variable.
4. Substitute the solution from Step 3 into the solved equation from Step 1.
5. Solve the equation from Step 4 for the remaining variable.

6. Write the proposed ordered pair solution using the solutions from Steps 3 and 5.
7. Check the proposed ordered pair solution from Step 6 in both original system equations.

Note: Any system of two linear equations in two variables can be solved using the substitution method.

B. Solve a system when one of the system equations is already solved for one of the variables.

EXAMPLE 11-7: Solve using the substitution method: $\begin{cases} y = x + 1 \\ x + 2y = 5 \end{cases}$

Solution:

$y = x + 1 \longleftarrow$ solved equation   *Think:* $y = x + 1$ is already solved for y.

$x + 2y = 5 \longleftarrow$ other system equation

$x + 2(x + 1) = 5$ Substitute $x + 1$ for y in $x + 2y = 5$ to eliminate the variable y.

$x + 2x + 2 = 5$ Solve for the remaining variable x.

$3x + 2 = 5$

$3x = 3$

$x = 1 \longleftarrow$ proposed x-coordinate of the system solution

$y = x + 1 \longleftarrow$ solved equation

$y = (1) + 1$ Substitute the proposed x-coordinate 1 in the solved equation to eliminate the variable x.

$y = 1 + 1$ Solve for the remaining variable y.

$y = 2 \longleftarrow$ proposed y-coordinate of the system solution

$x = 1$ and $y = 2$ means the proposed system solution is $(x, y) = (1, 2)$.

Check $(1, 2)$ in $y = x + 1$.   Check $(1, 2)$ in $x + 2y = 5$.

$\begin{array}{c|c} y = x + 1 \\ \hline 2 & 1 + 1 \\ 2 & 2 \longleftarrow \text{true} \end{array}$ $\begin{array}{c|c} x + 2y = 5 \\ \hline 1 + 2(2) & 5 \\ 1 + 4 & 5 \\ 5 & 5 \longleftarrow \text{true } [(1, 2) \text{ checks in both original system equations}] \end{array}$

$(x, y) = (1, 2)$ is the one and only solution of $\begin{cases} y = x + 1 \\ x + 2y = 5 \end{cases}$.

Note: In Example 11-7, substituting $x + 1$ for y in $x + 2y = 5$ resulted in an equation in which the variable y was eliminated: $x + 2(x + 1) = 5$. Whenever you first substitute one system equation into another to eliminate a variable, the system is being solved by the **substitution method.**

C. Solve when neither system equation is solved for a variable.

To solve a system using the substitution method when neither system equation is solved for a variable, you first solve one of the equations for one of the variables.

EXAMPLE 11-8: Solve using the substitution method: $\begin{cases} 2x + 3y = 4 \\ 3x - 5y = 1 \end{cases}$

Solution:

$$2x + 3y = 4 \qquad \text{Solve one of the equations for one of the variables [see the following } Note\text{]}.$$

$$2x = 4 - 3y$$

$$x = 2 - \tfrac{3}{2}y \longleftarrow \text{solved equation}$$

$$3x - 5y = 1 \longleftarrow \text{other system equation}$$

$$3(2 - \tfrac{3}{2}y) - 5y = 1 \qquad \text{Substitute } 2 - \tfrac{3}{2}y \text{ for } x \text{ in } 3x - 5y = 1 \text{ to eliminate the variable } x.$$

$$6 - \tfrac{9}{2}y - 5y = 1 \qquad \text{Solve for the remaining variable } y.$$

$$2(6 - \tfrac{9}{2}y - 5y) = 2(1)$$

$$12 - 9y - 10y = 2$$

$$12 - 19y = 2$$

$$-19y = -10$$

$$y = \tfrac{10}{19} \longleftarrow \text{proposed } y\text{-coordinate of the system solution}$$

$$x = 2 - \tfrac{3}{2}y \longleftarrow \text{solved equation}$$

$$x = 2 - \tfrac{3}{2}(\tfrac{10}{19}) \qquad \text{Substitute the proposed } y\text{-coordinate } \tfrac{10}{19} \text{ in the solved equation to eliminate the variable } y.$$

$$x = 2 - \tfrac{15}{19} \qquad \text{Solve for the remaining variable } x.$$

$$x = \tfrac{23}{19} \longleftarrow \text{proposed } x\text{-coordinate of the system solution}$$

$x = \tfrac{23}{19}$ and $y = \tfrac{10}{19}$ means the proposed system solution is $(x, y) = (\tfrac{23}{19}, \tfrac{10}{19})$.

$(x, y) = (\tfrac{23}{19}, \tfrac{10}{19})$ is the one and only solution of $\begin{cases} 2x + 3y = 4 \\ 3x - 5y = 1 \end{cases}$ [check as before].

Note: To find the solved equation in Example 11-8, you can
1. solve $2x + 3y = 4$ for x to get $x = 2 - \tfrac{3}{2}y$;
2. solve $2x + 3y = 4$ for y to get $y = \tfrac{4}{3} - \tfrac{2}{3}x$;
3. solve $3x - 5y = 1$ for x to get $x = \tfrac{5}{3}y + \tfrac{1}{3}$; or
4. solve $3x - 5y = 1$ for y to get $y = \tfrac{3}{5}x - \tfrac{1}{5}$.

D. Identify a dependent system using the substitution method.

A system that reduces to a true number sentence like $-6 = -6$ is a dependent system because it has infinitely many solutions.

EXAMPLE 11-9: Solve using the substitution method: $\begin{cases} -2x + y = 3 \\ 4x - 2y = -6 \end{cases}$

Solution:

$$-2x + y = 3 \qquad \text{Solve one of the equations for one of the variables.}$$

$$y = 2x + 3 \longleftarrow \text{solved equation}$$

$$4x - 2y = -6 \longleftarrow \text{other system equation}$$

$$4x - 2(2x + 3) = -6 \qquad \text{Substitute } 2x + 3 \text{ for } y \text{ in } 4x - 2y = -6 \text{ to eliminate the variable } y.$$

$$4x - 4x - 6 = -6 \qquad \text{Solve for the remaining variable } x.$$

$$0x - 6 = -6 \qquad Think: \ 4x - 4x = 0x$$

$$-6 = -6 \longleftarrow \text{the system reduces to a true number sentence}$$

A true number sentence like $-6 = -6$ means the given system has infinitely many solutions [see the following *Note*].

Check using slope-intercept forms:

$$-2x + y = 3 \qquad\qquad\qquad 4x - 2y = -6 \longleftarrow \text{given system equations}$$

$$-2x + 2x + y = 2x + 3 \qquad\qquad -2y = -4x - 6 \qquad \text{Solve for } y.$$

$$y = 2x + 3 \qquad\qquad\qquad y = 2x + 3 \longleftarrow \text{slope intercept forms}$$

Straight lines with the same slope and y-intercept are coinciding lines [see Example 11-3].

$\begin{cases} -2x + y = 3 \\ 4x - 2y = -6 \end{cases}$ has infinitely many solutions [a dependent system].

Note: To understand that "$-6 = -6$ means the given system has infinitely many solutions," consider the equation preceding $-6 = -6$ as follows:

For every real number x, $0x - 6 = -6$. That is, $0x - 6 = -6$ has infinitely many solutions.

E. Identify an inconsistent system using the substitution method.

A system that reduces to a false number sentence like $6 = 5$ is an inconsistent system because it has no solutions.

EXAMPLE 11-10: Solve using the substitution method: $\begin{cases} x + 2y = 3 \\ 2x + 4y = 5 \end{cases}$

Solution:

$$x + 2y = 3$$

$$x = -2y + 3 \longleftarrow \text{solved equation}$$

$$2x + 4y = 5 \longleftarrow \text{other system equation}$$

$$2(-2y + 3) + 4y = 5 \qquad \text{Substitute } -2y + 3 \text{ for } x \text{ in } 2x + 4y = 5 \text{ to eliminate the variable } x.$$

$$-4y + 6 + 4y = 5 \qquad \text{Solve for the remaining variable } y.$$

$$0y + 6 = 5 \qquad \textit{Think: } -4y + 4y = 0y$$

$$6 = 5 \longleftarrow \text{system reduces to a false number sentence.}$$

A false number sentence like $6 = 5$ means the given system has no solutions [see the following *Note*].

Check using slope-intercept forms:

$$x + 2y = 3 \qquad\qquad\qquad 2x + 4y = 5 \longleftarrow \text{given system equations}$$

$$2y = -x + 3 \qquad\qquad\qquad 4y = -2x + 5 \qquad \text{Solve for } y.$$

$$y = -\tfrac{1}{2}x + \tfrac{3}{2} \qquad\qquad\qquad y = -\tfrac{1}{2}x + \tfrac{5}{4} \longleftarrow \text{slope intercept forms}$$

Straight lines with the same slope and different y-intercepts are distinct parallel lines [see Example 11-4].

$\begin{cases} x + 2y = 3 \\ 2x + 4y = 5 \end{cases}$ has no solutions [an inconsistent system].

Note: To understand that "$6 = 5$ means the given system has no solutions," consider the equation that preceded $6 = 5$ as follows:

There are no real numbers y such that $0y + 6 = 5$. That is, $0y + 6 = 5$ has no solutions.

11-4. Solving Systems Using the Addition Method

A. Solve a system using the addition method.

To **solve a system using the addition method,** you

1. Write each system equation in standard form $[Ax + By = C]$ when necessary.
2. Multiply one or both system equations to get opposite like terms when necessary.
3. Add the system equations with opposite like terms to eliminate a variable.
4. Solve the equation from Step 3 for the remaining variable.
5. Substitute the solution from Step 4 into either of the original system equations.
6. Solve the equation from Step 5 for the remaining variable.
7. Write the proposed ordered pair solution using the solutions from Steps 4 and 6.
8. Check the proposed ordered pair solution from Step 7 in both original system equations.

Note: Any system of two linear equations in two variables can be solved using the addition method.

EXAMPLE 11-11: Solve using the addition method: $\begin{Bmatrix} 2x + y = 5 \\ 3x + 15 = y \end{Bmatrix}$

Solution:

$$2x + y = 5 \xrightarrow{\text{same}} 2x + y = 5$$

$$3x + 15 = y \xrightarrow[\text{write standard form}]{} 3x - y = -15$$

$2x + y = 5$ *Think:* y and $-y$ are opposite like terms: $y + (-y) = 0$

$\dfrac{3x - y = -15}{5x + 0 = -10}$ Add the system equations to eliminate the variable y.

$5x = -10$ Solve for the remaining variable x.

$x = -2 \longleftarrow$ proposed x-coordinate for system solution

$2x + y = 5 \longleftarrow$ either one of the original system equations

$2(-2) + y = 5$ Substitute the proposed x-coordinate -2 in $2x + y = 5$ to eliminate the variable x.

$-4 + y = 5$ Solve for the remaining variable y.

$y = 9 \longleftarrow$ proposed y-coordinate for the system solution

$x = -2$ and $y = 9$ means that the proposed system solution is $(x, y) = (-2, 9)$.

Check $(-2, 9)$ in $2x + y = 5$.

$$\dfrac{2x + y = 5}{2(-2) + 9 \; | \; 5}$$
$$-4 + 9 \; | \; 5$$
$$5 \; | \; 5 \longleftarrow \text{true}$$

Check $(-2, 9)$ in $3x + 15 = y$.

$$\dfrac{3x + 15 = y}{3(-2) + 15 \; | \; 9}$$
$$-6 + 15 \; | \; 9$$
$$9 \; | \; 9 \longleftarrow \text{true}$$

Substitute $(-2, 9)$ into both original system equations to see if you get two true number sentences.

$(x, y) = (-2, 9)$ is the one and only solution of $\begin{Bmatrix} 2x + y = 5 \\ 3x + 15 = y \end{Bmatrix}$.

Note: In Example 11-11, adding the system equations in standard form resulted in an equation in which the variable y was eliminated: $5x + 0 = -10$ or $5x = -10$. Whenever you first add system equations in standard form to eliminate a variable, the system is being solved by the addition method.

B. Solve a system using the addition method when one multiplier is needed.

Caution: To solve a system using the addition method when adding the system equations in standard form will not eliminate a variable, you must first multiply at least one of the system equations to get opposite like terms.

EXAMPLE 11-12: Solve using the addition method: $\begin{cases} 4x + 3y = -6 \\ 3x = 6y - 10 \end{cases}$

Solution:

$4x + 3y = -6 \xrightarrow{\text{same}} 4x + 3y = -6$

$3x = 6y - 10 \xrightarrow[\text{write standard form}]{} 3x - 6y = -10$

Think: Adding equations will not eliminate a variable:

$4x + 3x = 7x$ and

$3y - 6y = -3y.$

To get opposite y-terms, multiply $4x + 3y = -6$ by 2 [See the following *Note.*]
and leave $3x - 6y = -10$ as is.

$4x + 3y = -6 \xrightarrow{\text{multiply by 2}} 8x + 6y = -12$

$3x - 6y = -10 \xrightarrow[\text{same}]{} 3x - 6y = -10$

Think: Adding equations will eliminate the variable y: $6y - 6y = 0$

$$\begin{aligned} 8x + 6y &= -12 \\ 3x - 6y &= -10 \\ \hline 11x + 0 &= -22 \end{aligned}$$

Add equations to eliminate the y-terms.

$11x + 0 = -22 \longleftarrow$ y-terms eliminated

$11x = -22$ Solve for the remaining variable.

$x = -2 \longleftarrow$ proposed x-coordinate for system solution

$4x + 3y = -6 \longleftarrow$ either one of the original system equations

$4(-2) + 3y = -6$ Substitute the proposed x-coordinate -2 to eliminate the variable x.

$-8 + 3y = -6$ Solve for the remaining variable y.

$3y = 2$

$y = \frac{2}{3} \longleftarrow$ proposed y-coordinate for system solution

$x = -2$ and $y = \frac{2}{3}$ means that the proposed system solution is $(x, y) = (-2, \frac{2}{3})$.

$(x, y) = (-2, \frac{2}{3})$ is the one and only solution of $\begin{cases} 4x + 3y = -6 \\ 3x = 6y - 10 \end{cases}$ [check as before].

Note: To get opposite like terms when the system equations are in standard form and when a variable term in one equation divides the corresponding like variable term in the other equation evenly so that the quotient is an integer, you multiply the one equation by the opposite of that integer. Example 11-12 illustrates this: $-6y \div (3y) = -2$ means that you multiply $4x + 3y = -6$ by 2 [the opposite of -2] to get opposite y-terms.

C. Solve using the addition method when two multipliers are needed.

To get the opposite like terms when the system equations are in standard form and when neither pair of like terms divide evenly so that the quotient is an integer, you

1. multiply the first equation by the numerical coefficient of either variable term from the second equation, and then
2. multiply the second equation by the opposite numerical coefficient of the corresponding variable term from the first equation.

EXAMPLE 11-13: Solve using the addition method: $\begin{Bmatrix} 2y = 3 - 4x \\ 5 - 5x = 3y \end{Bmatrix}$

Solution:

$2y = 3 - 4x \xrightarrow{\text{write standard form}} 4x + 2y = 3$ *Think:* $4x$ will not divide $5x$ evenly and $2y$ will

$5 - 5x = 3y \xrightarrow[\text{write standard form}]{} 5x + 3y = 5$ not divide $3y$ evenly.

$4x + 2y = 3 \xrightarrow{\text{multiply by 3}} 12x + 6y = 9$ Multiply to get opposite y-terms

$5x + 3y = 5 \xrightarrow[\text{multiply by } -2]{} -10x - 6y = -10$ [See the following *Note 1*.]

$12x + 6y = 9$ Add equations to eliminate the y-terms.

$\dfrac{-10x - 6y = -10}{2x + \ \ 0 = -1} \longleftarrow$ *y-term eliminated*

$2x = -1$ Solve for the remaining variable x.

$x = -\dfrac{1}{2} \longleftarrow$ *proposed x-coordinate*

$2y = 3 - 4x \longleftarrow$ *either one of the original system equations*

$2y = 3 - 4\left(-\dfrac{1}{2}\right)$ Substitute the proposed x-coordinate $-\frac{1}{2}$ to eliminate the variable x.

$2y = 3 + 2$ Solve for the remaining variable y.

$2y = 5$

$y = \dfrac{5}{2} \longleftarrow$ *proposed y-coordinate for system solution*

$x = -\frac{1}{2}$ and $y = \frac{5}{2}$ means the proposed system solution is $(x, y) = (-\frac{1}{2}, \frac{5}{2})$.

$(x, y) = (-\frac{1}{2}, \frac{5}{2})$ is the one and only solution of $\begin{Bmatrix} 2y = 3 - 4x \\ 5 - 5x = 3y \end{Bmatrix}$ [check as before].

Note 1: To get opposite y-terms in Example 11-13, you
   1. multiply $4x + 2y = 3$ by 3 because 3 is the numerical coefficient of $3y$ in $5x + 3y = 5$, and then
   2. multiply $5x + 3y = 5$ by -2 because -2 is the opposite of the numerical coefficient of $2y$ in $4x + 2y = 3$.

Note 2: To get opposite x-terms in Example 11-13, you
   1. multiply $4x + 2y = 3$ by 5 because 5 is the numerical coefficient of $5x$ in $5x + 3y = 5$, and then
   2. multiply $5x + 3y = 5$ by -4 because -4 is the opposite of the numerical coefficient of $4x$ in $4x + 2y = 3$.

D. Identify a dependent system using the addition method.

Recall: A system that reduces to a true number sentence like $0 = 0$ is a dependent system because it has infinitely many solutions [see Example 11-9].

EXAMPLE 11-14: Solve using the addition method: $\begin{Bmatrix} 2x = 3y - 1 \\ 6y = 4x + 2 \end{Bmatrix}$

Solution:

$$2x = 3y - 1 \xrightarrow{\text{write standard form}} 2x - 3y = -1 \qquad \textit{Think: Multiply } 2x - 3y = -1 \text{ by 2 to get}$$
$$6y = 4x + 2 \xrightarrow[\text{write standard form}]{} -4x + 6y = 2 \qquad \text{opposite } x\text{-terms.}$$

$$2x - 3y = -1 \xrightarrow{\text{multiply by 2}} 4x - 6y = -2 \qquad \textit{Think: } 4x \text{ and } -4x \text{ are opposite like terms.}$$
$$-4x + 6y = 2 \xrightarrow[\text{same}]{} -4x + 6y = 2$$

$$\begin{array}{r} 4x - 6y = -2 \\ -4x + 6y = 2 \\ \hline 0x + 0y = 0 \end{array} \qquad \text{Add equations to eliminate a variable.}$$
$$0 + 0 = 0 \longleftarrow \text{ both variables are eliminated}$$
$$0 = 0 \longleftarrow \text{ system reduces to a true number sentence}$$

$\begin{cases} 2x = 3y - 1 \\ 6y = 4x + 2 \end{cases}$ has infinitely many solutions [a dependent system] because the system reduces to a true number sentence.

Note: To understand that the system in Example 11-14 "has infinitely many solutions because the system reduces to a true number sentence" like $0 = 0$, consider the equation preceding $0 = 0$ and $0 + 0 = 0$ as follows:

For every real number x and y, $0x + 0y = 0$. That is, $0x + 0y = 0$ has infinitely many solutions.

E. Identify an inconsistent system using the addition method.

Recall: A system that reduces to a false number sentence like $0 = 12$ is an inconsistent system because it has no solutions [see Example 11-10].

EXAMPLE 11-15: Solve using the addition method: $\begin{cases} 4y = 2 + 3x \\ 9x - 12y - 6 = 0 \end{cases}$

Solution:

$$4y = 2 + 3x \xrightarrow{\text{write standard form}} -3x + 4y = 2 \qquad \textit{Think: Multiply } -3x + 4y = 2$$
$$9x - 12y - 6 = 0 \xrightarrow[\text{write standard form}]{} 9x - 12y = 6 \qquad \text{by 3 to get opposite}$$
$$\qquad x\text{-terms.}$$

$$-3x + 4y = 2 \xrightarrow{\text{multiply by 3}} -9x + 12y = 6 \qquad \textit{Think: } -9x \text{ and } 9x \text{ are opposite like terms.}$$
$$9x - 12y = 6 \xrightarrow[\text{same}]{} 9x - 12y = 6$$

$$\begin{array}{r} -9x + 12y = 6 \\ 9x - 12y = 6 \\ \hline 0x + 0y = 12 \end{array} \qquad \text{Add equations to eliminate a variable.}$$
$$0 + 0 = 12 \longleftarrow \text{ both variables are eliminated}$$
$$0 = 12 \longleftarrow \text{ system reduces to a false number sentence}$$

$\begin{cases} 4y = 2 + 3x \\ 9x - 12y - 6 = 0 \end{cases}$ has no solutions [an inconsistent system] because the system reduces to a false number sentence.

Note: To understand that the system in Example 11-15 "has no solutions because the system reduces to a false number sentence" like $0 = 12$, consider the equation preceding $0 = 12$ and $0 + 0 = 12$ as follows:

There are no real numbers x and y so that $0x + 0y = 12$. That is, $0x + 0y = 12$ has no solutions.

RAISE YOUR GRADES

Can you . . . ?

☑ check a proposed system solution
☑ identify the three basic types of graphs for a system of two linear equations in two variables
☑ identify a dependent system without graphing
☑ identify an inconsistent system without graphing
☑ identify an independent-consistent system without graphing
☑ solve an independent-consistent system by graphing
☑ solve a system using the substitution method
☑ solve a system using the addition method

SUMMARY

1. Two or more equations considered at the same time are called a system of equations, or just a system.
2. Every ordered pair that is a common solution of each system equation is called a solution of the system or a system solution.
3. To graph a system, you graph each system equation on the same rectangular coordinate system.
4. Straight lines that lie one on top of the other are called coinciding lines.
5. Coinciding lines always have the same slope and *y*-intercept [or *x*-intercept].
6. A system that graphs as coinciding lines will have infinitely many solutions.
7. A system that has infinitely many solutions is called a dependent system.
8. A dependent system containing two equations will always graph as two coinciding lines.
9. Straight lines that never intersect are called distinct parallel lines.
10. Distinct parallel lines always have the same slope and different *y*-intercepts [or *x*-intercepts].
11. A system that graphs as distinct parallel lines will have no solutions.
12. A system that has no solutions is called an inconsistent system.
13. An inconsistent system containing two equations will always graph as two distinct parallel lines.
14. Straight lines that intersect at exactly one point are called intersecting lines.
15. Intersecting lines always have different slopes.
16. A system that graphs as intersecting lines will have exactly one solution.
17. A system that has exactly one solution is called an independent-consistent system.
18. An independent-consistent system containing two equations will always graph as two intersecting lines.
19. Given any two straight lines on a rectangular coordinate system, the two given lines must be coinciding lines, distinct parallel lines, or intersecting lines.
20. A system of two linear equations in two variables must have
    1. infinitely many solutions [a dependent system];
    2. no solutions [an inconsistent system]; or
    3. exactly one solution [an independent-consistent system].
21. To identify a dependent system, an inconsistent system, or an independent-consistent system without graphing, you can compare slope-intercept forms as follows:
    1. The same slope and the same *y*-intercept represent a dependent system [infinitely many solutions].
    2. The same slope and different *y*-intercepts represents an inconsistent system [no solutions].
    3. Different slopes represent an independent-consistent system [exactly one solution].
22. To solve a system, you find all the system solutions.
23. In this text, the words "solve a system" will mean *solve a system of two linear equations in two variables*.

24. To solve a system by graphing that has exactly one solution [an independent-consistent system], you
    (a) Write slope-intercept form for each system equation.
    (b) Graph each system equation using the slope-intercept method.
    (c) Identify the ordered pair that represents the point of intersection.
    (d) Check the proposed ordered pair from Step (c) in both original system equations.
25. If the one and only solution of an independent-consistent system does not have integral coordinates, then you will probably not be able to find the solution of the system by graphing.
26. In this text, the words "solve each independent-consistent system by graphing" will mean that the system solution has integral coordinates.
27. To solve a system using the substitution method, you
    (a) Solve one of the system equations for one of the variables to get a solved equation.
    (b) Substitute the solution from Step (a) into the other system equation to eliminate a variable.
    (c) Solve the equation from Step (b) for the remaining variable.
    (d) Substitute the solution from Step (c) into the solved equation from Step (a).
    (e) Solve the equation from Step (d) for the remaining variable.
    (f) Write the proposed ordered pair solution using the solutions from Steps (c) and (e).
    (g) Check the proposed ordered pair solution from Step (f) in both original system equations.
28. To solve a system using the substitution method when neither system equation is solved for a variable, you first solve one of the equations for one of the variables.
29. To solve a system using the addition method, you
    (a) Write the system equations in standard form $[Ax + By = C]$ when necessary.
    (b) Multiply one or both system equations to get opposite like terms when necessary.
    (c) Add the system equations with opposite like terms to eliminate a variable.
    (d) Solve the equation from Step (c) for the remaining variable.
    (e) Substitute the solution from Step (d) into either of the original system equations.
    (f) Solve the equation from Step (e) for the remaining variable.
    (g) Write the proposed ordered pair solution using the solutions from Steps (d) and (f).
    (h) Check the proposed ordered pair solution from Step (g) in both original system equations.
30. To solve a system using the addition method when adding the system equations in standard form will not eliminate a variable, you must first multiply one or both system equations to get opposite like terms.
31. To get opposite like terms when the system equations are in standard form and when a variable term in one equation divides the corresponding like variable term in the other equation evenly so that the quotient is an integer, you multiply the one equation by the opposite of that integer.
32. To get opposite like terms when the system equations are in standard form and when neither pair of like terms divide evenly so that the quotient is an integer, you
    (a) multiply the first equation by the numerical coefficient of either variable term from the second equation, and then
    (b) multiply the second equation by the opposite of the numerical coefficient of the corresponding variable term from the first equation.
33. A system that reduces to a true number sentence is a dependent system because it has infinitely many solutions.
34. A system that reduces to a false number sentence is an inconsistent system because it has no solutions.

SOLVED PROBLEMS

PROBLEM 11-1 Is (1, 1) a system solution for **(a)** $\left\{\begin{array}{l} 2x - 3y = -1 \\ -6x + 9y = 3 \end{array}\right\}$ or **(b)** $\left\{\begin{array}{l} 3x - 4y = -1 \\ 6x - 8y = 2 \end{array}\right\}$?

Solution: Recall that every ordered pair that is a common solution of each system equation is called a solution of the system, or system solution [see Example 11-1]:

(a) Check (1, 1) in $2x - 3y = -1$.

$$2x - 3y = -1$$

$2x - 3y$	-1
$2(1) - 3(1)$	-1
$2 - 3$	-1
-1	-1

$\longleftarrow$ (1, 1) checks in
$2x - 3y = -1$

Check (1, 1) in $-6x + 9y = 3$.

$$-6x + 9y = 3$$

$-6x + 9y$	3
$-6(1) + 9(1)$	3
$-6 + 9$	3
3	3

$\longleftarrow$ (1, 1) checks in
$-6x + 9y = 3$

(1, 1) is a system solution of $\begin{cases} 2x - 3y = -1 \\ -6x + 9y = 3 \end{cases}$ because (1, 1) is a solution of both system equations.

(b) Check (1, 1) in $3x - 4y = -1$.

$$3x - 4y = -1$$

$3x - 4y$	-1
$3(1) - 4(1)$	-1
$3 - 4$	-1
-1	-1

$\longleftarrow$ (1, 1) checks in
$3x - 4y = -1$

Check (1, 1) in $6x - 8y = 2$.

$$6x - 8y = 2$$

$6x - 8y$	2
$6(1) - 8(1)$	2
$6 - 8$	2
-2	2

$\longleftarrow$ (1, 1) does not check in
$6x - 8y = 2$

(1, 1) is not a solution of $\begin{cases} 3x - 4y = -1 \\ 6x - 8y = 2 \end{cases}$ because (1, 1) is not a solution of both the system equations.

PROBLEM 11-2   Match each graph in Column 1 with the correct description in Column 2, Column 3, and Column 4:

Column 1	Column 2	Column 3	Column 4

(a)

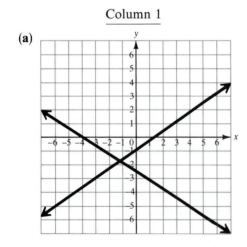

distinct parallel lines

inconsistent system

no solution

coinciding lines

independent-consistent system

exactly one solution

intersecting lines

dependent system

infinitely many solutions

(b)

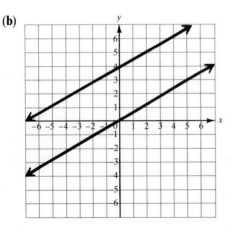

distinct parallel lines

inconsistent system

no solution

coinciding lines

independent-consistent system

exactly one solution

intersecting lines

dependent system

infinitely many solutions

	Column 1	Column 2	Column 3	Column 4

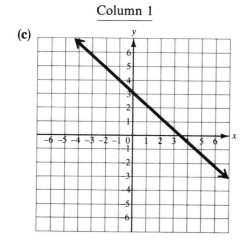

(c)

distinct
parallel
lines

inconsistent
system

no solution

coinciding
lines

independent-
consistent
system

exactly one solution

intersecting
lines

dependent
system

infinitely many
solutions

Solution: Recall that given any two straight lines on a rectangular coordinate system, the two given lines must be coinciding lines, distinct parallel lines, or intersecting lines [see Example 11-2]:
(a) intersecting lines, independent-consistent system, exactly one solution
(b) distinct parallel lines, inconsistent system, no solutions
(c) coinciding lines, dependent system, infinitely many solutions

PROBLEM 11-3 Identify each system as dependent, inconsistent, or independent-consistent without graphing:

(a) $\begin{cases} -3x + 4y = 1 \\ 6x + 8y = 2 \end{cases}$ **(b)** $\begin{cases} 2x - 4y = 1 \\ -6x + 12y = 3 \end{cases}$ **(c)** $\begin{cases} 2x - 3y = 1 \\ -4x + 6y = -2 \end{cases}$

Solution: Recall that to identify a dependent system, an inconsistent system, or an independent-consistent system without graphing, you compare slope-intercept form [see Examples 11-3, 11-4, and 11-5]:

(a) $-3x + 4y = 1$ $\qquad$ $6x + 8y = 2$

$\qquad 4y = 3x + 1$ $\qquad\qquad 8y = -6x + 2$

$\qquad y = \frac{3}{4}x + \frac{1}{4}$ $\qquad\qquad y = -\frac{3}{4}x + \frac{1}{4}$ ⟵ different slopes

$\begin{cases} -3x + 4y = 1 \\ 6x + 8y = 2 \end{cases}$ is an independent-consistent system [exactly one solution].

(b) $2x - 4y = 1$ $\qquad$ $-6x + 12y = 3$

$\qquad -4y = -2x + 1$ $\qquad\qquad 12y = 6x + 3$

$\qquad y = \frac{1}{2}x - \frac{1}{4}$ $\qquad\qquad y = \frac{1}{2}x + \frac{1}{4}$ ⟵ same slopes, different y-intercepts

$\begin{cases} 2x - 4y = 1 \\ -6x + 12y = 3 \end{cases}$ is an inconsistent system [no solutions].

(c) $2x - 3y = 1$ $\qquad$ $-4x + 6y = -2$

$\qquad -3y = -2x + 1$ $\qquad\qquad 6y = 4x - 2$

$\qquad y = \frac{2}{3}x - \frac{1}{3}$ $\qquad\qquad y = \frac{2}{3}x - \frac{1}{3}$ ⟵ same slopes, same y-intercepts

$\begin{cases} 2x - 3y = 1 \\ -4x + 6y = -2 \end{cases}$ is a dependent system [infinitely many solutions].

PROBLEM 11-4   Solve an independent-consistent system by graphing: $\begin{Bmatrix} x + y = 5 \\ y + 1 = x \end{Bmatrix}$

Solution: Recall that to solve a system by graphing that has exactly one solution [an independent-consistent system], you

1. Write slope-intercept form for each system equation.
2. Graph each system equation using the slope-intercept method.
3. Identify the ordered pair that represents the point of intersection from Step 2.
4. Check the proposed ordered pair from Step 3 in both original system equations.
   [See Example 11-6.]

$x + y = 5$	$y + 1 = x$ ⟵ given system equations
$y = -x + 5$	$y = x - 1$ Solve for y.
$y = -1x + 5$	$y = 1x - 1$ ⟵ slope-intercept form

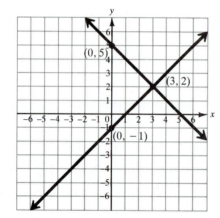

For $y = -1x + 5$: the slope $= -1$
and: the y-intercept $= (0, 5)$.

For $y = 1x - 1$: the slope $= 1$
and: the y-intercept $= (0, -1)$.

The two graphs appear to intersect at $(3, 2)$.

Check $(3, 2)$ in $x + y = 5$.

$x + y = 5$	
$3 + 2$	5
5	5 ⟵ $(3, 2)$ checks in $x + y = 5$

Check $(3, 2)$ in $y + 1 = x$.

$y + 1 = x$	
$2 + 1$	3
3	3 ⟵ $(3, 2)$ checks in $y + 1 = x$

$(x, y) = (3, 2)$ is the one and only solution of $\begin{Bmatrix} x + y = 5 \\ y + 1 = x \end{Bmatrix}$.

PROBLEM 11-5   Solve each system using the substitution method: **(a)** $\begin{Bmatrix} x - 3y = 1 \\ x = 5 - y \end{Bmatrix}$

(b) $\begin{Bmatrix} x - 2y = -1 \\ 2x + y = 1 \end{Bmatrix}$ **(c)** $\begin{Bmatrix} x - 2y + 1 = 0 \\ 2y = x + 1 \end{Bmatrix}$ **(d)** $\begin{Bmatrix} 4y + 3 = x \\ x + 3 = 4y \end{Bmatrix}$

Solution: Recall that to solve a system using the substitution method, you

1. Solve one of the system equations for one of the variables to get a solved equation.
2. Substitute the solution from Step 1 into the other system equation to eliminate a variable.
3. Solve the equation from Step 2 for the remaining variable.
4. Substitute the solution from Step 3 into the solved equation from Step 1.
5. Solve the equation from Step 4 for the remaining variable.
6. Write the proposed ordered pair solution using the solutions from Steps 3 and 5.
7. Check the proposed ordered pair solution from Step 6 in both original system equations.
   [See Examples 11-7, 11-8, 11-9, and 11-10.]

(a) $x = 5 - y$ ⟵ solved equation

$x - 3y = 1$ ⟵ other equation

$(5 - y) - 3y = 1$ Substitute.

$5 - y - 3y = 1$ Solve for y.

$5 - 4y = 1$

$-4y = -4$

$y = 1$ ⟵ proposed y-coordinate

$x = 5 - y$ ⟵ solved equation

$x = 5 - (1)$ Substitute.

$x = 5 - 1$ Solve for x.

$x = 4$ ⟵ proposed x-coordinate

$(x, y) = (4, 1)$ ⟵ solution [check as before]

(b) $x - 2y = -1$

$x = 2y - 1$ ⟵ solved equation

$2x + y = 1$ ⟵ other equation

$2(2y - 1) + y = 1$ Substitute.

$4y - 2 + y = 1$ Solve for y.

$5y - 2 = 1$

$5y = 3$

$y = \dfrac{3}{5}$ ⟵ proposed y-coordinate

$x = 2y - 1$ ⟵ solved equation

$x = 2\left(\dfrac{3}{5}\right) - 1$ Substitute.

$x = \dfrac{6}{5} - 1$ Solve for x.

$x = \dfrac{1}{5}$ ⟵ proposed x-coordinate

$(x, y) = \left(\dfrac{1}{5}, \dfrac{3}{5}\right)$ ⟵ solution [check as before]

(c) $x - 2y + 1 = 0$

$x = 2y - 1$ ⟵ solved equation

$2y = x + 1$ ⟵ other equation

$2y = (2y - 1) + 1$ Substitute.

$2y = 2y - 1 + 1$ Solve for y.

$2y = 2y$

$0 = 0$ ⟵ true number sentence

There are infinitely many solutions [a dependent system].

(d) $4y + 3 = x$ ⟵ solved equation

$x + 3 = 4y$ ⟵ other equation

$(4y + 3) + 3 = 4y$

$4y + 3 + 3 = 4y$

$4y + 6 = 4y$

$6 = 0$ ⟵ false number sentence

There are no solutions [an inconsistent system].

PROBLEM 11-6   Solve each system using the addition method:

(a) $\begin{cases} y = 5 - x \\ x = y - 1 \end{cases}$ **(b)** $\begin{cases} 2x + y + 3 = 0 \\ x = 1 - 3y \end{cases}$ **(c)** $\begin{cases} 3x = 4y - 7 \\ 5x + 3y = -2 \end{cases}$ **(d)** $\begin{cases} x = 1 - 2y \\ 6y = 3 - 3x \end{cases}$

(e) $\begin{cases} 2x - 3 = 5y \\ 4x - 10y + 6 = 0 \end{cases}$

Solution: Recall that to solve a system using the addition method, you

1. Write each system equation in standard form $[Ax + By = C]$ when necessary.
2. Multiply one or both system equations to get opposite like terms when necessary.
3. Add the system equations with opposite like terms to eliminate a variable.
4. Solve the equation from Step 3 for the remaining variable.
5. Substitute the solution from Step 4 into either of the original system equations.
6. Solve the equation from Step 5 for the remaining variable.
7. Write the proposed ordered pair solution using the solutions from Steps 4 and 6.
8. Check the proposed ordered pair solution from Step 7 in both original system equations.
   [See Examples 11-11, 11-12, 11-13, 11-14, and 11-15.]

(a) $y = 5 - x \longrightarrow$ $\overbrace{x + y = 5}^{\text{standard form}}$

$x = y - 1 \longrightarrow \underline{x - y = -1}$

$\qquad\qquad\qquad 2x + 0 = 4 \qquad$ Add.

$\qquad\qquad\qquad\quad 2x = 4 \qquad$ Solve for x.

$\qquad\qquad\qquad\quad\; x = 2$

$y = 5 - x \longleftarrow$ one of the system equations

$y = 5 - 2 \qquad$ Substitute.

$y = 3 \qquad\qquad$ Solve for y.

$(x, y) = (2, 3) \longleftarrow$ solution [check as before]

(b) $2x + y + 3 = 0 \longrightarrow$ $\overbrace{2x + \; y = -3}^{\text{standard form}}$ $\xrightarrow{\text{same}}$ $2x + \; y = -3 \quad\overset{\text{opposite like terms}}{\downarrow}$

$x = 1 - 3y \longrightarrow x + 3y = 1 \xrightarrow[\text{multiply by } -2]{} \underline{-2x - 6y = -2}$

$\qquad\qquad\qquad\qquad\qquad\qquad\qquad\qquad -5y = -5 \qquad$ Add.

$\qquad\qquad\qquad\qquad\qquad\qquad\qquad\qquad\quad\; y = 1 \qquad$ Solve for y.

$x = 1 - 3y \longleftarrow$ one of the system equations

$x = 1 - 3(1) \qquad$ Substitute.

$x = 1 - 3 \qquad\quad$ Solve for x.

$x = -2$

$(x, y) = (-2, 1) \longleftarrow$ solution [check as before]

(c) $3x = 4y - 7 \longrightarrow$ $\overbrace{3x - 4y = -7}^{\text{standard form}}$ $\xrightarrow{\text{multiply by } 3}$ $9x - 12y = -21 \quad\overset{\text{opposite like terms}}{\downarrow}$

$5x + 3y = -2 \longrightarrow 5x + 3y = -2 \xrightarrow[\text{multiply by } 4]{} \underline{20x + 12y = -8}$

$\qquad\qquad\qquad\qquad\qquad\qquad\qquad\qquad 29x + \; 0 \; = -29 \qquad$ Add.

$\qquad\qquad\qquad\qquad\qquad\qquad\qquad\qquad\quad 29x = -29 \qquad$ Solve for x.

$\qquad\qquad\qquad\qquad\qquad\qquad\qquad\qquad\qquad\; x = -1$

$3x - 4y = -7 \longleftarrow$ either one of the original system equations

$3(-1) - 4y = -7 \qquad$ Substitute.

$-3 - 4y = -7 \qquad$ Solve for y.

$-4y = -4$

$y = 1$

$(x, y) = (-1, 1) \longleftarrow$ solution [check as before]

(d) $x = 1 - 2y \longrightarrow$ $\overbrace{x + 2y = 1}^{\text{standard form}}$ $\xrightarrow{\text{multiply by } -3}$ $-3x - 6y = -3$

$6y = 3 - 3x \longrightarrow 3x + 6y = 3 \xrightarrow[\text{same}]{} \underline{3x + 6y = 3}$

$\qquad\qquad\qquad\qquad\qquad\qquad\qquad\qquad 0x + 0y = 0 \qquad$ Add.

$\qquad\qquad\qquad\qquad\qquad\qquad\qquad\qquad\quad\; 0 = 0 \longleftarrow$ true

There are infinitely many solutions [a dependent system].

(e)

standard form

$$2x - 3 = 5y \longrightarrow \overbrace{2x - 5y = 3} \xrightarrow{\text{multiply by } -2} -4x + 10y = -6$$
$$4x - 10y + 6 = 0 \longrightarrow 4x - 10y = -6 \xrightarrow{\text{same}} \underline{\quad 4x - 10y = -6}$$
$$0x + 0y = -12 \quad \text{Add.}$$
$$0 = -12 \longleftarrow \text{false}$$

There are no solutions [an inconsistent system].

Supplementary Exercises

PROBLEM 11-7 Is $(0, 1)$ a system solution for **(a)** $\begin{cases} x + y = 1 \\ x + 2y = 0 \end{cases}$ or **(b)** $\begin{cases} 3y - 2x = 3 \\ 3x - 2y = -2 \end{cases}$?

PROBLEM 11-8 Identify each system as either dependent, inconsistent, or independent-consistent without graphing: **(a)** $\begin{cases} 2x + 5y = 10 \\ 4x + 10y = 10 \end{cases}$ **(b)** $\begin{cases} 2x + 5y = 10 \\ 3x + 10y = 20 \end{cases}$ **(c)** $\begin{cases} 2x + 5y = 10 \\ 4x + 10y = 20 \end{cases}$

PROBLEM 11-9 Solve each independent-consistent system by graphing:

(a) $\begin{cases} x + y = 2 \\ x - y = 2 \end{cases}$ **(b)** $\begin{cases} x - 2y = -1 \\ x + 2y = 3 \end{cases}$ **(c)** $\begin{cases} 2x = 3y + 5 \\ 3x = -2y + 1 \end{cases}$ **(d)** $\begin{cases} 3y = 4x - 1 \\ -4y = 3x + 18 \end{cases}$

(e) $\begin{cases} x + y = 0 \\ x - y = 0 \end{cases}$ **(f)** $\begin{cases} x - 2y = 0 \\ x + 2y = 0 \end{cases}$ **(g)** $\begin{cases} 2x = 3y - 6 \\ 2y = 4 - 3x \end{cases}$ **(h)** $\begin{cases} 3y = 4x - 6 \\ -3x = 4y + 8 \end{cases}$

(i) $\begin{cases} x - y = 2 \\ x + y = 6 \end{cases}$ **(j)** $\begin{cases} x + y = 3 \\ x - y = 1 \end{cases}$ **(k)** $\begin{cases} 2x = y \\ x + y = 3 \end{cases}$ **(l)** $\begin{cases} x = 6 + y \\ 2x - y = 6 \end{cases}$

PROBLEM 11-10 Solve each system using the substitution method:

(a) $\begin{cases} x = y + 1 \\ y = 2x + 3 \end{cases}$ **(b)** $\begin{cases} x = 2 \\ y = 3 \end{cases}$ **(c)** $\begin{cases} x = 3y + 1 \\ 3x + 2y = 7 \end{cases}$ **(d)** $\begin{cases} y = 1 - 3x \\ 5x + 3y = 9 \end{cases}$

(e) $\begin{cases} 2x - 4y = 2 \\ y - 1 = 0 \end{cases}$ **(f)** $\begin{cases} x + 3y = 1 \\ x + y = 1 \end{cases}$ **(g)** $\begin{cases} x + 4y = -3 \\ x - y = 2 \end{cases}$ **(h)** $\begin{cases} 2x - 2y + 9 = 0 \\ 4x + y = 2 \end{cases}$

(i) $\begin{cases} y - x = -3 \\ x = y + 3 \end{cases}$ **(j)** $\begin{cases} 8x = 6y + 6 \\ 3y = 4x - 3 \end{cases}$ **(k)** $\begin{cases} 2y = x + 6 \\ x - 2y = 4 \end{cases}$ **(l)** $\begin{cases} 4x - 3y = 9 \\ 3y = 4x + 9 \end{cases}$

(m) $\begin{cases} 3x + 2y = -1 \\ x = 1 \end{cases}$ **(n)** $\begin{cases} 2y - 3x = -2 \\ y = -2 \end{cases}$ **(o)** $\begin{cases} y + 2 = x \\ x + 2 = y \end{cases}$ **(p)** $\begin{cases} y = 2 - x \\ 2x - y = 1 \end{cases}$

(q) $\begin{cases} 2y = x - 2 \\ x - 2y = 2 \end{cases}$ **(r)** $\begin{cases} x + y + 3 = 0 \\ x = y - 3 \end{cases}$ **(s)** $\begin{cases} x = 3 - y \\ x = y + 3 \end{cases}$ **(t)** $\begin{cases} x + y = -1 \\ y - 1 = x \end{cases}$

(u) $\begin{cases} y - 2 = x \\ x + 2 = y \end{cases}$ **(v)** $\begin{cases} x + y = 1 \\ x = y - 1 \end{cases}$ **(w)** $\begin{cases} 5y - 1 = 2x \\ 2y + 3 = 4x \end{cases}$ **(x)** $\begin{cases} 4x + 2y + 3 = 0 \\ 3x = 2y + 3 \end{cases}$

(y) $\begin{cases} 2x + 3y = 4 \\ 3x - 5y = 1 \end{cases}$ **(z)** $\begin{cases} 3x - 4y = -7 \\ 5x + 3y = -2 \end{cases}$

PROBLEM 11-11 Solve each system using the addition method:

(a) $\begin{cases} x - y = 4 \\ x + y = 5 \end{cases}$ **(b)** $\begin{cases} -x + y = -3 \\ x + y = 5 \end{cases}$ **(c)** $\begin{cases} x - y = 4 \\ y - 2x = 8 \end{cases}$

(d) $\begin{cases} 3y - x = -1 \\ x - y = 3 \end{cases}$

(e) $\begin{cases} 4x + 2y = 7 \\ 3x = 2y \end{cases}$

(f) $\begin{cases} 8y = 4x - 6 \\ 4x = 6 - 3y \end{cases}$

(g) $\begin{cases} 4x + 3y - 2 = 0 \\ 8y = 4x + 20 \end{cases}$

(h) $\begin{cases} -12y = 8 - 8x \\ 9x + 12y - 26 = 0 \end{cases}$

(i) $\begin{cases} 2x = 2y + 1 \\ 3x - 4y = 6 \end{cases}$

(j) $\begin{cases} 12x - 18y = -17 \\ 24x + 13 = -6y \end{cases}$

(k) $\begin{cases} 6x - 5y = 8 \\ y = 3x + 1 \end{cases}$

(l) $\begin{cases} 5x + y = -19 \\ 2x - 5y = 14 \end{cases}$

(m) $\begin{cases} 2x = 2y + 4 \\ x + 3y = 6 \end{cases}$

(n) $\begin{cases} 2x - y = -7 \\ 3x + 4y - 6 = 0 \end{cases}$

(o) $\begin{cases} y - 2x + 9 = 0 \\ 4y = x - 1 \end{cases}$

(p) $\begin{cases} 3x - y = 1 \\ 3y - 2x = 4 \end{cases}$

(q) $\begin{cases} 3y + 5 = 4x \\ 2y = 3x - 5 \end{cases}$

(r) $\begin{cases} 5x = 3y + 12 \\ 3x + 4y = 13 \end{cases}$

(s) $\begin{cases} 3x = 2y + 2 \\ 5y = 5x - 3 \end{cases}$

(t) $\begin{cases} 3y + 12 = 2x \\ 3x + 2y - 5 = 0 \end{cases}$

(u) $\begin{cases} 3x + 2y = 4 \\ 5y + 4x = 3 \end{cases}$

(v) $\begin{cases} 3x = 5y + 8 \\ -2 = 2x + 4y \end{cases}$

(w) $\begin{cases} 5x - 3y = 14 \\ 3x - 4y = 14 \end{cases}$

(x) $\begin{cases} 4x + 3y = 3 \\ 3x + 5y = 16 \end{cases}$

(y) $\begin{cases} 5x = 3y + 10 \\ 6y = 10x - 15 \end{cases}$

(z) $\begin{cases} 3x = 2y + 1 \\ 6x - 2 = 4y \end{cases}$

Answers to Supplementary Exercises

(11-7)   (a) no   (b) yes

(11-8)   (a) inconsistent   (b) independent-consistent   (c) dependent

(11-9)   (a) $(2, 0)$ (b) $(1, 1)$ (c) $(1, -1)$ (d) $(-2, -3)$ (e) $(0, 0)$ (f) $(0, 0)$
   (g) $(0, 2)$ (h) $(0, -2)$ (i) $(4, 2)$ (j) $(2, 1)$ (k) $(1, 2)$ (l) $(0, -6)$

(11-10)   (a) $(-4, -5)$ (b) $(2, 3)$ (c) $(\frac{23}{11}, \frac{4}{11})$ (d) $(-\frac{3}{2}, \frac{11}{2})$ (e) $(3, 1)$ (f) $(1, 0)$
   (g) $(1, -1)$ (h) $(-\frac{1}{2}, 4)$ (i) infinitely many solutions   (j) infinitely many solutions
   (k) no solutions   (l) no solutions   (m) $(1, -2)$ (n) $(-\frac{2}{3}, -2)$ (o) no solutions
   (p) $(1, 1)$ (q) infinitely many solutions   (r) $(-3, 0)$ (s) $(3, 0)$ (t) $(-1, 0)$
   (u) infinitely many solutions   (v) $(0, 1)$ (w) $(\frac{17}{16}, \frac{5}{8})$ (x) $(0, -\frac{3}{2})$
   (y) $(\frac{23}{19}, \frac{10}{19})$ (z) $(-1, 1)$

(11-11)   (a) $(\frac{9}{2}, \frac{1}{2})$ (b) $(4, 1)$ (c) $(-12, -16)$ (d) $(4, 1)$ (e) $(1, \frac{3}{2})$
   (f) $(\frac{3}{2}, 0)$ (g) $(-1, 2)$ (h) $(2, \frac{2}{3})$ (i) $(-4, -\frac{9}{2})$ (j) $(-\frac{2}{3}, \frac{1}{2})$
   (k) $(-\frac{13}{9}, -\frac{10}{3})$ (l) $(-3, -4)$ (m) $(3, 1)$ (n) $(-2, 3)$ (o) $(5, 1)$
   (p) $(1, 2)$ (q) $(5, 5)$ (r) $(3, 1)$ (s) $(\frac{4}{5}, \frac{1}{5})$ (t) $(3, -2)$ (u) $(2, -1)$
   (v) $(1, -1)$ (w) $(\frac{14}{11}, -\frac{28}{11})$ (x) $(-3, 5)$ (y) no solutions
   (z) infinitely many solutions

12 APPLICATIONS OF SYSTEMS

THIS CHAPTER IS ABOUT

☑ **Solving Number Problems Using Systems**
☑ **Solving Age Problems Using Systems**
☑ **Solving Digit Problems Using Systems**
☑ **Solving Mixture Problems Using Systems**
☑ **Solving Uniform Motion Problems Involving Opposite Rates Using Systems** $[d = rt]$

12-1. Solving Number Problems Using Systems

Agreement: In this chapter, the word "system" will mean *a system of two linear equations in two unknowns.*

To **solve a number problem using a system,** you

1. *Read* the problem very carefully several times.
2. *Identify* the unknown numbers.
3. *Decide* how to represent the unknown numbers using two variables.
4. *Translate* the problem to a system using key words.
5. *Solve* the system.
6. *Interpret* the solutions of the system with respect to each represented unknown number to find the proposed solutions of the original problem.
7. *Check* to see if the proposed solutions satisfy all the conditions of the original problem.

EXAMPLE 12-1: Solve the following number problem using a system.

Solution:

1. *Read:* The sum of two numbers is 84. The difference between the numbers is 48. What are the numbers?

2. *Identify:* The unknown numbers are $\begin{cases} \text{the larger number} \\ \text{the smaller number} \end{cases}$.

3. *Decide:* Let $x =$ the larger number
and $y =$ the smaller number [see the following *Note 1*].

4. *Translate:* The sum of two numbers is 84.
$$x + y = 84$$

The difference between the two numbers is 48.
$$x - y = 48 \qquad \text{[See the following *Note 2*.]}$$

5. *Solve:*

$$\left. \begin{array}{r} x + y = 84 \\ x - y = 48 \end{array} \right\} \longleftarrow \text{ system of two linear equations in two variables}$$

$$2x + 0 = 132 \qquad \text{Add to eliminate a variable [see Example 11-11].}$$

$$2x = 132 \qquad \text{Solve for } x.$$

$$x = 66$$

$$x + y = 84 \longleftarrow \text{ one of the system equations}$$

$$(66) + y = 84 \qquad \text{Substitute 66 for } x.$$

$$y = 84 - 66 \qquad \text{Solve for } y.$$

$$y = 18$$

6. *Interpret:* $x = 66$ means the larger number is 66.

$ y = 18$ means the smaller number is 18. $\Big\rangle$ proposed solutions

7. *Check:*   Did you find two numbers?   Yes: 66 and 18

Is the sum of the two numbers 84?   Yes: $66 + 18 = 84$

Is the difference between the two numbers 48?   Yes: $66 - 18 = 48$

Note 1: To solve a number problem using a system, you represent each different unknown number with a different variable.

Note 2: To represent the difference between two numbers x and y when x is the larger number, you write

(a) $x - y$ if the given difference is positive $[x - y = 48]$.

(b) $y - x$ if the given difference is negative $[y - x = -48]$.

12-2 Solving Age Problems Using Systems

To **solve an age problem using a system,** you

1. *Read* the problem very carefully several times.

2. *Identify* the unknown ages.

3. *Decide* how to represent the unknown current ages using two variables.

4. *Make a table* to help represent any unknown ages in the future or past.

5. *Translate* the problem to a system.

6. *Solve* the system.

7. *Interpret* the solutions of the system with respect to each represented unknown age to find the proposed solutions of the original problem.

8. *Check* to see if the proposed solutions satisfy all the conditions of the original problem.

EXAMPLE 12-2: Solve the following age problem using a system.

1. *Read:*   In four years, Bob will be three times the age that Chris is now. Three years ago, the sum of their ages was 10. How old was Bob three years ago? How old will Chris be in four years?

2. *Identify:*   The unknown ages are $\left\{ \begin{array}{l} \text{the current age of Bob} \\ \text{the current age of Chris} \\ \text{their ages in four years} \\ \text{their ages three years ago} \end{array} \right\}$.

3. *Decide:*   Let $b =$ the current age of Bob
and $c =$ the current age of Chris [see the following *Note 1*].

4. *Make a table:*

	now	in four years	three years ago	
Bob	b	$b + 4$	$b - 3$	[see the following *Note 2*.]
Chris	c	$c + 4$	$c - 3$	

5. *Translate:*   In four years, Bob will be three times the age that Chris is now.

$$b + 4 \quad = \quad 3 \quad \cdot \quad c$$

Three years ago, the sum of their ages was 10.

$$(b - 3) + (c - 3) \quad = \quad 10$$

6. *Solve:*

$b + 4 = 3c$ $\xrightarrow{\text{write standard form}}$ $b - 3c = -4$

$(b - 3) + (c - 3) = 10$ $\xrightarrow{\text{write standard form}}$ $b + c = 16$

$b + c = 16 \longleftarrow$ one system equation

$b = 16 - c \longleftarrow$ solved equation

$b - 3c = -4 \longleftarrow$ other system equation

$(16 - c) - 3c = -4$ Substitute to eliminate a variable [see Example 11-8].

$16 - c - 3c = -4$ Solve for c.

$16 - 4c = -4$

$-4c = -20$

$c = 5$

$b = 16 - c \longleftarrow$ solved equation

$b = 16 - (5)$ Substitute for c.

$b = 11$ Solve for b.

7. *Interpret:* $b = 11$ means that Bob is now 11 years old.
$c = 5$ means that Chris is now 5 years old.
$b - 3 = 11 - 3 = 8$ means that three years ago Bob was 8 years old.
$c + 4 = 5 + 4 = 9$ means that in four years Chris will be 9 years old.

8. *Check:*   Is the age of Bob in four years three times the age of Chris now?
Yes: $b + 4 = 11 + 4 = 15$ and $3c = 3(5) = 15$
Was the sum of their ages 10 three years ago?
Yes: $(b - 3) + (c - 3) = (11 - 3) + (5 - 3) = 8 + 2 = 10$

Note 1: To solve an age problem using a system, you represent each different unknown current age with a different variable.

Note 2: To represent any unknown ages that are in the
(a) future, you add to the current age variables;
(b) past, you subtract from the current age variables.

12-3. Solving Digit Problems Using Systems

A. Write expanded notation.

Every whole number can be written in **expanded notation.**

EXAMPLE 12-3: Write 429 in expanded notation.

Solution:

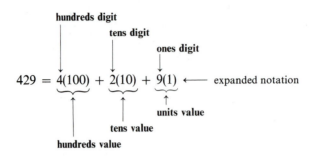

B. Reverse the digits of a number.

To reverse the digits of a given whole number, you write the last digit first, the next to last digit second, and so on.

EXAMPLE 12-4: Reverse the digits of **(a)** 38;   **(b)** 429.

Solution: **(a)** To reverse the digits of 38, you write 83.
(b) To reverse the digits of 429, you write 924.

reversed numbers

C. Solve a digit problem using a system.

To **solve a digit problem using a system,** you

1. *Read* the problem very carefully several times.
2. *Identify* the unknown digits and numbers.
3. *Decide* how to represent the unknown digits using two variables.
4. *Make a table* to help represent any unknown numbers using expanded notation.
5. *Translate* the problem to a system.
6. *Solve* the system.
7. *Interpret* the solutions of the system with respect to each represented unknown digit and number to find the proposed solutions of the original problem.
8. *Check* to see if the proposed solutions satisfy all the conditions of the original problem.

EXAMPLE 12-5: Solve the following digit problem using a system.

Solution:

1. *Read:*   The sum of the digits of a 2-digit number is 10. The reversed number is 54 less than the original number. What is the original number?

2. *Identify:*   The unknown digits and numbers are $\left\{\begin{array}{l}\text{the units digit}\\\text{the tens digit}\\\text{the original number}\\\text{the reversed number}\end{array}\right\}$.

3. *Decide:*   Let u = the units digit
                  and t = the tens digit [see the following *Note 1*].

4. *Make a table:*

	tens value	units value	expanded notation
original number	$t(10)$ or $10t$	$u(1)$ or u	$10t + u$
reversed number	$u(10)$ or $10u$	$t(1)$ or t	$10u + t$

[See the following *Note 2.*]

5. *Translate:*

The sum of the digits is 10.

$$u + t = 10$$

The reversed number is 54 less than the original number.

$$10u + t = (10t + u) - 54$$

6. *Solve:*

$$u + t = 10 \xrightarrow{\text{same}} u + t = 10 \quad \text{[See Example 11-12.]}$$

$$10u + t = 10t + u - 54 \xrightarrow[\text{write standard form}]{} 9u - 9t = -54$$

$$u + t = 10 \xrightarrow{\text{multiply by 9}} 9u + 9t = 90$$

$$9u - 9t = -54 \xrightarrow[\text{same}]{} 9u - 9t = -54$$

$$18u + 0 = 36 \quad \text{Add to eliminate a variable.}$$

$$18u = 36 \quad \text{Solve for } u.$$

$$u = 2$$

$$u + t = 10 \longleftarrow \text{one system equation}$$

$$(2) + t = 10 \quad \text{Substitute 2 for } u.$$

$$t = 8 \quad \text{Solve for } t.$$

7. *Interpret:* $t = 8$ means the tens digit is 8.

$u = 2$ means the units digit is 2.

The tens digit is 8 and the units digit is 2 means the original number is 82.

The units digit is 2 and the tens digit is 8 means the reversed number is 28.

8. *Check:* Is the sum of the digits 10? Yes: $8 + 2 = 10$

Is the reversed number 54 less than the original number? Yes: $82 - 54 = 28$

Note 1: To solve a digit problem using a system, you represent each different unknown digit with a different variable.

Note 2: To represent any unknown numbers in a digit problem, you use the chosen digit variables to write expanded notation.

12-4. Solving Mixture Problems Using Systems

To **solve a mixture problem using a system,** you

1. *Read* the problem very carefully several times.
2. *Identify* the unknown base amounts.
3. *Decide* how to represent the unknown base amounts using two variables.
4. *Make a table* to help represent the unknown amount of ingredient in each base amount.
5. *Translate* the problem to a system.
6. *Solve* the system.
7. *Interpret* the solutions of the system with respect to each represented unknown base amount to find the proposed solutions of the original problem.
8. *Check* to see if the proposed solutions satisfy all the conditions of the original problem.

EXAMPLE 12-6: Solve the following mixture problem using a system.

Solution:

1. *Read:*   How much antifreeze [ingredient] must be added to 12 gallons of a 20%-antifreeze solution [20% antifreeze and 80% water] to obtain a 60%-antifreeze solution?

2. *Identify:*   The unknown base amounts are $\begin{cases} \text{the amount of antifreeze to be added} \\ \text{the amount of 60\%-antifreeze solution} \end{cases}$.

3. *Decide:*   Let $x =$ the amount of antifreeze to be added
and $y =$ the amount of 60%-antifreeze solution [see the following *Note 1*].

4. *Make a table:*

	percent [of antifreeze]	base amount [in gallons]	amount of antifreeze [in gallons]
20% antifreeze solution	20%	12	20%(12) or 2.4
added antifreeze	100%	x	100%x or 1x
60% antifreeze solution	60%	y	60%y or 0.6y

[See the following Note 2.]

5. *Translate:*   Total amount of: 20% solution plus added antifreeze equals 60% solution.

$$12 \quad + \quad x \quad = \quad y$$

Amount of antifreeze in: 20% solution plus added antifreeze equals 60% solution.

$$2.4 \quad + \quad 1x \quad = \quad 0.6y$$

6. *Solve:* $12 + x = y \xrightarrow{\text{same}} 12 + x = y$

$2.4 + 1x = 0.6y \xrightarrow[\text{clear decimals}]{} 24 + 10x = 6y$ Multiply by 10.

$y = x + 12 \longleftarrow$ solved equation

$24 + 10x = 6y \longleftarrow$ other system equation

$24 + 10x = 6(12 + x)$ Substitute to eliminate a variable.

$24 + 10x = 72 + 6x$ Solve for x.

$10x = 48 + 6x$

$4x = 48$

$x = 12$

$y = x + 12 \longleftarrow$ solved equation

$y = (12) + 12$ Substitute 12 for x.

$y = 24$ Solve for y.

7. *Interpret:* $x = 12$ means the amount of antifreeze to be added is 12 gallons.
$y = 24$ means the amount of 60% antifreeze solution is 24 gallons.

8. *Check:*   Does the total amount of 20% solution plus added antifreeze equal the total amount of 60% solution?   Yes:

Total amount of: 20% solution	added antifreeze	60% solution
12	+ 12	= 24

Does the amount of antifreeze in the 20% solution and added antifreeze equal the amount of antifreeze in the 60% solution?   Yes:

Amount of antifreeze in: $\underline{20\% \text{ solution}}$ $\underline{\text{added antifreeze}}$ $\underline{60\% \text{ solution}}$

$$20\%(12) \quad + \quad 100\%(12) \quad = \quad 60\%(24)$$

$$2.4 \quad + \quad 12 \quad = \quad 14.4$$

Note 1: To solve a mixture problem using a system, you represent each different unknown base amount with a different variable.

Note 2: To represent an unknown amount of ingredient in an unknown base amount, you multiply the known percent of ingredient times the corresponding base-amount variable.

12-5. Solving Uniform Motion Problems Involving Opposite Rates Using Systems [*d = rt*]

A. Find the groundspeed of a plane with or against the wind given the airspeed and the wind velocity.

If $x =$ the **airspeed** of a plane [the rate of the plane in still air]
and $y =$ the **wind velocity** [the constant speed of the blowing wind],
then $x + y =$ the **groundspeed** of the plane [the **true speed** of the plane with respect to the ground] when flying with the wind
    $x - y =$ the groundspeed of the plane when flying against the wind.

Note: The phrases "with the wind" and "against the wind" describe **opposite wind rates.**

EXAMPLE 12-7: If the airspeed of a plane is 150 mph and the wind velocity is 50 mph, then what is the groundspeed [true speed] of the plane when flying (**a**) with the wind; and (**b**) against the wind?

Solution:

(**a**) airspeed of plane plus wind velocity equals groundspeed of plane with the wind

    150   +   50   =   200   *Think:* $150 + 50 = 200$

The groundspeed [true speed] of the plane flying with the wind is 200 mph.

(**b**) airspeed of plane minus wind velocity equals groundspeed of plane against wind

    150   −   50   =   100   *Think:* $150 - 50 = 100$

The groundspeed [true speed] of the plane flying against the wind is 100 mph.

B. Solve a uniform motion problem involving opposite wind rates using a system.

To **solve uniform motion problems involving opposite rates using a system,** you

1. *Read* the problem very carefully several times.
2. *Identify* the unknown rates.
3. *Decide* how to represent the unknown rates using two variables.
4. *Make a table* to help represent the true rates when appropriate.
5. *Translate* the problem to a system using the following formula: distance = (rate)(time), or $d = rt$.
6. *Solve* the system.
7. *Interpret* the solutions of the system with respect to each unknown constant rate and true speed to find the proposed solutions of the problem.
8. *Check* to see if the proposed solutions satisfy all the conditions of the original problem.

EXAMPLE 12-8: Solve this uniform motion problem involving opposite wind rates using a system.

Solution:

1. *Read:*   Flying at a constant airspeed, a pilot takes 2 hours to fly 660 miles with the wind, and then 3 hours to make the return trip against the same wind. Assuming the wind velocity was constant during the entire trip, what was the constant airspeed of the

plane and wind velocity? What was the groundspeed [true speed] of the plane with the wind and against the wind?

2. *Identify:*   The unknown rates are $\left\{\begin{array}{l}\text{the constant airspeed of the plane}\\\text{the constant wind velocity}\\\text{the groundspeed with the wind}\\\text{the groundspeed against the wind}\end{array}\right\}.$

3. *Decide:*   Let a = the constant airspeed of the plane
and w = the constant wind velocity [see the following *Note 1*].

4. *Make a table:*

	distance d [in miles]	rate r [groundspeed in mph]	time t [in hours]	
with the wind	660	$a + w$	2	[See the following *Note 2*.]
against the wind	660	$a - w$	3	

5. *Translate:*   With the wind, the distance equals the rate times the time.   *Think: $d = rt$*

$$660 = (a + w) \cdot 2$$

Against the wind, the distance equals the rate times the time.

$$660 = (a - w) \cdot 3$$

6. *Solve:*

$660 = (a + w)2 \xrightarrow{\text{divide by 2}} a + w = 330$

$660 = (a - w)3 \xrightarrow[\text{divide by 3}]{} a - w = 220$ ⟵ standard form

$$\overline{2a + 0 = 550} \quad \text{Add to eliminate a variable.}$$

$$2a = 550 \quad \text{Solve for } a.$$

$$a = 275$$

$a + w = 330$ ⟵ one system equation

$(275) + w = 330 \quad$ Substitute 275 for a.

$w = 55 \quad$ Solve for w.

7. *Interpret:* $a = 275$ means the airspeed of the plane is 275 mph.
$w = 55$ means the velocity of the wind is 55 mph.
$a + w = 275 + 55 = 330$ means the groundspeed with the wind is 330 mph.
$a - w = 275 - 55 = 220$ means the groundspeed against the wind is 220 mph.

8. *Check:*   Does it take 2 hours to fly 660 miles with the wind if the groundspeed is 330 mph?
Yes: $d = rt$, or $660 = (330)(2)$.

Does it take 3 hours to fly 660 miles against the wind if the groundspeed is 220 mph?
Yes: $d = rt$, or $660 = (220)(3)$.

Note 1: To solve a uniform motion problem involving opposite wind rates using a system, you represent the unknown constant airspeed and wind velocity with different variables.

Note 2: To represent the groundspeed [true speed] of a plane flying
  (a) with the wind [**tailwind**], you add the wind velocity to the airspeed;
  (b) against the wind [**headwind**], you subtract the wind velocity from the airspeed.

C. Find the landspeed of a boat with or against the current, given the waterspeed of the boat and the current rate.

  If x = the **waterspeed** of a boat [the speed of the boat in still water]
  and y = the **current rate** [the constant speed of the flowing river water],

then $x + y$ = the **landspeed** of the boat [the true speed of the boat with respect to the land on shore] while traveling **downstream** [with the current]

and $x - y$ = the landspeed of the boat while traveling **upstream** [against the current].

Note: The words "downstream" and "upstream" describe **opposite current rates.**

EXAMPLE 12-9: If the waterspeed of the boat is 10 mph and the current rate is 2 mph, then what is the landspeed [true speed] of the boat while traveling **(a)** downstream; or **(b)** upstream?

Solution: **(a)** waterspeed of boat plus current rate equals landspeed of boat going downstream

$$10 \quad + \quad 2 \quad = \quad 12 \qquad \textit{Think: } 10 + 2 = 12$$

The landspeed of the boat going downstream is 12 mph.

(b) waterspeed of boat minus current rate equals landspeed of boat going upstream

$$10 \quad - \quad 2 \quad = \quad 8 \qquad \textit{Think: } 10 - 2 = 8$$

The landspeed of the boat going upstream is 8 mph.

D. Solve a uniform motion problem involving opposite current rates using a system.

1. *Read:* Rowing at a constant rate, it takes Carol 3 hours to row 27 miles downstream and 9 hours to make the return trip. What is the waterspeed of the boat and the current rate? What is the landspeed of the boat going downstream and upstream?

2. *Identify:* The unknown rates are $\begin{cases} \text{the constant waterspeed of the boat} \\ \text{the current rate} \\ \text{the landspeed going downstream} \\ \text{the landspeed going upstream} \end{cases}$.

3. *Decide:* Let x = the constant waterspeed of the boat
and y = the constant current rate [see the following *Note 1*].

4. *Make a table:*

	distance d [in miles]	rate r [landspeed in mph]	time t [in hours]	
downstream	27	$x + y$	3	[See the following
upstream	27	$x - y$	9	*Note 2.*]

5. *Translate:* Rowing downstream, the distance equals the rate times the time.   *Think: $d = rt$*

$$27 \quad = \quad (x + y) \quad \cdot \quad 3$$

Rowing upstream, the distance equals the rate times the time.

$$27 \quad = \quad (x - y) \quad \cdot \quad 9$$

6. *Solve:*

$27 = (x + y)3 \xrightarrow{\text{divide by 3}} x + y = 9$

$27 = (x - y)9 \xrightarrow[\text{divide by 9}]{} x - y = 3$ standard form

$\overline{2x + 0 = 12}$ Add to eliminate a variable.

$2x = 12$ Solve for x.

$x = 6$

$x + y = 9 \longleftarrow$ one system equation

$(6) + y = 9$ Substitute 6 for x.

$y = 3$ Solve for y.

7. *Interpret:* $x = 6$ means the constant waterspeed of the boat is 6 mph.
$y = 3$ means the constant current rate is 3 mph.
$x + y = 6 + 3 = 9$ means the landspeed [true speed] downstream is 9 mph.
$x - y = 6 - 3 = 3$ means the landspeed [true speed] upstream is 3 mph.

8. *Check:*   Does it take 3 hours to row 27 miles downstream if the landspeed is 9 mph?
Yes: $d = rt$, or $27 = (9)(3)$.

Does it take 9 hours to row 27 miles upstream if the landspeed is 3 mph?
Yes: $d = rt$, or $27 = (3)(9)$.

Note 1: To solve a uniform motion problem involving opposite current rates using a system, you represent the unknown constant waterspeed of the boat and current rate with different variables.

Note 2: To represent the landspeed [true speed] of the boat moving
(a) with the current [downstream], you add the current rate to the waterspeed;
(b) against the current [upstream], you subtract the current rate from the waterspeed.

RAISE YOUR GRADES

Can you ... ?

☑ solve a number problem using a system
☑ solve an age problem using a system
☑ write expanded notation
☑ reverse the digits of a number
☑ solve a digit problem using a system
☑ solve a mixture problem using a system
☑ find the groundspeed of a plane with or against the wind given the airspeed and wind velocity
☑ solve a uniform motion problem involving opposite wind rates using a system
☑ find the landspeed of a boat with or against the current given the waterspeed of the boat and the current rate
☑ solve a uniform motion problem involving opposite current rates using a system

SUMMARY

1. To solve a number problem, age problem, digit problem, mixture problem, or uniform motion problem involving opposite rates using a system, you
   (a) *Read* the problem carefully several times.
   (b) *Identify* the unknowns.
   (c) *Decide* how to represent the unknowns using two variables.
   (d) *Make a table* to help represent the unknowns when appropriate.
   (e) *Translate* the problem to a system.
   (f) *Solve* the system.
   (g) *Interpret* the solutions of the system with respect to each represented unknown to find the proposed solutions of the original problem.
   (h) *Check* to see if the proposed solutions satisfy all the conditions of the original problem.
2. To solve a number problem using a system, you represent each different unknown number with a different variable.
3. To represent the difference between two numbers x and y where x is the larger number, you write
   (a) $x - y$ if the given difference is positive.
   (b) $y - x$ if the given difference is negative.
4. To solve an age problem using a system, you represent each different unknown current age with a different variable.

5. To represent any unknown ages that are in the
   (a) future, you add to the current age variables;
   (b) past, you subtract from the current age variables.
6. Every whole number can be written in expanded notation: $[429 = 4(100) + 2(10) + 9(1)]$.
7. To reverse the digits of a given whole number, you write the last digit first, then the next to last digit second, and so on.
8. To solve a digit problem using a system, you represent each different unknown digit with a different variable.
9. To represent any unknown numbers in a digit problem, you use the chosen digit variables to write expanded notation.
10. To solve a mixture problem using a system, you represent each different unknown base amount with a different variable.
11. To represent an unknown amount of ingredient in an unknown base amount, you multiply the known percent of ingredient times the corresponding base-amount variable.
12.   If x = the airspeed of a plane [the rate of the plane in still air]
       and y = the wind velocity [the constant speed of the blowing wind],
    then $x + y$ = the groundspeed of the plane [the true speed of the plane with respect to the ground] when flying with the wind
       and $x - y$ = the groundspeed of the plane when flying against the wind.
13. The phrases "with the wind" and "against the wind" describe opposite wind rates.
14. To solve a uniform motion problem involving opposite wind rates using a system, you represent the unknown constant airspeed and wind velocity with different variables.
15. To represent the groundspeed [true speed] of a plane flying
    (a) with the wind [tailwind], you add the wind velocity to the airspeed;
    (b) against the wind [headwind], you subtract the wind velocity from the airspeed.
16.   If x = the waterspeed of a boat [the speed of the boat in still water]
       and y = the current rate [the constant speed of the flowing river water],
    then $x + y$ = the landspeed of the boat [the true speed of the boat with respect to the land on shore] downstream [with the current]
       and $x - y$ = the landspeed of the boat upstream [against the current].
17. The words "downstream" and "upstream" describe opposite current rates.
18. To solve a uniform motion problem involving opposite current rates using a system, you represent the unknown constant waterspeed of the boat and current rate with different variables.
19. To represent landspeed [true speed] of a boat moving
    (a) with the current [downstream], you add the current to the waterspeed;
    (b) against the current [upstream], you subtract the current rate from the waterspeed.

SOLVED PROBLEMS

PROBLEM 12-1   Solve each number problem using a system:

(a) The difference between two numbers is -48. The larger number is four times the smaller number. Find the numbers.

(b) The sum of two numbers is 27. Two-thirds of the larger number added to one-fourth of the smaller number is 13. What are the numbers?

Solution: Recall that to solve a number problem using a system, you

1. *Read* the problem very carefully several times.
2. *Identify* the unknown numbers.
3. *Decide* how to represent the unknown numbers using two variables.
4. *Translate* the problem to a system using key words.
5. *Solve* the system.
6. *Interpret* the solutions of the system with respect to each represented unknown number to find the proposed solutions of the original problem.
7. *Check* to see if the proposed solutions satisfy all the conditions of the original problem. [See Example 12-1.]

(a) *Identify:*   The unknown numbers are $\begin{Bmatrix} \text{the larger number} \\ \text{the smaller number} \end{Bmatrix}$.

Decide: Let x = the larger number
and y = the smaller number.

Translate: The difference between two numbers is -48.

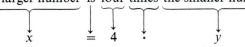

$$y - x \qquad\qquad = -48 \qquad \text{\textit{Think:} Write } y - x \text{ because the}$$
                                                given difference is negative.

The larger number is four times the smaller number.

$$x \qquad = \quad 4 \quad \cdot \qquad y$$

Solve: $x = 4y \longleftarrow$ solved equation

$y - x = -48 \longleftarrow$ other system equation

$y - (4y) = -48$ Substitute to eliminate a variable.

$y - 4y = -48$ Solve for y.

$-3y = -48$

$y = 16 \longleftarrow$ the smaller number

$x = 4y \longleftarrow$ solved equation

$x = 4(16)$ Substitute 16 for y and then solve for x.

$x = 64 \longleftarrow$ the larger number [check as before]

(b) *Identify:*   The unknown numbers are $\begin{Bmatrix} \text{the smaller number} \\ \text{the larger number} \end{Bmatrix}$.

Decide: Let x = the smaller number
and y = the larger number.

Translate: The sum of the two numbers is 27.

$$x + y \qquad\qquad = 27$$

Two-thirds of the larger number added to one-fourth of the smaller number is 13.

$$\frac{2}{3} \quad \cdot \qquad y \qquad + \qquad \frac{1}{4} \quad \cdot \qquad x \qquad = 13$$

Solve: $x + y = 27 \xrightarrow{\text{same}} x + y = 27$

$\dfrac{2}{3}y + \dfrac{1}{4}x = 13 \xrightarrow[\substack{\text{clear fractions and} \\ \text{write standard form}}]{} 3x + 8y = 156$ Multiply by the LCD 12.

$x + y = 27 \xrightarrow{\text{multiply by } -3} -3x - 3y = -81$

$3x + 8y = 156 \xrightarrow{\text{same}} \underline{3x + 8y = 156}$

$\qquad\qquad\qquad\qquad\qquad\qquad 0 + 5y = 75$ Add to eliminate a variable.

$5y = 75$ Solve for y.

$y = 15 \longleftarrow$ the larger number

$x + y = 27 \longleftarrow$ one of the system equations

$x + (15) = 27$ Substitute 15 for y and then solve for x.

$x = 12 \longleftarrow$ the smaller number [check as before]

PROBLEM 12-2 Solve each age problem using a system:

(a) Gary is 5 years older than Denise. The sum of their ages is 37. How old is each person?

(b) Diane is 6 years older than Nelson. In 5 years, Diane will be three times as old as Nelson was 3 years ago. How old will Diane be in 5 years? How old was Nelson 3 years ago?

Solution: Recall that to solve an age problem using a linear equation, you

1. *Read* the problem very carefully several times.
2. *Identify* the unknown ages.
3. *Decide* how to represent the unknown current ages using two variables.
4. *Make a table* to help represent any unknown ages in the future or past.
5. *Translate* the problem to a system.
6. *Solve* the system.
7. *Interpret* the solutions of the system with respect to each represented unknown age to find the proposed solutions of the original problem.
8. *Check* to see if the proposed solutions satisfy all the conditions of the original problem.

(a) *Identify:* The unknown ages are $\begin{cases} \text{the current age of Gary} \\ \text{the current age of Denise} \end{cases}$.

Decide: Let g = the current age of Gary
and d = the current age of Denise [see the following *Note*].

Translate: Gary is 5 years older than Denise.

$$g = 5 + d$$

The sum of their ages is 37.

$$g + d = 37$$

Solve: $g = d + 5 \longleftarrow$ solved equation

$g + d = 37 \longleftarrow$ other system equation

$(d + 5) + d = 37$ Substitute to eliminate a variable.

$d + 5 + d = 37$ Solve for d.

$2d + 5 = 37$

$2d = 32$

$d = 16 \longleftarrow$ the current age of Denise

$g = d + 5 \longleftarrow$ solved equation

$g = (16) + 5$ Substitute 16 for d and then solve for g.

$g = 21 \longleftarrow$ the current age of Gary [check as before]

Note: Because there are no future or past ages to represent, you do not need to make a table for Problem 12-2**(a)**.

(b) *Identify:* The unknown ages are $\begin{cases} \text{the current age of Diane} \\ \text{the current age of Nelson} \\ \text{their ages 5 years from now} \\ \text{their ages 3 years ago} \end{cases}$.

Decide: Let d = the current age of Diane
and n = the current age of Nelson.

Make a table:

	now	5 years from now	3 years ago
Diane	d	$d + 5$	$d - 3$
Nelson	n	$n + 5$	$n - 3$

Translate: Diane is 6 years older than Nelson.

$$d = 6 + n$$

In 5 years, Diane will be three times as old as Nelson was 3 years ago.

$$d + 5 = 3 \cdot (n - 3)$$

Solve:

$d = 6 + n \longleftarrow$ solved equation

$d + 5 = 3(n - 3) \longleftarrow$ other equation

$(6 + n) + 5 = 3(n - 3)$ Substitute to eliminate a variable.

$6 + n + 5 = 3n - 9$ Solve for n.

$n + 11 = 3n - 9$

$-2n + 11 = -9$

$-2n = -20$

$n = 10 \longleftarrow$ the current age of Nelson

$d = 6 + n \longleftarrow$ solved equation

$d = 6 + (10)$ Substitute 10 for n and then solve for d.

$d = 16 \longleftarrow$ the current age of Diane

$d + 5 = (16) + 5 = 21 \longleftarrow$ the age Diane will be 5 years from now

$n - 3 = (10) - 3 = 7 \longleftarrow$ the age Nelson was 3 years ago [check as before]

PROBLEM 12-3 Write expanded notation for **(a)** 5; **(b)** 63; **(c)** 827; and **(d)** 4019.

Solution: Recall that every whole number can be written in expanded notation [see Example 12-3]:

(a) $5 = 5(1)$ **(b)** $63 = 6(10) + 3(1)$ **(c)** $827 = 8(100) + 2(10) + 7(1)$

(d) $4019 = 4(1000) \overset{\text{optional}}{+ 0(100)} + 1(10) + 9(1)$ or $4(1000) + 1(10) + 9(1)$

PROBLEM 12-4 Reverse the digits of **(a)** 63; **(b)** 827; **(c)** 4019; and **(d)** 50.

Solution: Recall that to reverse the digits of a whole number, you write the last digit first, the next to last digit second, and so on [see Example 12-4]:

(a) To reverse the digits of 63, you write 36.
(b) To reverse the digits of 827, you write 728.
(c) To reverse the digits of 4019, you write 9104.
(d) To reverse the digits of 50, you write 05 or just 5.

PROBLEM 12-5 Solve each digit problem using a system:

(a) The sum of the digits of a 2-digit number is 10. The difference of the same two digits is 4. What are the two possible numbers?

(b) A 2-digit number is 5 less than 5 times the sum of the digits. The reversed number is 18 greater than the original number. Find the reversed number.

Solution: Recall that to solve a digit problem using a system, you

1. *Read* the problem very carefully several times.
2. *Identify* the unknown digits and numbers.
3. *Decide* how to represent the unknown digits using two variables.
4. *Make a table* to help represent any unknown numbers using expanded notation.
5. *Translate* the problem to a system.
6. *Solve* the system.
7. *Interpret* the solutions of the system with respect to each represented unknown digit and number to find the proposed solutions of the original problem.
8. *Check* to see if the proposed solutions satisfy all the conditions of the original problem. [See Example 12-5.]

(a) *Identify:*   The unknown digits and numbers are $\left\{ \begin{array}{l} \text{the units digit} \\ \text{the tens digit} \\ \text{the two possible original numbers} \end{array} \right\}.$

Decide: Let u = the units digit
and t = the tens digit [see the following *Note*].

Translate: The sum of the digits of a 2-digit number is 10.

$$t + u \qquad\qquad = 10$$

The difference of the same two digits is 4.

$$\text{Case 1: } t - u \qquad = 4 \text{ or Case 2: } u - t = 4$$

Caution: The difference between t and u can be interpreted as either $t - u$ or $u - t$.

Solve: Case 1:

$$\begin{array}{r} t + u = 10 \\ t - u = 4 \\ \hline 2t + 0 = 14 \end{array}$$

$$2t = 14$$
$$t = 7$$

$$t + u = 10$$
$$\downarrow$$
$$(7) + u = 10$$
$$u = 3$$

The original number in Case 1 is 73.

Case 2:

$$\begin{array}{r} t + u = 10 \longrightarrow u + t = 10 \\ u - t = 4 \longrightarrow u - t = 4 \\ \hline 2u + 0 = 14 \end{array}$$

$$2u = 14$$
$$u = 7$$

$$u + t = 10$$
$$\downarrow$$
$$(7) + t = 10$$
$$t = 3$$

The original number in Case 2 is 37.

The two possible numbers where the sum of the digits is 10 and difference of the digits is 4 are 73 and 37 [check as before].

Note: Because there are no numbers to represent in expanded notation, you do not need to make a table in Problem 12-5(a).

(b) *Identify:*   The unknown digits and numbers are $\left\{ \begin{array}{l} \text{the units digit} \\ \text{the tens digit} \\ \text{the original number} \\ \text{the reversed number} \end{array} \right\}.$

Decide: Let u = the units digit
and t = the tens digit.

Make a table:

	tens value	units value	expanded notation
original number	$t(10)$ or $10t$	$u(1)$ or u	$10t + u$
reversed number	$u(10)$ or $10u$	$t(1)$ or t	$10u + t$

Translate:

A 2-digit number is 5 less than 5 times the sum of the digits.

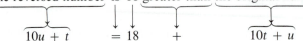

$$10t + u = 5 \cdot (t + u) - 5$$

The reversed number is 18 greater than the original number.

$$10u + t = 18 + 10t + u$$

Solve: $10t + u = 5(t + u) - 5$ $\xrightarrow{\text{write standard form}}$ $5t - 4u = -5$

$10u + t = 18 + 10t + u$ $\xrightarrow[\text{write standard form}]{}$ $-9t + 9u = 18$

$5t - 4u = -5$ $\xrightarrow{\text{same}}$ $5t - 4u = -5$ $\xrightarrow{\text{same}}$ $5t - 4u = -5$

$-9t + 9u = 18$ $\xrightarrow[\text{divide by 9}]{}$ $-t + u = 2$ $\xrightarrow[\text{multiply by 4}]{}$ $\dfrac{-4t + 4u = 8}{t + 0 = 3}$ Add to eliminate a variable.

$t = 3 \longleftarrow$ the tens digit

$5t - 4u = -5 \longleftarrow$ one of the system equations

$\downarrow$

$5(3) - 4u = -5$ Substitute 3 for t.

$15 - 4u = -5$ Solve for u.

$-4u = -20$

$u = 5 \longleftarrow$ the units digit

The original number is 35.
The reversed number is 53 [check as before].

PROBLEM 12-6   Solve each mixture using a system:

(a) The owner of a coffee store wants to make a blend of coffee that is worth $5.40 per pound [$5.40/lb] from two kinds of coffee– one kind worth $5 per pound and the other kind worth $7 per pound. How many pounds of each kind of coffee must be used to make 100 pounds of the blend?

(b) How many gallons of water must be added to 12 gallons of a 25% alcohol solution [25% alcohol and 75% water] in order to have a 15% alcohol solution?

Solution: Recall that to solve a mixture problem using a system, you

1. *Read* the problem very carefully several times.
2. *Identify* the unknown base amounts.
3. *Decide* how to represent the unknown base amounts using two variables.
4. *Make a table* to help represent the unknown amount of ingredient in each base amount.
5. *Translate* the problem to a system.
6. *Solve* the system.
7. *Interpret* the solutions of the system with respect to each represented unknown base amount to find the proposed solutions of the original problem.
8. *Check* to see if the proposed solutions satisfy all the conditions of the original problem.

(a) *Identify:*   The unknown base amounts are $\begin{cases} \text{the number of pounds of \$5/lb coffee} \\ \text{the number of pounds of \$7/lb coffee} \end{cases}$.

Decide: Let x = the number of pounds of $5/lb coffee
and y = the number of pounds of $7/lb coffee.

	cost per pound [in dollars]	base amount [in pounds]	value of base amount [in dollars]
$5/lb coffee	5	x	$5x$
$7/lb coffee	7	y	$7y$
$5.40/lb blend	5.40	100	5.40(100) or 540

Make a table: (label on left)

Translate: Total weight of: $5/lb coffee plus $7/lb coffee equals 100 pounds.

$$x \quad + \quad y \quad = \quad 100$$

Total value of: $5/lb coffee plus $7/lb coffee equals $540.

$$5x \quad + \quad 7y \quad = \quad 540$$

Solve:

$x + y = 100$ 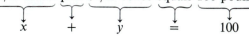 multiply by -5 → $-5x - 5y = -500$

$5x + 7y = 540$ ———same——→ $5x + 7y = 540$

$$0 + 2y = 40 \quad \text{Add to eliminate a variable.}$$

$$2y = 40 \quad \text{Solve for } y.$$

$$y = 20 \text{ (lb)} \longleftarrow \text{ number of pounds of } \$7/\text{lb coffee}$$

$x + y = 100 \longleftarrow$ one of the system equations

$x + (20) = 100$ Substitute 20 for y and then solve for x.

$x = 80 \text{ (lb)} \longleftarrow$ number of pounds of $5/lb coffee [check as before]

(b) *Identify:* The unknown base amounts are $\begin{cases} \text{the amount of water to be added} \\ \text{the amount of 15\% alcohol solution} \end{cases}$.

Decide: Let x = the amount of water to be added
and y = the amount of 15%-alcohol solution.

	percent of alcohol	base amount in gallons	amount of alcohol in gallons
25% alcohol solution	25%	12	25%(12) or 3
added water	0%	x	0%x or 0
15% alcohol solution	15%	y	15%y or 0.15y

Make a table: (labels on left)

Translate: Total amount of: 25% solution plus added water equals 15% solution.

$$12 \quad + \quad x \quad = \quad y$$

Amount of alcohol in: 25% solution plus added water equals 15% solution.

$$3 \quad + \quad 0 \quad = \quad 0.15y$$

Solve: $3 + 0 = 0.15y \longleftarrow$ one system equation

$3 = 0.15y$ Solve for y.

$y = 20 \text{ (gallons)} \longleftarrow$ the amount of 15% alcohol solution

$12 + x = y \longleftarrow$ other system equation

$12 + x = (20)$ Substitute 20 for y and then solve for x.

$x = 8 \text{ (gallons)} \longleftarrow$ the amount of water to be added [check as before]

PROBLEM 12-7 If the airspeed of a plane is 200 mph and the wind velocity is 25 mph, then what is the groundspeed of the plane when flying **(a)** with the wind; and **(b)** against the wind?

Solution: Recall that if $x =$ the airspeed of a plane

and $y =$ the wind velocity,

then $x + y =$ the groundspeed of the plane when flying with the wind

and $x - y =$ the groundspeed of the plane when flying against the wind.

[See Example 12-7.]

(a) airspeed of plane plus wind velocity equals groundspeed of plane with the wind

200   +   25   =   225   *Think: 200 + 25 = 225*

(b) airspeed of plane minus wind velocity equals groundspeed of plane against the wind

200   −   25   =   175   *Think: 200 − 25 = 175*

PROBLEM 12-8 Solve each uniform motion problem involving opposite wind rates using a system:

(a) At a constant airspeed, a pilot flew 180 miles in 3 hours against a headwind. At the same constant airspeed, the return trip took 2 hours with the help of a tailwind of the same velocity. What was the constant airspeed? What was the velocity of the wind?

(b) At a constant airspeed of 175 mph, a plane can fly 600 miles against a headwind in the same amount of time that it takes to fly 800 miles with a tailwind of the same velocity. How long is that amount of time? What is the velocity of the wind? What is the groundspeed of the plane with the tailwind?

Solution: Recall that to solve a uniform motion problem involving opposite wind rates using a system, you

1. *Read* the problem very carefully several times.
2. *Identify* the unknown air rates.
3. *Decide* how to represent the unknown air rates using two variables.
4. *Make a table* to help represent the true rates when appropriate.
5. *Translate* the problem to a system using the formula distance = (rate)(time), or $d = rt$.
6. *Solve* the system.
7. *Interpret* the solutions of the system with respect to each unknown constant air rate and true speed to find the proposed solutions of the problem.
8. *Check* to see if the proposed solutions satisfy all the conditions of the original problem. [See Example 12-8.]

(a) *Identify:*   The unknown rates are $\begin{cases} \text{the constant airspeed of the plane} \\ \text{the constant wind velocity} \end{cases}$.

Decide: Let $a =$ the constant airspeed of the plane

and $w =$ the constant wind velocity.

	distance d [in miles]	rate r [groundspeed in mph]	time t [in hours]
Make a table: against the wind	180	$a - w$	3
with the wind	180	$a + w$	2

Translate: Flying against the wind, the distance equals the rate times the time.

180   = $(a - w)$ ·   3

Flying with the wind, the distance equals the rate times the time.

180   = $(a + w)$ ·   2

Solve: $\quad 180 = (a - w)3 \xrightarrow{\text{divide by 3}} a - w = 60$

$180 = (a + w)2 \xrightarrow[\text{divide by 2}]{} a + w = 90$

$$2a + 0 = 150 \quad \text{Add to eliminate a variable.}$$

$$2a = 150 \quad \text{Solve for } a.$$

$$a = 75 \,(\text{mph}) \longleftarrow \text{constant airspeed}$$

$a + w = 90 \longleftarrow$ one of the system equations

$(75) + w = 90 \quad$ Substitute 75 for a and then solve for w.

$w = 15 \,(\text{mph}) \longleftarrow$ constant wind velocity [check as before]

(b) *Identify:* The unknown rates are $\begin{cases} \text{the constant velocity of the wind} \\ \text{"the same amount of time"} \\ \text{the groundspeed with the wind} \end{cases}$.

Decide: Let $w = $ the velocity of the wind
and $t = $ "the same amount of time."

Make a table:

	distance d [in miles]	rate r [groundspeed in mph]	time t [in hours]
against the wind	600	$175 - w$	t
with the wind	800	$175 + w$	t

Translate: Flying against the wind, the distance equals the rate times the time.

$$600 = (175 - w) \cdot t$$

Flying with the wind, the distance equals the rate times the time.

$$800 = (175 + w) \cdot t$$

Solve: $600 = (175 - w)t \xrightarrow{\text{write standard form}} 175t - wt = 600 \quad$ *Think:* $-wt$ and wt are opposite like terms.

$800 = (175 + w)t \xrightarrow[\text{form}]{\text{write standard}} 175t + wt = 800$

$$350t + 0 = 1400 \quad \text{Add to eliminate a variable.}$$

$$350t = 1400 \quad \text{Solve for } t.$$

$$t = 4 \,(\text{hours}) \longleftarrow \text{"the same amount of time"}$$

$175t + wt = 800 \longleftarrow$ one of the system equations

$175(4) + w(4) = 800 \quad$ Substitute 4 for t.

$700 + 4w = 800 \quad$ Solve for w.

$4w = 100$

$w = 25 \,(\text{mph}) \longleftarrow$ the constant wind velocity

$175 + w = 175 + (25) \quad$ Substitute 25 for w.

$= 200 \,(\text{mph}) \longleftarrow$ the groundspeed with the wind [check as before]

PROBLEM 12-9 If the waterspeed of a boat is 30 mph and the current rate is 5 mph, then what is the landspeed of the boat while traveling **(a)** downstream; and **(b)** upstream?

Solution: Recall that if $\quad x =$ the waterspeed of the boat
and $\quad y =$ the current rate,
then $x + y =$ the landspeed of the boat while traveling downstream
and $x - y =$ the landspeed of the boat while traveling upstream.
[See Example 12-9.]

(a) waterspeed of the boat plus current rate equals landspeed of boat downstream

$$30 \qquad + \qquad 5 \qquad = \qquad 35 \qquad \textit{Think: } 30 + 5 = 35$$

(b) waterspeed of the boat minus current rate equals landspeed of boat upstream

$$30 \qquad - \qquad 5 \qquad = \qquad 25 \qquad \textit{Think: } 30 - 5 = 25$$

PROBLEM 12-10 Solve each uniform motion problem involving opposite current rates using a system:

(a) A boat travels downstream at full speed of 14 km/h. On the return trip, the boat travels upstream at full speed of 6 km/h. What is the waterspeed of the boat and the current rate?

(b) With uniform effort Greg can row downstream 18 miles in 2 hours. With the same uniform effort, it takes him 6 hours to make the return trip upstream. How fast can Greg row in still water with the same uniform effort? What is the rate of the current? What is the landspeed of the boat downstream?

Solution: Recall that to solve a uniform motion problem involving opposite current rates using a system, you

1. *Read* the problem very carefully several times.
2. *Identify* the unknown water rates.
3. *Decide* how to represent the unknown water rates using two variables.
4. *Make a table* to help represent the true rates when appropriate.
5. *Translate* the problem to a system using the following formula: distance = (rate)(time), or $d = rt$.
6. *Solve* the system.
7. *Interpret* the solutions of the system with respect to each unknown constant water rate and true speed to find the proposed solutions of the problem.
8. *Check* to see if the proposed solutions satisfy all the conditions of the original problem.
   [See Example 12-10.]

(a) *Identify:* The unknown rates are $\begin{cases} \text{the constant waterspeed of the boat} \\ \text{the constant current rate} \end{cases}$.

    Decide: Let $x =$ the constant waterspeed of the boat
and $y =$ the constant current rate [see the following *Note*].

    Translate: The landspeed of the boat downstream equals 14 km/h

$$\underbrace{x + y} \qquad = \qquad 14$$

The landspeed of the boat upstream equals 6 km/h.

$$\underbrace{x - y} \qquad = \qquad 6$$

    Solve:
$$\begin{array}{l} x + y = 14 \qquad \textit{Think: } y \text{ and } -y \text{ are opposite like terms.} \\ \underline{x - y = 6} \\ 2x + 0 = 20 \qquad \text{Add to eliminate a variable.} \end{array}$$

$$2x = 20 \qquad \text{Solve for } x.$$

$$x = 10 \text{ (km/h)} \longleftarrow \text{the waterspeed of the boat}$$

$$x + y = 14 \longleftarrow \text{one of the system equations}$$

$$(10) + y = 14 \qquad \text{Substitute 10 for } x \text{ and then solve for } y.$$

$$y = 4 \text{ (km/h)} \longleftarrow \text{the current rate [check as before]}$$

Note: Because the landspeed [true speed] is given both with and against the current, you do not need to make a table in Problem 12-10 (**a**).

(b) *Identify:* The unknown rates are $\left\{\begin{array}{l}\text{the constant waterspeed of the boat}\\\text{the constant current rate}\\\text{the landspeed of the boat downstream}\end{array}\right\}$.

Decide: Let x = the constant waterspeed of the boat
and y = the constant current rate.

Make a table:

	distance d [in miles]	rate r [landspeed in mph]	time t [in hours]
downstream	18	$x + y$	2
upstream	18	$x - y$	6

Translate: Rowing downstream, the distance equals the rate times the time.

$$18 = (x + y) \cdot 2$$

Rowing upstream, the distance equals the rate times the time.

$$18 = (x - y) \cdot 6$$

Solve: $18 = (x + y)2 \xrightarrow{\text{divide by 2}} x + y = 9$
$18 = (x - y)6 \xrightarrow[\text{divide by 6}]{} x - y = 3$ standard form

$$\begin{array}{r} 2x + 0 = 12 \quad \text{Add to eliminate a variable.}\\ 2x = 12 \quad \text{Solve for } x.\\ x = 6 \, (\text{mph}) \longleftarrow \text{waterspeed of the boat (in still water)} \end{array}$$

$x + y = 9 \longleftarrow$ one of the system equations

$(6) + y = 9$ Substitute 6 for x and then solve for y.

$y = 3 \, (\text{mph}) \longleftarrow$ the current rate

$x + y = (6) + (3) = 9 \, (\text{mph}) \longleftarrow$ the landspeed of the boat downstream [check as before]

Supplementary Exercises

PROBLEM 12-11 Solve each number problem using a system:

(a) The sum of two numbers is 87. The difference between the numbers is 31. Find the numbers.

(b) The sum of two numbers is 150. The larger number is 5 times the smaller number. What is the difference between the numbers?

(c) The difference between two numbers is 12. One-third of the smaller number equals one-fourth of the larger number. What is the sum of the numbers?

(d) Two times the smaller of two numbers is two less than the difference between the two numbers. The sum of the larger number and one-half of the smaller number is equal to three times the smaller number increased by five. What are the numbers?

PROBLEM 12-12   Solve each age problem using a system:

(a) Myrle is 5 years older than Evelyn. The sum of their ages is 47. How old is each person?

(b) Arthur is 10 years older than Carol. Nine years ago, his age was three times her age. How old will each be 4 years from now?

(c) Five years ago, Chuck was twice as old as Jeanne will be 6 years from now. He is now one year older than 4 times her age. How old was Chuck 5 years ago? How old will Jeanne be 6 years from now?

(d) Greg is one-third as old as his father who is two years older than his mother. When Greg was born, the sum of the ages of his father and mother was 46. How old is Greg now?

PROBLEM 12-13   Solve each digit problem using a system:

(a) The sum of the digits of a two digit number is 10. The tens digit is one more than twice the units digit. What is the number?

(b) The sum of the digits of a two-digit number is 12. The difference between the digits is 6. What are the two possible numbers?

(c) The sum of the digits of a two digit number is 10. The number is 16 times the unit digit. Find the number.

(d) The difference between a two-digit number and the reversed number is 7 times the sum of the digits. The tens digit increased by 2 is 10 times the unit digit. What is the sum of the digits?

PROBLEM 12-14   Solve each mixture problem using a system:

(a) A nurse wants to strengthen 7 gallons of a 20% alcohol solution to a 40% alcohol solution. How much alcohol must he add?

(b) A janitor wants to dilute 7 gallons of 40% cleaner solution [40% cleaner and 60% water] to a 20% cleaner solution. How much water must she add?

(c) Tincture of arnica is 20% arnica and 80% alcohol. How much alcohol must be added to a pint of tincture of arnica to get a 5% arnica solution?

(d) How much water must be evaporated from 100 gallons of a 4% salt solution [4% salt and 96% water] to get a 5% salt solution?

(e) A grocer wants 100 pounds of a nut mixture to sell for $6.20/lb. She makes the mixture from two kinds of nuts—one kind that sells for $5/lb and one kind that sells for $8/lb. How many pounds of each kind does it take to make the nut mixture?

(f) Ring gold is usually 14 carat [14K] alloyed with copper. Coin gold is usually 90% gold and the rest copper. How many ounces of pure gold [24K] must be added to 24 ounces of ring gold to make an alloy of coin gold? [*Hint:* nK gold is $\frac{n}{24}$% gold.]

PROBLEM 12-15   Solve each uniform motion problem involving opposite rates using a system:

(a) An airplane flew 340 miles at a constant airspeed in 2 hours with the wind. At the same constant airspeed, the airplane made the return trip against the same wind in $2\frac{1}{2}$ hours. What was the constant airspeed and wind velocity? What was the groundspeed when flying with a tailwind?

(b) An airplane can travel 240 km against a head-wind at a constant airspeed in $1\frac{1}{2}$ hours. At the same constant airspeed the return trip takes 1 hour and 12 minutes with a tailwind of the same velocity. What is the constant airspeed and wind velocity? What is the groundspeed when flying against the wind?

(c) A motor boat can travel 36 miles downstream in 3 hours. If it takes twice as long to make the return trip upstream, what is the waterspeed of the boat and the current rate?

(d) A boat can travel downstream 6 miles in 45 minutes. The return trip takes $1\frac{1}{2}$ hours. What is the waterspeed of the boat and the current rate?

(e) At a constant airspeed of 180 km/h, an airplane can fly 500 km with a tailwind in the same amount of time that it takes to fly 400 km against a headwind of the same velocity. How long is that amount of time? What is the velocity of the wind? What is the groundspeed of the plane against a headwind?

(f) A boat can travel 8 miles upstream in the same time that it takes to travel 12 miles downstream. The waterspeed of the boat is 20 mph. How long is that time? What is the current rate and landspeed downstream?

Answers to Supplementary Exercises

(12-11)   **(a)** 59, 28   **(b)** 100   **(c)** 84   **(d)** 6, 20

(12-12)   **(a)** 26 years, 21 years   **(b)** 28 years, 18 years   **(c)** 28 years, 14 years   **(d)** 12 years

(12-13)   **(a)** 73   **(b)** 39 and 93   **(c)** 64   **(d)** 9

(12-14)   **(a)** $2\frac{1}{3}$ gallons   **(b)** 7 gallons   **(c)** 3 pints   **(d)** 20 gallons
(e) 60 lb of \$5/lb, 40 lb of \$8/lb   **(f)** 76 ounces

(12-15)   **(a)** 153 mph, 17 mph, 170 mph   **(b)** 180 km/h, 20 km/h, 160 km/h   **(c)** 9 mph, 3 mph
(d) 6 mph, 2 mph   **(e)** $2\frac{1}{2}$ hours, 20 km/h, 160 km/h   **(f)** 30 minutes, 4 mph, 24 mph

13 RATIONAL EXPRESSIONS AND EQUATIONS

THIS CHAPTER IS ABOUT

☑ **Simplifying Rational Expressions**
☑ **Combining Rational Expressions**
☑ **Multiplying and Dividing with Rational Expressions**
☑ **Simplifying Complex Fractions**
☑ **Solving Rational Equations and Formulas**

13-1. Simplifying Rational Expressions

A. Identify rational expressions.

If P and Q are polynomials $[Q \neq 0]$, then any algebraic expression that can be written in **fraction form** as $\dfrac{P}{Q}$ is called a **rational expression.**

EXAMPLE 13-1: Identify which of the following are rational expressions: **(a)** $\dfrac{2}{3}$ **(b)** -5 **(c)** $\dfrac{4}{z}$

(d) $\dfrac{m+n}{n}$ **(e)** $y^2 - 4$ **(f)** $\dfrac{|w|}{8}$ **(g)** $\dfrac{-b + \sqrt{b^2 - 4ac}}{2a}$ **(h)** 2^x

Solution: **(a)** $\dfrac{2}{3}$ is a rational expression because both 2 and 3 are polynomials.

(b) $-5 = \dfrac{-5}{1}$ is a rational expression because both -5 and 1 are polynomials.

(c) $\dfrac{4}{z}$ is a rational expression because both 4 and z are polynomials.

(d) $\dfrac{m+n}{n}$ is a rational expression because both $m + n$ and n are polynomials.

(e) $y^2 - 4 = \dfrac{y^2 - 4}{1}$ is a rational expression because both $y^2 - 4$ and 1 are polynomials.

(f) $\dfrac{|w|}{8}$ is not a rational expression because $|w|$ is not a polynomial.

(g) $\dfrac{-b + \sqrt{b^2 - 4ac}}{2a}$ is not a rational expression because $\sqrt{b^2 - 4ac}$ is not a polynomial.

(h) 2^x is not a rational expression because 2^x is not a polynomial.

B. Identify rational expressions in lowest terms.

A rational expression in fraction form for which the numerator and denominator do not share a common polynomial factor other than 1 or -1 is said to be in **lowest terms.**

EXAMPLE 13-2: Identify which are rational expressions in lowest terms: (a) $\dfrac{2}{3}$ (b) $\dfrac{6}{8}$ (c) $\dfrac{x}{5}$

(d) $\dfrac{w^2}{3w}$ (e) $\dfrac{y+1}{y-1}$ (f) $\dfrac{y+1}{y^2-1}$ (g) $\dfrac{m+n}{n}$

Solution: (a) $\dfrac{2}{3}$ is in lowest terms because 2 and 3 do not share a common polynomial factor other than 1 or -1.

(b) $\dfrac{6}{8}$ is not in lowest terms because 6 and 8 share a common polynomial factor of 2.

(c) $\dfrac{x}{5}$ is in lowest terms because x and 5 do not share a common polynomial factor other than 1 or -1.

(d) $\dfrac{w^2}{3w}$ is not in lowest terms because w^2 and $3w$ share a common polynomial factor of w.

(e) $\dfrac{y+1}{y-1}$ is in lowest terms because $y+1$ and $y-1$ do not share a common polynomial factor other than 1 or -1.

(f) $\dfrac{y+1}{y^2-1}$ is not in lowest terms because $y+1$ and $y^2-1 = (y+1)(y-1)$ share a common polynomial factor of $y+1$.

(g) $\dfrac{m+n}{n}$ is in lowest terms because $m+n$ and n do not share a common polynomial factor other than 1 or -1.

C. Simplify rational expressions.

To **simplify a rational expression** that is not in lowest terms, you reduce the rational expression to lowest terms using the following rule:

Fundamental Rule for Rational Expressions

If P, Q, R, and S are polynomials $[Q \neq 0$ and $R \neq 0]$, then $\dfrac{P \cdot R}{Q \cdot R} = \dfrac{P}{Q}$.

Note 1: By the Fundamental Rule for Rational Expressions, the value of a rational expression will not change when you divide both the numerator and denominator by the same nonzero polynomial.

Note 2: To simplify a rational expression, eliminate all common factors just as you would to simplify a fraction [see Example 1-31].

EXAMPLE 13-3: Simplify: (a) $\dfrac{20x}{15x}$ (b) $\dfrac{6a^2b}{-12ab^3}$ (c) $\dfrac{4y-8}{3y-6}$ (d) $\dfrac{m^2-n^2}{m^2+2mn+n^2}$

(e) $\dfrac{3-w}{w^2-9}$

Solution:

(a) $\dfrac{20x}{15x} = \dfrac{4 \cdot 5x}{3 \cdot 5x}$ Factor to get like polynomial factors.

$= \dfrac{4 \cdot \cancel{5x}}{3 \cdot \cancel{5x}}$ Eliminate the like polynomial factors using the Fundamental Rule for Rational Expressions.

$= \dfrac{4}{3}$ ⟵——— lowest terms

(b) $\dfrac{6a^2b}{-12ab^3} = \dfrac{a(6ab)}{-2b^2(6ab)}$

$\qquad = \dfrac{a}{-2b^2}$ ⟵ lowest terms

(c) $\dfrac{4y-8}{3y-6} = \dfrac{4(y-2)}{3(y-2)}$

$\qquad = \dfrac{4}{3}$ ⟵ lowest terms

(d) $\dfrac{m^2-n^2}{m^2+2mn+n^2} = \dfrac{(m+n)(m-n)}{(m+n)(m+n)}$

$\qquad = \dfrac{m-n}{m+n}$ ⟵ lowest terms

(e) $\dfrac{3-w}{w^2-9} = \dfrac{-(w-3)}{(w+3)(w-3)}$

$\qquad = \dfrac{-1}{w+3}$ ⟵ lowest terms

Note: To eliminate a factor in the numerator, you must also eliminate the same factor in the denominator. To eliminate a factor in the denominator, you must also eliminate the same factor in the numerator.

Caution: Never eliminate part of a sum or difference.

EXAMPLE 13-4: Show that you cannot eliminate part of a sum or difference.

Solution: **(a)** If $m = 3$ and $n = 2$, then $\dfrac{m+n}{n} = \dfrac{3+2}{2} = \dfrac{5}{2}$ ⟵

and $\dfrac{m+\cancel{n}}{\cancel{n}} = \dfrac{3+\cancel{2}}{\cancel{2}} = 3$ ⟵

$\qquad\qquad\qquad\qquad\qquad$ different

so $\dfrac{m+n}{n} \neq \dfrac{m+\cancel{n}}{\cancel{n}}$ $\qquad$ [Never eliminate part of a sum.]

(b) If $x = 5$ and $y = 2$, then $\dfrac{x}{x-y} = \dfrac{5}{5-2} = \dfrac{5}{3}$ ⟵

and $\dfrac{\cancel{x}}{\cancel{x}-y} = \dfrac{\cancel{5}}{\cancel{5}-2} = \dfrac{1}{-2}$ ⟵

$\qquad\qquad\qquad\qquad\qquad$ different

so $\dfrac{x}{x-y} \neq \dfrac{\cancel{x}}{\cancel{x}-y}$ $\qquad$ [Never eliminate part of a difference.]

13-2. Combining Rational Expressions

A. Combine like rational expressions.

Rational expressions with the same denominator are called **like rational expressions**. To **combine like rational expressions,** you can use the following rule:

Combine Like Rational Expressions Rule

If $\dfrac{P}{Q}$ and $\dfrac{R}{Q}$ are like rational expressions, then $\dfrac{P}{Q} + \dfrac{R}{Q} = \dfrac{P+R}{Q}$

and $\dfrac{P}{Q} - \dfrac{R}{Q} = \dfrac{P-R}{Q}$.

Note: To combine like rational expressions, add [or subtract] the numerators and then write the same denominator just as you would with like fractions [see Examples 1-36 and 1-37].

EXAMPLE 13-5: Combine like rational expressions:

(a) $\dfrac{x^2}{x+1} + \dfrac{3x}{x+1} + \dfrac{2}{x+1}$ (b) $\dfrac{2y-1}{y} - \dfrac{y+1}{y}$

Solution: (a) $\dfrac{x^2}{x+1} + \dfrac{3x}{x+1} + \dfrac{2}{x+1} = \dfrac{x^2+3x+2}{x+1}$ Add the numerators.
Write the same denominator.

$= \dfrac{(x+1)(x+2)}{x+1}$ Simplify when possible.

$= x + 2$ ⟵ lowest terms

(b) $\dfrac{2y-1}{y} - \dfrac{y+1}{y} = \dfrac{(2y-1)-(y+1)}{y}$ Subtract the numerators [see the following *Caution*]. Write the same denominator.

$= \dfrac{2y-1-y-1}{y}$ Clear parentheses.

$= \dfrac{y-2}{y}$ ⟵ lowest terms

Caution: To avoid an error, use parentheses in the numerator when subtracting rational expressions.

EXAMPLE 13-6: Show that not using parentheses when subtracting can cause an error.

wrong sign

Solution: $\dfrac{2y-1}{y} - \dfrac{y+1}{y} = \dfrac{2y-1-y+1}{y}$ No! Write $(2y-1)-(y+1)$ or $2y-1-(y+1)$.

$= \dfrac{y}{y}$

$= 1$ ⟵ wrong answer [see Example 13-5]

B. Combine unlike rational expressions.

Rational expressions with different denominators are called **unlike rational expressions.** To **combine unlike rational expressions,** you use the following rule:

Combine Unlike Rational Expressions Rule

1. Find the LCD for the given unlike rational expressions.
2. Build up to get like rational expressions using the LCD from Step 1.
3. Combine the like rational expressions from Step 2.
4. Simplify the rational expression result from Step 3 when possible.

Note: To combine unlike rational expressions, add [or subtract] as you would with unlike fractions [see Examples 1-39 and 1-40].

EXAMPLE 13-7: Combine unlike rational expressions:

(a) $\dfrac{1}{a} + \dfrac{1}{b}$ (b) $\dfrac{m}{m^2-n^2} - \dfrac{1}{m-n}$ (c) $\dfrac{x}{2x+2} + \dfrac{1}{x^2} + \dfrac{x-1}{x^2+x}$

Solution: **(a)** The LCD for $\dfrac{1}{a}$ and $\dfrac{1}{b}$ is ab [see the following *Note 1*].

building factor

$$\dfrac{1}{a} = \dfrac{1(b)}{a(b)} = \dfrac{b}{ab}$$

Build up to get like rational expressions using the LCD.

$$\dfrac{1}{b} = \dfrac{1(a)}{b(a)} = \dfrac{a}{ab}$$

LCD

$$\dfrac{1}{a} + \dfrac{1}{b} = \dfrac{b}{ab} + \dfrac{a}{ab}$$

Substitute the equal like rational expressions.

$$= \dfrac{b + a}{ab} \text{ or } \dfrac{a + b}{ab}$$

Combine like rational expressions.

(b) The LCD for $\dfrac{m}{m^2 - n^2}$ and $\dfrac{1}{m - n}$ is $(m + n)(m - n)$ [see the following *Note 2*].

$$\dfrac{m}{m^2 - n^2} = \dfrac{m}{(m + n)(m - n)}$$

$$\dfrac{1}{m - n} = \dfrac{1(m + n)}{(m + n)(m - n)} = \dfrac{m + n}{(m + n)(m - n)}$$

LCD

Build up.

$$\dfrac{m}{m^2 - n^2} - \dfrac{1}{m - n} = \dfrac{m}{(m + n)(m - n)} - \dfrac{m + n}{(m + n)(m - n)}$$

Substitute.

$$= \dfrac{m - (m + n)}{(m + n)(m - n)}$$

$$= \dfrac{m - m - n}{(m + n)(m - n)}$$

$$= \dfrac{-n}{(m + n)(m - n)} \text{ or } \dfrac{-n}{m^2 - n^2} \text{ or } \dfrac{n}{n^2 - m^2}$$

(c) The LCD for $\dfrac{x}{2x + 2}$, $\dfrac{1}{x^2}$, and $\dfrac{x - 1}{x^2 + x}$ is $2x^2(x + 1)$ [see the following *Note 3*].

$$\dfrac{x}{2x + 2} = \dfrac{x}{2(x + 1)} = \dfrac{x(x^2)}{2(x + 1)(x^2)} = \dfrac{x^3}{2x^2(x + 1)}$$

Build up.

$$\dfrac{1}{x^2} = \dfrac{1[2(x + 1)]}{x^2[2(x + 1)]} = \dfrac{2(x + 1)}{2x^2(x + 1)} = \dfrac{2x + 2}{2x^2(x + 1)}$$

LCD

$$\dfrac{x - 1}{x^2 + x} = \dfrac{x - 1}{x(x + 1)} = \dfrac{(x - 1)(2x)}{x(x + 1)(2x)} = \dfrac{2x^2 - 2x}{2x^2(x + 1)}$$

$$\dfrac{x}{2x + 2} + \dfrac{1}{x^2} + \dfrac{x - 1}{x^2 + x} = \dfrac{x^3}{2x^2(x + 1)} + \dfrac{2x + 2}{2x^2(x + 1)} + \dfrac{2x^2 - 2x}{2x^2(x + 1)}$$

Substitute.

$$= \dfrac{x^3 + 2x + 2 + 2x^2 - 2x}{2x^2(x + 1)}$$

$$= \dfrac{x^3 + 2x^2 + 2}{2x^2(x + 1)} \text{ or } \dfrac{x^3 + 2x^2 + 2}{2x^3 + 2x^2}$$

Note 1: The LCD for $\dfrac{1}{a}$ and $\dfrac{1}{b}$ is the product of the denominators ab, because the two denominators do not share a common polynomial factor other than 1 or -1 [see Example 1-38 (**c**)].

Note 2: The LCD for $\dfrac{m}{m^2 - n^2}$ and $\dfrac{1}{m - n}$ is the first denominator $m^2 - n^2 = (m + n)(m - n)$ because the other denominator $m - n$ divides into $m^2 - n^2$ evenly:

$$\frac{m^2 - n^2}{m - n} = \frac{(m + n)(m - n)}{m - n} = m + n \qquad \text{[see Example 1-38 (\textbf{b})]}.$$

Note 3: The LCD of $\dfrac{x}{2x + 2}, \dfrac{1}{x^2},$ and $\dfrac{x - 1}{x^2 + x}$ is $2x^2(x + 1)$ because

$$
\begin{aligned}
2x + 2 &= 2 (x + 1) \\
x^2 &= x^2 \\
x^2 + x &= x (x + 1)
\end{aligned}
$$

$$\text{LCD} = 2(x^2)(x + 1) \quad \longleftarrow \quad \text{Factoring Method for Finding the LCD [see Example 1-38 (\textbf{a})].}$$

13-3. Multiplying and Dividing with Rational Expressions

A. Multiply with rational expressions.

To **multiply with rational expressions,** you can use the following rule:

Multiply with Rational Expressions Rule

If $\dfrac{P}{Q}$ and $\dfrac{R}{S}$ are rational expressions, then $\dfrac{P}{Q} \cdot \dfrac{R}{S} = \dfrac{P \cdot R}{Q \cdot S}.$

Note: To multiply with rational expressions, eliminate all common factors, multiply the numerators, and then multiply the denominators as you would when multiplying with fractions [see Example 1-32].

EXAMPLE 13-8: Multiply $\dfrac{2x^2 + x - 6}{10x} \cdot \dfrac{6x - 12}{x^2 - 4}.$

Solution:
$$\frac{2x^2 + x - 6}{10x} \cdot \frac{6x - 12}{x^2 - 4} = \frac{(2x - 3)(x + 2)}{10x} \cdot \frac{6(x - 2)}{(x + 2)(x - 2)} \qquad \text{Factor each numerator and denominator}$$

$$= \frac{(2x + 3)(x + 2)}{2 \cdot 5x} \cdot \frac{2 \cdot 3(x - 2)}{(x + 2)(x - 2)} \qquad \text{Eliminate common factors.}$$

$$= \frac{(2x + 3) \cdot 3}{5x \cdot 1} \qquad \text{Multiply rational expressions.}$$

$$= \frac{6x + 9}{5x} \quad \longleftarrow \quad \text{lowest terms}$$

Note: If you first eliminate all common polynomial factors before multiplying, then the product will always be in lowest terms after multiplying.

B. Divide with rational expressions.

Recall: The reciprocal of a rational expression $\dfrac{R}{S}$ is $\dfrac{S}{R}$ $[R \neq 0]$.

To **divide with rational expressions,** you can use the following rule:

Divide with Rational Expressions Rule

If $\dfrac{P}{Q}$ and $\dfrac{R}{S}$ are rational expressions $[R \neq 0]$, then $\dfrac{P}{Q} \div \dfrac{R}{S} = \dfrac{P}{Q} \cdot \dfrac{S}{R}$.

Note: To divide with rational expressions, multiply by the reciprocal of the divisor as you would when dividing with fractions [see Example 1-35].

EXAMPLE 13-9: Divide $\dfrac{x-1}{x} \div \dfrac{x}{x+1}$.

Solution: $\dfrac{x-1}{x} \div \dfrac{x}{x+1} = \dfrac{x-1}{x} \cdot \dfrac{x+1}{x}$ Change to multiplication. Write the reciprocal of the divisor.

$$= \frac{(x-1)(x+1)}{x \cdot x}$$ Multiply rational expressions.

$$= \frac{x^2-1}{x^2}$$ ← lowest terms

13-4. Simplifying Complex Fractions

A. Identify complex fractions.

A rational expression with exactly one fraction bar is called a **simple rational expression,** or **simple fraction.** A rational expression with more than one fraction bar is called a **complex rational expression,** or **complex fraction.**

EXAMPLE 13-10: Identify which of the following are complex fractions: **(a)** $\dfrac{1}{2}$ **(b)** $\dfrac{3}{y}$

(c) $\dfrac{x}{\frac{5}{x}}$ **(d)** $\dfrac{\frac{1}{w}}{2}$ **(e)** $\dfrac{\frac{2}{m}}{\frac{m}{4}}$ **(f)** $\dfrac{1+\frac{1}{x}}{1-\frac{1}{x}}$
← numerator
← main fraction bar
← denominator

Solution: **(a)** $\dfrac{1}{2}$ is a simple fraction because it has exactly one fraction bar.

(b) $\dfrac{3}{y}$ is a simple fraction because it has exactly one fraction bar.

(c) $\dfrac{x}{\frac{5}{x}}$ is a complex fraction because it has more than one fraction bar.

(d) $\dfrac{\frac{1}{w}}{2}$ is a complex fraction because it has more than one fraction bar.

(e) $\dfrac{\frac{2}{m}}{\frac{m}{4}}$ is a complex fraction because it has more than one fraction bar.

(f) $\dfrac{1+\frac{1}{x}}{1-\frac{1}{x}}$ is a complex fraction because it has more than one fraction bar.

Note: In a complex fraction

1. The largest or thickest fraction bar is called the **main fraction bar.**
2. The rational expression above the main fraction bar is called the numerator.
3. The rational expression below the main fraction bar is called the denominator.

B. Simplify complex fractions using the division method.

To **simplify a complex fraction,** you rename the complex fraction as a simple fraction in lowest terms.

To simplify a complex fraction, you can use the following method:

Division Method for Simplifying a Complex Fraction

1. Rename to get simple fractions in both the numerator and denominator when necessary.

2. Rename the complex fraction from Step 1 as division using $\dfrac{a}{b} = a \div b$.

3. Divide the simple fractions from Step 2 to get a simple fraction in lowest terms.

EXAMPLE 13-11: Simplify using the division method: **(a)** $\dfrac{\frac{2}{m}}{\frac{m}{4}}$ **(b)** $\dfrac{1 + \frac{1}{x}}{1 - \frac{1}{x}}$

Solution:

(a) Because $\dfrac{2}{m}$ and $\dfrac{m}{4}$ are simple fractions, you first rename the complex fraction as division.

$$\frac{\frac{2}{m}}{\frac{m}{4}} = \frac{2}{m} \div \frac{m}{4} \qquad \text{Rename as division using: } \frac{a}{b} = a \div b.$$

$$= \frac{2}{m} \cdot \frac{4}{m} \qquad \text{Divide to get a simple fraction in lowest terms.}$$

$$= \frac{8}{m^2} \longleftarrow \text{simple fraction in lowest terms}$$

(b) Because $1 + \dfrac{1}{x}$ and $1 - \dfrac{1}{x}$ are not simple fractions, you first rename to get simple fractions.

$$\frac{1 + \frac{1}{x}}{1 - \frac{1}{x}} = \frac{\frac{x+1}{x}}{\frac{x-1}{x}} \qquad \text{Rename to get simple fractions } 1 + \frac{1}{x} = \frac{x}{x} + \frac{1}{x} = \frac{x+1}{x}$$

$$\text{and } 1 - \frac{1}{x} = \frac{x}{x} - \frac{1}{x} = \frac{x-1}{x}$$

$$= \frac{x+1}{x} \div \frac{x-1}{x} \qquad \text{Rename as division using } \frac{a}{b} = a \div b.$$

$$= \frac{x+1}{x} \cdot \frac{x}{x-1} \qquad \text{Divide to get a simple fraction in lowest terms.}$$

$$= \frac{x+1}{x-1} \longleftarrow \text{simple fraction in lowest terms}$$

Note: Every complex fraction can be simplified as a simple fraction in lowest terms.

C. Simplify complex fractions using the LCD method.

To simplify a complex fraction, you can also use the following method:

LCD Method for Simplifying a Complex Fraction

1. Find the LCD of all the terms contained in the numerator and denominator.
2. Multiply both the numerator and denominator by the LCD from Step 1.
3. Simplify the complex fraction from Step 2 to get a simple fraction in lowest terms.

EXAMPLE 13-12: Simplify using the LCD method: **(a)** $\dfrac{\dfrac{2}{m}}{\dfrac{m}{4}}$ **(b)** $\dfrac{1 + \dfrac{1}{x}}{1 - \dfrac{1}{x}}$

Solution: **(a)** The LCD of $\dfrac{2}{m}$ and $\dfrac{m}{4}$ is **4m.**

$$\frac{\dfrac{2}{m}}{\dfrac{m}{4}} = \frac{\left(\dfrac{2}{m}\right)4m}{\left(\dfrac{m}{4}\right)4m} \qquad \text{Multiply both the numerator and denominator by the LCD.}$$

$$= \frac{\dfrac{2}{m} \cdot \dfrac{4m}{1}}{\dfrac{m}{4} \cdot \dfrac{4m}{1}} \qquad \text{Simplify to get a simple fraction in lowest terms.}$$

$$= \frac{8}{m^2} \longleftarrow \text{same simple fraction in lowest terms found in Example 13-11 (a)}$$

(b) The LCD of $1, \dfrac{1}{x}, 1,$ and $-\dfrac{1}{x}$ is **x.**

$$\frac{1 + \dfrac{1}{x}}{1 - \dfrac{1}{x}} = \frac{\left(1 + \dfrac{1}{x}\right)x}{\left(1 - \dfrac{1}{x}\right)x} \qquad \text{Multiply both the numerator and denominator by the LCD.}$$

$$= \frac{1 \cdot x + \dfrac{1}{x} \cdot x}{1 \cdot x - \dfrac{1}{x} \cdot x} \qquad \text{Simplify to get a simple fraction in lowest terms.}$$

$$= \frac{x + 1}{x - 1} \longleftarrow \text{same simple fraction in lowest terms that was found in Example 13-11(b)}$$

Note: To simplify a complex fraction, you can use either the division method or the LCD method. You should use the method that is easiest for you.

13-5. Solving Rational Equations and Formulas

A. Identify rational equations.

An equation containing only rational expressions is called a **rational equation.**

EXAMPLE 13-13: Identify which of the following are rational equations:

(a) $2x + 3y = 5$ **(b)** $\dfrac{w}{w-1} = \dfrac{1}{w-1}$ **(c)** $\dfrac{1}{f} = \dfrac{1}{a} + \dfrac{1}{b}$ **(d)** $\sqrt{m} = 3$

(e) $y = 2x^2 - 3x + 5$ **(f)** $2^x = 4$ **(g)** $|x| = 3$

Solution:

(a) $2x + 3y = 5$ is a rational equation because both members are rational expressions.

(b) $\dfrac{w}{w - 1} = \dfrac{1}{w - 1}$ is a rational equation because both members are rational expressions.

(c) $\dfrac{1}{f} = \dfrac{1}{a} + \dfrac{1}{b}$ is a rational equation because both members are rational expressions.

(d) $\sqrt{m} = 3$ is not a rational equation because $\sqrt{m}$ is not a rational expression.

(e) $y = 2x^2 - 3x + 5$ is a rational equation because both members are rational expressions.

(f) $2^x = 4$ is not a rational equation because 2^x is not a rational expression.

(g) $|x| = 3$ is not a rational equation because $|x|$ is not a rational expression.

B. Find excluded values for rational equations.

An **excluded value of a rational equation** is any value of a variable that causes a denominator to become zero. To find the excluded value(s) of a given rational equation, you first set each denominator equal to zero and then solve each equation.

EXAMPLE 13-14: Find the excluded value(s) of: **(a)** $\dfrac{2}{3}x + \dfrac{1}{2} = \dfrac{3}{4}$ **(b)** $y - \dfrac{3}{y} = 2$

(c) $\dfrac{w^2 - 4}{w^2 + 4} = \dfrac{4}{w^2 + 4}$ **(d)** $\dfrac{3w}{w^2 + 3w} + \dfrac{2}{w - 1} = \dfrac{5w}{w^2 + 5w + 6}$

Solution:

(a) $\dfrac{2}{3}x + \dfrac{1}{2} = \dfrac{3}{4}$ has no excluded values because the denominators 3, 2, and 4 are never zero.

(b) $y - \dfrac{3}{y} = 2$ has an excluded value of 0 because the denominator y is zero when $y = 0$.

(c) $\dfrac{w^2 - 4}{w^2 + 4} = \dfrac{4}{w^2 + 4}$ has no excluded values because $w^2 + 4$ is never less than 4.

(d) $\dfrac{3w}{w^2 + 3w} + \dfrac{2}{w - 1} = \dfrac{5w}{w^2 + 5w + 6}$ has excluded values of 1, 0, -2, and -3 because

First Denominator $= 0$	Second Denominator $= 0$	Third Denominator $= 0$
$w^2 + 3w = 0$	$w - 1 = 0$	$w^2 + 5w + 6 = 0$
$w(w + 3) = 0$	$w = 1$	$(w + 2)(w + 3) = 0$
$w = 0$ or $w + 3 = 0$		$w + 2 = 0$ or $w + 3 = 0$
$w = 0$ or $w = -3$		$w = -2$ or $w = -3$

Note: To solve an equation like $w(w + 3) = 0$ or $(w + 2)(w + 3) = 0$, you use the following property:

Zero-product Property

If $ab = 0$, then $a = 0$ or $b = 0$ [see the previous Example 13-14(d)].

C. Solve rational equations.

To **solve a rational equation,** you

1. Find the LCD of all the rational expressions contained in the given equation.
2. Clear fractions by multiplying each member of the equation by the LCD from Step 1.
3. Solve the equation from Step 2 to find the proposed solutions.
4. Find the excluded values of the original equation.

5. Compare each proposed solution from Step 3 with the excluded value(s) from Step 4.
6. Reject any proposed solution that is also an excluded value.
7. Check each proposed solution that is not an excluded value in the original equation.

Note: If a rational equation has no excluded values and there were no errors made in finding the proposed solution(s), then the proposed solution(s) will always check in the original equation.

EXAMPLE 13-15: Solve $\frac{2}{3}x + \frac{1}{2} = \frac{3}{4}$.   *Think:* $\frac{2}{3}x + \frac{1}{2} = \frac{3}{4}$ has no excluded values.

Solution: The LCD of $\frac{2}{3}, \frac{1}{2},$ and $\frac{3}{4}$ is 12 [see Example 1-38].

$$12(\tfrac{2}{3}x + \tfrac{1}{2}) = 12(\tfrac{3}{4}) \qquad \text{Clear fractions by multiplying each term by the LCD.}$$

$$12(\tfrac{2}{3}x) + 12(\tfrac{1}{2}) = 12(\tfrac{3}{4})$$

$$8x + 6 = 9 \longleftarrow \text{fractions cleared}$$

$$8x = 3 \qquad \text{Solve for } x.$$

$$x = \tfrac{3}{8} \longleftarrow \text{proposed solution}$$

$$
\begin{array}{c|c}
Check: \ \frac{2}{3}x + \frac{1}{2} = \frac{3}{4} & \longleftarrow \text{original equation} \\
\frac{2}{3}(\frac{3}{8}) + \frac{1}{2} \ \Big| \ \frac{3}{4} & \text{Substitute the proposed solution in the original} \\
\frac{1}{4} + \frac{1}{2} \ \Big| \ \frac{3}{4} & \text{equation to see if you get a true number sentence.} \\
\frac{3}{4} \ \Big| \ \frac{3}{4} & \longleftarrow x = \frac{3}{8} \text{ checks [see the following } Note.]
\end{array}
$$

Note: In Example 13-15, $\frac{2}{3}x + \frac{1}{2} = \frac{3}{4}$ has no excluded values and there were no errors made in finding the proposed solution $x = \frac{3}{8}$, so the proposed solution $x = \frac{3}{8}$ checks in the original equation.

Caution: If a rational equation does have excluded values, then a proposed solution may or may not check in the original equation even if there were no errors made in finding that proposed solution. That is, if a proposed solution of a rational equation is also an excluded value of the rational equation, then that proposed solution must be rejected because it will cause division by zero when checked in the original equation.

EXAMPLE 13-16: Solve $\dfrac{w}{w-1} = \dfrac{1}{w-1}$.   *Think:* $\dfrac{w}{w-1} = \dfrac{1}{w-1}$ has an excluded value of 1.

Solution: The LCD of $\dfrac{w}{w-1} = \dfrac{1}{w-1}$ is $w - 1$.

$$(w-1) \cdot \dfrac{w}{w-1} = (w-1) \cdot \dfrac{1}{w-1} \qquad \begin{array}{l}\text{Clear fractions by multiplying each term by} \\ \text{the LCD.}\end{array}$$

$$w = 1 \longleftarrow \text{proposed solution}$$

$$
\begin{array}{c|c}
Check: \ \dfrac{w}{w-1} = \dfrac{1}{w-1} & \longleftarrow \text{original equation} \\[2mm]
\hline
\dfrac{1}{1-1} \ \Big| \ \dfrac{1}{1-1} & \text{Substitute 1 for } w \text{ to see if you get a true number sentence.} \\[3mm]
\dfrac{1}{0} \ \Big| \ \dfrac{1}{0} & \textbf{Stop!} \quad \begin{array}{l}\text{The proposed solution } w = 1 \text{ must be rejected because} \\ \frac{1}{0} \text{ is not defined [see the following } Notes \ 1 \text{ and } 2].\end{array}
\end{array}
$$

$\dfrac{w}{w-1} = \dfrac{1}{w-1}$ has no solutions because the only proposed solution is also an excluded value.

Note 1: In Example 13-16, there were no errors made in finding the proposed solution $w = 1$, but the proposed solution $w = 1$ did not check in the original equation because $w = 1$ was also an excluded value of the original equation.

Note 2: In Example 13-16, it was not necessary to check the proposed solution $w = 1$ in the original equation to see that it should be rejected. To see that the proposed solution $w = 1$ should be rejected, it is only necessary to observe that $w = 1$ is both a proposed solution and an excluded value of the original equation. A proposed solution of a given equation must always be rejected if it is also an excluded value of that equation.

D. Solve rational formulas for a given variable.

A formula that contains only rational expressions is called a **rational formula.**

To **solve a rational formula for a given variable,** you

1. Find the LCD of all the rational expressions contained in the given formula.
2. Clear fractions by multiplying each member of the formula by the LCD from Step 1.
3. Solve the literal equation from Step 2 by isolating the given variable in one member. [See Section 5-5.]

EXAMPLE 13-17: Solve for a: $\dfrac{1}{f} = \dfrac{1}{a} + \dfrac{1}{b}$ [optics formula]

Solution: The LCD of $\dfrac{1}{f}, \dfrac{1}{a}$, and $\dfrac{1}{b}$ is fab.

$$fab\left(\frac{1}{f}\right) = fab\left(\frac{1}{a} + \frac{1}{b}\right) \qquad \text{Clear fractions by multiplying each term by the LCD.}$$

$$fab\left(\frac{1}{f}\right) = fab\left(\frac{1}{a}\right) + fab\left(\frac{1}{b}\right)$$

$$ab = fb + fa \longleftarrow \text{fractions cleared}$$

$$ab - fa = fb + fa - fa \qquad \text{Isolate the given variable } a.$$

$$a(b - f) = fb$$

$$a \text{ is isolated} \longrightarrow a = \frac{fb}{b - f} \text{ or } \frac{bf}{b - f} \longleftarrow \text{solution}$$

RAISE YOUR GRADES
Can you . . . ?

☑ identify rational expressions
☑ identify rational expressions in lowest terms
☑ simplify rational expressions
☑ combine like rational expressions
☑ combine unlike rational expressions
☑ multiply with rational expressions
☑ divide with rational expressions
☑ identify complex fractions
☑ simplify complex fractions using the division method
☑ simplify complex fractions using the LCD method
☑ identify rational equations

☑ find excluded values of rational equations
☑ solve rational equations
☑ solve rational formulas for a given variable

SUMMARY

1. If P and Q are polynomials $[Q \neq 0]$, then any algebraic expression that can be written in the fraction form $\dfrac{P}{Q}$ is called a rational expression.

2. A rational expression in fraction form for which the numerator and denominator do not share a common polynomial factor other than 1 or -1 is said to be in lowest terms.

3. To simplify a rational expression that is not in lowest terms, you reduce the rational expression to lowest terms using the following rule:
   Fundamental Rule for Rational Expressions:

   If P, Q, and R are polynomials $[Q \neq 0$ and $R \neq 0]$, then $\dfrac{P \cdot R}{Q \cdot R} = \dfrac{P}{Q}$.

4. By the Fundamental Rule for Rational Expressions, the value of a rational expression will not change when you divide both the numerator and denominator by the same nonzero polynomial.

5. To simplify a rational expression, eliminate all common factors just as you would to simplify fractions.

6. Rational expressions with the same denominator are called like rational expressions.

7. Combine Like Rational Expressions Rule:

   If $\dfrac{P}{Q}$ and $\dfrac{R}{Q}$ are like rational expressions, then $\dfrac{P}{Q} + \dfrac{R}{Q} = \dfrac{P + R}{Q}$

   and $\dfrac{P}{Q} - \dfrac{R}{Q} = \dfrac{P - R}{Q}$.

8. To combine like rational expressions, add [or subtract] the numerators and then write the same denominator just as you would with like fractions.

9. To avoid an error, use parentheses in the numerator when subtracting rational expressions.

10. Rational expressions with different denominators are called unlike rational expressions.

11. Combine Unlike Rational Expressions Rule:
    (a) Find the LCD for the given unlike rational expressions.
    (b) Build up to get like rational expressions using the LCD from Step (a).
    (c) Combine the like rational expressions from Step (b).
    (d) Simplify the rational expressions from Step (c) when possible.

12. To combine unlike rational expressions, add [or subtract] as you would with unlike fractions.

13. When the denominators in two or more rational expressions do not share a common polynomial factor other than 1 or -1, the product of the denominators is the LCD.

14. When one denominator divides into the other denominator evenly, the other denominator is the LCD.

15. To find the LCD when no other method works, you can use the Factoring Method for Finding the LCD.

16. Multiply with Rational Expressions Rule:
    If $\dfrac{P}{Q}$ and $\dfrac{R}{S}$ are rational expressions, then $\dfrac{P}{Q} \cdot \dfrac{R}{S} = \dfrac{P \cdot R}{Q \cdot S}$.

17. To multiply with rational expressions, eliminate all common factors, multiply the numerators, and then multiply the denominators as you would when multiplying with fractions.

18. If you first eliminate all common polynomial factors before multiplying, then the product will always be in lowest terms after multiplying.

19. The reciprocal of a rational expression $\dfrac{R}{S}$ is $\dfrac{S}{R}$ $[R \neq 0]$.

20. Divide with Rational Expressions Rule:

   If $\dfrac{P}{Q}$ and $\dfrac{R}{S}$ are rational expressions $[R \neq 0]$, then $\dfrac{P}{Q} \div \dfrac{R}{S} = \dfrac{P}{Q} \cdot \dfrac{S}{R}$.

21. To divide with rational expressions, multiply by the reciprocal of the divisor as you would with fractions.

22. A rational expression with exactly one fraction bar is called a simple rational expression, or simple fraction.

23. A rational expression with more than one fraction bar is called a complex rational expression, or complex fraction.

24. To simplify a complex fraction, you rename the complex fraction as a simple fraction in lowest terms.

25. Division Method for Simplifying a Complex Fraction:
   (a) Rename to get simple fractions in both the numerator and denominator when necessary.
   (b) Rename the complex fraction from Step (a) as division using $\dfrac{a}{b} = a \div b$.
   (c) Divide the simple fractions from Step (b) to get a simple fraction in lowest terms.

26. LCD Method for Simplifying a Complex Fraction:
   (a) Find the LCD of all of the terms contained in the numerator and denominator.
   (b) Multiply both the numerator and denominator by the LCD from Step (a).
   (c) Simplify the complex fraction from Step (b) to get a simple fraction in lowest terms.

27. An equation containing only rational expressions is called a rational equation.

28. An excluded value of a rational equation is any value of a variable that causes a denominator to become zero.

29. To find the excluded value(s) of a given rational equation, you first set each denominator equal to zero and then solve each equation.

30. To solve an equation like $w(w + 3) = 0$ or $(w + 2)(w + 3) = 0$, you use the following property:
   Zero-Product Property:
   If $ab = 0$, then $a = 0$ or $b = 0$.

31. To solve a rational equation, you
   (a) Find the LCD of all the rational expressions contained in the given equation.
   (b) Clear fractions by multiplying each member of the equation by the LCD from Step (a).
   (c) Solve the equation from Step (b) to find the proposed solutions.
   (d) Find the excluded values of the original equation.
   (e) Compare each proposed solution from Step (c) with the excluded value(s) from Step (d).
   (f) Reject any proposed solution that is also an excluded value.
   (g) Check each proposed solution that is not an excluded value in the original equation.

32. If a rational equation has no excluded values and there were no errors made in finding the proposed solution(s), then the proposed solution(s) will always check in the original equation.

33. If a rational equation does have excluded values and there were no errors made in finding the proposed solution(s), then the proposed solution(s) may or may not check in the original equation.

34. If a proposed solution of a rational equation is also an excluded value of the rational equation, then that proposed solution must be rejected because it will cause division by zero when checked in the original equation.

35. A formula that contains only rational expressions is called a rational formula.

36. To solve a rational formula for a given variable, you
   (a) Find the LCD of all the rational expressions contained in the given formula.
   (b) Clear fractions by multiplying each member of the formula by the LCD from Step (a).
   (c) Solve the literal equation from Step (b) by isolating the given variable in one member.

SOLVED PROBLEMS

PROBLEM 13-1   Simplify: (a) $\dfrac{12}{18}$ (b) $\dfrac{24x^2}{18x^3}$ (c) $\dfrac{y^2 + 2y}{y^2 - y - 6}$ (d) $\dfrac{m^2 + mn}{m^2 - n^2}$

(e) $\dfrac{6a - 6b}{2b - 2a}$ (f) $\dfrac{6u^2}{4v}$

Solution: Recall that to simplify a rational expression, eliminate all common factors just as you would to simplify fractions [see Example 13-3]:

(a) $\dfrac{12}{18} = \dfrac{2 \cdot \cancel{6}}{3 \cdot \cancel{6}}$

$\quad = \dfrac{2}{3}$

(b) $\dfrac{24x^2}{18x^3} = \dfrac{4(\cancel{6x^2})}{3x(\cancel{6x^2})}$

$\quad = \dfrac{4}{3x}$

(c) $\dfrac{y^2 + 2y}{y^2 - y - 6} = \dfrac{y\cancel{(y + 2)}}{\cancel{(y + 2)}(y - 3)}$

$\quad = \dfrac{y}{y - 3}$

(d) $\dfrac{m^2 + mn}{m^2 - n^2} = \dfrac{m\cancel{(m + n)}}{\cancel{(m + n)}(m - n)}$

$\quad = \dfrac{m}{m - n}$

(e) $\dfrac{6a - 6b}{2b - 2a} = \dfrac{2 \cdot 3(a - b)}{2(b - a)}$

$\quad = \dfrac{\cancel{2} \cdot 3\cancel{(a - b)}}{-1 \cdot \cancel{2}\cancel{(b - a)}}$

$\quad = -3$

(f) $\dfrac{6u^2}{4v} = \dfrac{\cancel{2} \cdot 3u^2}{2 \cdot \cancel{2}v}$

$\quad = \dfrac{3u^2}{2v}$

PROBLEM 13-2   Combine like rational expressions: (a) $\dfrac{3}{8} + \dfrac{1}{8}$ (b) $\dfrac{4x - 3}{3x - 2} + \dfrac{3x + 2}{3x - 2}$

(c) $\dfrac{y^2}{y - 1} - \dfrac{2y - 1}{y - 1}$ (d) $\dfrac{-3w}{w^2 - 4} + \dfrac{2w^2}{w^2 - 4} - \dfrac{2}{w^2 - 4}$

Solution: Recall that to combine like rational expressions, add [or subtract] the numerators and then write the same denominator just as you would with like fractions [see Examples 13-5 and 13-6]:

(a) $\dfrac{3}{8} + \dfrac{1}{8} = \dfrac{3 + 1}{8}$

$\quad = \dfrac{4}{8}$

$\quad = \dfrac{\cancel{4}}{2 \cdot \cancel{4}}$

$\quad = \dfrac{1}{2}$

(b) $\dfrac{4x - 3}{3x - 2} + \dfrac{3x + 2}{3x - 2} = \dfrac{4x - 3 + 3x + 2}{3x - 2}$

$\quad = \dfrac{7x - 1}{3x - 2}$

(c) $\dfrac{y^2}{y - 1} - \dfrac{2y - 1}{y - 1} = \dfrac{y^2 - (2y - 1)}{y - 1}$

$\quad = \dfrac{y^2 - 2y + 1}{y - 1}$

$\quad = \dfrac{(y - 1)\cancel{(y - 1)}}{\cancel{y - 1}}$

$\quad = y - 1$

(d) $\dfrac{-3w}{w^2 - 4} + \dfrac{2w^2}{w^2 - 4} - \dfrac{2}{w^2 - 4} = \dfrac{-3w + 2w^2 - 2}{w^2 - 4}$

$\quad = \dfrac{2w^2 - 3w - 2}{w^2 - 4}$

$\quad = \dfrac{(2w + 1)\cancel{(w - 2)}}{(w + 2)\cancel{(w - 2)}}$

$\quad = \dfrac{2w + 1}{w + 2}$

PROBLEM 13-3 Combine unlike rational expressions: **(a)** $\dfrac{2}{3} + \dfrac{3}{4}$ **(b)** $\dfrac{x}{x-3} + \dfrac{2x+6}{x^2-9}$

(c) $\dfrac{5}{2y} - \dfrac{3}{y^2}$ **(d)** $\dfrac{m^2}{m-n} - m$ **(e)** $\dfrac{4w-2}{w^2-w-6} - \dfrac{5w+2}{w^2-4} + \dfrac{2w+7}{w^2-5w+6}$

Solution: Recall that to combine unlike rational expressions, you

1. Find the LCD for the given unlike rational expressions.
2. Build up to get like rational expressions using the LCD from Step 1.
3. Combine the like rational expressions from Step 2.
4. Simplify the rational expression result from Step 3 when possible.
    [See Example 13-7.]

(a) The LCD of $\dfrac{2}{3}$ and $\dfrac{3}{4}$ is 12.

$$\frac{2}{3} = \frac{2(4)}{3(4)} = \frac{8}{12} \longleftarrow$$

$$\frac{3}{4} = \frac{3(3)}{4(3)} = \frac{9}{12} \longleftarrow \quad \text{LCD}$$

$$\frac{2}{3} + \frac{3}{4} = \frac{8}{12} + \frac{9}{12}$$

$$= \frac{8+9}{12}$$

$$= \frac{17}{12} \text{ or } 1\frac{5}{12}$$

(b) The LCD of $\dfrac{x}{x-3}$ and $\dfrac{2x+6}{(x+3)(x-3)}$ is $(x+3)(x-3)$.

$$\frac{x}{x-3} = \frac{x(x+3)}{(x-3)(x+3)} = \frac{x^2+3x}{(x+3)(x-3)} \longleftarrow$$

$$\frac{2x+6}{(x+3)(x-3)} \xrightarrow{\text{same}} \frac{2x+6}{(x+3)(x-3)} \longleftarrow \quad \text{LCD}$$

$$\frac{x}{x-3} + \frac{2x+6}{(x+3)(x-3)} = \frac{x^2+3x}{(x+3)(x-3)} + \frac{2x+6}{(x+3)(x-3)}$$

$$= \frac{x^2+3x+2x+6}{(x+3)(x-3)}$$

$$= \frac{x^2+5x+6}{(x+3)(x-3)}$$

$$= \frac{(x+2)\cancel{(x+3)}}{\cancel{(x+3)}(x-3)}$$

$$= \frac{x+2}{x-3}$$

(c) The LCD of $\dfrac{5}{2y}$ and $\dfrac{3}{y^2}$ is $2y^2$.

$$\frac{5}{2y} = \frac{5(y)}{2y(y)} = \frac{5y}{2y^2} \longleftarrow$$

$$\frac{3}{y^2} = \frac{3(2)}{y^2(2)} = \frac{6}{2y^2} \longleftarrow \quad \text{LCD}$$

$$\frac{5}{2y} - \frac{3}{y^2} = \frac{5y}{2y^2} - \frac{6}{2y^2}$$

$$= \frac{5y-6}{2y^2}$$

(d) The LCD of $\dfrac{m^2}{m-n}$ and $\dfrac{m}{1}$ is $m-n$.

$$\frac{m^2}{m-n} \xrightarrow{\text{same}} \frac{m^2}{m-n} \longleftarrow$$

$$\frac{m}{1} = \frac{m(m-n)}{1(m-n)} = \frac{m^2-mn}{m-n} \longleftarrow \quad \text{LCD}$$

$$\frac{m^2}{m-n} - m = \frac{m^2}{m-n} - \frac{m}{1}$$

$$= \frac{m^2}{m-n} - \frac{m^2-mn}{m-n}$$

$$= \frac{m^2-(m^2-mn)}{m-n}$$

$$= \frac{m^2-m^2+mn}{m-n}$$

$$= \frac{mn}{m-n}$$

(e) The LCD of $\dfrac{4w-2}{(w+2)(w-3)}$, $\dfrac{5w+2}{(w+2)(w-2)}$, and $\dfrac{2w+7}{(w-2)(w-3)}$ is $(w+2)(w-2)(w-3)$.

$$\frac{4w-2}{(w+2)(w-3)}=\frac{(4w-2)(w-2)}{(w+2)(w-3)(w-2)}=\frac{4w^2-10w+4}{(w+2)(w-2)(w-3)}\leftarrow$$

$$\frac{5w+2}{(w+2)(w-2)}=\frac{(5w+2)(w-3)}{(w+2)(w-2)(w-3)}=\frac{5w^2-13w-6}{(w+2)(w-2)(w-3)}\leftarrow \quad \text{LCD}$$

$$\frac{2w+7}{(w-2)(w-3)}=\frac{(2w+7)(w+2)}{(w-2)(w-3)(w+2)}=\frac{2w^2+11w+14}{(w+2)(w-2)(w-3)}\leftarrow$$

$$\frac{4w-2}{(w+2)(w-3)}-\frac{5w+2}{(w+2)(w-2)}+\frac{2w+7}{(w-2)(w-3)}$$

$$=\frac{4w^2-10w+4}{(w+2)(w-2)(w-3)}-\frac{5w^2-13w-6}{(w+2)(w-2)(w-3)}+\frac{2w^2+11w+14}{(w+2)(w-2)(w-3)}$$

$$=\frac{4w^2-10w+4-(5w^2-13w-6)+2w^2+11w+14}{(w+2)(w-2)(w-3)}$$

$$=\frac{4w^2-10w+4-5w^2+13w+6+2w^2+11w+14}{(w+2)(w-2)(w-3)}$$

$$=\frac{w^2+14w+24}{(w+2)(w-2)(w-3)}$$

$$=\frac{(w+2)(w+12)}{(w+2)(w-2)(w-3)}$$

$$=\frac{w+12}{(w-2)(w-3)}$$

PROBLEM 13-4 Multiply: **(a)** $\dfrac{1}{2}\cdot\dfrac{3}{4}$ **(b)** $\dfrac{3}{5m^2}\cdot\dfrac{10m}{2}$ **(c)** $\dfrac{n}{n+1}\cdot\dfrac{3}{n-1}$

(d) $\dfrac{x^2+xy-2y^2}{x^2y^3}\cdot\dfrac{x^3y}{x^2-xy-6y^2}$ **(e)** $\dfrac{w^2-5w}{w^2+3w+2}\cdot\dfrac{5w+10}{15}\cdot\dfrac{3w+15}{25-w^2}$

Solution: Recall that to multiply with rational expressions, eliminate all common factors, multiply the numerators, and then multiply the denominators as you would when multiplying with fractions [see Example 13-8]:

(a) $\dfrac{1}{2}\cdot\dfrac{3}{4}=\dfrac{1\cdot3}{2\cdot4}$ **(b)** $\dfrac{3}{5m^2}\cdot\dfrac{10m}{2}=\dfrac{3}{5mm}\cdot\dfrac{2\cdot5m}{2}$ **(c)** $\dfrac{n}{n+1}\cdot\dfrac{3}{n-1}=\dfrac{n(3)}{(n+1)(n-1)}$

$\qquad\qquad =\dfrac{3}{8}$ $\qquad\qquad\qquad\quad =\dfrac{3}{m}$ $\qquad\qquad\qquad\qquad =\dfrac{3n}{n^2-1}$

(d) $\dfrac{x^2+xy-2y^2}{x^2y^3}\cdot\dfrac{x^3y}{x^2-xy-6y^2}=\dfrac{(x+2y)(x-y)}{x^2y^2y}\cdot\dfrac{xx^2y}{(x+2y)(x-3y)}$

$$=\frac{x-y}{y^2}\cdot\frac{x}{x-3y}$$

$$=\frac{(x-y)x}{y^2(x-3y)}$$

$$=\frac{x^2-xy}{xy^2-3y^3}$$

(e) $\dfrac{w^2 - 5w}{w^2 + 3w + 2} \cdot \dfrac{5w + 10}{15} \cdot \dfrac{3w + 15}{25 - w^2} = \dfrac{w(w - 5)}{(w + 1)(w + 2)} \cdot \dfrac{5(w + 2)}{3(5)} \cdot \dfrac{3(w + 5)}{(5 + w)(5 - w)}$

$$= \dfrac{w(w - 5)}{(w + 1)(5 - w)}$$

$$= -1 \cdot \dfrac{w(w - 5)}{(w + 1)(5 - w)} \qquad Think: \dfrac{w - 5}{5 - w} = -1$$

$$= -\dfrac{w}{w + 1}$$

PROBLEM 13-5   Divide: **(a)** $\dfrac{1}{2} \div \dfrac{2}{3}$ **(b)** $\dfrac{2x}{3} \div \dfrac{5x}{6}$ **(c)** $\dfrac{1}{a} \div \dfrac{1}{b}$

(d) $\dfrac{2y^2 + 4y}{1 - y^2} \div \dfrac{4 - y^2}{y^2 - 3y + 2}$ **(e)** $\dfrac{3w - w^2}{2w^2 + 11w + 5} \div \dfrac{3w^2 - 10w + 3}{25 - w^2} \div \dfrac{2w - 10}{6w^2 + w - 1}$

Solution: Recall that to divide with rational expressions, multiply by the reciprocal of the divisor as you would when dividing with fractions [See Example 13-9]:

(a) $\dfrac{1}{2} \div \dfrac{2}{3} = \dfrac{1}{2} \cdot \dfrac{3}{2}$ **(b)** $\dfrac{2x}{3} \div \dfrac{5x}{6} \cdot = \dfrac{2x}{3} \cdot \dfrac{6}{5x}$ **(c)** $\dfrac{1}{a} \div \dfrac{1}{b} = \dfrac{1}{a} \cdot \dfrac{b}{1}$

$$= \dfrac{1(3)}{2(2)} \qquad\qquad\qquad = \dfrac{2x}{3} \cdot \dfrac{2 \cdot 3}{5x} \qquad\qquad = \dfrac{b}{a}$$

$$= \dfrac{3}{4} \qquad\qquad\qquad\quad = \dfrac{4}{5}$$

(d) $\dfrac{2y^2 + 4y}{1 - y^2} \div \dfrac{4 - y^2}{y^2 - 3y + 2} = \dfrac{2y^2 + 4y}{1 - y^2} \cdot \dfrac{y^2 - 3y + 2}{4 - y^2}$

$$= \dfrac{2y(y + 2)}{(1 + y)(1 - y)} \cdot \dfrac{(y - 1)(y - 2)}{(2 + y)(2 - y)}$$

$$= -1 \cdot \dfrac{2y}{(1 + y)(1 - y)} \cdot \dfrac{(y - 1)(y - 2)}{2 - y} \qquad Think: \dfrac{y - 1}{1 - y} = -1$$

$$= -1 \cdot \dfrac{2y}{y + 1} \cdot (-1) \cdot \dfrac{y - 2}{2 - y} \qquad Think: \dfrac{y - 2}{2 - y} = -1$$

$$= \dfrac{2y}{y + 1} \qquad Think: -1(-1) = +1$$

(e) $\dfrac{3w - w^2}{2w^2 + 11w + 5} \div \dfrac{3w^2 - 10w + 3}{25 - w^2} \div \dfrac{2w - 10}{6w^2 + w - 1}$

$$= \dfrac{3w - w^2}{2w^2 + 11w + 5} \cdot \dfrac{25 - w^2}{3w^2 - 10w + 3} \cdot \dfrac{6w^2 + w - 1}{2w - 10}$$

$$= \dfrac{w(3 - w)}{(2w + 1)(w + 5)} \cdot \dfrac{(5 + w)(5 - w)}{(3w - 1)(w - 3)} \cdot \dfrac{(3w - 1)(2w + 1)}{2(w - 5)}$$

$$= -1(-1) \cdot \dfrac{w(3 - w)}{1} \cdot \dfrac{5 - w}{w - 3} \cdot \dfrac{1}{2(w - 5)}$$

$$= \dfrac{w}{2}$$

PROBLEM 13-6 Simplify using the division method: **(a)** $\dfrac{\frac{1}{2}}{\frac{2}{3}}$ **(b)** $\dfrac{\frac{2}{3}}{5}$ **(c)** $\dfrac{1}{\frac{7}{8}}$

(d) $\dfrac{1 + \frac{1}{3}}{1 - \frac{1}{3}}$ **(e)** $\dfrac{\frac{4}{x}}{\frac{8}{x}}$ **(f)** $\dfrac{\frac{1}{y+1}}{y-1}$ **(g)** $\dfrac{2}{\frac{a}{a+b}}$ **(h)** $\dfrac{\frac{1}{m} + \frac{1}{n}}{\frac{1}{m} - \frac{1}{n}}$ **(i)** $\dfrac{u + \frac{u}{v}}{1 + \frac{1}{v}}$

Solution: Recall that to simplify a complex fraction using the division method, you

1. Rename to get simple fractions in both the numerator and denominator when necessary.

2. Rename the complex fraction from Step 1 as division using $\dfrac{a}{b} = a \div b$.

3. Divide the simple fractions from Step 2 to get a simple fraction in lowest terms.
[See Example 13-11.]

(a)
$$\frac{\frac{1}{2}}{\frac{2}{3}} = \frac{1}{2} \div \frac{2}{3}$$
$$= \frac{1}{2} \cdot \frac{3}{2}$$
$$= \frac{3}{4}$$

(b)
$$\frac{\frac{2}{3}}{5} = \frac{2}{3} \div 5$$
$$= \frac{2}{3} \div \frac{5}{1}$$
$$= \frac{2}{3} \cdot \frac{1}{5}$$
$$= \frac{2}{15}$$

(c)
$$\frac{1}{\frac{7}{8}} = 1 \div \frac{7}{8}$$
$$= \frac{1}{1} \cdot \frac{8}{7}$$
$$= \frac{8}{7} \text{ or } 1\frac{1}{7}$$

(d)
$$\frac{1 + \frac{1}{3}}{1 - \frac{1}{3}} = \frac{\frac{3}{3} + \frac{1}{3}}{\frac{3}{3} - \frac{1}{3}}$$
$$= \frac{\frac{4}{3}}{\frac{2}{3}}$$
$$= \frac{4}{3} \div \frac{2}{3}$$
$$= \frac{2(2)}{\not{3}} \cdot \frac{\not{3}}{\not{2}}$$
$$= 2$$

(e)
$$\frac{\frac{4}{x}}{\frac{8}{x}} = \frac{4}{x} \div \frac{8}{x}$$
$$= \frac{4}{x} \cdot \frac{x}{8}$$
$$= \frac{\not{4}}{x} \cdot \frac{x}{\not{4}(2)}$$
$$= \frac{1}{2}$$

(f)
$$\frac{\frac{1}{y+1}}{y-1} = \frac{1}{y+1} \div (y-1)$$
$$= \frac{1}{y+1} \div \frac{y-1}{1}$$
$$= \frac{1}{y+1} \cdot \frac{1}{y-1}$$
$$= \frac{1}{y^2 - 1}$$

(g)
$$\frac{2}{\frac{a}{a+b}} = 2 \div \frac{a}{a+b}$$
$$= \frac{2}{1} \cdot \frac{a+b}{a}$$
$$= \frac{2(a+b)}{a}$$
$$= \frac{2a + 2b}{a}$$

(h)
$$\frac{\frac{1}{m} + \frac{1}{n}}{\frac{1}{m} - \frac{1}{n}} = \frac{\frac{n+m}{mn}}{\frac{n-m}{mn}}$$
$$= \frac{n+m}{mn} \div \frac{n-m}{mn}$$
$$= \frac{n+m}{\not{mn}} \cdot \frac{\not{mn}}{n-m}$$
$$= \frac{n+m}{n-m}$$

(i)
$$\frac{u + \frac{u}{v}}{1 + \frac{1}{v}} = \frac{\frac{uv+u}{v}}{\frac{v+1}{v}}$$
$$= \frac{uv+u}{v} \div \frac{v+1}{v}$$
$$= \frac{u(v+1)}{\not{v}} \cdot \frac{\not{v}}{v+1}$$
$$= u$$

PROBLEM 13-7   Simplify using the LCD method: **(a)** $\dfrac{\frac{1}{2}}{\frac{2}{3}}$ **(b)** $\dfrac{\frac{2}{3}}{5}$ **(c)** $\dfrac{1}{\frac{7}{8}}$

(d) $\dfrac{1+\frac{1}{3}}{1-\frac{1}{3}}$ **(e)** $\dfrac{\frac{4}{x}}{\frac{8}{x}}$ **(f)** $\dfrac{\frac{1}{y+1}}{\frac{y-1}{1}}$ **(g)** $\dfrac{2}{\frac{a}{a+b}}$ **(h)** $\dfrac{u+\frac{u}{v}}{1+\frac{1}{v}}$ **(i)** $\dfrac{w+1}{1+\frac{2}{w}+\frac{1}{w^2}}$

Solution: Recall that to simplify a complex fraction using the LCD method, you

1. Find the LCD of all the terms contained in the numerator and denominator.
2. Multiply both the numerator and denominator by the LCD from Step 1.
3. Simplify the complex fraction from Step 2 to get a simple fraction in lowest terms [See Example 13-12.]

(a) The LCD of $\frac{1}{2}$ and $\frac{2}{3}$ is 6.   **(b)** The LCD of $\frac{2}{3}$ and $\frac{5}{1}$ is 3.   **(c)** The LCD of $\frac{1}{1}$ and $\frac{7}{8}$ is 8.

$$\frac{\frac{1}{2}}{\frac{2}{3}}=\frac{\left(\frac{1}{2}\right)6}{\left(\frac{2}{3}\right)6}$$
$$=\frac{3}{4}$$

$$\frac{\frac{2}{3}}{5}=\frac{\left(\frac{2}{3}\right)3}{(5)3}$$
$$=\frac{2}{15}$$

$$\frac{1}{\frac{7}{8}}=\frac{(1)8}{\left(\frac{7}{8}\right)8}$$
$$=\frac{8}{7}\text{ or }1\frac{1}{7}$$

(d) The LCD of $\frac{1}{1}$ and $\frac{1}{3}$ is 3.   **(e)** The LCD of $\frac{4}{x}$ and $\frac{8}{x}$ is x.

$$\frac{1+\frac{1}{3}}{1-\frac{1}{3}}=\frac{\left(1+\frac{1}{3}\right)3}{\left(1-\frac{1}{3}\right)3}$$
$$=\frac{3+1}{3-1}$$
$$=\frac{4}{2}$$
$$=2$$

$$\frac{\frac{4}{x}}{\frac{8}{x}}=\frac{\left(\frac{4}{x}\right)x}{\left(\frac{8}{x}\right)x}$$
$$=\frac{4}{8}$$
$$=\frac{1}{2}$$

(f) The LCD of $\dfrac{1}{y+1}$ and $\dfrac{y-1}{1}$ is $y+1$.   **(g)** The LCD of $\dfrac{2}{1}$ and $\dfrac{a}{a+b}$ is $a+b$.

$$\frac{\frac{1}{y+1}}{\frac{y-1}{1}}=\frac{\left(\frac{1}{y+1}\right)(y+1)}{(y-1)(y+1)}$$
$$=\frac{1}{y^2-1}$$

$$\frac{2}{\frac{a}{a+b}}=\frac{2(a+b)}{\left(\frac{a}{a+b}\right)(a+b)}$$
$$=\frac{2a+2b}{a}$$

(h) The LCD of $\dfrac{u}{1}, \dfrac{u}{v}, \dfrac{1}{1}$, and $\dfrac{1}{v}$ is v. **(i)** The LCD of $\dfrac{w}{1}, \dfrac{1}{1}, \dfrac{2}{w}$, and $\dfrac{1}{w^2}$ is w^2.

$$\dfrac{u + \dfrac{u}{v}}{1 + \dfrac{1}{v}} = \dfrac{\left(u + \dfrac{u}{v}\right)v}{\left(1 + \dfrac{1}{v}\right)v}$$

$$= \dfrac{(u)v + \left(\dfrac{u}{v}\right)\!\!\not{v}}{(1)v + \left(\dfrac{1}{v}\right)\!\!\not{v}}$$

$$= \dfrac{uv + u}{v + 1}$$

$$= \dfrac{u(\not{v+1})}{\not{v+1}}$$

$$= u$$

$$\dfrac{w + 1}{1 + \dfrac{2}{w} + \dfrac{1}{w^2}} = \dfrac{(w + 1)w^2}{\left(1 + \dfrac{2}{w} + \dfrac{1}{w^2}\right)w^2}$$

$$= \dfrac{(w + 1)w^2}{(1)w^2 + \left(\dfrac{2}{w}\right)w^2 + \left(\dfrac{1}{w^2}\right)w^2}$$

$$= \dfrac{(w + 1)w^2}{w^2 + 2w + 1}$$

$$= \dfrac{(w+1)w^2}{(w+1)(w + 1)}$$

$$= \dfrac{w^2}{w + 1}$$

PROBLEM 13-8 Solve: **(a)** $\dfrac{3}{4}x - \dfrac{1}{2} = 1$ **(b)** $\dfrac{4}{y} = \dfrac{3}{y - 1}$ **(c)** $w - \dfrac{3}{w} = 2$

(d) $\dfrac{5z}{z - 3} - 2 = \dfrac{2z + 9}{z - 3}$ **(e)** $\dfrac{3}{m} = \dfrac{1}{m} + \dfrac{2}{m}$ **(f)** $\dfrac{5}{n - 3} = \dfrac{4}{n - 3}$

Solution: Recall that to solve a rational equation, you

1. Find the LCD of all the rational expressions contained in the given equation.
2. Clear fractions by multiplying each member of the equation by the LCD from Step 1.
3. Solve the equation from Step 2 to find the proposed solutions.
4. Find the excluded values of the original equation.
5. Compare each proposed solution from Step 3 with the excluded values from Step 4.
6. Reject any proposed solution that is also an excluded value.
7. Check each proposed solution that is not an excluded value in the original equation.
   [See Examples 13-15 and 13-16.]

(a) The LCD of $\dfrac{3}{4}, \dfrac{1}{2}$, and $\dfrac{1}{1}$ is 4.

$$4\left(\dfrac{3}{4}x - \dfrac{1}{2}\right) = 4(1)$$

$$4\left(\dfrac{3}{4}x\right) - 4\left(\dfrac{1}{2}\right) = 4(1)$$

$$3x - 2 = 4$$

$$3x = 6$$

$$x = 2$$

$\dfrac{3}{4}x - \dfrac{1}{2} = 1$ has no excluded values.

Check: $\dfrac{3}{4}x - \dfrac{1}{2} = 1$

$$\begin{array}{c|c} \dfrac{3}{4}(2) - \dfrac{1}{2} & 1 \\ \hline \dfrac{3}{2} - \dfrac{1}{2} & 1 \\ \hline 1 & 1 \end{array} \longleftarrow x = 2 \text{ checks}$$

(b) The LCD of $\dfrac{4}{y}$ and $\dfrac{3}{y - 1}$ is $y(y - 1)$.

$$\not{y}(y - 1) \cdot \dfrac{4}{\not{y}} = y(\not{y-1}) \cdot \dfrac{3}{\not{y-1}}$$

$$(y - 1)4 = y \cdot 3$$

$$4y - 4 = 3y$$

$$y - 4 = 0$$

$$y = 4$$

$\dfrac{4}{y} = \dfrac{3}{y - 1}$ has excluded values of 0 and 1.

Check: $\dfrac{4}{y} = \dfrac{3}{y - 1}$

$$\begin{array}{c|c} \dfrac{4}{4} & \dfrac{3}{4 - 1} \\ \hline 1 & 1 \end{array} \longleftarrow y = 4 \text{ checks}$$

(c) The LCD of $\frac{w}{1}$, $\frac{3}{w}$, and $\frac{2}{1}$ is w.

$$w\left(w - \frac{3}{w}\right) = w(2)$$

$$w(w) - w\left(\frac{3}{w}\right) = w(2)$$

$$w^2 - 3 = 2w$$

$$w^2 - 2w - 3 = 0$$

$$(w + 1)(w - 3) = 0$$

$$w + 1 = 0 \text{ or } w - 3 = 0$$

$$w = -1 \text{ or } \quad w = 3$$

$w - \dfrac{3}{w} = 2$ has an excluded value of 0.

Because the proposed solutions $w = -1$ and 3 are not excluded values, both proposed solutions should be checked in the original equation.

$w - \dfrac{3}{w} = 2$ has solutions of

$w = -1$ or 3.

(e) The LCD of $\frac{3}{m}$, $\frac{1}{m}$, and $\frac{2}{m}$ is m.

$$m\left(\frac{3}{m}\right) = m\left(\frac{1}{m} + \frac{2}{m}\right)$$

$$m\left(\frac{3}{m}\right) = m\left(\frac{1}{m}\right) + m\left(\frac{2}{m}\right)$$

$$3 = 1 + 2$$

$$3 = 3 \quad \text{Stop!}$$

An equation that simplifies as a true number sentence like $3 = 3$ has every real number as a solution except, of course, excluded values.

$\dfrac{3}{m} = \dfrac{1}{m} + \dfrac{2}{m}$ has every real number except the excluded value 0 as a solution.

(d) The LCD of $\frac{5z}{z-3}$, $\frac{2}{1}$, and $\frac{2z+9}{z-3}$ is $z - 3$

$$(z - 3)\left(\frac{5z}{z-3} - 2\right) = (z - 3)\left(\frac{2z+9}{z-3}\right)$$

$$(z - 3)\left(\frac{5z}{z-3}\right) - (z-3)(2) = (z-3)\left(\frac{2z+9}{z-3}\right)$$

$$5z - 2z + 6 = 2z + 9$$

$$3z + 6 = 2z + 9$$

$$z + 6 = 9$$

$$z = 3$$

$\dfrac{5z}{z-3} - 2 = \dfrac{2z+9}{z-3}$ has an excluded value of 3.

Because the proposed solution $z = 3$ is also an excluded value, the proposed solution $z = 3$ must be rejected. Therefore

$\dfrac{5z}{z-3} - 2 = \dfrac{2z+9}{z-3}$ has no solutions.

(f) The LCD of $\dfrac{5}{n-3}$ and $\dfrac{4}{n-3}$ is $n - 3$.

$$(n - 3)\frac{5}{n-3} = (n - 3)\frac{4}{n-3}$$

$$5 = 4 \quad \text{Stop!}$$

An equation that simplifies as a false number sentence like $5 = 4$ has no solutions.

$\dfrac{5}{n-3} = \dfrac{4}{n-3}$ has no solutions.

PROBLEM 13-9   Solve for n: $I = \dfrac{E}{R + \dfrac{r}{n}}$ [electricity formula]

Solution: Recall that to solve a rational formula for a given variable, you

1. Find the LCD of all the rational expressions contained in the given formula.
2. Clear fractions by multiplying each member of the formula by the LCD from Step 1.
3. Solve the literal equation from Step 2 by isolating the given variable in one member.
   [See Example 13-17.]

Solution 1

The LCD of $\dfrac{I}{1}$ and $\dfrac{E}{R + \dfrac{r}{n}}$ is $R + \dfrac{r}{n}$.

$$\left(R + \frac{r}{n}\right)I = \left(R + \frac{r}{n}\right)\frac{E}{R + \frac{r}{n}}$$

$$RI + \frac{rI}{n} = E$$

The LCD of $\dfrac{RI}{1}, \dfrac{rI}{n}$, and $\dfrac{E}{1}$ is n.

$$n\left(RI + \frac{rI}{n}\right) = n(E)$$

$$n(RI) + n\left(\frac{rI}{n}\right) = n(E)$$

$$nRI + rI = nE \quad \longleftarrow \text{ fractions cleared}$$

$$nRI - nE = -rI$$

$$n(RI - E) = -rI$$

$$n = \frac{-rI}{RI - E} \text{ or } \frac{rI}{E - RI}$$

Solution 2

$$I = \frac{E}{R + \dfrac{r}{n}}$$

$$I = \frac{E}{\dfrac{nR + r}{n}}$$

$$I = \frac{E(n)}{\left(\dfrac{nR + r}{n}\right)(n)}$$

$$I = \frac{En}{nR + r}$$

First simplify the complex fraction $\dfrac{E}{R + \dfrac{r}{n}}$.

The LCD of $\dfrac{I}{1}$ and $\dfrac{En}{nR + r}$ is $nR + r$.

$$(nR + r)I = (nR + r)\left(\frac{nE}{nR + r}\right)$$

$$nRI + rI = nE \quad \longleftarrow \text{ fractions cleared}$$

$$nRI - nE = -rI$$

$$n(RI - E) = -rI$$

$$n = \frac{-rI}{RI - E} \text{ or } \frac{rI}{E - RI}$$

Supplementary Exercises

PROBLEM 13-10 Simplify: **(a)** $-\dfrac{12x^6}{18x^4}$ **(b)** $\dfrac{20y^2}{32y^7}$ **(c)** $\dfrac{6a - 12}{6}$ **(d)** $-\dfrac{b^2 + b}{b}$

(e) $\dfrac{w + 3}{w + 3}$ **(f)** $-\dfrac{2 - m}{2 - m}$ **(g)** $-\dfrac{4n + 12}{2n + 6}$ **(h)** $\dfrac{r^2 - 2r}{8r - 16}$ **(i)** $\dfrac{s^2 - 1}{s^2 + 2s - 3}$

(j) $-\dfrac{2u^2 + 3u - 2}{4u^2 - 1}$ **(k)** $\dfrac{v - 4}{4 - v}$ **(l)** $\dfrac{4z - 20}{10 - 2z}$ **(m)** $\dfrac{2h - h^2}{2h - 4}$ **(n)** $\dfrac{6 - k - k^2}{k^2 + 2k - 8}$

(o) $\dfrac{2c + 2}{2}$ **(p)** $-\dfrac{3d - 6}{3}$ **(q)** $-\dfrac{x^2 - x}{x}$ **(r)** $\dfrac{2y}{y^2 + y}$ **(s)** $\dfrac{w^2 - 9}{3 - w}$ **(t)** $-\dfrac{z^2 - 16}{z + 4}$

(u) $-\dfrac{a^2 + 4a + 3}{a^2 - 1}$ **(v)** $\dfrac{b^2 - 4}{b^2 - 5b + 6}$ **(w)** $\dfrac{3 - 4c - 4c^2}{6c^2 + c - 2}$ **(x)** $-\dfrac{4d^2 + 11d - 3}{4d^2 + 7d - 2}$

(y) $-\dfrac{x^2 - xy}{x^2 - y^2}$ **(z)** $\dfrac{m^2 - n^2}{m^2 - 2nm + n^2}$

PROBLEM 13-11 Combine rational expressions: **(a)** $\dfrac{6}{x + 3} + \dfrac{2x}{x + 3}$ **(b)** $\dfrac{2y + 1}{4y^2} + \dfrac{2y - 1}{4y^2}$

(c) $\dfrac{w + 1}{w - 2} + \dfrac{w}{2 - w}$ (d) $\dfrac{2z^2}{1 - z^2} + \dfrac{2z}{z^2 - 1}$ (e) $\dfrac{3a}{a - 2} - \dfrac{6}{a - 2}$ (f) $\dfrac{4b + 3}{2b^2} - \dfrac{3 - 4b}{2b^2}$

(g) $\dfrac{c + 2}{c^2 - 1} - \dfrac{2}{1 - c^2}$ (h) $\dfrac{3}{1 - d^2} - \dfrac{2d + 1}{d^2 - 1}$ (i) $\dfrac{m + 1}{m} - \dfrac{m - 1}{m} + \dfrac{1}{m}$

(j) $\dfrac{u}{u - v} - \dfrac{v}{v - u} + \dfrac{u}{v - u}$ (k) $\dfrac{n + 5}{4n^2 - 1} + \dfrac{n}{4n^2 - 1} - \dfrac{4}{4n^2 - 1}$ (l) $\dfrac{r^2}{r - s} - \dfrac{2rs}{r - s} - \dfrac{s^2}{s - r}$

(m) $\dfrac{1}{h^2 - 1} + \dfrac{1}{h - 1}$ (n) $\dfrac{k}{k^2 + 6k + 8} + \dfrac{2}{k + 4}$ (o) $\dfrac{3}{c - 5} - \dfrac{4}{c}$ (p) $\dfrac{x}{x^2 - y^2} - \dfrac{1}{x - y}$

(q) $\dfrac{b^2}{a - b} + b$ (r) $m + n - \dfrac{m^2 - 4n^2}{m - n}$ (s) $\dfrac{u}{2v} + \dfrac{3u}{v} - \dfrac{2}{v^2}$ (t) $\dfrac{1}{d^2 - 7d + 12} + \dfrac{1}{d - 4} - \dfrac{1}{3 - d}$

(u) $\dfrac{r}{2s} + \dfrac{3}{2s - 1}$ (v) $\dfrac{2\pi r}{\pi r + \pi h} + \dfrac{2\pi r}{\pi r - \pi h}$ (w) $\dfrac{x + 2}{x^2 - 9} - \dfrac{x + 4}{x^2 + 7x + 12}$

(x) $\dfrac{y + 3}{y^2 + 6y + 8} + \dfrac{y + 5}{y + 2}$ (y) $\dfrac{w + 2}{w + 3} + \dfrac{w + 6}{w + 5} + \dfrac{2}{w^2 + 8w + 15}$

(z) $\dfrac{a - 3}{a + 1} - \dfrac{a^2}{1 - a^2} + \dfrac{a + 5}{a - 1}$

PROBLEM 13-12 Multiply and/or divide: (a) $\dfrac{24a^2b}{27ab^4} \cdot \dfrac{54a^4b^3}{8a^3b^2}$ (b) $\dfrac{9x^3y^2}{35x^5y^2} \cdot \dfrac{21x^4y^2}{18xy^4}$

(c) $\dfrac{3m^2 + m}{3m^2 - 15m} \cdot \dfrac{6m^3 - 30m}{6m^2 + 2m}$ (d) $\dfrac{n^2 - 4n - 5}{n^2 - 5n + 6} \cdot \dfrac{n^2 - 2n - 3}{n^2 + n - 6}$ (e) $\dfrac{u - 1}{u^2 - 4} \cdot \dfrac{2 - u}{1 - u^2}$

(f) $\dfrac{(v + 1)^2}{v^2 + 1} \cdot \dfrac{(v - 1)^2}{1 - v^2}$ (g) $\dfrac{c^2 - 12c + 32}{c^2 - 2c - 48} \cdot \dfrac{30 + c - c^2}{c^2 - 10c + 24}$

(h) $\dfrac{20 + 7d - 6d^2}{8d^2 - 14d + 3} \cdot \dfrac{16d - 15 - 4d^2}{12d^2 - 11d + 2}$ (i) $\dfrac{12r^2s^3}{15r^2s^6} \div \dfrac{8r^4s}{6rs^3}$ (j) $\dfrac{6h^4k^2}{14hk^4} \div \dfrac{15h^5k}{35h^2k^3}$

(k) $\dfrac{x^3 + 6x^2}{2x^2 - 4x} \div \dfrac{3x^2 + 18x}{6x - 12}$ (l) $\dfrac{y^2 - 10y + 16}{y^2 - 8y - 20} \div \dfrac{y^2 - 7y - 8}{10 + 9y - y^2}$ (m) $\dfrac{a}{b} \div \dfrac{b}{a} \cdot \dfrac{b^2}{a^2}$

(n) $\dfrac{a}{b} \div \left(\dfrac{b}{a} \cdot \dfrac{b^2}{a^2}\right)$ (o) $\dfrac{a}{b} \cdot \dfrac{b}{a} \div \dfrac{b^2}{a^2}$ (p) $\dfrac{a}{b} \cdot \dfrac{b}{a} \div \dfrac{b}{a} \cdot \dfrac{b}{b}$ (q) $\dfrac{1}{x} \cdot \dfrac{2}{y}$ (r) $\dfrac{5}{u^2 + 2u + 1} \cdot \dfrac{2}{u}$

(s) $\dfrac{3}{r} \div \dfrac{r^2 - r}{2}$ (t) $\dfrac{v + 2}{v} \div \dfrac{v^2 + 5v + 4}{2}$

(u) $\dfrac{m^2 + 12m + 36}{m^2 + 13m + 42} \cdot \dfrac{m^2 + 2m - 35}{m^2 + 11m + 30} \cdot \dfrac{m^2 + 13m + 40}{m^2 + 10m + 16}$

(v) $\dfrac{8n^2 - 26n + 21}{4n^2 + 25n - 56} \cdot \dfrac{3n^2 - 11n - 42}{3n^2 - 20n + 12} \div \dfrac{21 - 5n - 6n^2}{3n^2 + 16n - 12}$ (w) $\dfrac{9 - c^2}{3c - 6} \div \dfrac{c^2 - 2c - 3}{2 - c} \cdot \dfrac{3c^2 + 3c}{c + 3}$

(x) $\dfrac{d^2 - 2d + 1}{3d^2 - 3} \div \dfrac{3 - 3d}{12d + 12}$ (y) $\dfrac{3}{w + 2} \div \dfrac{w}{w - 2}$ (z) $(x^2 - y^2) \div (y - x)$

PROBLEM 13-13 Simplify: (a) $\dfrac{3}{\dfrac{4}{5}}$ (b) $\dfrac{\dfrac{3}{4}}{5}$ (c) $\dfrac{-\dfrac{1}{2}}{\dfrac{1}{2}}$ (d) $\dfrac{\dfrac{4}{9}}{\dfrac{2}{3}}$ (e) $\dfrac{2}{2 - \dfrac{1}{3}}$

(f) $\dfrac{3 - \dfrac{2}{3}}{3}$ (g) $\dfrac{2 - \dfrac{1}{4}}{2 + \dfrac{1}{4}}$ (h) $\dfrac{\dfrac{5}{8} + \dfrac{1}{10}}{\dfrac{5}{6} - \dfrac{1}{8}}$ (i) $\dfrac{x}{\dfrac{x}{3}}$ (j) $\dfrac{\dfrac{y}{4}}{y}$ (k) $\dfrac{\dfrac{3}{w}}{\dfrac{4}{w}}$ (l) $\dfrac{\dfrac{a+1}{b-1}}{\dfrac{b-1}{a}}$

(m) $\dfrac{1 - \dfrac{2}{c}}{c + \dfrac{1}{c}}$ (n) $\dfrac{d + \dfrac{1}{d}}{d - \dfrac{1}{d}}$ (o) $\dfrac{1 - m}{\dfrac{1}{m^2} - 1}$ (p) $\dfrac{1 - \dfrac{1}{n-2}}{1 + \dfrac{1}{n-2}}$ (q) $\dfrac{\dfrac{x}{y}}{\dfrac{1}{y}}$ (r) $\dfrac{1 + \dfrac{a}{b}}{1 - \dfrac{a}{b}}$

(s) $\dfrac{h + 1}{1 + \dfrac{1}{h}}$ (t) $\dfrac{\dfrac{1}{x} + \dfrac{1}{y}}{\dfrac{1}{x} - \dfrac{1}{y}}$ (u) $\dfrac{\dfrac{1}{m+n}}{\dfrac{1}{m-n}}$ (v) $\dfrac{\dfrac{k}{k+1} - 1}{k + 1}$ (w) $\dfrac{r + \dfrac{s}{t}}{r - \dfrac{s}{t}}$ (x) $\dfrac{\dfrac{u}{4} - \dfrac{4}{u}}{\dfrac{u}{2} - \dfrac{2}{u}}$

(y) $\dfrac{\dfrac{4ab^2}{a^2 + ab - 2b^2}}{\dfrac{2a^2b}{a^2 - 4b^2}}$ (z) $\dfrac{\dfrac{w^2}{w^2 + w - 6}}{\dfrac{w^3}{w^2 - 9}}$

PROBLEM 13-14 Solve: (a) $\dfrac{1}{4} + \dfrac{1}{6} = \dfrac{1}{x}$ (b) $\dfrac{1}{y - 2} = \dfrac{2}{y}$ (c) $\dfrac{4}{w + 1} + 2 = \dfrac{6}{w + 1}$

(d) $\dfrac{2}{z - 2} - 2 = \dfrac{2}{z - 2}$ (e) $\dfrac{3}{a - 7} = \dfrac{2}{a - 5}$ (f) $\dfrac{4}{b - 3} = -2$ (g) $1 - \dfrac{c}{c + 2} = \dfrac{1}{c + 5}$

(h) $\dfrac{-2}{5} = \dfrac{d}{-20}$ (i) $\dfrac{-4}{h} = \dfrac{16}{20}$ (j) $\dfrac{2}{m - 2} - \dfrac{12}{m^2 + 2m - 8} = \dfrac{5}{m + 4}$

(k) $\dfrac{1}{n + 1} + \dfrac{1}{n - 1} = \dfrac{1}{n^2 - 1}$ (l) $\dfrac{r - 3}{r^2 - 4} = \dfrac{r}{r^2 - r - 6}$ (m) $\dfrac{u - 3}{2u + 1} + 1 = \dfrac{u - 6}{2u^2 - 5u - 3}$

(n) $\dfrac{2}{s^2 - 1} = \dfrac{1}{s^2 - 1} + \dfrac{1}{s^2 - 1}$ (o) for w: $\dfrac{w}{W} = \dfrac{L}{l}$ [lever formula]   (p) for W: $\dfrac{w}{W} = \dfrac{L}{l}$

(q) for v: $a = \dfrac{v - v_0}{t}$ [acceleration formula]   (r) for t: $a = \dfrac{v - v_0}{t}$

(s) for E: $I = \dfrac{E}{R_1 + R_2}$ [electricity formula]   (t) for R_2: $I = \dfrac{E}{R_1 + R_2}$

(u) for a: $s = \dfrac{a}{1 - r}$ [series formula]   (v) for r: $s = \dfrac{a}{1 - r}$

(w) for t: $f = \dfrac{1}{1 - m + mt}$ [finance formula]   (x) for m: $f = \dfrac{1}{1 - m + mt}$

(y) for c^2: $v = \dfrac{v_1 + v_2}{1 + \dfrac{v_1 v_2}{c^2}}$ [Einstein's velocity formula]   (z) for v_1: $v = \dfrac{v_1 + v_2}{1 + \dfrac{v_1 v_2}{c^2}}$

Answers to Supplementary Exercises

(13-10) (a) $-\dfrac{2x^2}{3}$ (b) $\dfrac{5}{8y^5}$ (c) $a - 2$ (d) $-b - 1$ (e) 1 (f) -1 (g) -2

(h) $\dfrac{r}{8}$ (i) $\dfrac{s + 1}{s + 3}$ (j) $-\dfrac{u + 2}{2u + 1}$ (k) -1 (l) -2 (m) $-\dfrac{h}{2}$

(n) $-\dfrac{k+3}{k+4}$ **(o)** $c+1$ **(p)** $2-d$ **(q)** $1-x$ **(r)** $\dfrac{2}{y+1}$ **(s)** $-w-3$

(t) $4-z$ **(u)** $-\dfrac{a+3}{a-1}$ **(v)** $\dfrac{b+2}{b-3}$ **(w)** $-\dfrac{2c+3}{3c+2}$ **(x)** $-\dfrac{d+3}{d+2}$

(y) $-\dfrac{x}{x+y}$ **(z)** $\dfrac{m+n}{m-n}$

(13-11) **(a)** 2 **(b)** $\dfrac{1}{y}$ **(c)** $\dfrac{1}{w-2}$ **(d)** $-\dfrac{2z}{z+1}$ **(e)** 3 **(f)** $\dfrac{4}{b}$ **(g)** $\dfrac{c+4}{c^2-1}$

(h) $\dfrac{2d+4}{1-d^2}$ **(i)** $\dfrac{3}{m}$ **(j)** $\dfrac{v}{u-v}$ **(k)** $\dfrac{1}{2n-1}$ **(l)** $r-s$ **(m)** $\dfrac{h+2}{h^2-1}$

(n) $\dfrac{3k+4}{k^2+6k+8}$ **(o)** $\dfrac{20-c}{c^2-5c}$ **(p)** $-\dfrac{y}{x^2-y^2}$ **(q)** $\dfrac{ab}{a-b}$ **(r)** $\dfrac{3n^2}{m-n}$

(s) $\dfrac{7uv-4}{2v^2}$ **(t)** $\dfrac{2}{d-4}$ **(u)** $\dfrac{2rs-r+6s}{4s^2-2s}$ **(v)** $\dfrac{4r^2}{r^2-h^2}$ **(w)** $\dfrac{5}{x^2-9}$

(x) $\dfrac{y^2+10y+23}{y^2+6y+8}$ **(y)** 2 **(z)** $\dfrac{3a^2+2a+8}{a^2-1}$

(13-12) **(a)** $\dfrac{6a^2}{b^2}$ **(b)** $\dfrac{3x}{10y^2}$ **(c)** $\dfrac{m^2-5}{m-5}$ **(d)** $\dfrac{n^3-3n^2-9n-5}{n^3-n^2-8n+12}$ **(e)** $\dfrac{1}{u^2+3u+2}$

(f) $\dfrac{1-v^2}{1+v^2}$ **(g)** $-\dfrac{c+5}{c+6}$ **(h)** $\dfrac{12d^3-44d^2-5d+100}{48d^3-56d^2+19d-2}$ **(i)** $\dfrac{3}{5r^3s}$ **(j)** 1

(k) 1 **(l)** $-\dfrac{y-2}{y+2}$ **(m)** 1 **(n)** $\dfrac{a^4}{b^4}$ **(o)** $\dfrac{a^2}{b^2}$ **(p)** $\dfrac{a}{b}$ **(q)** $\dfrac{2}{xy}$

(r) $\dfrac{10}{u^3+2u^2+u}$ **(s)** $\dfrac{6}{r^3-r^2}$ **(t)** $\dfrac{2v+4}{v^3+5v^2+4v}$ **(u)** $\dfrac{m-5}{m+2}$

(v) $-\dfrac{n+6}{n+8}$ **(w)** c **(x)** $-\dfrac{4}{3}$ **(y)** $\dfrac{3w-6}{w^2+2w}$ **(z)** $-x-y$

(13-13) **(a)** $\dfrac{15}{4}$ **(b)** $\dfrac{3}{20}$ **(c)** -1 **(d)** $\dfrac{2}{3}$ **(e)** $\dfrac{6}{5}$ **(f)** $\dfrac{7}{9}$ **(g)** $\dfrac{7}{9}$ **(h)** $\dfrac{87}{85}$ **(i)** 3

(j) $\dfrac{1}{4}$ **(k)** $\dfrac{3}{4}$ **(l)** $\dfrac{a^2+a}{b^2-2b+1}$ **(m)** $\dfrac{c-2}{c^2+1}$ **(n)** $\dfrac{d^2+1}{d^2-1}$ **(o)** $\dfrac{m^2}{m+1}$

(p) $\dfrac{n-3}{n-1}$ **(q)** x **(r)** $\dfrac{b+a}{b-a}$ **(s)** h **(t)** $\dfrac{y+x}{y-x}$ **(u)** $\dfrac{m-n}{m+n}$

(v) $-\dfrac{1}{k^2+2k+1}$ **(w)** $\dfrac{rt+s}{rt-s}$ **(x)** $\dfrac{u^2-16}{2u^2-8}$ **(y)** $\dfrac{2ab-4b^2}{a^2-ab}$ **(z)** $\dfrac{w-3}{w^2-2w}$

(13-14) **(a)** $\dfrac{12}{5}$ **(b)** 4 **(c)** 0 **(d)** no solutions   **(e)** 1 **(f)** 1 **(g)** -8 **(h)** 8

(i) -5 **(j)** no solutions   **(k)** $\dfrac{1}{2}$ **(l)** $\dfrac{9}{4}$ **(m)** 2

(**n**) every real number except 1 and -1 (**o**) $w = \dfrac{WL}{l}$ (**p**) $W = \dfrac{wl}{L}$

(**q**) $v = at + v_0$ (**r**) $t = \dfrac{v - v_0}{a}$ (**s**) $E = I(R_1 + R_2)$ (**t**) $R_2 = \dfrac{E}{I} - R_1$

(**u**) $a = s(1 - r)$ (**v**) $r = 1 - \dfrac{a}{s}$ (**w**) $t = \dfrac{1 - f + fm}{fm}$ (**x**) $m = \dfrac{1 - f}{ft - f}$

(**y**) $c^2 = \dfrac{v_1 v_2 v}{v_1 + v_2 - v}$ (**z**) $v_1 = \dfrac{c^2(v - v_2)}{c^2 - vv_2}$

14 APPLICATIONS OF RATIONAL EQUATIONS

THIS CHAPTER IS ABOUT

☑ **Solving Number Problems Using Rational Equations**
☑ **Solving Proportion Problems Using Rational Equations**
☑ **Solving Work Problems Using Rational Equations** $[w = rt]$
☑ **Solving Uniform Motion Problems Using Rational Equations** $[d = rt]$

14-1. Solving Number Problems Using Rational Equations

Agreement: In this chapter, the term "rational equation" will mean *a rational equation in one variable.*

To **solve a number problem using a rational equation,** you

1. *Read* the problem very carefully several times.
2. *Identify* the unknown numbers.
3. *Decide* how to represent the unknown numbers using one variable.
4. *Translate* the problem to a rational equation using key words.
5. *Solve* the rational equation.
6. *Interpret* the solution of the rational equation with respect to each represented unknown number to find the proposed solutions of the original problem.
7. *Check* to see if the proposed solutions satisfy all the conditions of the original problem.

EXAMPLE 14-1: Solve the following number problem using a rational equation.

Solution:

1. *Read:* The larger of two numbers is twice the smaller number. The sum of their reciprocals is 1. What are the numbers?

2. *Identify:* Let n = the smaller number
then $2n$ = the larger number [see the following *Note*].

4. *Translate:* The sum of their reciprocals is 1.

$$\frac{1}{n} + \frac{1}{2n} \stackrel{?}{=} 1$$

5. *Solve:* The LCD of $\frac{1}{n}, \frac{1}{2n}$, and $\frac{1}{1}$ is $2n$ [see Example 13-7].

$$2n\left(\frac{1}{n} + \frac{1}{2n}\right) = 2n(1)$$

$$2n\left(\frac{1}{n}\right) + 2n\left(\frac{1}{2n}\right) = 2n(1)$$

$$2 + 1 = 2n$$

$$3 = 2n$$

$$\frac{3}{2} = n$$

6. *Interpret:* $n = \frac{3}{2}$ means the smaller number is $\frac{3}{2}$.

$2n = 2(\frac{3}{2}) = 3$ means the larger number is 3.

7. *Check:*   Did you find two numbers?   Yes: $\frac{3}{2}$ and 3.

Is the larger of the two numbers twice the smaller number?   Yes: $3 = 2(\frac{3}{2})$

Is the sum of their reciprocals 1?   Yes: $\dfrac{1}{\frac{3}{2}} + \dfrac{1}{3} = \dfrac{2}{3} + \dfrac{1}{3} = 1$

Note: To represent two numbers so that the larger number is twice the smaller number, you can

let $n =$ the smaller number   let $n =$ the larger number
or
then $2n =$ the larger number   then $\frac{1}{2}n =$ the smaller number

because $2n$ is twice n, and n is twice $\frac{1}{2}n$.

14-2. Solving Proportion Problems Using Rational Equations

A. Identify the three most common measurement families.

A **measure** is made up of a real-number amount and a **unit of measure.** The three most common **measurement families** are **length measures, capacity measures,** and **weight measures.**

EXAMPLE 14-2: Name one of each of the following measures: **(a)** length measure; **(b)** capacity measure;   **(c)** weight measure.

Solution:

amount
unit of measure

(a) 5 miles ⟵———— length measure
(b) 1.5 gallons ⟵——— capacity measure
(c) $3\frac{1}{2}$ pounds ⟵——— weight measure

B. Write a ratio of two given like measures in lowest terms.

Measures that have the same unit of measure are called **like measures.** When a fraction is formed using two like measures it is called a **ratio.** To find the ratio of two given measures in lowest terms, you

1. Write the first given measure as the numerator.
2. Write the second given measure as the denominator.
3. Rename to get like units of measure when necessary.
4. Eliminate the like unit of measure to get a ratio [fraction].
5. Reduce the ratio [fraction] to lowest terms.

EXAMPLE 14-3: Find the ratio of 18 hours to 1 day in lowest terms.

Solution: The ratio of 18 hours to 1 day $= \dfrac{18\text{ hours}}{1\text{ day}}$ ⟵— first given measure / second given measure

$= \dfrac{18\text{ hours}}{24\text{ hours}}$ Rename to get like measures.

$= \dfrac{18\,\cancel{\text{hours}}}{24\,\cancel{\text{hours}}}$ Eliminate the like unit of measure.

$= \dfrac{18}{24}$ ⟵— ratio [fraction]

$= \dfrac{3(6)}{4(6)}$ Reduce the ratio to lowest terms.

$= \dfrac{3}{4}$ ⟵— ratio in lowest terms

Note: If a person is awake 18 hours out of 1 day, then the ratio of hours awake to total hours per day for that person is $\frac{3}{4}$ or 3 to 4. That is, a person that is awake 18 hours out of 24 hours is awake 3 out of every 4 hours that day.

C. Determine if two given ratios are directly proportional.

When one given ratio is equal to another given ratio, the two given ratios are said to be **directly proportional.** To determine if two given ratios are directly proportional, you

1. Find the **cross products** of the two given ratios.
2. Compare the cross products from Step 1 as follows:
   (a) If the cross products are equal, the two given ratios are directly proportional.
   (b) If the cross products are not equal, the two given ratios are not directly proportional.

EXAMPLE 14-4: Determine which ratios are directly proportional: (a) $\frac{18}{24}$ and $\frac{3}{4}$ (b) $\frac{18}{24}$ and $\frac{4}{3}$

Solution:

(a) $\frac{18}{24}$ and $\frac{3}{4}$ are directly proportional because the cross products of $\frac{18}{24}$ and $\frac{3}{4}$ are equal:

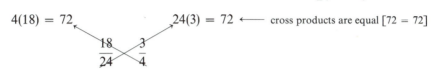

$$4(18) = 72 \qquad\qquad 24(3) = 72 \longleftarrow \text{cross products are equal } [72 = 72]$$

(b) $\frac{18}{24}$ and $\frac{4}{3}$ are not directly proportional because the cross products of $\frac{18}{24}$ and $\frac{4}{3}$ are not equal:

$$3(18) = 54 \qquad\qquad 24(4) = 96 \longleftarrow \text{cross products are not equal } [54 \neq 96]$$

Note: When two ratios are directly proportional, the ratios are always equal.

D. Determine if two given ratios are indirectly proportional.

When one given ratio is equal to the reciprocal of another given ratio, the two given ratios are said to be **indirectly proportional.** To determine if two given ratios are indirectly proportional, you

1. Write the reciprocal of one of the two given ratios.
2. Find the cross products of the reciprocal-ratio from Step 1 and the other given ratio.
3. Compare the cross products from Step 2 as follows:
   (a) If the cross products are equal, then the two given ratios are indirectly proportional.
   (b) If the cross products are not equal, then the two given ratios are not indirectly proportional.

EXAMPLE 14-5: Determine which ratios are indirectly proportional: (a) $\frac{18}{24}$ and $\frac{3}{4}$ (b) $\frac{18}{24}$ and $\frac{4}{3}$

Solution:

(a) $\frac{18}{24}$ and $\frac{3}{4}$ are not indirectly proportional because the cross products of $\frac{18}{24}$ and $\frac{4}{3}$ $\left[\text{the reciprocal of } \frac{3}{4}\right]$ are not equal: $3(18) = 54$ and $24(4) = 96$

(b) $\frac{18}{24}$ and $\frac{4}{3}$ are indirectly proportional because the cross products of $\frac{18}{24}$ and $\frac{3}{4}$ $\left[\text{the reciprocal of } \frac{4}{3}\right]$ are equal: $4(18) = 72$ and $24(3) = 72$

Note 1: When two different ratios are indirectly proportional, the ratios are never equal.

Note 2: When two ratios are indirectly proportional, the reciprocal of one ratio is always equal to the other ratio.

E. **Write a proportion given two ratios that are directly proportional.**

When an equation is used to show the equality of two equal ratios, the equation is called a **proportion.** To write a proportion given two ratios that are directly proportional, you just set one ratio equal to the other ratio.

EXAMPLE 14-6: Write a proportion given two ratios that are directly proportional: $\dfrac{18}{24}$ and $\dfrac{3}{4}$

Solution: $\dfrac{18}{24}$ and $\dfrac{3}{4}$ are directly proportional means $\overbrace{\dfrac{18}{24} = \dfrac{3}{4}}^{\text{proportion}}$.

Check: Is $\dfrac{18}{24} = \dfrac{3}{4}$ a proportion? Yes: $\dfrac{18}{24} = \dfrac{3(6)}{4(6)} = \dfrac{3}{4}$

F. **Write a proportion given two ratios that are indirectly proportional.**

To write a proportion given two ratios that are indirectly proportional, you set one ratio equal to the reciprocal of the other ratio.

EXAMPLE 14-7: Write a proportion given two ratios that are indirectly proportional: $\dfrac{18}{24}$ and $\dfrac{4}{3}$

Solution: $\dfrac{18}{24}$ and $\dfrac{4}{3}$ are indirectly proportional means $\overbrace{\dfrac{18}{24} = \dfrac{3}{4}}^{\text{proportion}}$ ⟵ reciprocal of $\dfrac{4}{3}$

or $\dfrac{24}{18} = \dfrac{4}{3}$. ⟵ reciprocal of $\dfrac{18}{24}$

Check: Is $\dfrac{18}{24} = \dfrac{3}{4}$ a proportion? Yes: $\dfrac{18}{24} = \dfrac{3(6)}{4(6)} = \dfrac{3}{4}$

Is $\dfrac{24}{18} = \dfrac{4}{3}$ a proportion? Yes: $\dfrac{24}{18} = \dfrac{4(6)}{3(6)} = \dfrac{4}{3}$

Note: Given two ratios that are indirectly proportional, it is always possible to write two different proportions.

G. **Solve a proportion problem using a rational equation.**

To **solve a proportion problem using a rational equation,** you

1. *Read* the problem very carefully several times.
2. *Identify* the unknown measure.
3. *Decide* how to represent the unknown measure using one variable.
4. *Make a table* to help form two ratios that are either directly proportional or indirectly proportional.
5. *Translate* the problem to a proportion using the ratios from step 4.
6. *Solve* the proportion.
7. *Interpret* the solution of the proportion with respect to the unknown measure to find the proposed solution of the original problem.
8. *Check* to see if the proposed solution satisfies all the conditions of the original problem.

When the two given units of measure involved in a proportion problem act in the same way, the ratios that are formed when solving the proportion problem are directly proportional. That is, the ratios that are formed when solving a proportion problem are directly proportional if one of the two given units of measure increases when the other unit of measure increases, or if one of the two given units of measure decreases when the other unit of measure decreases.

EXAMPLE 14-8: Solve this proportion problem using a rational equation.

Solution:

1. *Read:* A certain car can travel 120 miles on 10 gallons of gas. At that rate, how far can the car travel on 25 gallons of gas?

2. *Identify:* The unknown measure is the distance the car can travel on 25 gallons of gas.

3. *Decide:* Let d = the distance the car can travel on 25 gallons of gas.

4. *Make a table:*

	miles ratio	gallons ratio
120 miles on 10 gallons of gas	$\dfrac{120}{d}$	$\dfrac{10}{25}$
d miles on 25 gallons of gas		

Think: $\dfrac{120}{d}$ and $\dfrac{10}{25}$ are directly proportional [see the following *Note 1*]

5. *Translate:* The miles ratio equals the gallons ratio [see the following *Note 2*].

$$\frac{120}{d} = \frac{10}{25} \leftarrow \text{rational equation}$$

6. *Solve:* The LCD of $\dfrac{120}{d}$ and $\dfrac{10}{25}$ is $25d$.

$$25d\left(\frac{120}{d}\right) = 25d\left(\frac{10}{25}\right)$$
$$25(120) = d(10)$$
$$3000 = 10d$$
$$300 = d$$

7. *Interpret:* $d = 300$ means on 25 gallons of gas the car can travel 300 miles.

8. *Check:* Do $\dfrac{120}{300}$ and $\dfrac{10}{25}$ form a proportion? Yes: $\dfrac{120}{300} = \dfrac{2(60)}{5(60)} = \dfrac{2}{5}$ and $\dfrac{10}{25} = \dfrac{2(5)}{5(5)} = \dfrac{2}{5}$

Note 1: In Example 14-8, the miles ratio $\dfrac{120}{d}$ and the gallons ratio $\dfrac{10}{25}$ are directly proportional because the two units of measure (miles and gallons) act in the same way. That is, the number of miles increases as the number of gallons increases, and the number of miles decreases as the number of gallons decreases.

Note 2: To form a proportion given two ratios that are directly proportional, you set one ratio equal to the other ratio.

When the two given units of measure involved in a proportion problem act in opposite ways, the two ratios that are formed when solving the proportion problem are indirectly proportional. That is, the two ratios that are formed when solving a proportion problem are indirectly proportional if one of the two given units of measure increases when the other unit of measure decreases, or if one of the two given units of measure decreases when the other unit increases.

EXAMPLE 14-9: Solve the following proportion problem using a rational equation.

Solution:

1. *Read:* At $4\frac{1}{2}$ miles per hour [mph], it takes 5 minutes to walk around a certain track. How fast must you jog to get around the track in 3 minutes?

2. *Identify:* The unknown measure is the rate to jog around the track in 3 minutes.

3. *Decide:*   Let r = the rate to jog around the track in 3 minutes.

4. *Make a table:* $4\frac{1}{2}$ mph takes **5** minutes

r mph takes **3** minutes

mph ratio	minutes ratio
$4\frac{1}{2}$	$\dfrac{5}{3}$
r	3

Think: $\dfrac{4\frac{1}{2}}{r}$ and $\dfrac{5}{3}$ are indirectly proportional [see *Note 1*].

5. *Translate:*   The mph ratio equals the reciprocal of the minutes ratio [see the following *Note 2*].

$$\frac{4\frac{1}{2}}{r} = \frac{3}{5} \longleftarrow \text{rational equation}$$

6. *Solve:*   The LCD of $\dfrac{4\frac{1}{2}}{r}$ and $\dfrac{3}{5}$ is $5r$.

$$5r\left(\frac{4\frac{1}{2}}{r}\right) = 5r\left(\frac{3}{5}\right)$$

$$5(4\tfrac{1}{2}) = r(3)$$

$$22\tfrac{1}{2} = 3r$$

$$7\tfrac{1}{2} = r$$

7. *Interpret:* $r = 7\frac{1}{2}$ means the rate needed to jog around the track in 3 minutes is $7\frac{1}{2}$ mph.

8. *Check:*   Do $\dfrac{4\frac{1}{2}}{7\frac{1}{2}}$ and $\dfrac{3}{5}$ form a proportion?   Yes: $\dfrac{4\frac{1}{2}}{7\frac{1}{2}} = \dfrac{\frac{9}{2}}{\frac{15}{2}} = \dfrac{9}{2} \div \dfrac{15}{2} = \dfrac{9}{2} \cdot \dfrac{2}{15} = \dfrac{9}{15} = \dfrac{3}{5}$

Note 1: In Example 14-9, the mph ratio $\dfrac{4\frac{1}{2}}{r}$ and the minutes ratio $\dfrac{5}{3}$ are indirectly proportional because the two units of measure (mph and minutes) act in opposite ways. That is, the number of mph increases as the number of minutes decreases, and the number of mph decreases as the number of minutes increases.

Note 2: To form a proportion given two ratios that are indirectly proportional, you set one ratio equal to the reciprocal of the other ratio $\left[\dfrac{4\frac{1}{2}}{r} = \dfrac{3}{5} \text{ or } \dfrac{r}{4\frac{1}{2}} = \dfrac{5}{3}\right]$.

14-3. Solving Work Problems Using Rational Equations [$w = rt$]

A. Find the unit rate.

The unit rate indicates the fractional part of a job that can be completed in one unit of a given time period. For example, the **unit rate** for a person who completes a certain job [1 job] in n hours is $\dfrac{1}{n}$ job per hour.

EXAMPLE 14-10: Find the unit rate for a person who completes a certain job in: **(a)** 4 hours; **(b)** 4 days;   **(c)** 4 weeks.

Solution:

(a) The unit rate for 1 job in 4 hours is $\dfrac{1 \text{ job}}{4 \text{ hours}} = \dfrac{1}{4}\dfrac{\text{ job}}{\text{ hour}}$ or $\dfrac{1}{4}$ job per hour.

(b) The unit rate for 1 job in 4 days is $\dfrac{1 \text{ job}}{4 \text{ days}} = \dfrac{1}{4}\dfrac{\text{ job}}{\text{ day}}$ or $\dfrac{1}{4}$ job per day.

(c) The unit rate for 1 job in 4 weeks is $\dfrac{1 \text{ job}}{4 \text{ weeks}} = \dfrac{1}{4}\dfrac{\text{ job}}{\text{ week}}$ or $\dfrac{1}{4}$ job per week.

B. Find the amount of work completed with respect to the whole job.

To find the amount of **work** completed with respect to the whole job, you evaluate the **work formula:**
$$\text{work} = (\text{unit rate})(\text{time period}), \text{ or } w = rt.$$

EXAMPLE 14-11: If a given person can complete a certain job in 8 hours, then how much of the job can that person complete in 6 hours?

Solution: The unit rate r for 1 job [the whole job] in 8 hours is $r = \dfrac{1}{8}$ [job per hour]. The time period t spent towards completing the whole 8-hour job is $t = 6$ [hours]. The amount of work w completed in 6 hours with respect to the whole 8-hour job is

$$\text{work} = (\text{unit rate})(\text{time period}) \text{ or } w = rt$$

$$= \left(\frac{1}{8}\frac{\text{job}}{\text{hour}}\right)(6 \text{ hours}) \qquad = \left(\frac{1}{8}\right)(6)$$

$$= \frac{6}{8}\text{job} \qquad\qquad = \frac{6}{8}$$

$$= \frac{3}{4}\text{job} \qquad\qquad = \frac{3}{4} \text{ [of the whole 8-hour job]}$$

Note: If a given person can complete a certain job in 8 hours, then in 6 hours that same person can complete $\dfrac{3}{4}$ of the same job.

C. Solve a work problem using a rational equation.

To **solve a work problem using a rational equation,** you

1. *Read* the problem very carefully several times.
2. *Identify* the unknown time periods.
3. *Decide* how to represent the unknown time periods using one variable.
4. *Make a table* to help represent individual amounts of work using

$$\text{work} = (\text{unit rate})(\text{time period}), \text{ or } w = rt.$$

5. *Translate* the problem to a rational equation.
6. *Solve* the rational equation.
7. *Interpret* the solution of the rational equation with respect to each unknown time period to find the proposed solutions of the original problem.
8. *Check* to see if the proposed solutions satisfy all the conditions of the original problem.

EXAMPLE 14-12: Solve the following problem using a rational equation.

Solution:

1. *Read:*	Shari can paint a certain house in 3 hours. Shari and Gary working together can paint the same house in 2 hours. How long would it take Gary working alone to paint the house?
2. *Identify:*	The unknown time period is the time needed for Gary to paint the house alone.
3. *Decide:*	Let $x =$ the time needed for Gary to paint the house alone.

4. *Make a table:*

	unit rate r [per hour]	time period t [in hours]	work $[w = rt]$ [portion of whole job]	
Shari	$\dfrac{1}{3}$	2	$\dfrac{1}{3}(2)$ or $\dfrac{2}{3}$	[See the following *Note 1.*]
Gary	$\dfrac{1}{x}$	2	$\dfrac{1}{x}(2)$ or $\dfrac{2}{x}$	

5. *Translate:*   Shari's work plus Gary's work equals the total work [see the following *Note 2*].

$$\frac{2}{3} \quad + \quad \frac{2}{x} \quad = \quad 1 \longleftarrow \text{rational equation}$$

6. *Solve:*   The LCD of $\frac{2}{3}$, $\frac{2}{x}$, and $\frac{1}{1}$ is $3x$.

$$3x\left(\frac{2}{3} + \frac{2}{x}\right) = 3x(1)$$

$$3x\left(\frac{2}{3}\right) + 3x\left(\frac{2}{x}\right) = 3x(1)$$

$$2x + 6 = 3x$$

$$6 = x$$

7. *Interpret:* $x = 6$ means the time needed for Gary to paint the house alone is 6 hours.

8. *Check:*   Does Shari's unit rate [per hour] plus Gary's unit rate equal their unit rate together?   Yes: $\frac{1}{3} + \frac{1}{6} = \frac{2}{6} + \frac{1}{6} = \frac{3}{6} = \frac{1}{2}$

Note 1: In Example 14-12, Shari's portion of the total work is known $\left[\frac{2}{3} \text{ of the whole job}\right]$ while

Gary's portion of the total work is unknown $\left[\frac{2}{x} \text{ of the whole job}\right]$.

Note 2: The **total work** is the whole completed job [painting the house]:
total work $= 1$ [complete job].

14-4. Solving Uniform Motion Problems Using Rational Equations $[d = rt]$

To represent an unknown time t given the distance d and the rate r, you solve the distance formula $[d = rt]$ for t to get the **time formula** $t = \dfrac{d}{r}$.

To represent an unknown rate r given the distance d and the time t, you solve the distance formula $[d = rt]$ for r to get the **rate formula** $r = \dfrac{d}{t}$.

To **solve a uniform motion problem using a rational equation,** you

1. *Read* the problem very carefully several times.
2. *Identify* the unknown distances, rates, and times.
3. *Decide* how to represent the unknown distances, rates, and/or times using one variable.
4. *Make a table* to help represent the unknown rates [or times] using $r = \dfrac{d}{t}\left[\text{or } t = \dfrac{d}{r}\right]$.
5. *Translate* the problem to a rational equation.
6. *Solve* the rational equation.
7. *Interpret* the solution of the rational equation with respect to each unknown distance, rate, and time to find the proposed solutions of the original problem.
8. *Check* to see if the proposed solutions satisfy all the conditions of the original problem.

EXAMPLE 14-13: Solve the following uniform motion problem using a rational equation.

Solution:

1. *Read:*   It takes a boat the same amount of time to go 30 km up a certain river as it does to go 50 km down the same river. The speed of the boat in still water [waterspeed]

is 20 km/h. What is the rate of the current? What was the rate of the boat going down the river? How long did it take the boat to go up the river?

2. *Identify:*   The unknown rates and times are $\left\{\begin{array}{l}\text{the rate of the current}\\ \text{the rate of the boat up the river}\\ \text{the rate of the boat down the river}\\ \text{the time to go up the river}\\ \text{the time to return down the river}\end{array}\right\}$.

3. *Decide:*

Let $\quad x =$ the rate of the current

then $20 - x =$ the rate of the boat up the river

and $20 + x =$ the rate of the boat down the river [see Example 12-9].

4. *Make a table:*

	distance d [in kilometers]	rate r [km/h]	time $\left[t = \dfrac{d}{r}\right]$ [in hours]
up the river	30	$20 - x$	$\dfrac{30}{20 - x}$
down the river	50	$20 + x$	$\dfrac{50}{20 + x}$

5. *Translate:*

The time to go up the river equals the time to go down the river.

$$\frac{30}{20 - x} \qquad = \qquad \frac{50}{20 + x}$$

6. *Solve:*   The LCD of $\dfrac{30}{20 - x}$ and $\dfrac{50}{20 + x}$ is $(20 - x)(20 + x)$.

$$(20 - x)(20 + x) \cdot \frac{30}{20 - x} = (20 - x)(20 + x) \cdot \frac{50}{20 + x}$$

$$(20 + x)30 = (20 - x)50$$

$$(20 + x)3 = (20 - x)5$$

$$60 + 3x = 100 - 5x$$

$$8x = 40$$

$$x = 5$$

7. *Interpret:*

$x = 5$ means the rate of the current is 5 km/h.

$20 - x = 20 - 5 = 15$ means the rate of the boat up the river is 15 km/h.

$20 + x = 20 + 5 = 25$ means the rate of the boat down the river is 25 km/h.

$$\frac{30}{20 - x} = \frac{30}{15} = 2 \text{ means the time to go up the river is 2 hours.}$$

8. *Check:*   Does it take 2 hours to go 30 km up the river or 50 km down the river?   Yes:

$$\frac{50}{20 + x} = \frac{50}{25} = 2 \text{ means the time to go down the river is also 2 hours.}$$

RAISE YOUR GRADES

Can you . . . ?

☑ solve a number problem using a rational equation in one variable
☑ identify the three most common measurement families
☑ write the ratio of two given like measures in lowest terms

☑ determine if two given ratios are directly proportional
☑ determine if two given ratios are indirectly proportional
☑ write a proportion given two ratios that are directly proportional
☑ write a proportion given two ratios that are indirectly proportional
☑ solve a proportion problem using a rational equation
☑ find the unit rate
☑ find the amount of work completed with respect to a whole job
☑ solve a work problem using a rational equation
☑ solve a uniform motion problem using a rational equation

SUMMARY

1. To solve a number problem, proportion problem, work problem, or uniform motion problem using a rational equation, you
   (a) *Read* the problem very carefully several times.
   (b) *Identify* the unknown values.
   (c) *Decide* how to represent the unknown values using one variable.
   (d) *Make a table* to help represent the unknown values when appropriate.
   (e) *Translate* the problem to a rational equation.
   (f) *Solve* the rational equation.
   (g) *Interpret* the solution of the rational equation with respect to each represented unknown to find the proposed solutions of the original problem.
   (h) *Check* to see if the proposed solutions satisfy all the conditions of the original problem.
2. A measure is made up of a real-number amount and a unit of measure.
3. The three most common measurement families are length measures, capacity measures, and weight measures.
4. Measures that have the same unit of measure are called like measures.
5. When a fraction is formed using two like measures, it is called a ratio.
6. To find the ratio of two given measures in lowest terms, you
   (a) Write the first given measure as the numerator.
   (b) Write the second given measure as the denominator.
   (c) Rename to get like units of measure when necessary.
   (d) Eliminate the like unit of measure to get a ratio [fraction].
   (e) Reduce the ratio [fraction] to lowest terms when possible.
7. When one given ratio is equal to another given ratio, the two given ratios are said to be directly proportional.
8. To determine if two given ratios are directly proportional, you
   (a) Find the cross products of the two given ratios.
   (b) Compare the cross products from Step (a) as follows:
      (*i*) If the cross products are equal, then the two given ratios are directly proportional.
      (*ii*) If the cross products are not equal, then the two given ratios are not directly proportional.
9. When two given ratios are directly proportional, the ratios are always equal.
10. When one given ratio is equal to the reciprocal of another given ratio, the two given ratios are said to be indirectly proportional.
11. To determine if two given ratios are indirectly proportional, you
   (a) Write the reciprocal of one of the two given ratios.
   (b) Find the cross products of the reciprocal-ratio from Step (a) and the other given ratio.
   (c) Compare the cross products from Step (b) as follows:
      (*i*) If the cross products are equal, then the two given ratios are indirectly proportional.
      (*ii*) If the cross products are not equal, then the two given ratios are not indirectly proportional.
12. When two different ratios are indirectly proportional
   (a) the ratios are never equal.
   (b) the reciprocal of one ratio is always equal to the other ratio.

13. When an equation is used to show the equality of two equal ratios, the equation is called a proportion.
14. To write a proportion given two ratios that are directly proportional, you just set one ratio equal to the other ratio.
15. To write a proportion given two ratios that are indirectly proportional, you set one ratio equal to the reciprocal of the other ratio.
16. When the two given units of measure involved in a proportion problem act in the same way, the two ratios that are formed when solving the proportion problem are directly proportional. That is, the two ratios that are formed when solving a proportion problem are directly proportional if one of the two given units of measure increases when the other unit of measure increases, or if one of the two given units of measure decreases when the other unit of measure decreases.
17. When the two given units of measure involved in a proportion problem act in opposite ways, the two ratios that are formed when solving the proportion problem are indirectly proportional. That is, the two ratios that are formed when solving a proportion problem are indirectly proportional if one of the two given units of measure increases when the other unit of measure decreases, or if one of the two given units of measure decreases when the other unit of measure increases.
18. The unit rate indicates the fractional part of a job that can be completed in one unit of a given time period.
19. The unit rate for a person who completes a certain job [1 job] in n hours is $\frac{1}{n}$ job per hour.
20. To find the amount of work completed with respect to the whole job, you evaluate the work formula: work = (unit rate)(time period), or $w = rt$.
21. The total work is the whole completed job, or: total work = 1 [completed job].
22. To represent an unknown time t given the distance d and the rate r, you use $t = \frac{d}{r}$.
23. To represent an unknown rate r given the distance d and time t, you use $r = \frac{d}{t}$.

SOLVED PROBLEMS

PROBLEM 14-1 Solve each number problem using a rational equation:

(a) The denominator of a certain fraction is 1 less than 3 times the numerator. If 6 is added to both the numerator and denominator, the result is equal to $\frac{1}{2}$. What is the fraction?

(b) One number is 3 times another number. The product of their reciprocals equals the reciprocal of their sum. Find the numbers.

Solution: Recall that to solve a number problem using a rational equation, you

1. *Read* the problem very carefully several times.
2. *Identify* the unknown numbers.
3. *Decide* how to represent the unknown numbers using one variable.
4. *Translate* the problem to a rational equation using key words.
5. *Solve* the rational equation.
6. *Interpret* the solution of the rational equation with respect to each represented unknown number to find the proposed solutions of the original problem.
7. *Check* to see if the proposed solutions satisfy all the conditions of the original problem.
   [See Example 14-1.]

(a) *Identify:* The unknown numbers are $\begin{cases} \text{the numerator of the fraction} \\ \text{the denominator of the fraction} \end{cases}$.

    Decide: Let x = the numerator of the fraction
              then $3x - 1$ = the denominator of the fraction
              because $3x - 1$ is 1 less than 3 times x.

Translate: <u>If 6 is added to both the numerator and denominator</u> <u>the new fraction</u> <u>equals $\frac{1}{2}$.</u>

$$\frac{x + (6)}{3x - 1 + (6)} \qquad\qquad = \qquad \frac{1}{2}$$

Solve: The LCD of $\dfrac{x + 6}{3x + 5}$ and $\dfrac{1}{2}$ is $2(3x + 5)$.

$$2(3x + 5) \cdot \frac{x + 6}{3x + 5} = 2(3x + 5) \cdot \frac{1}{2}$$

$$2(x + 6) = 3x + 5$$

$$2x + 12 = 3x + 5$$

$$7 = x$$

Interpret: $x = 7$ means the numerator of the fraction is 7.
$3x - 1 = 3(7) - 1 = 20$ means the denominator of the fraction is 20.
The numerator is 7 and the denominator is 20 means the fraction is $\frac{7}{20}$.

Check: Is the denominator 1 less than 3 times the numerator?   Yes: $20 = 3(7) - 1$
If 6 is added to both the numerator and denominator of $\frac{7}{20}$, is the result equal to $\frac{1}{2}$?

$$\text{Yes: } \frac{7 + 6}{20 + 6} = \frac{13}{26} = \frac{1(13)}{2(13)} = \frac{1}{2}$$

(b) *Identify:*   The unknown numbers are $\begin{cases} \text{the smaller number} \\ \text{the larger number} \end{cases}$.

Decide: Let $n =$ the smaller number
then $3n =$ the larger number
because $3n$ is 3 times n.

Translate: <u>The product of their reciprocals</u> <u>equals</u> <u>the reciprocal of their sum.</u>

$$\frac{1}{n} \cdot \frac{1}{3n} \qquad\qquad = \qquad\qquad \frac{1}{n + 3n}$$

Solve: The LCD of $\dfrac{1}{3n^2}$ and $\dfrac{1}{4n}$ is $12n^2$.

$$4(3n^2) \cdot \frac{1}{3n^2} = 3n(4n) \cdot \frac{1}{4n}$$

$$4 = 3n$$

$$\frac{4}{3} = n$$

Interpret: $n = \frac{4}{3}$ means the smaller number is $\frac{4}{3}$.
$3n = 3 \cdot \frac{4}{3} = 4$ means the larger number is 4.

Check: Is 4 three times $\frac{4}{3}$?   Yes: $4 = 3(\frac{4}{3})$
Does the product of their reciprocals equal the reciprocal of their sum?   Yes:

$$\frac{1}{\frac{4}{3}} \cdot \frac{1}{4} = \frac{3}{4} \cdot \frac{1}{4} = \frac{3}{16} \text{ and } \frac{1}{4 + \frac{4}{3}} = \frac{1}{\frac{12}{3} + \frac{4}{3}} = \frac{1}{\frac{16}{3}} = \frac{3}{16}$$

PROBLEM 14-2   Find each ratio in lowest terms of: **(a)** 75 cents to 2 dollars;
(b) 120 students to 4 instructors.

Solution: Recall that to find the ratio of two given like measures in lowest terms, you

1. Write the first given measure as the numerator.

2. Write the second given measure as the denominator.
3. Rename to get like units of measure when necessary.
4. Eliminate the like unit of measure to get a ratio [fraction].
5. Reduce the ratio [fraction] to lowest terms [see Example 14-3]:

(a) 75 cents to 2 dollars $= \dfrac{75 \text{ cents}}{2 \text{ dollars}}$

$$= \dfrac{75 \cancel{\text{ cents}}}{200 \cancel{\text{ cents}}}$$

$$= \dfrac{75}{200}$$

$$= \dfrac{3(25)}{8(25)}$$

$$= \dfrac{3}{8}$$

(b) 120 students to 4 instructors $= \dfrac{120 \text{ students}}{4 \text{ instructors}}$

$$= \dfrac{120 \cancel{\text{ people}}}{4 \cancel{\text{ people}}}$$

$$= \dfrac{120}{4}$$

$$= \dfrac{30(4)}{1(4)}$$

$$= \dfrac{30}{1}$$

Caution: Do not rename the ratio $\dfrac{30}{1}$ as 30. A ratio is a fraction, not a whole number.

PROBLEM 14-3   Determine which ratios are directly or indirectly proportional:

(a) $\dfrac{12}{18}$ and $\dfrac{10}{15}$ **(b)** $\dfrac{8}{6}$ and $\dfrac{3}{4}$ **(c)** $\dfrac{7}{8}$ and $\dfrac{15}{16}$

Solution: Recall that two ratios are directly proportional if the cross products of one ratio and the other ratio are equal. Two ratios are indirectly proportional if the reciprocal of one ratio and the other ratio are equal. [See Examples 14-4 and 14-5.]

(a) $\dfrac{12}{18}$ and $\dfrac{10}{15}$ are directly proportional because $15(12) = 18(10) = 180$ means $\dfrac{12}{18} = \dfrac{10}{15}$.

(b) $\dfrac{8}{6}$ and $\dfrac{3}{4}$ are indirectly proportional because $3(8) = 6(4) = 24$ means $\dfrac{8}{6} = \dfrac{4}{3}$ or $\dfrac{6}{8} = \dfrac{3}{4}$.

(c) $\dfrac{7}{8}$ and $\dfrac{15}{16}$ are not directly or indirectly proportional because

$16(7) = 112$ and $8(15) = 120$ means $\dfrac{7}{8} \neq \dfrac{15}{16}$ and

$15(7) = 105$ and $8(16) = 128$ means $\dfrac{7}{8} \neq \dfrac{16}{15}$.

PROBLEM 14-4   Write a proportion given each of the following:

(a) $\dfrac{12}{18}$ and $\dfrac{10}{15}$ are directly proportional;   **(b)** $\dfrac{8}{6}$ and $\dfrac{3}{4}$ are indirectly proportional.

Solution: Recall that to write a proportion given two ratios that are directly proportional, you set one ratio equal to the other ratio. To write a proportion given two ratios that are indirectly proportional, you set one ratio equal to the reciprocal of the other ratio.
[See Examples 14-6 and 14-7.]

(a) $\dfrac{12}{18}$ and $\dfrac{10}{15}$ are directly proportional means $\overbrace{\dfrac{12}{18} = \dfrac{10}{15}}^{\text{proportion}}$.

proportion

(b) $\dfrac{8}{6}$ and $\dfrac{3}{4}$ are indirectly proportional means $\overbrace{\dfrac{8}{6} = \dfrac{4}{3}}$ ⟵ reciprocal of $\dfrac{3}{4}$

$\qquad\qquad\qquad\qquad$ or $\dfrac{6}{8} = \dfrac{3}{4}$ ⟵ reciprocal of $\dfrac{8}{6}$.

PROBLEM 14-5   Solve each proportion problem using a rational equation:

(a) Five pounds of ham will feed 20 people. At that rate, how much ham is needed to feed 6 people?

(b) Five people take 3 hours to unload a certain truck. At that rate, how long would it take to unload the same truck with 2 people?

Solution: Recall that to solve a proportion problem using a rational equation, you

1. *Read* the problem very carefully several times.
2. *Identify* the unknown measure.
3. *Decide* how to represent the unknown measure using one variable.
4. *Make a table* to help form two ratios that are either directly proportional or indirectly proportional.
5. *Translate* the problem to a proportion using the ratios from Step 4.
6. *Solve* the proportion.
7. *Interpret* the solution of the proportion with respect to the unknown measure to find the proposed solution of the original problem.
8. *Check* to see if the proposed solution satisfies all the conditions of the original problem. [See Example 14-8.]

(a) *Identify:*   The unknown measure is the amount of ham needed to feed 6 people.

$\qquad$ *Decide:*   Let a = the amount of ham needed to feed 6 people.

Make a table:

	pounds ratio	people ratio
5 pounds feeds 20 people	$\dfrac{5}{a}$	$\dfrac{20}{6}$
a pounds feeds 6 people		

Think: $\dfrac{5}{a}$ and $\dfrac{20}{6}$ are directly proportional because the number of pounds increases [or decreases] as the number of people increases [or decreases].

$\qquad$ *Translate:*   The pounds ratio equals the people ratio.

$$\dfrac{5}{a} \qquad = \qquad \dfrac{20}{6}$$

Think: $\dfrac{5}{a}$ and $\dfrac{20}{6}$ are directly proportional means $\dfrac{5}{a} = \dfrac{20}{6}$.

$\qquad$ *Solve:*   The LCD of $\dfrac{5}{a}$ and $\dfrac{20}{6}$ is $6a$.

$$6a \cdot \dfrac{5}{a} = 6a \cdot \dfrac{20}{6}$$

$$6(5) = a(20)$$

$$30 = 20a$$

$$\dfrac{3}{2} = a$$

$\qquad$ *Interpret:* $a = \dfrac{3}{2}$ or $1\dfrac{1}{2}$ means to feed 6 people it takes $1\dfrac{1}{2}$ pounds of ham.

$\qquad$ *Check:*   Do $\dfrac{5}{\frac{3}{2}}$ and $\dfrac{20}{6}$ form a proportion?   Yes: $\dfrac{5}{\frac{3}{2}} = \dfrac{10}{3}$ and $\dfrac{20}{6} = \dfrac{10(2)}{3(2)} = \dfrac{10}{3}$

(b) *Identify:*   The unknown measure is the time for 2 people to unload the truck.

      Decide: Let t = the time for 2 people to unload the truck.

      Make a table:

	people ratio	hour ratio	
5 people take 3 hours	$\dfrac{5}{2}$	$\dfrac{3}{t}$	
2 people take t hours			

Think: $\dfrac{5}{2}$ and $\dfrac{3}{t}$ are indirectly proportional because the number of people increases [decreases] as the number of hours decreases [increases].

      Translate: The people ratio equals the reciprocal of the hour ratio.

$$\frac{5}{2} \qquad = \qquad \frac{t}{3}$$

Think: $\dfrac{5}{2}$ and $\dfrac{3}{t}$ are indirectly proportional means

$$\frac{5}{2} = \frac{t}{3} \text{ or }$$

$$\frac{2}{5} = \frac{3}{t}.$$

      Solve: The LCD of $\dfrac{5}{2}$ and $\dfrac{t}{3}$ is 2(3) or 6.

$$2(3) \cdot \frac{5}{2} = 2(3) \cdot \frac{t}{3}$$

$$3(5) = 2(t)$$

$$15 = 2t$$

$$\frac{15}{2} = t$$

      Interpret: $t = \frac{15}{2}$ or $7\frac{1}{2}$ means to unload the truck with 2 people would take $7\frac{1}{2}$ hours.

      Check: Do $\dfrac{5}{2}$ and $\dfrac{\frac{15}{2}}{3}$ form a proportion?   Yes: $\dfrac{\frac{15}{2}}{3} = \dfrac{15}{2} \div \dfrac{3}{1} = \dfrac{3 \cdot 5}{2} \cdot \dfrac{1}{3} = \dfrac{5}{2}$

PROBLEM 14-6   Find the unit rate for a person who completes a certain job in **(a)** 3 hours; **(b)** 5 days;   **(c)** 2 weeks;   **(d)** 10 minutes.

Solution: Recall that the unit rate for a person who completes a certain job [1 job] in n hours is $\dfrac{1}{n}$ job per hour [see Example 14-10]:

(a) The unit rate for 1 job in 3 hours is $\dfrac{1 \text{ job}}{3 \text{ hours}} = \dfrac{1}{3} \dfrac{\text{job}}{\text{hour}}$ or $\dfrac{1}{3}$ job per hour.

(b) The unit rate for 1 job in 5 days is $\dfrac{1 \text{ job}}{5 \text{ days}} = \dfrac{1}{5} \dfrac{\text{job}}{\text{day}}$ or $\dfrac{1}{5}$ job per day.

(c) The unit rate for 1 job in 2 weeks is $\dfrac{1 \text{ job}}{2 \text{ weeks}} = \dfrac{1}{2} \dfrac{\text{job}}{\text{week}}$ or $\dfrac{1}{2}$ job per week.

(d) The unit rate for 1 job in 10 minutes is $\dfrac{1 \text{ job}}{10 \text{ minutes}} = \dfrac{1}{10} \dfrac{\text{job}}{\text{minute}}$ or $\dfrac{1}{10}$ job per minute.

PROBLEM 14-7   If a person can complete a job in 6 days, then how much of the job can that person complete in 4 days?

Solution: Recall that to find the amount of work completed with respect to the whole job, you evaluate the work formula: work = (unit rate)(time period), or $w = rt$ [see Example 14-11]:

The unit rate r for 1 job in 6 days is $r = \frac{1}{6}$ [job per day].
The time period t spent towards completing the whole 6-day job is $t = 4$ [days].
The amount of work w completed in 4 days with respect to the whole 6-day job is

$$w = rt$$
$$= (\tfrac{1}{6})(4)$$
$$= \tfrac{4}{6}$$
$$= \tfrac{2}{3} \text{ [of the whole job]}$$

PROBLEM 14-8   Solve each work problem using a rational equation:

(a) Kristina can paint a certain room with a brush in 8 hours. Stephanie can paint the same room with a roller in 2 hours. How long will it take for them to paint the room together?

(b) An outlet pipe can empty a pool in 24 hours. An inlet pipe can fill the same pool in 16 hours. How long will it take to fill the pool using the inlet pipe if the outlet pipe is un-intentionally left open?

Solution: Recall that to solve a work problem using a rational equation, you

1. *Read* the problem very carefully several times.
2. *Identify* the unknown time periods.
3. *Decide* how to represent the unknown time periods using one variable.
4. *Make a table* to help represent individual amounts of work using

$$\text{work} = (\text{unit rate})(\text{time period}), \text{ or } w = rt.$$

5. *Translate* the problem to a rational equation.
6. *Solve* the rational equation.
7. *Interpret* the solution of the rational equation with respect to each unknown time period to find the proposed solutions of the original problem.
8. *Check* to see if the proposed solutions satisfy all the conditions of the original problem [see Example 14-12]:

(a) *Identify:*   The unknown time period is the time needed to paint the room together.

Decide: Let x = the time needed to paint the room together.

	unit rate r [per hour]	time period t [in hours]	work [$w = rt$] [portion of whole job]
Make a table: Kristina	$\dfrac{1}{8}$	x	$\dfrac{1}{8}x \text{ or } \dfrac{x}{8}$
Stephanie	$\dfrac{1}{2}$	x	$\dfrac{1}{2}x \text{ or } \dfrac{x}{2}$

Translate: Kristina's work plus Stephanie's work equals the total work.

$$\frac{x}{8} \qquad + \qquad \frac{x}{2} \qquad = \qquad 1$$

Solve: The LCD of $\dfrac{x}{8}$, $\dfrac{x}{2}$, and $\dfrac{1}{1}$ is 8.

$$(8)\cdot\frac{x}{8} + 2(4)\cdot\frac{x}{2} = (8)1$$

$$x + 4x = 8$$

$$5x = 8$$

$$x = \frac{8}{5} \text{ or } 1\frac{3}{5}$$

Interpret: $x = 1\frac{3}{5}$ means Kristina and Stephanie can paint the room together in $1\frac{3}{5}$ hours.

Check: Does Kristina's unit rate [per hour] plus Stephanie's unit rate equal their unit rate together?   Yes: $\dfrac{1}{8} + \dfrac{1}{2} = \dfrac{1}{8} + \dfrac{4}{8} = \dfrac{5}{8} = \dfrac{1}{\frac{8}{5}} = \dfrac{1}{1\frac{3}{5}}$

(b) *Identify:*   The unknown time period is the time to fill the pool with both pipes open.

Decide: Let $x =$ the time to fill the pool with both pipes open.

Make a table:

	unit rate r [per hour]	time period t [in hours]	work $[w = rt]$ [portion of whole job]
outlet	$\dfrac{1}{24}$	x	$\dfrac{1}{24}x = \dfrac{x}{24}$
inlet	$\dfrac{1}{16}$	x	$\dfrac{1}{16}x = \dfrac{x}{16}$

Translate: Water from inlet pipe minus water from outlet pipe equals total work.

$$\frac{x}{16} \qquad - \qquad \frac{x}{24} \qquad = \qquad 1$$

Solve: The LCD of $\dfrac{x}{16}$, $\dfrac{x}{24}$, and $\dfrac{1}{1}$ is 48.

$$3(16)\cdot\frac{x}{16} - 2(24)\cdot\frac{x}{24} = 48(1)$$

$$3x - 2x = 48$$

$$x = 48$$

Interpret: $x = 48$ means the pool can be filled with both pipes open in 48 hours.

Check: Does the unit rate [per hour] of the inlet pipe minus the unit rate of the outlet pipe equal their unit rate together?   Yes: $\dfrac{1}{16} - \dfrac{1}{24} = \dfrac{3}{48} - \dfrac{2}{48} = \dfrac{1}{48}$

PROBLEM 14-9   Solve each uniform motion problem using a rational equation:

(a) One car is traveling at 60 mph. Another car is traveling at 55 mph. How long will it take the faster car to catch up to the slower car if the faster car is 12 miles behind the slower car?

(b) An airplane has a 7-hour fuel supply. If the pilot flies as far as possible at 150 mph and then straight back again at 200 mph using all the fuel the plane has, how long did each part of the trip take?

Solution: Recall that to solve a uniform motion problem using a rational equation, you

1. *Read* the problem very carefully several times.
2. *Identify* the unknown distances, rates, and times.
3. *Decide* how to represent the unknown distances, rates, and/or times using one variable.
4. *Make a table* to help represent the unknown rates [or times] using $r = \dfrac{d}{t}$ $\left[\text{or } t = \dfrac{d}{r}\right]$.
5. *Translate* the problem to a rational equation.
6. *Solve* the rational equation.
7. *Interpret* the solution of the rational equation with respect to each unknown distance, rate, and time to find the proposed solutions of the original problem.
8. *Check* to see if the proposed solutions satisfy all the conditions of the original problem. [See Example 14-13.]

(a) *Identify:*   The unknown time and distances are $\begin{cases} \text{the distance for the slower car} \\ \text{the distance for the faster car} \\ \text{the time for the faster car} \end{cases}$.

Decide: Let d = the distance for the slower car
then $d + 12$ = the distance for the faster car.

Make a table:

	distance d [in miles]	rate r [mph]	time $\left[t = \dfrac{d}{r}\right]$ [in hours]
faster car	$d + 12$	60	$\dfrac{d + 12}{60}$
slower car	d	55	$\dfrac{d}{55}$

Translate: The time for slower car equals the time for faster car.

$$\frac{d}{55} \quad = \quad \frac{d + 12}{60}$$

Solve: The LCD for $\dfrac{d}{55}$ and $\dfrac{d + 12}{60}$ is 660.

$$12(55) \cdot \frac{d}{55} = 11(60) \cdot \frac{d + 12}{60}$$

$$12(d) = 11(d + 12)$$

$$12d = 11d + 132$$

$$d = 132$$

Interpret: $d = 132$ means the distance for the slower car is 132 miles.
$d + 12 = 132 + 12 = 144$ means the distance for the faster car is 144 miles.

$$\frac{d + 12}{60} = \frac{144}{60} = 2.4 \text{ means the time for the faster car is 2.4 hours.}$$

Check: Is the time for the slower car to travel 132 miles at 55 mph equal to the 2.4 hours it took the faster car to travel 144 miles at 60 mph?

$$\text{Yes: } t = \frac{d}{r} = \frac{132}{55} = 2.4 \text{ [hours]}$$

(b) *Identify:*   The unknown distance and times are $\begin{cases} \text{the distance out [or back]} \\ \text{the time to fly out} \\ \text{the time to fly back} \end{cases}$.

Decide: Let d = the distance out [or back]

		distance d [in miles]	rate r [mph]	time $\left[t = \dfrac{d}{r}\right]$ [in hours]
Make a table:	out	d	150	$\dfrac{d}{150}$
	back	d	200	$\dfrac{d}{200}$

Translate: The time to fly out plus the time to fly back is 7 hours.

$$\dfrac{d}{150} \quad + \quad \dfrac{d}{200} \quad = \quad 7$$

Solve: The LCD for $\dfrac{d}{150}$, $\dfrac{d}{200}$, and $\dfrac{7}{1}$ is 600.

$$4(\cancel{150}) \cdot \dfrac{d}{\cancel{150}} + (3)(\cancel{200}) \cdot \dfrac{d}{\cancel{200}} = 600(7)$$

$$4d + 3d = 4200$$

$$7d = 4200$$

$$d = 600$$

Interpret: $d = 600$ means the distance out [or back] is 600 miles.

$$\dfrac{d}{150} = \dfrac{600}{150} = 4 \text{ means the time to fly out is 4 hours.}$$

$$\dfrac{d}{200} = \dfrac{600}{200} = 3 \text{ means the time to fly back is 3 hours.}$$

Check: Is the time for the round trip 7 hours?   Yes: $4 + 3 = 7$

Supplementary Exercises

PROBLEM 14-10   Solve each number problem using a rational equation.

(a) The denominator of a fraction is 3 greater than the numerator. If both the numerator and denominator are increased by 1, the resulting fraction equals $\frac{1}{2}$. What is the original fraction?

(b) The numerator of a fraction is 5 less than the denominator. When 1 is added to the numerator and 1 is subtracted from the denominator, the resulting fraction equals $\frac{1}{2}$. Find the resulting fraction.

(c) When the same number is added to both the numerator and denominator of $\frac{13}{17}$, the resulting fraction equals $\frac{9}{11}$. What is the number?

(d) The sum of the reciprocal of twice a number and twice the number's reciprocal is $\frac{11}{3}$. Find the number.

PROBLEM 14-11   Solve each proportion problem using a rational equation:

(a) A photo is 8 inches long and 5 inches wide. If a reduced copy is made so that the width is 3 inches, then what will be the length of the copy?

(b) A woman earns $560 in 15 days. At that rate, how much will she earn in a week (7 days), to the nearest cent?

(c) An 8-ounce jar of instant coffee contains 12 servings. At that rate, how many servings will a 22-ounce jar of the same coffee contain?

(d) At the rate given in problem **(c)**, how much instant coffee is needed for 22 servings?

(e) Concrete is made of cement, sand, and rock. The ratio of sand to rock in concrete is 2:4. How much sand is needed for 125 pounds of rocks?

(f) At the rate given in problem **(e)**, how much rock is needed for 125 tons of sand?

(g) It takes 6 men 14 days to do a job. At that rate, how long will it take 8 men to do the same job?

(h) At the rate given in problem **(g)**, what is the fewest number of men needed to do the same job in 9 days?

(i) A driven pulley with a diameter of 10 inches turns at 500 rpm. How many revolutions per minute does the driving pulley turn at if its diameter is 15 inches? [*Hint:* As the diameter of a pulley increases, the rpm decreases. As the diameter of a pulley decreases, the rpm increases.]

(j) At the rate given in problem **(i)**, what is the diameter of the driving pulley if it turns at 200 rpm?

(k) A map has a scale of $1\frac{1}{4}$ inch to 10 miles. If the map distance between two cities is $3\frac{1}{2}$ inches, then what is the actual distance between the two cities?

(l) Using the scale given in problem **(k)**, if the actual distance between two cities is 500 miles, what is the map distance between the two cities?

(m) The average 150-pound person should eat 1800 calories each day to maintain that weight. How many calories should be eaten by a 110-pound person to maintain his weight?

(n) Using the rate given in problem **(m)**, how much should you weigh to eat 2000 calories a day so that you will maintain your weight?

(o) Three men take 5 hours to unload a train car. At that rate, how long will it take 10 men to unload the same train car?

(p) At the rate given in problem **(o)**, how many men will be needed to unload the same train car in a maximum of 2 hours?

(q) Apples are on sale for $1.50 for 3 pounds. At that rate, how much will 5 pounds of apples cost?

(r) At the rate given in problem **(q)**, how many pounds of apples can be purchased for $5?

(s) To feed 12 people, it takes 5 pounds of canned ham. At that rate, how many people can be given a full serving from 3 pounds of canned ham?

(t) At the rate given in problem **(s)**, how many pounds of canned ham will it take to feed 20 people?

(u) A certain car travels 500 miles in 8 hours. At that rate, how far can the car travel in 12 hours?

(v) At the rate given in problem **(u)**, how many hours will it take the car to travel 300 miles?

(w) A driving gear with 36 teeth turns at 90 rpm. How many teeth are in the driven gear if it turns at 60 rpm? [*Hint:* As the number of teeth in a gear increases, the rpm decreases. As the number of teeth in a gear decreases, the rpm increases.

(x) Using the same driving gear in problem **(w)**, how fast is the driven gear turning if it has 20 teeth?

(y) Two windows are to be constructed so that they are proportional. The height and width of one window is 5 feet by 3 feet, respectively. What should the height of the other window be if its width is 5 feet?

(z) Given the dimensions of the window in problem **(y)**, what should the width of the other window be if its height is 3 feet?

PROBLEM 14-12   Solve each work problem using a rational equation.

(a) Ryan can complete a certain job in 6 hours. Sean can complete the same job in 8 hours. How long will it take to complete the same job with both boys working together?

(b) Mr. Ivanhoe and Mr. Tippy can complete a certain job in 3 days working together. Mr. Ivanhoe can complete the same job in 5 days working alone. How long would it take Mr. Tippy to complete the job working alone?

(c) An inlet pipe can fill a certain tank in 15 hours. An outlet pipe can drain the same tank in 20 hours. How long will it take to fill the tank if both pipes are left open?

(d) A certain sink can be filled in 15 minutes using the faucet. The same sink can be drained in 10 minutes by pulling the plug. How long will it take to empty a full sink by pulling the plug if the faucet is left on?

PROBLEM 14-13   Solve each uniform motion problem using a rational equation.

(a) A certain row boat can travel twice as fast downstream as it can upstream. It takes the boat 3 hours to travel 5 miles downstream and 2 miles upstream. What is the rate of the current? What is the rate of the boat upstream? How long did the 5-mile trip downstream take?

(b) It takes a certain car 2 hours longer to travel 330 km than to travel 220 km at a constant rate. Find the constant rate of the car. Find the time needed to make the entire trip.

(c) It takes a motor boat twice as long to go 60 km upstream as it does to go 45 km downstream at a constant waterspeed of 30 km/h. What is the rate of the current? What is the landspeed of the boat upstream? How long does the trip downstream take?

(d) Elizabeth can run around a certain track in 1 minute. It takes Carlos 20 seconds longer than Elizabeth to complete one lap around the same track. How long will it take Elizabeth to gain one full lap ahead of Carlos if they both start at the same place?

Answers to Supplementary Exercises

(14-10)   **(a)** $\frac{2}{5}$ **(b)** $\frac{3}{6}$ **(c)** 5   **(d)** $\frac{15}{22}$

(14-11)   **(a)** $4\frac{4}{5}$ in.   **(b)** \$261.33   **(c)** 33 servings   **(d)** $14\frac{2}{3}$ oz   **(e)** $62\frac{1}{2}$ lb
   (f) 250 T   **(g)** $10\frac{1}{2}$ days   **(h)** 10 men   **(i)** $333\frac{1}{3}$ rpm   **(j)** 25 in.   **(k)** 28 mi
   (l) $62\frac{1}{2}$ in.   **(m)** 1320 calories   **(n)** $166\frac{2}{3}$ lb   **(o)** $1\frac{1}{2}$ hr   **(p)** 8 men   **(q)** \$2.50
   (r) 10 lb   **(s)** 7 people   **(t)** $8\frac{1}{3}$ lb   **(u)** 750 mi   **(v)** $4\frac{4}{5}$ hr   **(w)** 54 teeth
   (x) 162 rpm   **(y)** $8\frac{1}{3}$ ft   **(z)** $1\frac{4}{5}$ ft

(14-12)   **(a)** $3\frac{3}{7}$ hr   **(b)** $7\frac{1}{2}$ days   **(c)** 60 hr   **(d)** 30 min

(14-13)   **(a)** $\frac{3}{4}$ mph, $1\frac{1}{2}$ mph, $1\frac{2}{3}$ hr   **(b)** 55 mph, 10 hr   **(c)** 6 km/h, 24 km/h, $1\frac{1}{4}$ hr
   (d) 4 min.

15 RADICAL EXPRESSIONS AND EQUATIONS

THIS CHAPTER IS ABOUT

- ☑ **Simplifying Radical Expressions**
- ☑ **Combining Radical Expressions**
- ☑ **Multiplying Radical Expressions**
- ☑ **Dividing Radical Expressions**
- ☑ **Solving Radical Equations and Formulas**

15-1. Simplifying Radical Expressions

A. Find the positive and negative square root of a positive real number.

Recall: If $a = b^2$, then a is called the square of b and b is called a square root of a [see Examples 2-34 and 2-36].

Every positive real number has two square roots, one positive and one negative.

EXAMPLE 15-1: Find the square roots of the following real numbers: **(a)** 9   **(b)** $\frac{1}{4}$ **(c)** 0.01   **(d)** 0   **(e)** -1.

Solution:
(a) The two square roots of 9 are 3 and -3 because $9 = (3)^2$ and $9 = (-3)^2$.
(b) The two square roots of $\frac{1}{4}$ are $\frac{1}{2}$ and $-\frac{1}{2}$ because $\frac{1}{4} = (\frac{1}{2})^2$ and $\frac{1}{4} = (-\frac{1}{2})^2$.
(c) The two square roots of 0.01 are 0.1 and -0.1 because $0.01 = (0.1)^2$ and $0.01 = (-0.1)^2$.
(d) The only square root of 0 is 0 because $0 = (0)^2$ only.
(e) There are no real-number square roots of -1 because the square of a real number is never negative. That is, $-1 = b^2$ is not possible for any real number b.

If a is a positive real number, then the positive square root or **principal square root** of a is denoted by $\sqrt{a}$ and the negative square root of a is denoted by $-\sqrt{a}$. The $\sqrt{}$ in $\sqrt{a}$ or $-\sqrt{a}$ is called a **radical symbol** or **radical sign**. The a in $\sqrt{a}$ or $-\sqrt{a}$ is called the **radicand**. Both $\sqrt{a}$ and $-\sqrt{a}$ are called **radicals**.

EXAMPLE 15-2: Find the following positive and negative square roots: **(a)** $\sqrt{9}$ **(b)** $-\sqrt{9}$ **(c)** $\sqrt{\frac{1}{4}}$ **(d)** $\sqrt{0}$ **(e)** $-\sqrt{0}$ **(f)** $\sqrt{-9}$ **(g)** $\sqrt{3^2}$ **(h)** $\sqrt{(-3)^2}$ **(i)** $\sqrt{x^2}$ **(j)** $(\sqrt{9})^2$ **(k)** $(\sqrt{-9})^2$ **(l)** $(\sqrt{x})^2$.

Solution:

positive square root of 9
$\downarrow$

(a) $\sqrt{9} = 3$ because $9 = (3)^2$.

negative square root of 9
$\downarrow$

(b) $-\sqrt{9} = -(3) = -3$ because $9 = (-3)^2$.
(c) $\sqrt{\frac{1}{4}} = \frac{1}{2}$ because $\frac{1}{4} = (\frac{1}{2})^2$.

(d) $\sqrt{0} = 0$ because $0 = (0)^2$.

(e) $-\sqrt{0} = -(0) = 0$ because $0 = (0)^2$.

(f) $\sqrt{-9}$ is not a real number because $-9 < 0$ [see the following *Note*].

(g) $\sqrt{3^2} = \sqrt{9} = 3$ or $\sqrt{3^2} = |3| = 3$

(h) $\sqrt{(-3)^2} = \sqrt{9} = 3$ or $\sqrt{(-3)^2} = |-3| = 3$

(i) $\sqrt{x^2} = |x| = \begin{cases} x \text{ if } x \geq 0 & [\text{see the previous part (g)}] \\ -x \text{ if } x < 0 & [\text{see the previous part (h)}] \end{cases}$.

(j) $(\sqrt{9})^2 = (3)^2 = 9$ or $(\sqrt{9})^2 = 9$

(k) $(\sqrt{-9})^2$ is not a real number because the radicand -9 is negative.

(l) $(\sqrt{x})^2 = \begin{cases} x \text{ if } x \geq 0 \ [\text{see the previous part (j)}] \\ \text{is not a real number if } x < 0 \ [\text{see the previous part (k)}] \end{cases}$.

Note: If a is any negative real number, then $\sqrt{a}$ is not a real number because the square of a real number is never negative.

B. Identify radical expressions.

An algebraic expression that contains a radical is called a **radical expression.**

EXAMPLE 15-3: Identify which of the following are radical expressions: **(a)** 5   **(b)** $\sqrt{5}$

(c) $\sqrt{-5}$ **(d)** $\dfrac{m + n}{n}$ **(e)** $\dfrac{-b + \sqrt{b^2 - 4ac}}{2a}$ **(f)** $y^2 - 4$ **(g)** $|w|$ **(h)** 2^x.

Solution: **(a)** 5 is not a radical expression because it does not contain a radical.

   (b) $\sqrt{5}$ is a radical expression.

   (c) $\sqrt{-5}$ is not a radical expression because $\sqrt{-5}$ is not a real number.

   (d) $\dfrac{m + n}{n}$ is not a radical expression.

   (e) $\dfrac{-b + \sqrt{b^2 - 4ac}}{2a}$ is a radical expression.

   (f) $y^2 - 4$ is not a radical expression.

   (g) $|w|$ is not a radical expression.

   (h) 2^x is not a radical expression.

C. Identify radical expressions that are in simplest radical form.

A radical expression is in **simplest radical form** if

1. There are no square polynomial factors in any radicand other than 1.

2. There are no fractions in any radicand.

3. There are no radicals in any denominator.

EXAMPLE 15-4: Identify which radical expressions are in simplest radical form:

(a) $\sqrt{18}$ **(b)** $3\sqrt{2}$ **(c)** $\sqrt{\dfrac{5}{16}}$ **(d)** $\dfrac{\sqrt{3}}{2}$ **(e)** $\dfrac{3}{\sqrt{2}}$ **(f)** $\sqrt{x^2}$ **(g)** $y^2\sqrt{y}$ **(h)** $\dfrac{w + \sqrt{w}}{w}$.

Solution:

(a) $\sqrt{18}$ is not in simplest radical form because the radicand 18 has a square polynomial factor other than 1: $\sqrt{18} = \sqrt{9(2)}$ [9 is a square polynomial factor because: $9 = 3^2$].

(b) $3\sqrt{2}$ is in simplest radical form.

(c) $\sqrt{\dfrac{5}{16}}$ is not in simplest radical form because the radical contains the fraction $\dfrac{5}{16}$.

(d) $\dfrac{\sqrt{3}}{2}$ is in simplest radical form.

(e) $\dfrac{3}{\sqrt{2}}$ is not in simplest radical form because the radical $\sqrt{2}$ is in the denominator.

(f) $\sqrt{x^2}$ is not in simplest radical form because the radicand x^2 is the square of a polynomial [$x^2 = (x)^2$ and x is a polynomial].

(g) $y^2\sqrt{y}$ is in simplest radical form.

(h) $\dfrac{w + \sqrt{w}}{w}$ is in simplest radical form.

D. Simplify radicands that have square polynomial factors other than 1.

To **simplify a radicand that has a square polynomial factor other than 1,** you use the reverse of the following rule:

Product Rule for Radicals

If a and b are nonnegative, then $\sqrt{a}\sqrt{b} = \sqrt{ab}$.

EXAMPLE 15-5: Simplify radicands that contain a square polynomial factor other than 1. Assume x and y are both nonnegative: **(a)** $\sqrt{18}$ **(b)** $\sqrt{x^2}$ **(c)** $\sqrt{12y^3}$

Solution:

(a) $\sqrt{18} = \sqrt{9(2)}$

$\qquad = \sqrt{9}\sqrt{2}$

$\qquad = \sqrt{3^2}\sqrt{2}$

$\qquad = 3\sqrt{2}$

(b) $\sqrt{x^2} = |x| = x$
[See the following *Note 1.*]

(c) $\sqrt{12y^3} = \sqrt{4y^2(3y)}$

$\qquad = \sqrt{4y^2}\sqrt{3y}$

$\qquad = \sqrt{4}\sqrt{y^2}\sqrt{3y}$

$\qquad = 2|y|\sqrt{3y}$

$\qquad = 2y\sqrt{3y}$

[See the following *Note 2.*]

Note 1: In Example 15-5(b), $\sqrt{x^2} = |x| = x$ because x is assumed to be nonnegative [$x \geq 0$].

Note 2: In Example 15-5(c), $\sqrt{12y^3} = 2|y|\sqrt{3y} = 2y\sqrt{3y}$ because y is assumed to be nonnegative.

E. Simplify radicands given in fraction form.

To **simplify a radicand that is given in fraction form,** you use the reverse of the following rule:

Quotient Rule for Radicals

If a is nonnegative and b is positive, then $\dfrac{\sqrt{a}}{\sqrt{b}} = \sqrt{\dfrac{a}{b}}$.

EXAMPLE 15-6: Simplify radicands given in fraction form. Assume all variables are positive:

(a) $\sqrt{\dfrac{5}{16}}$ **(b)** $\sqrt{\dfrac{4x^7}{25}}$ **(c)** $\sqrt{\dfrac{24}{y^4}}$

Solution:

(a) $\sqrt{\dfrac{5}{16}} = \dfrac{\sqrt{5}}{\sqrt{16}}$

$\qquad = \dfrac{\sqrt{5}}{4}$

(b) $\sqrt{\dfrac{4x^7}{25}} = \dfrac{\sqrt{4x^7}}{\sqrt{25}}$

$\qquad = \dfrac{\sqrt{4(x^6)x}}{5}$

$\qquad = \dfrac{2|x^3|\sqrt{x}}{5}$

$\qquad = \dfrac{2x^3\sqrt{x}}{5}$

(c) $\sqrt{\dfrac{24}{y^4}} = \dfrac{\sqrt{4(6)}}{\sqrt{y^4}}$

$\qquad = \dfrac{2\sqrt{6}}{y^2}$

[because $x > 0$]

15-2. Combining Radical Expressions

A. Combine like radicals.

Two or more terms in a radical expression that have the same radical are called **like radicals**. To **combine like radicals,** you use the distributive properties to factor out the like radical.

EXAMPLE 15-7: Combine like radicals. Assume $y + 2$ is nonnegative: **(a)** $x\sqrt{2} + 3x\sqrt{2}$
(b) $5\sqrt{y + 2} - 2\sqrt{y + 2}$

Solution: **(a)** $x\sqrt{2} + 3x\sqrt{2} = (x + 3x)\sqrt{2}$ **(b)** $5\sqrt{y + 2} - 2\sqrt{y + 2} = (5 - 2)\sqrt{y + 2}$
$$= 4x\sqrt{2} \qquad\qquad\qquad\qquad\qquad\qquad = 3\sqrt{y + 2}$$

B. Simplify and then combine like radicals.

Two or more terms in a radical expression that have different radicals are called **unlike radicals**. Sometimes unlike radicals can be combined by first simplifying the unlike radicals to get like radicals.

EXAMPLE 15-8: Simplify and then combine like radicals. Assume x is nonnegative:
(a) $\sqrt{8} + \sqrt{18}$ **(b)** $3\sqrt{x^3} - 2\sqrt{4x}$

Solution:
(a) $\sqrt{8} + \sqrt{18} = \sqrt{4(2)} + \sqrt{9(2)}$ **(b)** $3\sqrt{x^3} - 2\sqrt{4x} = 3\sqrt{x^2(x)} - 2\sqrt{4(x)}$
$$\qquad\qquad = \sqrt{4}\sqrt{2} + \sqrt{9}\sqrt{2} \qquad\qquad\qquad\qquad = 3\sqrt{x^2}\sqrt{x} - 2\sqrt{4}\sqrt{x}$$
$$\qquad\qquad = 2\sqrt{2} + 3\sqrt{2} \qquad\qquad\qquad\qquad\qquad = 3|x|\sqrt{x} - 2(2)\sqrt{x}$$
$$\qquad\qquad = (2 + 3)\sqrt{2} \qquad\qquad\qquad\qquad\qquad = 3x\sqrt{x} - 4\sqrt{x} \quad [\text{because } x \geq 0]$$
$$\qquad\qquad = 5\sqrt{2} \qquad\qquad\qquad\qquad\qquad\qquad = (3x - 4)\sqrt{x}$$

15-3. Multiplying Radical Expressions

A. Multiply radical expressions using the Product Rule for Radicals.

To **multiply radicals,** you

1. Multiply using the Product Rule for Radicals: $\sqrt{a}\sqrt{b} = \sqrt{ab}\ [a \geq 0 \text{ and } b \geq 0]$
2. Simplify the product from Step 1 when possible.

EXAMPLE 15-9: Multiply using the Product Rule for Radicals. Assume x is nonnegative:
(a) $\sqrt{2}\sqrt{5}$ **(b)** $\sqrt{2}\sqrt{6}$ **(c)** $\sqrt{3x}\sqrt{6x}$

Solution: **(a)** $\sqrt{2}\sqrt{5} = \sqrt{2(5)}$ **(b)** $\sqrt{2}\sqrt{6} = \sqrt{2(6)}$ **(c)** $\sqrt{3x}\sqrt{6x} = \sqrt{3x(6x)}$
$$= \sqrt{10} \qquad\qquad\qquad = \sqrt{12} \qquad\qquad\qquad\qquad = \sqrt{18x^2}$$
$$\qquad\qquad\qquad\qquad\qquad = \sqrt{4(3)} \qquad\qquad\qquad\qquad = \sqrt{9x^2(2)}$$
$$\qquad\qquad\qquad\qquad\qquad = \sqrt{4}\sqrt{3} \qquad\qquad\qquad\qquad = \sqrt{9x^2}\sqrt{2}$$
$$\qquad\qquad\qquad\qquad\qquad = 2\sqrt{3} \qquad\qquad\qquad\qquad\quad = \sqrt{9}\sqrt{x^2}\sqrt{2}$$
$$\qquad\qquad\qquad\qquad\qquad\qquad\qquad\qquad\qquad\qquad\qquad = 3|x|\sqrt{2}$$
$$\qquad\qquad\qquad\qquad\qquad\qquad\qquad\qquad\qquad\qquad\qquad = 3x\sqrt{2} \quad [\text{because } x \geq 0]$$

B. Multiply radical expressions using the distributive properties.

To **multiply a radical expression containing two or more terms by a radical,** you

1. Multiply using the distributive properties.
2. Simplify the product from Step 1 when possible.

EXAMPLE 15-10: Multiply using the distributive properties. Assume x is nonnegative:
(a) $\sqrt{2}(\sqrt{5} + \sqrt{2})$ **(b)** $3\sqrt{x}(2 - \sqrt{2x})$

Solution: **(a)** $\sqrt{2}(\sqrt{5} + \sqrt{2}) = \sqrt{2}\sqrt{5} + \sqrt{2}\sqrt{2}$

$$= \sqrt{2(5)} + \sqrt{2(2)}$$

$$= \sqrt{10} + \sqrt{4}$$

$$= \sqrt{10} + 2$$

(b) $3\sqrt{x}(2 - \sqrt{2x}) = 3\sqrt{x}(2) - 3\sqrt{x}\sqrt{2x}$

$$= 6\sqrt{x} - 3\sqrt{x(2x)}$$

$$= 6\sqrt{x} - 3\sqrt{x^2(2)}$$

$$= 6\sqrt{x} - 3\sqrt{x^2}\sqrt{2}$$

$$= 6\sqrt{x} - 3x\sqrt{2}$$

C. Multiply radical expressions using the FOIL Method.

To **multiply two radical expressions each containing two terms,** you

1. Multiply using the FOIL Method [see Example 8-17].
2. Simplify the product from Step 1 when possible.

EXAMPLE 15-11: Multiply using the FOIL Method. Assume x is nonnegative:
(a) $(\sqrt{2} + 3)(5 + \sqrt{6})$ **(b)** $(\sqrt{x} - 2)(\sqrt{x} + 3)$

$$\qquad\qquad\qquad\qquad\qquad \overset{F}{\quad} \quad \overset{O}{\quad} \quad \overset{I}{\quad} \quad \overset{L}{\quad}$$

Solution: **(a)** $(\sqrt{2} + 3)(5 + \sqrt{6}) = 5\sqrt{2} + \sqrt{2}\sqrt{6} + 3(5) + 3\sqrt{6}$ Multiply using FOIL.

$$= 5\sqrt{2} + \sqrt{12} + 15 + 3\sqrt{6} \qquad \text{Simplify.}$$

$$= 5\sqrt{2} + 2\sqrt{3} + 15 + 3\sqrt{6} \qquad \textit{Think: } \sqrt{12} = \sqrt{4(3)}$$

$$\qquad\qquad\qquad\qquad\quad \overset{F}{\quad} \quad \overset{O}{\quad} \quad \overset{I}{\quad} \quad \overset{L}{\quad}$$

(b) $(\sqrt{x} - 2)(\sqrt{x} + 3) = \sqrt{x}\sqrt{x} + 3\sqrt{x} - 2\sqrt{x} - 2(3)$ Multiply using FOIL.

$$= \sqrt{x}\sqrt{x} + (3 - 2)\sqrt{x} - 2(3) \qquad \text{Combine like terms.}$$

$$= (\sqrt{x})^2 + 1\sqrt{x} - 6 \qquad \text{Simplify.}$$

$$= x + \sqrt{x} - 6 \qquad \textit{Think: } (\sqrt{x})^2 = x \text{ if } x \geq 0.$$

Note: If x is nonnegative, then $(\sqrt{x})^2 = x$ $[\sqrt{2}\sqrt{2} = (\sqrt{2})^2 = 2]$.

15-4. Dividing Radical Expressions

A. Divide radical expressions using the Quotient Rule for Radicals.

To **divide radicals,** you

1. Divide using the Quotient Rule for Radicals: $\dfrac{\sqrt{a}}{\sqrt{b}} = \sqrt{\dfrac{a}{b}}$ $[a \geq 0 \text{ and } b > 0]$

2. Simplify the quotient from Step 1 when possible.

EXAMPLE 15-12: Divide using the Quotient Rule for Radicals. Assume x is positive:

(a) $\dfrac{\sqrt{10}}{\sqrt{5}}$ **(b)** $\dfrac{\sqrt{24}}{\sqrt{3}}$ **(c)** $\dfrac{\sqrt{6x}}{\sqrt{2x^3}}$

Solution:

(a) $\dfrac{\sqrt{10}}{\sqrt{5}} = \sqrt{\dfrac{10}{5}}$

$$= \sqrt{2}$$

(b) $\dfrac{\sqrt{24}}{\sqrt{3}} = \sqrt{\dfrac{24}{3}}$

$$= \sqrt{8}$$

$$= \sqrt{4(2)}$$

$$= 2\sqrt{2}$$

(c) $\dfrac{\sqrt{6x}}{\sqrt{2x^3}} = \sqrt{\dfrac{6x}{2x^3}}$

$$= \sqrt{\dfrac{3}{x^2}}$$

$$= \dfrac{\sqrt{3}}{\sqrt{x^2}}$$

$$= \dfrac{\sqrt{3}}{x}$$

B. Rationalize denominators that contain one term.

Recall: For a radical expression to be in simplest radical form, there can be no radicals in any denominator of the radical expression.

To **rationalize the denominator** of a fraction containing a radical in the denominator, you **clear the radical** to get an equal fraction that does not contain a radical in the denominator.

To **rationalize a denominator that contains one term,** you

1. Identify the **rationalizing factor** as the radical in the denominator.
2. Multiply both the numerator and denominator by the rationalizing factor from Step 1.
3. Simplify the result from Step 2 when possible.

EXAMPLE 15-13: Rationalize denominators that contain one term. Assume x is positive:

(a) $\dfrac{3}{5\sqrt{2}}$ (b) $\dfrac{x}{\sqrt{3x}}$

Solution:

(a) $\dfrac{3}{5\sqrt{2}} = \dfrac{3}{5\sqrt{2}} \cdot \dfrac{\sqrt{2}}{\sqrt{2}}$ $\rangle$ rationalizing factor

$= \dfrac{3\sqrt{2}}{5(\sqrt{2})^2}$

$= \dfrac{3\sqrt{2}}{5(2)}$ ⟵ rational denominator

$= \dfrac{3\sqrt{2}}{10}$

(b) $\dfrac{x}{\sqrt{3x}} = \dfrac{x}{\sqrt{3x}} \cdot \dfrac{\sqrt{3x}}{\sqrt{3x}}$ $\rangle$ rationalizing factor

$= \dfrac{x\sqrt{3x}}{(\sqrt{3x})^2}$

$= \dfrac{x\sqrt{3x}}{3x}$ ⟵ rational denominator

$= \dfrac{\sqrt{3x}}{3}$

Note: If $\sqrt{a}$ is a radical expression, then $(\sqrt{a})^2 = a$ is always a rational expression.

C. Rationalize denominators that contain two terms.

To **rationalize a denominator that contains two terms,** you

1. Identify the rationalizing factor as the denominator with the opposite middle sign.
2. Multiply both the numerator and denominator by the rationalizing factor from Step 1.
3. Simplify the result from Step 2 when possible.

EXAMPLE 15-14: Rationalize denominators that contain two terms. Assume x is positive:

(a) $\dfrac{2}{\sqrt{2} + 1}$ (b) $\dfrac{\sqrt{x} + 2}{\sqrt{x} - 3}$

Solution: (a) $\dfrac{2}{\sqrt{2} + 1} = \dfrac{2}{\sqrt{2} + 1} \cdot \dfrac{\sqrt{2} - 1}{\sqrt{2} - 1}$ $\rangle$ rationalizing factor

⎿ change middle sign of the denominator $\sqrt{2} + 1$

$= \dfrac{2(\sqrt{2} - 1)}{(\sqrt{2} + 1)(\sqrt{2} - 1)}$ ⟵ difference of two squares

$= \dfrac{2(\sqrt{2} - 1)}{(\sqrt{2})^2 - (1)^2}$ *Think:* $(a + b)(a - b) = a^2 - b^2$

$= \dfrac{2(\sqrt{2} - 1)}{2 - 1}$ ⟵ rational denominator

$= 2(\sqrt{2} - 1)$ or $2\sqrt{2} - 2$ Simplify.

(b) $\dfrac{\sqrt{x} + 2}{\sqrt{x} - 3} = \dfrac{\sqrt{x} + 2}{\sqrt{x} - 3} \cdot \dfrac{\sqrt{x} + 3}{\sqrt{x} + 3}$ ⟵ rationalizing factor

$$= \dfrac{(\sqrt{x} + 2)(\sqrt{x} + 3)}{(\sqrt{x} - 3)(\sqrt{x} + 3)}$$ ⟵ difference of two squares

$$= \dfrac{\overset{F}{\sqrt{x}\sqrt{x}} + \overset{O}{3\sqrt{x}} + \overset{I}{2\sqrt{x}} + \overset{L}{6}}{(\sqrt{x})^2 - (3)^2}$$ *Think:* $(a - b)(a + b) = a^2 - b^2$

$$= \dfrac{x + 5\sqrt{x} + 6}{x - 9}$$ Simplify.

⟵ rational denominator

Note: If $a + b$ or $[a - b]$ is a radical expression, then $(a + b)(a - b) = a^2 - b^2$ is always a rational expression.

D. Divide, simplify, and rationalize radical expressions.

To **divide radical expressions,** you

1. *Divide* using the Quotient Rule for Radicals: $\dfrac{\sqrt{a}}{\sqrt{b}} = \sqrt{\dfrac{a}{b}}$

2. *Simplify* using the Quotient Rule for Radicals in reverse: $\sqrt{\dfrac{a}{b}} = \dfrac{\sqrt{a}}{\sqrt{b}}$

3. *Rationalize* the denominator when it contains a radical.

EXAMPLE 15-15: Divide, simplify, and rationalize. Assume x is positive:

(a) $8\sqrt{15} \div (2\sqrt{40})$ **(b)** $\dfrac{2x\sqrt{x}}{x\sqrt{24x^3}}$

Solution: **(a)** $8\sqrt{15} \div (2\sqrt{40}) = \dfrac{8\sqrt{15}}{2\sqrt{40}}$ Rename as division.

$$= \dfrac{8}{2}\sqrt{\dfrac{15}{40}}$$ Divide.

$$= \dfrac{4}{1}\sqrt{\dfrac{3}{8}}$$

$$= \dfrac{4\sqrt{3}}{\sqrt{8}}$$ Simplify.

$$= \dfrac{2(2)\sqrt{3}}{2\sqrt{2}}$$

$$= \dfrac{2\sqrt{3}}{\sqrt{2}}$$

$$= \dfrac{2\sqrt{3}}{\sqrt{2}} \cdot \dfrac{\sqrt{2}}{\sqrt{2}}$$ Rationalize.

$$= \dfrac{2\sqrt{3}\sqrt{2}}{(\sqrt{2})^2}$$

$$= \dfrac{2\sqrt{6}}{2}$$

$$= \sqrt{6}$$

(b) $\dfrac{2x\sqrt{x}}{x\sqrt{24x^3}} = \dfrac{2x}{x}\sqrt{\dfrac{x}{24x^3}}$ Divide.

$= 2\sqrt{\dfrac{1}{24x^2}}$

$= \dfrac{2\sqrt{1}}{\sqrt{4x^2(6)}}$ Simplify.

$= \dfrac{2(1)}{2x\sqrt{6}}$

$= \dfrac{1}{x\sqrt{6}}$

$= \dfrac{1}{x\sqrt{6}} \cdot \dfrac{\sqrt{6}}{\sqrt{6}}$ Rationalize.

$= \dfrac{1\sqrt{6}}{x(\sqrt{6})^2}$

$= \dfrac{\sqrt{6}}{6x}$

15-5. Solving Radical Equations and Formulas

A. Identify radical equations.

An equation that contains a variable in a radicand is called a **radical equation.**

EXAMPLE 15-16: Identify which of the following are radical equations: **(a)** $2x + 3y = 5$

(b) $\dfrac{w}{w-1} = \dfrac{1}{w-1}$ **(c)** $\sqrt{m} = 3m$ **(d)** $2x^2 - 3x + 5 = 0$ **(e)** $x = \dfrac{-b + \sqrt{b^2 - 4ac}}{2a}$

(f) $2^x = 4$ **(g)** $|x| = 3$ **(h)** $m = \sqrt{3}$

Solution:
(a) $2x + 3y = 5$ is not a radical equation because neither member contains a variable in a radicand.

(b) $\dfrac{w}{w-1} = \dfrac{1}{w-1}$ is not a radical equation because neither member contains a variable in a radicand.

(c) $\sqrt{m} = 3m$ is a radical equation because the left member contains a variable in a radicand.
(d) $2x^2 - 3x + 5 = 0$ is not a radical equation because neither member contains a variable in a radicand.

(e) $x = \dfrac{-b + \sqrt{b^2 - 4ac}}{2a}$ is a radical equation because the right member contains a variable in a radicand.

(f) $2^x = 4$ is not a radical equation because neither member contains a variable in a radicand.
(g) $|x| = 3$ is not a radical equation because neither member contains a variable in a radicand.
(h) $m = \sqrt{3}$ is not a radical equation because neither member contains a variable in a radicand.

B. Solve radical equations when the radical is isolated in one member.

To **solve a radical equation** when the radical is isolated in one member of the equation, you first clear the radical using the following rule:

Squaring Rule for Equations
If $a = b$, then $a^2 = b^2$.

EXAMPLE 15-17: Solve $\sqrt{m} = 3$. Assume m is nonnegative.

Solution: $\quad \sqrt{m} = 3$ ⟵ given equation

$\qquad (\sqrt{m})^2 = (3)^2 \qquad$ Clear the radical using the Squaring Rule for Equations.

$\qquad\qquad m = 3^2$ ⟵ radical cleared

$\qquad\qquad m = 9$ ⟵ proposed solution

Check: $\sqrt{m} = 3$ ⟵ original equation

$\qquad \dfrac{\sqrt{9}\ \Big|\ 3}{\ } \qquad$ Substitute the proposed solution into the original equation to see if you get a true number sentence.

$\qquad 3 \ \Big|\ 3$ ⟵ $m = 9$ checks [See the following *Caution*.]

Caution: Using the Squaring Rule for Equations may introduce proposed solutions that are not solutions of the original equation.

EXAMPLE 15-18: Solve $\sqrt{m} = -3$. Assume m is nonnegative.

Solution: $\quad \sqrt{m} = -3$ ⟵ given equation

$\qquad (\sqrt{m})^2 = (-3)^2 \qquad$ Clear the radical using the Squaring Rule for Equations.

$\qquad\qquad m = (-3)^2$ ⟵ radical cleared

$\qquad\qquad m = 9$ ⟵ proposed solution

Check: $\sqrt{m} = -3$ ⟵ original equation

$\qquad \dfrac{\sqrt{9}\ \Big|\ -3}{3\ \Big|\ -3} \qquad$ Substitute.

$\qquad\qquad\qquad$ ⟵ $m = 9$ does not check

$\sqrt{m} = -3$ has no real-number solutions because the only proposed real-number solution does not check.

Note: If a is negative, then $\sqrt{x} = a$ has no real-number solutions.

C. Solve radical equations when the radical is not isolated in one member.

To clear radicals using the Squaring Rule for Equations, the radical must first be isolated in one member of the equation.

EXAMPLE 15-19: Solve $\sqrt{2w} - 3 = 1$ using the Squaring Rule. Assume w is nonnegative.

Solution: $\quad \sqrt{2w} - 3 = 1$ ⟵ given equation $\qquad$ *Think:* $\sqrt{2w}$ is not isolated.

$\qquad \sqrt{2w} - 3 + 3 = 1 + 3 \qquad$ Isolate the radical in one member.

$\qquad\qquad \sqrt{2w} = 4$ ⟵ radical is isolated

$\qquad\qquad (\sqrt{2w})^2 = 4^2 \qquad$ Clear the radical using the Squaring Rule for Equations.

$\qquad\qquad\qquad 2w = 16$ ⟵ radical cleared

$\qquad\qquad\qquad w = 8$ ⟵ proposed solution

Check: $\sqrt{2w} - 3 = 1$ ⟵ original equation

$\qquad \dfrac{\sqrt{2(8)} - 3\ \Big|\ 1}{\ }$ $\qquad$ Substitute.

$\qquad \sqrt{16} - 3\ \Big|\ 1$

$\qquad\qquad 4 - 3\ \Big|\ 1$

$\qquad\qquad\qquad 1\ \Big|\ 1$ ⟵ $w = 8$ checks

Caution: If the radical is not isolated in one member of the equation, then using the Squaring Rule for Equations will not clear the radical.

EXAMPLE 15-20: Show that using the Squaring Rule for Equations on $\sqrt{2w} - 3 = 1$ without first isolating the radical in one member of the equation will not clear the radical.

Solution:

$$\sqrt{2w} - 3 = 1 \longleftarrow \text{given equation}$$

$$(\sqrt{2w} - 3)^2 = 1^2 \qquad \text{Try to clear the radical using the Squaring Rule for Equations without first isolating the radical.}$$

$$(\sqrt{2w} - 3)(\sqrt{2w} - 3) = 1 \qquad \textit{Think: } (a - b)^2 = (a - b)(a - b)$$

$$\overset{F}{\underline{\sqrt{2w}\,\sqrt{2w}}} \overset{O}{\underline{- 3\sqrt{2w}}} \overset{I}{\underline{- 3\sqrt{2w}}} \overset{L}{\underline{+ 9}} = 1$$

$$2w - 6\sqrt{2w} + 9 = 1 \qquad \text{Stop! The radical is not cleared.}$$

Note: To solve an equation like $\sqrt{2w} - 3 = 1$, you first isolate the radical in one member of the equation as shown in Example 15-19.

D. Solve radical formulas for a given variable.

A formula that contains a variable in a radicand is called a **radical formula.**
To **solve a radical formula for a given variable,** you

1. Isolate the radical in one member of the formula.
2. Clear the radical using the Squaring Rule for Equations.
3. Solve the rational formula from Step 2 by isolating the given variable in one member of the formula [see Example 13-17].

EXAMPLE 15-21: Solve for g: $T = 2\pi \sqrt{\dfrac{L}{g}}$ [pendulum formula]

Solution:

$$T = 2\pi \sqrt{\frac{L}{g}} \longleftarrow \text{given formula}$$

$$\frac{1}{2\pi} \cdot T = \frac{1}{2\pi} \cdot 2\pi \sqrt{\frac{L}{g}} \qquad \text{Isolate the radical in one member.}$$

$$\frac{T}{2\pi} = \sqrt{\frac{L}{g}} \longleftarrow \text{radical is isolated}$$

$$\left(\frac{T}{2\pi}\right)^2 = \left(\sqrt{\frac{L}{g}}\right)^2 \qquad \text{Clear the radical using the Squaring Rule for Equations to get a rational formula.}$$

$$\frac{T^2}{4\pi^2} = \frac{L}{g} \longleftarrow \text{rational formula}$$

The LCD of $\dfrac{T^2}{4\pi^2}$ and $\dfrac{L}{g}$ is $4\pi^2 g$.   Solve the rational formula [see Example 13-17].

$$4\pi^2 g \cdot \frac{T^2}{4\pi^2} = 4\pi^2 g \cdot \frac{L}{g}$$

$$gT^2 = 4\pi^2 L$$

$$gT^2 \cdot \frac{1}{T^2} = 4\pi^2 L \cdot \frac{1}{T^2} \qquad \text{Isolate the given variable in one member.}$$

$$g \text{ is isolated} \longrightarrow g = \frac{4\pi^2 L}{T^2} \longleftarrow \text{solution}$$

RAISE YOUR GRADES

Can you . . . ?

☑ find the positive and negative square root of a positive real number
☑ identify radical expressions
☑ identify radical expressions that are in simplest radical form

☑ simplify radicands that have square factors other than 1
☑ simplify radicands given in fraction form
☑ combine like radicals
☑ simplify and then combine like radicals
☑ multiply radical expressions using the Product Rule for Radicals
☑ multiply radical expressions using the distributive properties
☑ multiply radical expressions using the FOIL method
☑ divide radical expressions using the Quotient Rule for Radicals
☑ rationalize a denominator that contains one term
☑ rationalize a denominator that contains two terms
☑ divide, simplify, and rationalize radical expressions
☑ identify radical equations
☑ solve radical equations when the radical is isolated in one member
☑ solve radical equations when the radical is not isolated in one member
☑ solve radical formulas for a given variable

SUMMARY

1. If $a = b^2$, then a is called the square of b and b is called the square root of a.
2. Every positive real number has two square roots, one positive and one negative.
3. If a is a positive real number, then the positive square root or principal square root of a is denoted by $\sqrt{a}$ and the negative square root of a is denoted by $-\sqrt{a}$.
4. The $\sqrt{}$ in $\sqrt{a}$ or $-\sqrt{a}$ is called a radical symbol or radical sign.
5. The a in $\sqrt{a}$ or $-\sqrt{a}$ is called the radicand.
6. Both $\sqrt{a}$ and $-\sqrt{a}$ are called radicals.
7. $\sqrt{x^2} = |x| = \begin{cases} x \text{ if } x \geq 0 \\ -x \text{ if } x < 0 \end{cases}$
8. $(\sqrt{x})^2 = \begin{cases} x \text{ if } x \geq 0 \\ \text{is not a real number if } x < 0 \end{cases}$
9. If a is a negative real number, then $\sqrt{a}$ is not a real number.
10. An algebraic expression that contains a radical is called a radical expression.
11. A radical expression is in simplest radical form if
    (a) There are no square polynomial factors in any radicand other than 1.
    (b) There are no fractions in any radicand.
    (c) There are no radicals in any denominator.
12. To simplify a radicand that has a square polynomial factor other than 1, you use the reverse of the Product Rule for Radicals:
    If a and b are nonnegative, then $\sqrt{a}\sqrt{b} = \sqrt{ab}$.
13. To simplify a radicand that is given in fraction form, you use the reverse of the Quotient Rule for Radicals:

    If a is nonnegative and b is positive, then $\dfrac{\sqrt{a}}{\sqrt{b}} = \sqrt{\dfrac{a}{b}}$.

14. Two or more terms in a radical expression that have the same radicand are called like radicals.
15. To combine like radicals, you use the distributive properties to factor out the like radical.
16. Two or more terms in a radical expression that have different radicals are called unlike radicals.
17. Sometimes unlike radicals can be combined by first simplifying the unlike radicals to get like radicals.
18. To multiply two or more radicals, you
    (a) Multiply using the Product Rule for Radicals: $\sqrt{a}\sqrt{b} = \sqrt{ab}$ [$a \geq 0$ and $b \geq 0$]
    (b) Simplify the product from Step (a) when possible.

19. To multiply a radical expression containing two or more terms by a radical, you
    (a) Multiply using the distributive properties.
    (b) Simplify the product from Step (a) when possible.
20. To multiply two radical expressions each containing two terms, you
    (a) Multiply using the FOIL Method.
    (b) Simplify the product from Step (a) when possible.
21. To divide two radicals, you
    (a) Divide using the Quotient Rule for Radicals: $\dfrac{\sqrt{a}}{\sqrt{b}} = \sqrt{\dfrac{a}{b}}\ [a \geq 0 \text{ and } b > 0]$
    (b) Simplify the quotient from Step (a) when possible.
22. To rationalize the denominator of a fraction containing a radical in the denominator, you clear the radical to get an equal fraction that does not contain a radical in the denominator.
23. To rationalize a denominator that contains one term, you
    (a) Identify the rationalizing factor as the radical in the denominator.
    (b) Multiply both the numerator and denominator by the rationalizing factor from Step (a).
    (c) Simplify the result from Step (b) when possible.
24. If $\sqrt{a}$ is a radical expression, then $(\sqrt{a})^2 = a$ is always a rational expression.
25. To rationalize a denominator that contains two terms, you
    (a) Identify the rationalizing factor as the denominator with the opposite middle sign.
    (b) Multiply both the numerator and denominator by the rationalizing factor from Step (a).
    (c) Simplify the result from Step (b) when possible.
26. If $a + b$ [or $a - b$] is a radical expression, then $(a + b)(a - b) = a^2 - b^2$ is always a rational expression.
27. To divide radical expressions, you:
    (a) Divide using the Quotient Rule for Radicals: $\dfrac{\sqrt{a}}{\sqrt{b}} = \sqrt{\dfrac{a}{b}}$

    (b) Simplify using the Quotient Rule for Radicals in reverse: $\sqrt{\dfrac{a}{b}} = \dfrac{\sqrt{a}}{\sqrt{b}}$

    (c) Rationalize the denominator when it contains a radical.
28. An equation that contains a variable in a radicand is called a radical equation.
29. To solve a radical equation, you first clear the radical using the Squaring Rule for Equations: If $a = b$, then $a^2 = b^2$.
30. Using the Squaring Rule for Equations may introduce proposed solutions that are not solutions of the original equation.
31. If a is negative, then $\sqrt{x} = a$ has no real-number solutions.
32. To clear radicals using the Squaring Rule for Equations, the radical must first be isolated in one member of the equation.
33. A formula that contains a variable in a radicand is called a radical formula.
34. To solve a radical formula for a given variable, you
    (a) Isolate the radical in one member of the formula.
    (b) Clear the radical using the Squaring Rule for Equations.
    (c) Solve the rational formula from Step (b) by isolating the given variable in one member of the formula.

SOLVED PROBLEMS

PROBLEM 15-1   Find the square roots of the following: (a) 25   (b) $\frac{9}{16}$ (c) 0.36   (d) 0   (e) -2

Solution: Recall that every positive real number has two square roots, one positive and one negative [see Example 15-1]:

(a) The two square roots of 25 are 5 and -5 because $25 = (5)^2$ and $25 = (-5)^2$.
(b) The two square roots of $\frac{9}{16}$ are $\frac{3}{4}$ and $-\frac{3}{4}$ because $\frac{9}{16} = (\frac{3}{4})^2$ and $\frac{9}{16} = (-\frac{3}{4})^2$.

(c) The two square roots of 0.36 are 0.6 and -0.6 because $0.36 = (0.6)^2$ and $0.36 = (-0.6)^2$.
(d) The only square root of 0 is 0 because $0 = (0)^2$ only.
(e) There are no real-number square roots of -2 because the square of a real number is never negative.

PROBLEM 15-2 Find each square root: **(a)** $\sqrt{1}$ **(b)** $-\sqrt{1}$ **(c)** $\sqrt{\dfrac{4}{25}}$ **(d)** $-\sqrt{\dfrac{1}{100}}$

(e) $\sqrt{0}$ **(f)** $\sqrt{-1}$ **(g)** $\sqrt{4^2}$ **(h)** $\sqrt{(-4)^2}$ **(i)** $\sqrt{w^2}$ **(j)** $(\sqrt{4})^2$ **(k)** $(\sqrt{-4})^2$
(l) $(\sqrt{w})^2$ [see Example 15-2.]

Solution:
(a) $\sqrt{1} = 1$ because $1 = (1)^2$.
(b) $-\sqrt{1} = -(1) = -1$ because $1 = (1)^2$.
(c) $\sqrt{\frac{4}{25}} = \frac{2}{5}$ because $\frac{4}{25} = (\frac{2}{5})^2$.
(d) $-\sqrt{\frac{1}{100}} = -(\frac{1}{10}) = -\frac{1}{10}$ because $\frac{1}{100} = (\frac{1}{10})^2$.
(e) $\sqrt{0} = 0$ because $0 = (0)^2$.
(f) $\sqrt{-1}$ is not a real number because $-1 < 0$.
(g) $\sqrt{4^2} = \sqrt{16} = 4$ or $\sqrt{4^2} = |4| = 4$.
(h) $\sqrt{(-4)^2} = \sqrt{16} = 4$ or $\sqrt{(-4)^2} = |-4| = 4$.
(i) $\sqrt{w^2} = |w| = \begin{cases} w \text{ if } w \geq 0 \text{ [see the previous part (g)]} \\ -w \text{ if } w < 0 \text{ [see the previous part (h)]} \end{cases}$.
(j) $(\sqrt{4})^2 = (2)^2 = 4$ or $(\sqrt{4})^2 = 4$.
(k) $(\sqrt{-4})^2$ is not a real number because $-4 < 0$.
(l) $(\sqrt{w})^2 = \begin{cases} w \text{ if } w \geq 0 \text{ [see the previous part (j)]} \\ \text{is not a real number if } w < 0 \text{ [see the previous part (k)]} \end{cases}$.

PROBLEM 15-3 Simplify radicands that contain a square polynomial factor other than 1. Assume all variables are nonnegative: **(a)** $\sqrt{24}$ **(b)** $2\sqrt{180}$ **(c)** $\sqrt{w^2}$ **(d)** $x\sqrt{x^5}$ **(e)** $\sqrt{8y^4}$
(f) $\sqrt{27mn^3}$

Solution: Recall that to simplify a radicand that has a square polynomial factor other than 1, you use the reverse of the Product Rule for Radicals:

$$\text{If } a \text{ and } b \text{ are nonnegative, then } \sqrt{a}\sqrt{b} = \sqrt{ab} \text{ [see Example 15-5].}$$

(a) $\sqrt{24} = \sqrt{4(6)}$ **(b)** $2\sqrt{180} = 2\sqrt{36(5)}$ **(c)** $\sqrt{w^2} = |w|$
$= \sqrt{4}\sqrt{6}$ $= 2\sqrt{36}\sqrt{5}$ $= w$ [because $w \geq 0$]
$= 2\sqrt{6}$ $= 2(6)\sqrt{5}$
       $= 12\sqrt{5}$

(d) $x\sqrt{x^5} = x\sqrt{x^4(x)}$ **(e)** $\sqrt{8y^4} = \sqrt{4y^4(2)}$ **(f)** $\sqrt{27mn^3} = \sqrt{9n^2(3mn)}$
$= x\sqrt{x^4}\sqrt{x}$ $= \sqrt{4y^4}\sqrt{2}$ $= \sqrt{9n^2}\sqrt{3mn}$
$= x(x^2)\sqrt{x}$ $= 2y^2\sqrt{2}$ $= 3|n|\sqrt{3mn}$
$= x^3\sqrt{x}$ $= 3n\sqrt{3mn}$ [because $n \geq 0$]

PROBLEM 15-4 Simplify radicands in fraction form. Assume all variables are positive:

(a) $\sqrt{\dfrac{9}{16}}$ **(b)** $\dfrac{3}{4}\sqrt{\dfrac{12}{25}}$ **(c)** $\sqrt{\dfrac{x^3}{4}}$ **(d)** $\dfrac{y^2}{5}\sqrt{\dfrac{20}{y^2}}$ **(e)** $n\sqrt{\dfrac{49m^5}{36n^4}}$

Solution: Recall that to simplify a radicand that is given in fraction form, you use the reverse of the Quotient Rule for Radicals:

$$\text{If } a \text{ is nonnegative and } b \text{ is positive, then } \frac{\sqrt{a}}{\sqrt{b}} = \sqrt{\frac{a}{b}} \quad \text{[see Example 15-6].}$$

(a) $\sqrt{\dfrac{9}{16}} = \dfrac{\sqrt{9}}{\sqrt{16}}$

$\qquad = \dfrac{3}{4}$

(b) $\dfrac{3}{4}\sqrt{\dfrac{12}{25}} = \dfrac{3}{4} \cdot \dfrac{\sqrt{12}}{\sqrt{25}}$

$\qquad = \dfrac{3}{4} \cdot \dfrac{\sqrt{4(3)}}{5}$

$\qquad = \dfrac{3}{2(2)} \cdot \dfrac{2\sqrt{3}}{5}$

$\qquad = \dfrac{3}{2} \cdot \dfrac{\sqrt{3}}{5} = \dfrac{3\sqrt{3}}{10}$

(c) $\sqrt{\dfrac{x^3}{4}} = \dfrac{\sqrt{x^3}}{\sqrt{4}}$

$\qquad = \dfrac{\sqrt{x^2(x)}}{2}$

$\qquad = \dfrac{|x|\sqrt{x}}{2}$

$\qquad = \dfrac{x\sqrt{x}}{2}$ [because $x > 0$]

(d) $\dfrac{y^2}{5}\sqrt{\dfrac{20}{y^2}} = \dfrac{y^2}{5} \cdot \dfrac{\sqrt{20}}{\sqrt{y^2}}$

$\qquad = \dfrac{y^2}{5} \cdot \dfrac{\sqrt{4(5)}}{y}$ [because $y > 0$]

$\qquad = \dfrac{y(y)}{5} \cdot \dfrac{2\sqrt{5}}{y}$

$\qquad = \dfrac{2y\sqrt{5}}{5}$

(e) $n\sqrt{\dfrac{49m^5}{36n^4}} = \dfrac{n\sqrt{49m^4(m)}}{\sqrt{36n^4}}$

$\qquad = \dfrac{n(7m^2)\sqrt{m}}{6n^2}$

$\qquad = \dfrac{n(7m^2)\sqrt{m}}{6nn}$

$\qquad = \dfrac{7m^2\sqrt{m}}{6n}$

PROBLEM 15-5   Combine like radicals. Assume all radicands are nonnegative:

(a) $\sqrt{5} - 2\sqrt{5}$ **(b)** $\dfrac{\sqrt{2x}}{3} + \sqrt{2x}$ **(c)** $\dfrac{\sqrt{12}}{4} + \dfrac{2\sqrt{12}}{3} - \dfrac{3\sqrt{12}}{12}$.

Solution: Recall that to combine like radicals, you use the distributive properties to factor out the like radical [see Example 15-7]:

(a) $\sqrt{5} - 2\sqrt{5} = 1\sqrt{5} - 2\sqrt{5}$

$\qquad = (1 - 2)\sqrt{5}$

$\qquad = -1\sqrt{5}$

$\qquad = -\sqrt{5}$

(b) $\dfrac{\sqrt{2x}}{3} + \sqrt{2x} = \dfrac{\sqrt{2x}}{3} + \dfrac{3\sqrt{2x}}{3}$ Build up to get like fractions using the LCD 3.

$\qquad = \dfrac{1\sqrt{2x} + 3\sqrt{2x}}{3}$ Add like fractions.

$\qquad = \dfrac{(1 + 3)\sqrt{2x}}{3}$ Combine like radicals.

$\qquad = \dfrac{4\sqrt{2x}}{3}$

(c) $\dfrac{\sqrt{12}}{4} + \dfrac{2\sqrt{12}}{3} - \dfrac{3\sqrt{12}}{12} = \dfrac{3\sqrt{12}}{3(4)} + \dfrac{4(2)\sqrt{12}}{4(3)} - \dfrac{3\sqrt{12}}{12}$ Build up to get like fractions using the LCD 12.

$\qquad = \dfrac{3\sqrt{12}}{12} + \dfrac{8\sqrt{12}}{12} - \dfrac{3\sqrt{12}}{12}$ ⟵ like fractions

$\qquad = \dfrac{3\sqrt{12} + 8\sqrt{12} - 3\sqrt{12}}{12}$ Add and subtract like fractions.

$\qquad = \dfrac{(3 + 8 - 3)\sqrt{12}}{12}$ Combine like radicals.

$\qquad = \dfrac{8\sqrt{12}}{12}$ ⟵ like radicals are combined

$\qquad = \dfrac{4(2\sqrt{12})}{4(3)} = \dfrac{2\sqrt{12}}{3} = \dfrac{2(2\sqrt{3})}{3}$

$\qquad = \dfrac{4\sqrt{3}}{3}$ ⟵ simplest form

PROBLEM 15-6 Simplify and then combine like radicals. Assume each radicand is nonnegative:

(a) $\sqrt{27} + \sqrt{12}$ **(b)** $\sqrt{2x^3} + \sqrt{8x}$ **(c)** $5\sqrt{y^5} - y\sqrt{9y}$ **(d)** $m\sqrt{m} + \sqrt{m^3} + \sqrt{4m^2} - \sqrt{2m^2}$

Solution: Recall that sometimes unlike radicals can be combined by first simplifying the unlike radicals to get like radicals [see Example 15-8]:

(a) $\sqrt{27} + \sqrt{12} = \sqrt{9(3)} + \sqrt{4(3)}$ **(b)** $\sqrt{2x^3} + \sqrt{8x} = \sqrt{x^2(2x)} + \sqrt{4(2x)}$

$$= 3\sqrt{3} + 2\sqrt{3}$$
$$\qquad\qquad\qquad = |x|\sqrt{2x} + 2\sqrt{2x}$$
$$= 5\sqrt{3}$$
$$\qquad\qquad\qquad = x\sqrt{2x} + 2\sqrt{2x} \qquad \text{[because } x \geq 0\text{]}$$
$$\qquad\qquad\qquad = (x + 2)\sqrt{2x}$$

(c) $5\sqrt{y^5} - y\sqrt{9y} = 5\sqrt{y^4(y)} - y\sqrt{9(y)}$

$$= 5(y^2)\sqrt{y} - y(3)\sqrt{y}$$
$$= 5y^2\sqrt{y} - 3y\sqrt{y}$$
$$= (5y^2 - 3y)\sqrt{y}$$

(d) $m\sqrt{m} + \sqrt{m^3} + \sqrt{4m^2} - \sqrt{2m^2} = m\sqrt{m} + \sqrt{m^2(m)} + \sqrt{4(m^2)} - \sqrt{m^2(2)}$

$$= m\sqrt{m} + |m|\sqrt{m} + 2|m| - |m|\sqrt{2}$$
$$= m\sqrt{m} + m\sqrt{m} + 2m - m\sqrt{2} \qquad \text{[because } m \geq 0\text{]}$$
$$= 2m\sqrt{m} + 2m - m\sqrt{2}$$

PROBLEM 15-7 Multiply using the Product Rule for Radicals. Assume each radicand is nonnegative:

(a) $\sqrt{3}\sqrt{5}$ **(b)** $\sqrt{2}\sqrt{10}$ **(c)** $\sqrt{x}\sqrt{2x}\sqrt{2}$ **(d)** $\sqrt{w + 1}\sqrt{w - 1}$ **(e)** $2\sqrt{3}(4\sqrt{6})$

(f) $\dfrac{8}{9}\sqrt{\dfrac{m^2}{2}}\left(\dfrac{3}{4}\sqrt{\dfrac{m}{2}}\right)$

Solution: Recall that to multiply two or more radicals, you

(a) Multiply using the Product Rule for Radicals: $\sqrt{a}\sqrt{b} = \sqrt{ab}$ $[a \geq 0$ and $b \geq 0]$
(b) Simplify the product from Step **(a)** when possible [see Example 15-9]:

(a) $\sqrt{3}\sqrt{5} = \sqrt{3(5)}$ **(b)** $\sqrt{2}\sqrt{10} = \sqrt{2(10)}$ **(c)** $\sqrt{x}\sqrt{2x}\sqrt{2} = \sqrt{x(2x)2}$

$$= \sqrt{15}$$
$$\qquad = \sqrt{20}$$
$$\qquad\qquad = \sqrt{4x^2}$$
$$\qquad = \sqrt{4(5)}$$
$$\qquad\qquad = \sqrt{4}\sqrt{x^2}$$
$$\qquad = 2\sqrt{5}$$
$$\qquad\qquad = 2|x|$$
$$\qquad\qquad = 2x \qquad \text{[because } x \geq 0\text{]}$$

(d) $\sqrt{w + 1}\sqrt{w - 1} = \sqrt{(w + 1)(w - 1)}$ **(e)** $2\sqrt{3}(4\sqrt{6}) = 2(4)\sqrt{3(6)}$

$$= \sqrt{w^2 - 1}$$
$$\qquad\qquad = 8\sqrt{18}$$
$$\qquad\qquad = 8\sqrt{9(2)}$$
$$\qquad\qquad = 8(3)\sqrt{2} = 24\sqrt{2}$$

(f) $\dfrac{8}{9}\sqrt{\dfrac{m^2}{2}}\left(\dfrac{3}{4}\sqrt{\dfrac{m}{2}}\right) = \dfrac{8}{9}\left(\dfrac{3}{4}\right)\sqrt{\dfrac{m^2}{2}\left(\dfrac{m}{2}\right)}$

$$= \dfrac{2}{3}\sqrt{\dfrac{m^3}{4}}$$
$$= \dfrac{2|m|\sqrt{m}}{3(2)}$$
$$= \dfrac{m\sqrt{m}}{3} \qquad \text{[because } m \geq 0\text{]}$$

PROBLEM 15-8 Multiply using the distributive properties. Assume each radicand is nonnegative:

(a) $2\sqrt{3}(\sqrt{3} - \sqrt{2})$ (b) $\sqrt{w}(\sqrt{w} + 5)$

Solution: Recall that to multiply a radical expression containing two or more terms by a radical, you

(a) Multiply using the distributive properties.
(b) Simplify the product from Step (a) when possible [see Example 15-10]:

(a) $2\sqrt{3}(\sqrt{3} - \sqrt{2}) = 2\sqrt{3}\sqrt{3} - 2\sqrt{3}\sqrt{2}$ (b) $\sqrt{w}(\sqrt{w} + 5) = \sqrt{w}\sqrt{w} + \sqrt{w}(5)$

$\qquad\qquad = 2\sqrt{3^2} - 2\sqrt{3(2)}$ $\qquad\qquad = \sqrt{w^2} + 5\sqrt{w}$

$\qquad\qquad = 2(3) - 2\sqrt{6}$ $\qquad\qquad = |w| + 5\sqrt{w}$

$\qquad\qquad = 6 - 2\sqrt{6}$ $\qquad\qquad = w + 5\sqrt{w}$ [because $w \geq 0$]

PROBLEM 15-9 Multiply using the FOIL Method. Assume each radicand is nonnegative:

(a) $(\sqrt{2} + \sqrt{3})(\sqrt{2} - \sqrt{3})$ (b) $(1 - \sqrt{5})^2$ (c) $(\sqrt{x} + \sqrt{y})^2$ (d) $(\sqrt{m} + 2)(m - \sqrt{3})$

Solution: Recall that to multiply two radical expressions each containing two terms, you

(a) Multiply using the FOIL method or special products.
(b) Simplify the product from Step (a) when possible [see Example 15-11]:

$\overbrace{\qquad\qquad}^{\text{difference of two squares}}$

(a) $(\sqrt{2} + \sqrt{3})(\sqrt{2} - \sqrt{3}) = (\sqrt{2})^2 - (\sqrt{3})^2$ *Think:* $(a + b)(a - b) = a^2 - b^2$

$\qquad\qquad\qquad\qquad\qquad = 2 - 3$ *Think:* $(\sqrt{x})^2 = x$ if $x \geq 0$.

$\qquad\qquad\qquad\qquad\qquad = -1$

$\overbrace{\qquad\qquad}^{\text{perfect square trinomial}}$

(b) $(1 - \sqrt{5})^2 = (1)^2 - 2(1)(\sqrt{5}) + (\sqrt{5})^2$ *Think:* $(a - b)^2 = a^2 - 2ab + b^2$

$\qquad\qquad\quad = 1 - 2\sqrt{5} + 5$ *Think:* $(\sqrt{x})^2 = x$ if $x \geq 0$.

$\qquad\qquad\quad = 6 - 2\sqrt{5}$

$\overbrace{\qquad\qquad}^{\text{perfect square trinomial}}$

(c) $(\sqrt{x} + \sqrt{y})^2 = (\sqrt{x})^2 + 2\sqrt{x}\sqrt{y} + (\sqrt{y})^2$ *Think:* $(a + b)^2 = a^2 + 2ab + b^2$

$\qquad\qquad\quad = x + 2\sqrt{xy} + y$ *Think:* $(\sqrt{x})^2 = x$ and $(\sqrt{y})^2 = y$ because both x and y are assumed to be nonnegative [$x \geq 0$ and $y \geq 0$].

$\qquad\qquad\qquad\quad\text{F}\qquad\text{O}\qquad\text{I}\qquad\text{L}$

(d) $(\sqrt{m} + 2)(m - \sqrt{3}) = m\sqrt{m} - \sqrt{3}\sqrt{m} + 2m - 2\sqrt{3}$

$\qquad\qquad\qquad\qquad = m\sqrt{m} - \sqrt{3m} + 2m - 2\sqrt{3}$

PROBLEM 15-10 Divide using the Quotient Rule for Radicals. Assume each radicand is positive:

(a) $\dfrac{\sqrt{4}}{4}$ (b) $\dfrac{\sqrt{24}}{\sqrt{3}}$ (c) $\dfrac{\sqrt{w}}{\sqrt{w^3}}$ (d) $\dfrac{\sqrt{2xy}}{\sqrt{x}}$ (e) $\dfrac{\sqrt{6}}{\sqrt{75}}$ (f) $\dfrac{2\sqrt{6}}{5\sqrt{8}}$ (g) $2\sqrt{m^3} \div (m\sqrt{4m})$

Solution: Recall that to divide two radicals, you

(a) Divide using the Quotient Rule for Radicals: $\dfrac{\sqrt{a}}{\sqrt{b}} = \sqrt{\dfrac{a}{b}}$ [$a \geq 0$ and $b > 0$]

(b) Simplify the quotient from Step **(a)** when possible [see Example 15-12]:

(a) $\dfrac{\sqrt{4}}{4} = \dfrac{2}{4}$ **(b)** $\dfrac{\sqrt{24}}{\sqrt{3}} = \sqrt{\dfrac{24}{3}}$ **(c)** $\dfrac{\sqrt{w}}{\sqrt{w^3}} = \sqrt{\dfrac{w}{w^3}}$ **(d)** $\dfrac{\sqrt{2xy}}{\sqrt{x}} = \sqrt{\dfrac{2xy}{x}}$

$\qquad = \dfrac{1}{2}$ $\qquad\qquad = \sqrt{8}$ $\qquad\qquad\quad = \sqrt{\dfrac{1}{w^2}}$ $\qquad\qquad\quad = \sqrt{\dfrac{2y}{1}}$

$\qquad\qquad\qquad\quad = \sqrt{4(2)}$ $\qquad\qquad = \dfrac{\sqrt{1}}{\sqrt{w^2}}$ $\qquad\qquad\quad = \sqrt{2y}$

$\qquad\qquad\qquad\quad = 2\sqrt{2}$

$\qquad\qquad\qquad\qquad\qquad\qquad\quad = \dfrac{1}{|w|}$

$\qquad\qquad\qquad\qquad\qquad\qquad\quad = \dfrac{1}{w}\qquad$ [because $w > 0$]

(e) $\dfrac{\sqrt{6}}{\sqrt{75}} = \sqrt{\dfrac{6}{75}}$ **(f)** $\dfrac{2\sqrt{6}}{5\sqrt{8}} = \dfrac{2}{5}\sqrt{\dfrac{6}{8}}$ **(g)** $2\sqrt{m^3} \div (m\sqrt{4m}) = \dfrac{2\sqrt{m^3}}{m\sqrt{4m}}$

$\qquad = \sqrt{\dfrac{2}{25}}$ $\qquad\qquad = \dfrac{2}{5}\sqrt{\dfrac{3}{4}}$ $\qquad\qquad\qquad\qquad = \dfrac{2}{m}\sqrt{\dfrac{m^3}{4m}}$

$\qquad = \dfrac{\sqrt{2}}{\sqrt{25}}$ $\qquad\qquad = \dfrac{2\sqrt{3}}{5\sqrt{4}}$ $\qquad\qquad\qquad\qquad = \dfrac{2}{m}\sqrt{\dfrac{m^2}{4}}$

$\qquad = \dfrac{\sqrt{2}}{5}$ $\qquad\qquad\quad = \dfrac{2\sqrt{3}}{5(2)}$ $\qquad\qquad\qquad\qquad = \dfrac{2\sqrt{m^2}}{m\sqrt{4}}$

$\qquad\qquad\qquad\quad = \dfrac{\sqrt{3}}{5}$ $\qquad\qquad\qquad\qquad = \dfrac{2|m|}{m(2)}$

$\qquad\qquad\qquad\qquad\qquad\qquad\qquad\quad = \dfrac{2m}{2m}\qquad$ [because $m > 0$]

$\qquad\qquad\qquad\qquad\qquad\qquad\qquad\quad = 1$

PROBLEM 15-11   Rationalize denominators that contain one term. Assume each radicand is positive: **(a)** $\dfrac{1}{\sqrt{2}}$ **(b)** $\dfrac{3\sqrt{2}}{2\sqrt{3}}$ **(c)** $\dfrac{2x}{\sqrt{2x^3}}$

Solution: Recall that to rationalize a denominator that contains one term, you

(a) Identify the rationalizing factor as the radical in the denominator.
(b) Multiply both the numerator and denominator by the rationalizing factor from Step **(a)**.
(c) Simplify the result from Step **(b)** when possible [see Example 15-13]:

(a) $\dfrac{1}{\sqrt{2}} = \dfrac{1}{\sqrt{2}} \cdot \dfrac{\sqrt{2}}{\sqrt{2}}$ **(b)** $\dfrac{3\sqrt{2}}{2\sqrt{3}} = \dfrac{3\sqrt{2}}{2\sqrt{3}} \cdot \dfrac{\sqrt{3}}{\sqrt{3}}$ **(c)** $\dfrac{2x}{\sqrt{2x^3}} = \dfrac{2x}{\sqrt{x^2(2x)}}$

$\qquad = \dfrac{1\sqrt{2}}{(\sqrt{2})^2}$ $\qquad\qquad = \dfrac{3\sqrt{2}\sqrt{3}}{2(\sqrt{3})^2}$ $\qquad\qquad\qquad = \dfrac{2x}{|x|\sqrt{2x}}$

$\qquad = \dfrac{\sqrt{2}}{2}$ $\qquad\qquad\quad = \dfrac{3\sqrt{2}\sqrt{3}}{2(3)}$ $\qquad\qquad\qquad = \dfrac{2x}{x\sqrt{2x}}\qquad$ [because $x > 0$]

$\qquad\qquad\qquad\quad = \dfrac{3\sqrt{6}}{2(3)}$ $\qquad\qquad\qquad = \dfrac{2x}{x\sqrt{2x}} \cdot \dfrac{\sqrt{2x}}{\sqrt{2x}}$

$\qquad\qquad\qquad\quad = \dfrac{\sqrt{6}}{2}$ $\qquad\qquad\qquad\quad = \dfrac{2x\sqrt{2x}}{x(\sqrt{2x})^2}$

$\qquad\qquad\qquad\qquad\qquad\qquad\qquad = \dfrac{2x\sqrt{2x}}{x(2x)} = \dfrac{\sqrt{2x}}{x}$

PROBLEM 15-12 Rationalize denominators that contain two terms. Assume each radicand is positive and each denominator is nonzero:

(a) $\dfrac{3\sqrt{2} - 1}{\sqrt{2} + 3}$ **(b)** $\dfrac{\sqrt{x} + \sqrt{y}}{\sqrt{x} - \sqrt{y}}$

Solution: Recall that to rationalize a denominator that contains two terms, you

(a) Identify the rationalizing factor as the denominator with the opposite middle sign.
(b) Multiply both the numerator and denominator by the rationalizing factor from Step **(a)**.
(c) Simplify the result from Step **(b)** when possible [see Example 15-14]:

(a) $\dfrac{3\sqrt{2} - 1}{\sqrt{2} + 3} = \dfrac{3\sqrt{2} - 1}{\sqrt{2} + 3} \cdot \dfrac{\sqrt{2} - 3}{\sqrt{2} - 3}$

$= \dfrac{(3\sqrt{2} - 1)(\sqrt{2} - 3)}{(\sqrt{2} + 3)(\sqrt{2} - 3)}$

$= \dfrac{3(\sqrt{2})^2 - 9\sqrt{2} - \sqrt{2} + 3}{(\sqrt{2})^2 - (3)^2}$

$= \dfrac{6 - 10\sqrt{2} + 3}{2 - 9}$

$= \dfrac{9 - 10\sqrt{2}}{-7}$ or $\dfrac{10\sqrt{2} - 9}{7}$

(b) $\dfrac{\sqrt{x} + \sqrt{y}}{\sqrt{x} - \sqrt{y}} = \dfrac{\sqrt{x} + \sqrt{y}}{\sqrt{x} - \sqrt{y}} \cdot \dfrac{\sqrt{x} + \sqrt{y}}{\sqrt{x} + \sqrt{y}}$

$= \dfrac{(\sqrt{x} + \sqrt{y})(\sqrt{x} + \sqrt{y})}{(\sqrt{x} - \sqrt{y})(\sqrt{x} + \sqrt{y})}$

$= \dfrac{(\sqrt{x})^2 + 2\sqrt{x}\sqrt{y} + (\sqrt{y})^2}{(\sqrt{x})^2 - (\sqrt{y})^2}$

$= \dfrac{x + 2\sqrt{xy} + y}{x - y}$

PROBLEM 15-13 Divide, simplify, and then rationalize. Assume each radicand is positive:

(a) $\dfrac{6\sqrt{14}}{\sqrt{24}}$ **(b)** $x\sqrt{x} \div \sqrt{8x}$

Solution: Recall that to divide radical expressions, you

(a) Divide using the Quotient Rule for Radicals: $\dfrac{\sqrt{a}}{\sqrt{b}} = \sqrt{\dfrac{a}{b}}$

(b) Simplify using the Quotient Rule for Radicals in reverse: $\sqrt{\dfrac{a}{b}} = \dfrac{\sqrt{a}}{\sqrt{b}}$

(c) Rationalize the denominator when it contains a radical [see Example 15-15]:

(a) $\dfrac{6\sqrt{14}}{\sqrt{24}} = 6\sqrt{\dfrac{14}{24}}$ Divide.

$= 6\sqrt{\dfrac{7}{12}}$

$= \dfrac{6\sqrt{7}}{\sqrt{12}}$ Simplify.

$= \dfrac{2(3)\sqrt{7}}{2\sqrt{3}}$

$= \dfrac{3\sqrt{7}}{\sqrt{3}} \cdot \dfrac{\sqrt{3}}{\sqrt{3}}$ Rationalize.

$= \dfrac{3\sqrt{7}\sqrt{3}}{(\sqrt{3})^2}$

$= \dfrac{3\sqrt{21}}{3}$

$= \sqrt{21}$

(b) $x\sqrt{x} \div \sqrt{8x} = \dfrac{x\sqrt{x}}{\sqrt{8x}}$ Rename as division

$= x\sqrt{\dfrac{x}{8x}}$ Divide.

$= x\sqrt{\dfrac{1}{8}}$

$= \dfrac{x\sqrt{1}}{\sqrt{8}}$ Simplify.

$= \dfrac{x}{2\sqrt{2}}$

$= \dfrac{x}{2\sqrt{2}} \cdot \dfrac{\sqrt{2}}{\sqrt{2}}$ Rationalize.

$= \dfrac{x\sqrt{2}}{2(\sqrt{2})^2}$

$= \dfrac{x\sqrt{2}}{2(2)} = \dfrac{x\sqrt{2}}{4}$

PROBLEM 15-14 Solve each radical equation with the radical isolated in one member. Assume each radicand is nonnegative:

(a) $\sqrt{m} = 2$ (b) $\sqrt{2x + 5} = 1$ (c) $\sqrt{3y - 5} = -2$ (d) $\sqrt{x} = x$

Solution: Recall that to solve a radical equation when the radical is isolated in one member of the equation, you first clear the radical using the following rule:

Squaring Rule for Equations:

If $a = b$, then: $a^2 = b^2$ [see Examples 15-17 and 15-18]:

(a) $\sqrt{m} = 2$

    $(\sqrt{m})^2 = (2)^2$ Clear radicals.

      $m = 4$

Check: $\sqrt{m} = 2$

    $\dfrac{\sqrt{4}\ \ \ |\ \ 2}{2\ \ \ \ |\ \ 2}$ ⟵ $m = 4$ checks

(b) $\sqrt{2x + 5} = 1$

    $(\sqrt{2x + 5})^2 = (1)^2$ Clear radicals.

      $2x + 5 = 1$

         $2x = -4$

          $x = -2$

Check: $\sqrt{2x + 5} = 1$

    $\dfrac{\sqrt{2(-2) + 5}\ \ |\ \ 1}{}$

        $\dfrac{\sqrt{1}\ \ |\ \ 1}{1\ \ \ |\ \ 1}$ ⟵ $x = -2$ checks

(c) $\sqrt{3y - 5} = -2$ Stop!

$\sqrt{3y - 5} = -2$ has no real-number solutions because the principal square root of a real number is never negative.

(d) $\sqrt{x} = x$

    $(\sqrt{x})^2 = (x)^2$ Clear radicals.

      $x = x^2$

      $0 = x^2 - x$ Make one member zero.

      $0 = x(x - 1)$ Factor.

   $x = 0$ or $x - 1 = 0$ *Think:*
                       $ab = 0$ means
                       $a = 0$ or $b = 0$

   $x = 0$ or $x = 1$ ⟵ proposed solutions

Check: $\sqrt{x} = x$ | $\sqrt{x} = x$

    $\dfrac{\sqrt{0}\ \ |\ \ 0}{0\ \ \ |\ \ 0}$ | $\dfrac{\sqrt{1}\ \ |\ \ 1}{1\ \ \ |\ \ 1}$ ⟵ $x = 0$ or 1 checks

PROBLEM 15-15 Solve each radical equation where the radical is not isolated in one member. Assume each radicand is nonnegative: (a) $\sqrt{3x} - 2 = 1$ (b) $\sqrt{3x} + 2 = 1$ (c) $2\sqrt{y} = 6$ (d) $4\sqrt{w + 5} + 3 = 11$

Solution: Recall that to clear radicals using the Squaring Rule for Equations, the radical must first be isolated in one member of the equation [see Examples 15-19 and 15-20]:

(a) $\sqrt{3x} - 2 = 1$

    $\sqrt{3x} = 3$ ⟵ radical is isolated

    $(\sqrt{3x})^2 = (3)^2$

      $3x = 9$

       $x = 3$

(b) $\sqrt{3x} + 2 = 1$

    $\sqrt{3x} = -1$ Stop!

$\sqrt{3x} + 2 = 1$ has no real-number solutions because the principal square root of a real number is never negative.

Check: $\dfrac{\sqrt{3x} - 2 = 1}{\sqrt{3(3)} - 2 \mid 1}$
$\begin{array}{c|c} 3 - 2 & 1 \\ 1 & 1 \longleftarrow x = 3 \text{ checks} \end{array}$

(c) $2\sqrt{y} = 6$

$\sqrt{y} = 3 \longleftarrow$ radical is isolated

$(\sqrt{y})^2 = (3)^2$

$y = 9$

Check: $2\sqrt{y} = 6$
$\begin{array}{c|c} 2\sqrt{9} & 6 \\ 2(3) & 6 \\ 6 & 6 \longleftarrow y = 9 \text{ checks} \end{array}$

(d) $4\sqrt{w + 5} + 3 = 11$

$4\sqrt{w + 5} = 8$

$\sqrt{w + 5} = 2 \longleftarrow$ radical is isolated

$(\sqrt{w + 5})^2 = (2)^2$

$w + 5 = 4$

$w = -1$

Check: $\dfrac{4\sqrt{w + 5} + 3 = 11}{\begin{array}{c|c} 4\sqrt{(-1) + 5} + 3 & 11 \\ 4\sqrt{4} + 3 & 11 \\ 4(2) + 3 & 11 \\ 8 + 3 & 11 \\ 11 & 11 \longleftarrow w = -1 \\ & \text{checks} \end{array}}$

PROBLEM 15-16 Solve for h: $v = r\sqrt{\dfrac{g}{r + h}}$ [orbital velocity formula]

Solution: Recall that to solve a radical formula for a given variable, you

(a) Isolate the radical in one member of the formula.
(b) Clear the radical using the Squaring Rule for Equations.
(c) Solve the rational formula from Step (b) by isolating the given variable in one member of the formula [see Example 15-21]:

$$v = r\sqrt{\dfrac{g}{r + h}}$$

$$\dfrac{v}{r} = \sqrt{\dfrac{g}{r + h}} \qquad \text{Isolate the radical in one member.}$$

$$\left(\dfrac{v}{r}\right)^2 = \left(\sqrt{\dfrac{g}{r + h}}\right)^2 \qquad \begin{array}{l}\text{Clear the radical using the Squaring Rule for Equations} \\ \text{to get a rational formula.}\end{array}$$

$$\dfrac{v^2}{r^2} = \dfrac{g}{r + h} \longleftarrow \text{rational formula}$$

The LCD of $\dfrac{v^2}{r^2}$ and $\dfrac{g}{r + h}$ is $r^2(r + h)$.   Solve the rational formula.

$$r^2(r + h) \cdot \dfrac{v^2}{r^2} = r^2(r + h) \cdot \dfrac{g}{r + h}$$

$$(r + h)v^2 = r^2 g$$

$$rv^2 + hv^2 = r^2 g$$

$$hv^2 = r^2 g - rv^2$$

h is isolated $\longrightarrow h = \dfrac{r^2 g - rv^2}{v^2} \longleftarrow$ solution

Supplementary Exercises

PROBLEM 15-17   Find each square root: **(a)** $\sqrt{4}$ **(b)** $-\sqrt{16}$ **(c)** $\sqrt{\frac{1}{9}}$ **(d)** $-\sqrt{\frac{25}{64}}$
(e) $\sqrt{0}$ **(f)** $\sqrt{-2}$ **(g)** $\sqrt{5^2}$ **(h)** $\sqrt{(-5)^2}$ **(i)** $\sqrt{y^2}$ **(j)** $(\sqrt{5})^2$ **(k)** $(\sqrt{-5})^2$
(l) $(\sqrt{y})^2$ **(m)** $\sqrt{x^3}$ **(n)** $\sqrt{w^4}$ **(o)** $\sqrt{z^5}$ **(p)** $\sqrt{a^6}$ **(q)** $\sqrt{b^7}$ **(r)** $\sqrt{36}$ **(s)** $-\sqrt{49}$
(t) $\sqrt{\frac{81}{100}}$ **(u)** $-\sqrt{\frac{121}{144}}$ **(v)** $\sqrt{-169}$ **(w)** $\sqrt{(-169)^2}$ **(x)** $(\sqrt{-169})^2$ **(y)** $\sqrt{m^8}$
(z) $\sqrt{n^9}$

PROBLEM 15-18   Simplify. Assume each variable is positive: **(a)** $\sqrt{28}$ **(b)** $\sqrt{80}$ **(c)** $3\sqrt{50}$

(d) $2\sqrt{63}$ **(e)** $\sqrt{x^8 y^5}$ **(f)** $ab\sqrt{8a^3 b^{12}}$ **(g)** $\sqrt{\frac{3}{16}}$ **(h)** $\sqrt{\frac{2}{25}}$ **(i)** $\sqrt{\frac{28}{9}}$ **(j)** $\sqrt{\frac{20}{49}}$

(k) $\frac{1}{3}\sqrt{\frac{27}{100}}$ **(l)** $\frac{2}{3}\sqrt{\frac{45}{4}}$ **(m)** $\sqrt{\frac{x^2}{16}}$ **(n)** $\sqrt{\frac{1}{y^4}}$ **(o)** $\sqrt{\frac{2w^5}{49}}$ **(p)** $\sqrt{\frac{3z^{10}}{4}}$ **(q)** $\sqrt{\frac{8a^4}{36}}$

(r) $\sqrt{\frac{18b^3}{25}}$ **(s)** $\sqrt{\frac{1}{64c^2}}$ **(t)** $\sqrt{\frac{20}{100d^8}}$ **(u)** $\sqrt{\frac{12x^3}{25y^2}}$ **(v)** $\sqrt{\frac{24m^6}{81n^4}}$ **(w)** $-\sqrt{45}$ **(x)** $-\sqrt{52}$

(y) $-\sqrt{\frac{20m^2}{81}}$ **(z)** $\sqrt{\frac{80u^3}{9n^6}}$

PROBLEM 15-19   Combine radical expressions. Assume each radicand is nonnegative:

(a) $2\sqrt{3} + \sqrt{3}$ **(b)** $3\sqrt{2} - \sqrt{2}$ **(c)** $\sqrt{5} + \frac{3\sqrt{5}}{4}$ **(d)** $\frac{2}{3}\sqrt{7} - \sqrt{7}$ **(e)** $\sqrt{5} + \sqrt{20}$

(f) $\sqrt{18} - \sqrt{32}$ **(g)** $\frac{3\sqrt{32}}{2} + 3\sqrt{50}$ **(h)** $2\sqrt{36} - \frac{5\sqrt{64}}{4}$ **(i)** $3\sqrt{12} - \sqrt{12} + 2\sqrt{12}$

(j) $\sqrt{75} - \sqrt{3} + \sqrt{18}$ **(k)** $2\sqrt{a} + 3\sqrt{a}$ **(l)** $5\sqrt{b^2} - 3\sqrt{b^2}$ **(m)** $x\sqrt{4x} + 3\sqrt{x^3}$
(n) $y\sqrt{y+1} - \sqrt{y+1}$ **(o)** $w^2\sqrt{w} + w\sqrt{w^3} - \sqrt{w}$ **(p)** $4 + 2\sqrt{3} - 2\sqrt{3} + 3$
(q) $\frac{3\sqrt{32}}{2} + \frac{3\sqrt{8}}{4}$ **(r)** $\frac{2x\sqrt{24x}}{3} - \frac{5\sqrt{54x^3}}{9}$

PROBLEM 15-20   Multiply radical expressions. Assume each radicand is nonnegative.

(a) $\sqrt{7}\sqrt{3}$ **(b)** $\sqrt{12}\sqrt{2}$ **(c)** $\sqrt{\frac{2}{3}}\sqrt{\frac{3}{8}}$ **(d)** $\sqrt{6}\sqrt{8}\sqrt{3}$ **(e)** $3\sqrt{5}(2\sqrt{5})$ **(f)** $3\sqrt{12}(4\sqrt{6})$
(g) $\frac{3}{8}\sqrt{\frac{1}{2}}(\frac{2}{3}\sqrt{\frac{5}{8}})$ **(h)** $5\sqrt{3}(\sqrt{6} + \sqrt{3})$ **(i)** $(3\sqrt{8} - 2\sqrt{2})2\sqrt{2}$ **(j)** $(\sqrt{5} - \sqrt{2})(\sqrt{5} + \sqrt{2})$
(k) $(\sqrt{3} - \sqrt{8})(\sqrt{3} + \sqrt{2})$ **(l)** $(\sqrt{6} + \sqrt{2})^2$ **(m)** $(\sqrt{8} - \sqrt{3})^2$ **(n)** $\sqrt{a^3}\sqrt{a^5}$
(o) $\sqrt{12b}\sqrt{8b^3}\sqrt{18b}$ **(p)** $\sqrt{2c}(\sqrt{c} + \sqrt{2})$ **(q)** $\sqrt{x}(\sqrt{x^7} - \sqrt{x})$ **(r)** $(\sqrt{y} - 1)(\sqrt{y} + 1)$
(s) $(\sqrt{u} + \sqrt{v})^2$ **(t)** $(3 - \sqrt{w})^2$ **(u)** $x\sqrt{3x^7}\sqrt{6x^3}$ **(v)** $y^2\sqrt{10y^2}\sqrt{2y^5}$ **(w)** $\sqrt{8ab^4}\sqrt{2a^4b^3}$
(x) $\sqrt{3mn}\sqrt{mn}$ **(y)** $\sqrt{3r}\sqrt{2s}$ **(z)** $\sqrt{abc}\sqrt{a^2b^5}$

PROBLEM 15-21   Divide radical expressions. Assume each radicand and denominator is positive:

(a) $\frac{\sqrt{18}}{\sqrt{2}}$ **(b)** $\frac{\sqrt{6}}{\sqrt{27}}$ **(c)** $\frac{\sqrt{40}}{\sqrt{2}}$ **(d)** $\frac{\sqrt{2a^4}}{\sqrt{a}}$ **(e)** $\frac{\sqrt{b^2 + 2b + 1}}{\sqrt{b+1}}$

(f) $\frac{2\sqrt{8}}{\sqrt{2}}$ **(g)** $\frac{5\sqrt{18}}{\sqrt{3}}$ **(h)** $\frac{\sqrt{2}}{2\sqrt{8}}$ **(i)** $\frac{2\sqrt{54}}{3\sqrt{3}}$ **(j)** $\frac{4\sqrt{x^3 + x^2}}{x\sqrt{x+1}}$

(k) $\frac{1}{\sqrt{3}}$ **(l)** $\frac{2}{\sqrt{8}}$ **(m)** $\sqrt{\frac{3}{28}}$ **(n)** $\frac{2\sqrt{3}}{3\sqrt{2}}$ **(o)** $\frac{y}{\sqrt{y}}$ **(p)** $\frac{2}{\sqrt{5}-1}$ **(q)** $\frac{\sqrt{2}-\sqrt{3}}{\sqrt{2}+\sqrt{3}}$

(r) $\dfrac{\sqrt{w}}{\sqrt{w}+\sqrt{z}}$ (s) $\dfrac{\sqrt{40}}{\sqrt{6}}$ (t) $4\sqrt{3}\div(2\sqrt{8})$ (u) $\dfrac{18\sqrt{15}}{3\sqrt{54}}$ (v) $\dfrac{c\sqrt{5}}{\sqrt{c}}$

(w) $\dfrac{6\sqrt{d}}{\sqrt{3d}}$ (x) $\dfrac{3m\sqrt{2}}{2\sqrt{m}}$ (y) $\dfrac{3}{\sqrt{7}-2}$ (z) $\dfrac{\sqrt{a}-b}{\sqrt{a}+b}$

PROBLEM 15-22 Solve radical equations and formulas. Assume each radicand is nonnegative:

(a) $\sqrt{x}=3$ (b) $\sqrt{y}=-2$ (c) $\sqrt{2w}=4$ (d) $\sqrt{z+1}=1$ (e) $\sqrt{3a+9}=2$
(f) $\sqrt{2b}-1=5$ (g) $1-3\sqrt{c-3}=2$ (h) $\sqrt{3m-5}+5=7$ (i) $\sqrt{3n}=n$
(j) $\sqrt{3x-2}=x$ (k) $\sqrt{y+3}=y-3$ (l) $\sqrt{x^2}=4$

(m) for L: $f=\dfrac{1}{2\pi\sqrt{LC}}$ [electronics formula]   (n) for L: $T=0.4\sqrt{L}$ [wave period formula]

(o) for h: $r=\dfrac{590}{\sqrt{h}}$ [pulse rate formula]   (p) for d: $v=\sqrt{\dfrac{E}{d}}$ [motion formula]

(q) for R: $E=\sqrt{RP}$ [electricity formula]   (r) for m: $T=2\pi\sqrt{\dfrac{m}{k}}$ [spring period formula]

(s) for c: $y=\dfrac{-b+\sqrt{b^2-4ac}}{2a}$ [quadratic formula]

(t) for c^2: $M=\dfrac{m}{\sqrt{1-\dfrac{v^2}{c^2}}}$ [relativity formula]

Answers to Supplementary Exercises

(15-17) (a) 2   (b) -4 (c) $\frac{1}{3}$ (d) $-\frac{5}{8}$ (e) 0   (f) not a real number   (g) 5

(h) 5   (i) $|y|$ (j) 5   (k) not a real number   (l) $\begin{cases} y \text{ if } y \geq 0 \\ \text{not a real number if } y < 0 \end{cases}$

(m) $\begin{cases} x\sqrt{x} \text{ if } x \geq 0 \\ \text{not a real number if } x < 0 \end{cases}$ (n) w^2 (o) $\begin{cases} z^2\sqrt{z} \text{ if } z \geq 0 \\ \text{not a real number if } z < 0 \end{cases}$

(p) $|a^3|$ (q) $\begin{cases} b^3\sqrt{b} \text{ if } b \geq 0 \\ \text{not a real number if } b < 0 \end{cases}$ (r) 6   (s) -7 (t) $\frac{9}{10}$ (u) $-\frac{11}{12}$

(v) not a real number   (w) 169   (x) not a real number   (y) m^4

(z) $\begin{cases} n^4\sqrt{n} \text{ if } n \geq 0 \\ \text{not a real number if } n < 0 \end{cases}$

(15-18) (a) $2\sqrt{7}$ (b) $4\sqrt{5}$ (c) $15\sqrt{2}$ (d) $6\sqrt{7}$ (e) $x^4y^2\sqrt{y}$ (f) $2a^2b^7\sqrt{2a}$

(g) $\dfrac{\sqrt{3}}{4}$ (h) $\dfrac{\sqrt{2}}{5}$ (i) $\dfrac{2\sqrt{7}}{3}$ (j) $\dfrac{2\sqrt{5}}{7}$ (k) $\dfrac{\sqrt{3}}{10}$ (l) $\sqrt{5}$ (m) $\dfrac{x}{4}$ (n) $\dfrac{1}{y^2}$

(o) $\dfrac{w^2\sqrt{2w}}{7}$ (p) $\dfrac{z^5\sqrt{3}}{2}$ (q) $\dfrac{a^2\sqrt{2}}{3}$ (r) $\dfrac{3b\sqrt{2b}}{5}$ (s) $\dfrac{1}{8c}$ (t) $\dfrac{\sqrt{5}}{5d^4}$ (u) $\dfrac{2x\sqrt{3x}}{5y}$

(v) $\dfrac{2m^3\sqrt{6}}{9n^2}$ (w) $-3\sqrt{5}$ (x) $-2\sqrt{13}$ (y) $-\dfrac{2m\sqrt{5}}{9}$ (z) $\dfrac{4u\sqrt{5u}}{3n^3}$

(15-19) **(a)** $3\sqrt{3}$ **(b)** $2\sqrt{2}$ **(c)** $\dfrac{7\sqrt{5}}{4}$ **(d)** $-\dfrac{\sqrt{7}}{3}$ **(e)** $3\sqrt{5}$ **(f)** $-\sqrt{2}$ **(g)** $21\sqrt{2}$

(h) 2 **(i)** $8\sqrt{3}$ **(j)** $4\sqrt{3} + 3\sqrt{2}$ **(k)** $5\sqrt{a}$ **(l)** $2|b|$ **(m)** $5x\sqrt{x}$

(n) $(y-1)\sqrt{y+1}$ **(o)** $(2w^2 - 1)\sqrt{w}$ **(p)** 7 **(q)** $\dfrac{15\sqrt{2}}{2}$ **(r)** $-\dfrac{x\sqrt{6x}}{3}$

(15-20) **(a)** $\sqrt{21}$ **(b)** $2\sqrt{6}$ **(c)** $\frac{1}{2}$ **(d)** 12 **(e)** 30 **(f)** $72\sqrt{2}$ **(g)** $\dfrac{\sqrt{5}}{16}$

(h) $15 + 15\sqrt{2}$ **(i)** 16 **(j)** 3 **(k)** $-1 - \sqrt{6}$ **(l)** $8 + 4\sqrt{3}$ **(m)** $11 - 4\sqrt{6}$
(n) a^4 **(o)** $24b^2\sqrt{3b}$ **(p)** $c\sqrt{2} + 2\sqrt{c}$ **(q)** $x^4 - x$ **(r)** $y - 1$
(s) $u + 2\sqrt{uv} + v$ **(t)** $9 - 6\sqrt{w} + w$ **(u)** $3x^6\sqrt{2}$ **(v)** $2y^5\sqrt{5y}$ **(w)** $4a^2b^3\sqrt{ab}$
(x) $mn\sqrt{3}$ **(y)** $\sqrt{6rs}$ **(z)** $|a|b^3\sqrt{ac}$

(15-21) **(a)** 3 **(b)** $\dfrac{\sqrt{2}}{3}$ **(c)** $2\sqrt{5}$ **(d)** $a\sqrt{2a}$ **(e)** $\sqrt{b+1}$ **(f)** 4 **(g)** $5\sqrt{6}$ **(h)** $\frac{1}{4}$

(i) $2\sqrt{2}$ **(j)** 4 **(k)** $\dfrac{\sqrt{3}}{3}$ **(l)** $\dfrac{\sqrt{2}}{2}$ **(m)** $\dfrac{\sqrt{21}}{14}$ **(n)** $\dfrac{\sqrt{6}}{3}$ **(o)** $\sqrt{y}$ **(p)** $\dfrac{\sqrt{5}+1}{2}$

(q) $-5 + 2\sqrt{6}$ **(r)** $\dfrac{w - \sqrt{wz}}{w - z}$ **(s)** $\dfrac{2\sqrt{15}}{3}$ **(t)** $\dfrac{\sqrt{6}}{2}$ **(u)** $\sqrt{10}$ **(v)** $\sqrt{5c}$

(w) $2\sqrt{3}$ **(x)** $\dfrac{3\sqrt{2m}}{2}$ **(y)** $\sqrt{7} + 2$ **(z)** $\dfrac{a - 2b\sqrt{a} + b^2}{a - b^2}$

(15-22) **(a)** 9 **(b)** no real-number solutions **(c)** 8 **(d)** 0 **(e)** $-\frac{5}{3}$ **(f)** 18
(g) no real-number solutions **(h)** 3 **(i)** 3 or 0 **(j)** 1 or 2 **(k)** 6

(l) 4 or -4 **(m)** $L = \dfrac{1}{4\pi^2 Cf^2}$ **(n)** $L = \dfrac{T^2}{0.16}$ or $L = 6.25T^2$ **(o)** $h = \dfrac{348,100}{r^2}$

(p) $d = \dfrac{E}{v^2}$ **(q)** $R = \dfrac{E^2}{P}$ **(r)** $m = \dfrac{kT^2}{4\pi^2}$ **(s)** $c = -ay^2 - by$ or $c = -y(ay + b)$

(t) $c^2 = \dfrac{v^2}{1 - \dfrac{m^2}{M^2}}$ or $c^2 = \dfrac{v^2 M^2}{M^2 - m^2}$

16 QUADRATIC EQUATIONS

THIS CHAPTER IS ABOUT

- ☑ **Solving Quadratic Equations Using the Zero-Product Property**
- ☑ **Solving Quadratic Equations Using the Square Root Rule**
- ☑ **Solving Quadratic Equations by Completing the Square**
- ☑ **Solving Quadratic Equations Using the Quadratic Formula**
- ☑ **Solving Quadratic Equations Using the Easiest Method**

16-1. Solving Quadratic Equations Using the Zero-Product Property

A. Identify quadratic equations.

If a, b, and c are real numbers [$a \neq 0$] and x is any variable, then any equation that can be written in the form $ax^2 + bx + c = 0$ is called a **quadratic equation in one variable.**

EXAMPLE 16-1: Identify which of the following are quadratic equations in one variable:
(a) $w^2 + 5w - 6 = 0$ **(b)** $2m^2 - 3m + 5 = 0$ **(c)** $4x^2 + 3x = 0$ **(d)** $y^2 - 2 = 0$
(e) $2x + 3 = 0$ **(f)** $y = 2x^2 - 3x + 5$ **(g)** $2^x = 4$ **(h)** $|x| = 3$

Solution: **(a)** $w^2 + 5w - 6 = 0$ or $1w^2 + 5w + (-6) = 0$ is a quadratic equation in one variable.
 (b) $2m^2 - 3m + 5 = 0$ or $2m^2 + (-3)m + 5 = 0$ is a quadratic equation in one variable.
 (c) $4x^2 + 3x = 0$ or $4x^2 + 3x + 0 = 0$ is a quadratic equation in one variable.
 (d) $y^2 - 2 = 0$ or $1y^2 + 0y + (-2) = 0$ is a quadratic equation in one variable.
 (e) $2x + 3 = 0$ is not a quadratic equation in one variable because the 2nd-degree term [quadratic term] is missing. That is, $a = 0$ in $0x^2 + 2x + 3 = 0$.
 (f) $y = 2x^2 - 3x + 5$ is not a quadratic equation in one variable because there are two different variables x and y.
 (g) $2^x = 4$ is not a quadratic equation in one variable because there is a variable used as an exponent.
 (h) $|x| = 3$ is not a quadratic equation in one variable because there is a variable contained in an absolute value symbol.

Agreement: For the remainder of this chapter, the phrase "quadratic equation(s)" will mean *quadratic equation(s) in one variable.*

B. Write quadratic equations in standard form.

If $a > 0$, then the quadratic equation $ax^2 + bx + c = 0$ is in **standard form.**

EXAMPLE 16-2: Write each quadratic equation in standard form:
(a) $2x + 3 + 5x^2 = 0$ **(b)** $3y = 1 - 4y^2$ **(c)** $3w - 2w^2 - 4 = 0$ **(d)** $5 = 4m^2$
(e) $8n - 2n^2 = 0$

Solution: **(a)** $2x + 3 + 5x^2 = 0$ Write descending powers to get standard form.
$$\mathbf{5x^2} + ? + ? = 0$$
$$5x^2 + \mathbf{2x} + ? = 0$$
$$5x^2 + 2x + \mathbf{3} = 0 \longleftarrow \text{standard form}$$

379

(b) $3y = 1 - 4y^2$ Make one member zero while writing the other

         $4y^2 + 3y = 1$ member in descending powers to get standard form.

    $4y^2 + 3y - 1 = 0$ ⟵ standard form

(c) $3w - 2w^2 - 4 = 0$

      $-2w^2 + 3w - 4 = 0$ Write descending powers and make *a* positive

   $-1(-2w^2 + 3w - 4) = -1(0)$ $[a > 0]$ to get standard form.

     $2w^2 - 3w + 4 = 0$ ⟵ standard form

(d) $5 = 4m^2$ Make one member zero while writing descending

      $0 = 4m^2 - 5$ powers to get standard form.

    $4m^2 - 5 = 0$ ⟵ standard form

(e) $8n - 2n^2 = 0$

      $-2n^2 + 8n = 0$ Write descending powers and make *a* positive

   $-1(-2n^2 + 8n) = -1(0)$ $[a > 0]$ to get standard form.

     $2n^2 - 8n = 0$ ⟵ standard form

C. Write quadratic equations in simplest form and then identify *a*, *b*, and *c*.

A quadratic equation in standard form $ax^2 + bx + c = 0 \, [a > 0]$ is in **simplest form** if *a*, *b*, and *c* are all integers and the greatest common factor [GCF] of *a*, *b*, and *c* is 1 or -1.

EXAMPLE 16-3: Write each quadratic equation in simplest form and then identify *a*, *b*, and *c*:
(a) $4x^2 - 6x - 10 = 0$ **(b)** $\frac{2}{3} + y = \frac{1}{6}y^2$

Solution: **(a)** In $4x^2 - 6x - 10 = 0$: $a = 4$, $b = -6$, $c = -10$.
        The GCF of 4, -6, and -10 is 2 or -2 [see Example 9-10].

$$\frac{4x^2 - 6x - 10}{2} = \frac{0}{2} \qquad \text{Divide both members by the positive GCF 2.}$$

$$\frac{4x^2}{2} - \frac{6x}{2} - \frac{10}{2} = \frac{0}{2}$$

$2x^2 - 3x - 5 = 0$ ⟵ simplest form [The GCF of 2, -3, and -5 is 1 or -1.]

In $2x^2 - 3x - 5 = 0$: $a = 2$, $b = -3$, $c = -5$.

(b) The LCD of $\frac{2}{3}$ and $\frac{1}{6}$ is 6.

$$6(\tfrac{2}{3} + y) = 6 \cdot \tfrac{1}{6}y^2 \qquad \text{Clear fractions.}$$

$$6 \cdot \tfrac{2}{3} + 6(y) = 6 \cdot \tfrac{1}{6}y^2$$

$4 + 6y = y^2$ ⟵ fractions cleared

$0 = y^2 - 6y - 4$

$y^2 - 6y - 4 = 0$ ⟵ simplest form [The GCF of 1, -6, and -4 is 1 or -1.]

In $y^2 - 6y - 4 = 0$: $a = 1$, $b = -6$, $c = -4$.

Note: For a quadratic equation to be in simplest form, all five of the following conditions must be satisfied:

1. The right member is zero [0].
2. The left member is in descending powers $[ax^2 + bx + c]$.
3. In $ax^2 + bx + c = 0$, *a*, *b*, and *c* are all integers.

4. In $ax^2 + bx + c = 0$, a is positive $[a > 0]$.
5. In $ax^2 + bx + c = 0$, the GCF of a, b, and c is 1 or -1.

D. Solve quadratic equations that factor using the zero-product property.

To **solve a quadratic equation that factors in simplest form,** you can use the following method:

Zero-Product Property Method

1. Write the given quadratic equation in simplest form.
2. Factor the left member of the quadratic equation from Step 1.
3. Use the zero-product property to set both factors from Step 2 equal to zero.
4. Solve both linear equations from Step 3 to find the proposed solutions of the original quadratic equation.
5. Check by substituting each proposed solution from Step 4 in the original equation to see if you get a true number sentence.

EXAMPLE 16-4: Solve using the zero-product property method: **(a)** $8x^2 + 6x = 0$ **(b)** $y^2 = 4$
(c) $3w + 3w^2 = 18$ **(d)** $m^2 - 2 = 0$

Solution:

(a) $8x^2 + 6x = 0$ *Think:* $8x^2 + 6x = 0$ is not in simplest form because the GCF of 8 and 6 is 2 or -2.

$\dfrac{8x^2}{2} + \dfrac{6x}{2} = \dfrac{0}{2}$ Divide each term by the positive GCF 2.

$4x^2 + 3x = 0$ ⟵ simplest form

$x(4x + 3) = 0$ Factor [see Example 9-11].

$x = 0$ or $4x + 3 = 0$ Use the zero-product property: $ab = 0$ means $a = 0$ or $b = 0$.

$x = 0$ or $4x = -3$ Solve each linear equation.

$x = 0$ or $x = -\dfrac{3}{4}$ ⟵ proposed solutions

Check: $\begin{array}{c|c} 8x^2 + 6x = 0 & \text{⟵ original equation} \\ \hline 8(0)^2 + 6(0) & 0 \qquad \text{Substitute 0 for } x. \\ 0 + 0 & 0 \\ 0 & 0 \text{ ⟵ } x = 0 \text{ checks} \end{array}$ $\begin{array}{c|c} 8x^2 + 6x = 0 & \text{⟵ original equation} \\ \hline 8(-\frac{3}{4})^2 + 6(-\frac{3}{4}) & 0 \qquad \text{Substitute } -\frac{3}{4} \text{ for } x. \\ \frac{9}{2} + (-\frac{9}{2}) & 0 \\ 0 & 0 \text{ ⟵ } x = -\frac{3}{4} \text{ checks} \end{array}$

(b) $y^2 = 4$ *Think:* $y^2 = 4$ is not in simplest form because the right member is not 0.

$y^2 - 4 = 0$ ⟵ simplest form

$(y + 2)(y - 2) = 0$ Factor [see Example 9-36].

$y + 2 = 0$ or $y - 2 = 0$ Use the zero-product property: $ab = 0$ means $a = 0$ or $b = 0$.

$y = -2$ or $y = 2$ ⟵ proposed solutions

Check: $\begin{array}{c|c} y^2 = 4 & \text{⟵ original equation} \\ \hline (2)^2 & 4 \qquad \text{Substitute 2 for } y. \\ 4 & 4 \text{ ⟵ } y = 2 \text{ checks} \end{array}$ $\begin{array}{c|c} y^2 = 4 & \text{⟵ original equation} \\ \hline (-2)^2 & 4 \qquad \text{Substitute } -2 \text{ for } y. \\ 4 & 4 \text{ ⟵ } y = -2 \text{ checks} \end{array}$

(c) $3w + 3w^2 = 18$ *Think:* $3w + 3w^2 = 18$ is not in simplest form.

$3w^2 + 3w - 18 = 0$ ⟵ standard form [The GCF of 3, 3, and −18 is 3 or −3.]

$\dfrac{3w^2}{3} + \dfrac{3w}{3} - \dfrac{18}{3} = \dfrac{0}{3}$ Divide each term by the positive GCF 3.

$w^2 + w - 6 = 0$ ⟵ simplest form

$(w + 3)(w - 2) = 0$ Factor [see Example 9-22].

$w + 3 = 0$ or $w - 2 = 0$ Use the zero-product property: $ab = 0$ means $a = 0$
                                                                    or $b = 0$.

$w = -3$ or $w = 2$ ⟵ proposed solutions

Check:

$3w + 3w^2 = 18$ ⟵ original equation			$3w + 3w^2 = 18$ ⟵ original equation	
$3(-3) + 3(-3)^2$	18	Substitute -3 for w.	$3(2) + 3(2)^2$	18	Substitute 2 for w.
$-9 + 27$	18		$6 + 12$	18
18	18 ⟵ $w = -3$ checks		18	18 ⟵ $w = 2$ checks

(d) $m^2 - 2 = 0$ cannot be solved using the zero-product property method because $m^2 - 2$ will not factor over the integers.

16-2. Solving Quadratic Equations Using the Square Root Rule

A. Solve quadratic equations with form $x^2 = d$ using the Square Root Rule.

To **solve quadratic equations with form $x^2 = d$,** you can use the following rule:

Square Root Rule

If $x^2 = d$, then $x = \pm\sqrt{d}$.

Note: $x = \pm\sqrt{d}$ means $x = \sqrt{d}$ or $-\sqrt{d}$.

EXAMPLE 16-5: Solve using the Square Root Rule: **(a)** $y^2 = 4$ **(b)** $m^2 = 2$

Solution: **(a)** $y^2 = 4$ ⟵ $x^2 = d$ form

$y = \pm\sqrt{4}$ Use the Square Root Rule: $x^2 = d$ means $x = \pm\sqrt{d}$.

$y = \pm 2$ ⟵ same solutions as found in Example 16-4(b)

(b) $m^2 = 2$ ⟵ $x^2 = d$ form

$m = \pm\sqrt{2}$ ⟵ proposed solutions

Check:

$m^2 = 2$			$m^2 = 2$	
$(\sqrt{2})^2$	2		$(-\sqrt{2})^2$	2
2	2 ⟵ $m = \sqrt{2}$ checks		2	2 ⟵ $m = -\sqrt{2}$ checks

B. Solve quadratic equations with form $ax^2 + h = k$ using the Square Root Rule.

To **solve a quadratic equation with form $ax^2 + h = k$,** you can use the following method:

Square-Root Rule Method

1. Isolate x^2 in one member to get $x^2 = d$ form.
2. Solve the equation from Step 1 using the Square Root Rule.

3. Check by substituting each proposed solution from Step 2 in the original equation to see if you get a true number sentence.

EXAMPLE 16-6: Solve using the square-root rule method: **(a)** $2w^2 - 1 = 0$ **(b)** $4x^2 + 5 = 2$

Solution:
(a) $2w^2 - 1 = 0$ ⟵ $ax^2 + h = k$ form

$\quad\quad 2w^2 = 1$ Isolate w^2.

$\quad\quad w^2 = \dfrac{1}{2}$ ⟵ $x^2 = d$ form

$\quad\quad w = \pm\sqrt{\dfrac{1}{2}}$ Use the Square Root Rule: $x^2 = d$ means $x = \pm\sqrt{d}$.

$\quad\quad w = \pm\dfrac{\sqrt{1}}{\sqrt{2}}$ Simplify the radical when possible [see Example 15-13].

$\quad\quad w = \pm\dfrac{1}{\sqrt{2}} \cdot \dfrac{\sqrt{2}}{\sqrt{2}}$

$\quad\quad w = \pm\dfrac{\sqrt{2}}{2}$ ⟵ proposed solutions

Check:

$$2w^2 - 1 = 0$$
$$2\left(\dfrac{\sqrt{2}}{2}\right)^2 - 1 \;\Big|\; 0$$
$$2\left(\dfrac{2}{4}\right) - 1 \;\Big|\; 0$$
$$1 - 1 \;\Big|\; 0$$
$$0 \;\Big|\; 0 \;\longleftarrow\; w = \dfrac{\sqrt{2}}{2} \text{ checks}$$

$$2w^2 - 1 = 0$$
$$2\left(-\dfrac{\sqrt{2}}{2}\right)^2 - 1 \;\Big|\; 0$$
$$2\left(\dfrac{2}{4}\right) - 1 \;\Big|\; 0$$
$$1 - 1 \;\Big|\; 0$$
$$0 \;\Big|\; 0 \;\longleftarrow\; w = -\dfrac{\sqrt{2}}{2} \text{ checks}$$

(b) $4x^2 + 5 = 2$ ⟵ $ax^2 + h = k$ form

$\quad\quad 4x^2 = -3$ Isolate x^2.

$\quad\quad x^2 = -\dfrac{3}{4}$ Stop! The square of a real number is never negative.

$4x^2 + 5 = 2$ has no real-number solutions because $x^2 = -\frac{3}{4}$ has no real-number solutions.

Note: If a and c are both positive [or both negative], then $ax^2 + c = 0$ has no real-number solutions.

B. Solve quadratic equations with form $(x + h)^2 = k$ using the Square Root Rule.

To **solve quadratic equations with form $(x + h)^2 = k$,** you

1. Use the Square Root Rule to get: $x + h = \pm\sqrt{k}$
2. Simplify the radical $\sqrt{k}$ when possible.
3. Add the opposite of h to both members to find the proposed solutions of $(x + h)^2 = k$.
4. Check by substituting each proposed solution from Step 3 in $(x + h)^2 = k$ to see if you get a true number sentence.

EXAMPLE 16-7: Solve using the Square Root Rule: **(a)** $\left(m - \dfrac{1}{2}\right)^2 = \dfrac{3}{4}$ **(b)** $(x + 2)^2 = -1$

Solution:

(a) $\left(m - \dfrac{1}{2}\right)^2 = \dfrac{3}{4}$ ⟵ $(x + h)^2 = k$ form with $x = m$, $h = -\dfrac{1}{2}$, and $k = \dfrac{3}{4}$

$m - \dfrac{1}{2} = \pm\sqrt{\dfrac{3}{4}}$ Use the Square Root Rule: $(x + h)^2 = k$ means $x + h = \pm\sqrt{k}$.

$m - \dfrac{1}{2} = \pm\dfrac{\sqrt{3}}{2}$ Simplify the radical when possible.

$m = \dfrac{1}{2} \pm \dfrac{\sqrt{3}}{2}$ Add $\dfrac{1}{2}\left[\text{the opposite of } -\dfrac{1}{2}\right]$ to both members.

$m = \dfrac{1}{2} \pm \dfrac{\sqrt{3}}{2}$ or $\dfrac{1 \pm \sqrt{3}}{2}$ ⟵ proposed solutions [See the following *Note*.]

Check:

$$\left(m - \dfrac{1}{2}\right)^2 = \dfrac{3}{4}$$

$\left[\left(\dfrac{1}{2} + \dfrac{\sqrt{3}}{2}\right) - \dfrac{1}{2}\right]^2$	$\dfrac{3}{4}$
$\left(\dfrac{\sqrt{3}}{2}\right)^2$	$\dfrac{3}{4}$
$\dfrac{(\sqrt{3})^2}{2^2}$	$\dfrac{3}{4}$
$\dfrac{3}{4}$	$\dfrac{3}{4}$ ⟵ $m = \dfrac{1}{2} + \dfrac{\sqrt{3}}{2}$ checks

$$\left(m - \dfrac{1}{2}\right)^2 = \dfrac{3}{4}$$

$\left[\left(\dfrac{1}{2} - \dfrac{\sqrt{3}}{2}\right) - \dfrac{1}{2}\right]^2$	$\dfrac{3}{4}$
$\left(-\dfrac{\sqrt{3}}{2}\right)^2$	$\dfrac{3}{4}$
$\dfrac{(-\sqrt{3})^2}{2^2}$	$\dfrac{3}{4}$
$\dfrac{3}{4}$	$\dfrac{3}{4}$ ⟵ $m = \dfrac{1}{2} - \dfrac{\sqrt{3}}{2}$ checks

(b) $(x + 2)^2 = -1$ Stop! The square of a real number is never negative.

$(x + 2)^2 = -1$ has no real-number solutions.

Note: $x = \dfrac{1 \pm \sqrt{3}}{2}$ means $x = \dfrac{1 + \sqrt{3}}{2}$ or $\dfrac{1 - \sqrt{3}}{2}$.

16-3. Solving Quadratic Equations by Completing the Square

A. Complete the square for a binomial with form $x^2 + bx$.

Recall: $x^2 + 6x + 9$ is a perfect square trinomial because $x^2 + 6x + 9 = (x + 3)^2$.

Note: In $x^2 + 6x + 9$, the square of one-half 6 is equal to 9: $\left(\dfrac{6}{2}\right)^2 = (3)^2 = 9$

The previous note is true for any perfect square trinomial with form $x^2 + bx + c$ as stated in the following rule:

Perfect Square Trinomial Rule for $x^2 + bx + c$

If $x^2 + bx + c$ is a perfect square trinomial, then the square of one-half b is equal to c:

$$c = \left(\frac{1}{2}b\right)^2 \text{ or } \left(\frac{b}{2}\right)^2.$$

To make a binomial with form $x^2 + bx$ a perfect square trinomial with form $x^2 + bx + c$, you **complete the square** for $x^2 + bx$ by adding $c = \left(\frac{1}{2}b\right)^2$ or $\left(\frac{b}{2}\right)^2$ to it.

EXAMPLE 16-8: Complete the square for $x^2 + 3x$.

Solution: To complete the square for $x^2 + bx$, you add $\left(\frac{b}{2}\right)^2$ to it.

$$x^2 + 3x + \left(\frac{b}{2}\right)^2 = x^2 + 3x + \left(\frac{3}{2}\right)^2 \qquad \textit{Think: } \text{In } x^2 + 3x, b = 3.$$

$$= x^2 + 3x + \frac{3^2}{2^2} \qquad \text{Simplify.}$$

$$= x^2 + 3x + \frac{9}{4} \longleftarrow \text{proposed solution}$$

Check: Is $x^2 + 3x + \frac{9}{4}$ a perfect square trinomial?   Yes:

$$x^2 + 3x + \frac{9}{4} = \left(x + \frac{3}{2}\right)\left(x + \frac{3}{2}\right) \qquad \text{Factor [see Example 9-39].}$$

$$= \left(x + \frac{3}{2}\right)^2 \longleftarrow x^2 + 3x + \frac{9}{4} \text{ checks}$$

Note: The completed square of $x^2 + 3x$ is $x^2 + 3x + \frac{9}{4}$ because $x^2 + 3x + \frac{9}{4}$ is a perfect square trinomial: $x^2 + 3x + \frac{9}{4} = (x + \frac{3}{2})^2$

B. Solve quadratic equations by completing the square.

Every quadratic equation can be solved using the following method:

Complete-the-Square Method

1. Write the given quadratic equation in $Ax^2 + Bx + C = 0$ form.
2. If $A \neq 1$ in $Ax^2 + Bx + C = 0$ from Step 1, then divide both members by A to get $x^2 + bx + c = 0$ form [$a = 1$].
3. Isolate $x^2 + bx$ in one member of the equation from Step 2.
4. Complete the square for $x^2 + bx$ by adding $\left(\frac{1}{2}b\right)^2$ or $\left(\frac{b}{2}\right)^2$ to both members of the equation from Step 3.
5. Rename the equation from Step 4 to get $(x + h)^2 = k$ form.
6. Solve $(x + h)^2 = k$ from Step 5 for x using the Square Root Rule to find the proposed solutions of the original quadratic equation.
7. Check by substituting each proposed solution from Step 6 in the original equation to see if you get a true number sentence.

EXAMPLE 16-9: Solve using the complete-the-square method:
(a) $4x^2 = 16x - 14$ (b) $y^2 + 2y + 3 = 0$

Solution:
(a) $\qquad\qquad\qquad 4x^2 = 16x - 14$ ⟵ given quadratic equation

$\qquad\qquad 4x^2 - 16x + 14 = 0$ ⟵ standard form

$\qquad\dfrac{4x^2 - 16x + 14}{4} = \dfrac{0}{4}$ $\qquad$ Divide both members by $a = 4$ to get $x^2 + bx + c = 0$ form.

$\qquad\qquad x^2 - 4x + \dfrac{7}{2} = 0$ ⟵ $x^2 + bx + c = 0$ form

$\qquad\qquad\quad x^2 - 4x = -\dfrac{7}{2}$ $\qquad$ Isolate $x^2 + bx$ in one member.

$\qquad x^2 - 4x + \left(\dfrac{-4}{2}\right)^2 = -\dfrac{7}{2} + \left(\dfrac{-4}{2}\right)^2$ $\qquad$ Complete the square for $x^2 + bx$ by adding $\left(\dfrac{b}{2}\right)^2$ to both members [see the following *Caution*].

$\qquad\quad x^2 - 4x + 4 = -\dfrac{7}{2} + 4$ $\qquad$ Rename to get $(x + h)^2 = k$ form.

$\qquad\qquad\quad (x - 2)^2 = \dfrac{1}{2}$ ⟵ $(x + h)^2 = k$ form

$\qquad\qquad\quad x - 2 = \pm\sqrt{\dfrac{1}{2}}$ $\qquad$ Solve using the square root rule.

$\qquad\qquad\quad x - 2 = \pm\dfrac{\sqrt{2}}{2}$

$\qquad\qquad\qquad x = 2 \pm \dfrac{\sqrt{2}}{2}$ or $\dfrac{4 \pm \sqrt{2}}{2}$ ⟵ proposed solutions [see the following *Note.*]

Check: $\qquad\qquad\qquad 4x^2 = 16x - 14$

$\qquad\qquad 4\left(2 + \dfrac{\sqrt{2}}{2}\right)^2 \quad\Big|\quad 16\left(2 + \dfrac{\sqrt{2}}{2}\right) - 14$

$\qquad\qquad 4\left(4 + 2\sqrt{2} + \dfrac{1}{2}\right) \quad\Big|\quad 32 + 8\sqrt{2} - 14$

$\qquad\qquad 16 + 8\sqrt{2} + 2 \quad\Big|\quad 18 + 8\sqrt{2}$

$\qquad\qquad 18 + 8\sqrt{2} \quad\Big|\quad 18 + 8\sqrt{2}$ ⟵ $x = 2 + \dfrac{\sqrt{2}}{2}$ checks $\left[\text{check } x = 2 - \dfrac{\sqrt{2}}{2} \text{ in the same manner}\right]$

(b) $\qquad y^2 + 2y + 3 = 0$

$\qquad\qquad y^2 + 2y = -3$

$\qquad y^2 + 2y + \left(\dfrac{2}{2}\right)^2 = -3 + \left(\dfrac{2}{2}\right)^2$

$\qquad\quad y^2 + 2y + 1 = -3 + 1$

$\qquad\qquad (y + 1)^2 = -2$ $\qquad$ Stop! The square of a real number is never negative.

$y^2 + 2y + 3 = 0$ has no real-number solutions because $y^2 + 2y + 3 = 0$ and $(y + 1)^2 = -2$ have the same solutions.

Note: In Example 16-9 (a), $4x^2 = 16x - 14$ and $(x - 2)^2 = \frac{1}{2}$ have the same solutions:

$x = 2 \pm \dfrac{\sqrt{2}}{2}$ or $\dfrac{4 \pm \sqrt{2}}{2}$

Caution: To complete the square in Example 16-9(a) and at the same time not change the solutions of the original equation, you must add $\left(\dfrac{-4}{2}\right)^2$ to both members of $x^2 - 4x = -\dfrac{7}{2}$.

16-4. Solving Quadratic Equations Using the Quadratic Formula

A. Solve quadratic equations using the quadratic formula.

The solutions of a quadratic equation in $ax^2 + bx + c = 0$ form are given by the

$$\textbf{quadratic formula: } x = \frac{-b \pm \sqrt{b^2 - 4ac}}{2a}.$$

Every quadratic equation can be solved using the following method:

Quadratic-Formula Method

1. Write the given quadratic equation in simplest form $ax^2 + bx + c = 0$.
2. Identify a, b, and c in $ax^2 + bx + c = 0$ from Step 1.
3. Evaluate the quadratic formula $x = \dfrac{-b \pm \sqrt{b^2 - 4ac}}{2a}$ for a, b, and c from Step 2 to find the proposed solutions of the original quadratic equation.
4. Check by substituting each proposed solution from Step 3 in the original equation to see if you get a true number sentence.

EXAMPLE 16-10: Solve using the quadratic-formula method: (a) $4x^2 = 16x - 14$ (b) $y^2 + 2y + 3 = 0$.

Solution:

(a)
$$4x^2 = 16x - 14 \longleftarrow \text{given quadratic equation}$$

$$4x^2 - 16x + 14 = 0 \longleftarrow \text{standard form [see Example 16-2]}$$

$$\frac{4x^2 - 16x + 14}{2} = \frac{0}{2} \quad \text{Divide both members by the positive GCF 2 [see the following } \textit{Caution]}.$$

$$2x^2 - 8x + 7 = 0 \longleftarrow \text{simplest form [see Example 16-3]}$$

In $2x^2 - 8x + 7 = 0$: $a = 2, b = -8, c = 7.$ Identify a, b, and c.

$$x = \frac{-b \pm \sqrt{b^2 - 4ac}}{2a} \longleftarrow \text{quadratic formula}$$

$$x = \frac{-(-8) \pm \sqrt{(-8)^2 - 4(2)(7)}}{2(2)} \quad \text{Evaluate for } a = 2, b = -8, \text{ and } c = 7 \text{ to find the proposed solutions.}$$

$$x = \frac{8 \pm \sqrt{64 - 56}}{4}$$

$$x = \frac{8 \pm \sqrt{8}}{4}$$

$$x = \frac{8 \pm 2\sqrt{2}}{4} \quad \text{Simplify the radical when possible [see Section 15-1].}$$

$$x = \frac{2(4 \pm \sqrt{2})}{2(2)} \quad \text{Eliminate common factors when possible.}$$

$$= \frac{4 \pm \sqrt{2}}{2} \text{ or } 2 \pm \frac{\sqrt{2}}{2} \longleftarrow \text{same solutions as found in Example 16-9(a) [check as before]}$$

(b) In $y^2 + 2y + 3 = 0$: $a = 1$, $b = 2$, $c = 3$.   Identify a, b, and c.

$$y = \frac{-b \pm \sqrt{b^2 - 4ac}}{2a} \longleftarrow \text{quadratic formula}$$

$$y = \frac{-(2) \pm \sqrt{(2)^2 - 4(1)(3)}}{2(1)} \qquad \text{Evaluate for } a = 1, b = 2, \text{ and } c = 3.$$

$$y = \frac{-2 \pm \sqrt{4 - 12}}{2}$$

$$y = \frac{-2 \pm \sqrt{-8}}{2} \qquad \text{Stop!} \qquad \sqrt{-8} \text{ is not a real number [see the following } \textit{Note}].$$

$y^2 + 2y + 3 = 0$ has no real-number solutions. $\longleftarrow$ same result as found in Example 16-9(**b**)

Note: When the **discriminant** $b^2 - 4ac$ is negative in $\dfrac{-b \pm \sqrt{b^2 - 4ac}}{2a}$, the given quadratic equation has no real-number solutions [see Example 16-10(**b**)].

Caution: To find the solutions of $ax^2 + bx + c = 0$ when the positive GCF of a, b, and c is not 1, you should divide both members by the positive GCF before substituting into the quadratic formula to avoid computing with such large numbers.

B. Approximate solutions that contain radicals.

To **approximate a solution that contains a radical,** you can use Appendix Table 2 or a calculator.

EXAMPLE 16-11: Approximate the solutions of Examples 16-9(a) and 16-10(a): $\dfrac{4 \pm \sqrt{2}}{2}$

Solution: $\dfrac{4 \pm \sqrt{2}}{2} = \dfrac{4 + \sqrt{2}}{2}$ or $\dfrac{4 - \sqrt{2}}{2} \longleftarrow$ exact solutions

$\approx \dfrac{4 + 1.414}{2}$ or $\dfrac{4 - 1.414}{2}$ Approximate the radical using Appendix Table 2 or a calculator.

$= \dfrac{5.414}{2}$ or $\dfrac{2.586}{2}$ Simplify when possible.

$= 2.707$ or 1.293 $\longleftarrow$ approximate solutions in simplest form

16-5. Solving Quadratic Equations Using the Easiest Method

Every quadratic equation can be identified as one of the following three types:

Type 1: $ax^2 + bx + c = 0$ $[b \neq 0$ and $c \neq 0]$
Type 2: $ax^2 + bx = 0$ $[b \neq 0]$
Type 3: $ax^2 + c = 0$

A. Solve $ax^2 + bx + c = 0$ $[b \neq 0$ and $c \neq 0]$ using the easiest method when $ax^2 + bx + c$ does not factor.

Recall: Every quadratic equation can be solved using either the quadratic-formula method or the complete-the-square method.

To solve a quadratic equation with form $ax^2 + bx + c = 0$ $[b \neq 0$ and $c \neq 0]$ when $ax^2 + bx + c$ does not factor as the product of two binomials, you should find it easier to use the quadratic-formula method instead of the complete-the-square method.

EXAMPLE 16-12: Show that the quadratic-formula method is easier than the complete-the-square method to solve: $5x^2 - 3x - 4 = 0$

Solution:

Quadratic-Formula Method	**Complete-the-Square Method**

Quadratic-Formula Method

In $5x^2 - 3x - 4 = 0$, $a = 5$, $b = -3$, and $c = -4$.

$$x = \frac{-b \pm \sqrt{b^2 - 4ac}}{2a}$$

$$x = \frac{-(-3) \pm \sqrt{(-3)^2 - 4(5)(-4)}}{2(5)}$$

$$x = \frac{3 \pm \sqrt{89}}{10}$$

Complete-the-Square Method

$$5x^2 - 3x - 4 = 0$$

$$x^2 - \frac{3}{5}x - \frac{4}{5} = 0$$

$$x^2 - \frac{3}{5}x = \frac{4}{5}$$

$$x^2 - \frac{3}{5}x + \left[\frac{1}{2}\left(-\frac{3}{5}\right)\right]^2 = \frac{4}{5} + \left[\frac{1}{2}\left(-\frac{3}{5}\right)\right]^2$$

$$x^2 - \frac{3}{5}x + \frac{9}{100} = \frac{4}{5} + \frac{9}{100}$$

$$\left(x - \frac{3}{10}\right)^2 = \frac{89}{100}$$

$$x - \frac{3}{10} = \pm\sqrt{\frac{89}{100}}$$

$$x - \frac{3}{10} = \frac{\pm\sqrt{89}}{10}$$

$$x = \frac{3}{10} \pm \frac{\sqrt{89}}{10}$$

$$x = \frac{3 \pm \sqrt{89}}{10}$$

B. Solve $ax^2 + bx + c = 0$ [$b \neq 0$ and $c \neq 0$] using the easiest method when $ax^2 + bx + c$ does factor.

To solve a quadratic equation with form $ax^2 + bx + c = 0$ [$b \neq 0$ and $c \neq 0$] when $ax^2 + bx + c$ does factor as the product of two binomials, you should find it easier to use the zero-product property method instead of the quadratic-formula method.

EXAMPLE 16-13: Show that the zero-product property method is easier than the quadratic-formula method to solve: $5x^2 - 2x - 3 = 0$

Solution:

Zero-Product Property Method

$$5x^2 - 2x - 3 = 0$$

$$(5x + 3)(x - 1) = 0$$

$$5x + 3 = 0 \quad \text{or} \quad x - 1 = 0$$

$$5x = -3 \quad \text{or} \quad x = 1$$

$$x = -\frac{3}{5} \quad \text{or} \quad x = 1$$

Quadratic-Formula Method

In $5x^2 - 2x - 3 = 0$, $a = 5$, $b = -2$, and $c = -3$.

$$x = \frac{-b \pm \sqrt{b^2 - 4ac}}{2a}$$

$$x = \frac{-(-2) \pm \sqrt{(-2)^2 - 4(5)(-3)}}{2(5)}$$

$$x = \frac{+2 \pm \sqrt{64}}{10}$$

$$x = \frac{2 \pm 8}{10}$$

$$x = \frac{2 + 8}{10} \quad \text{or} \quad \frac{2 - 8}{10}$$

$$x = 1 \quad \text{or} \quad -\frac{3}{5}$$

C. Solve $ax^2 + bx = 0 \; [b \neq 0]$ using the easiest method.

To solve a quadratic equation with form $ax^2 + bx = 0 \, [b \neq 0]$, you should find it easier to use the zero-product property method instead of the quadratic-formula method.

EXAMPLE 16-14: Show that the zero-product property method is easier than the quadratic-formula method to solve: $2x^2 + 3x = 0$

Solution:

<table>
<tr><th>Zero-Product Property Method</th><th>Quadratic-Formula Method</th></tr>
<tr><td>

$2x^2 + 3x = 0$

$x(2x + 3) = 0$

$x = 0$ or $2x + 3 = 0$

$x = 0$ or $\quad 2x = -3$

$x = 0$ or $\quad x = -\dfrac{3}{2}$

</td><td>

In $2x^2 + 3x = 0$, $a = 2$, $b = 3$, and $c = 0$.

$x = \dfrac{-b \pm \sqrt{b^2 - 4ac}}{2a}$

$x = \dfrac{-(3) \pm \sqrt{(3)^2 - 4(2)(0)}}{2(2)}$

$x = \dfrac{-3 \pm \sqrt{9}}{4}$

$x = \dfrac{-3 \pm 3}{4}$

$x = \dfrac{-3 + 3}{4}$ or $\dfrac{-3 - 3}{4}$

$x = 0 \qquad$ or $-\dfrac{3}{2}$

</td></tr>
</table>

Note: The two solutions of $ax^2 + bx = 0$ are $x = 0$ and $x = -\dfrac{b}{a}$.

D. Solve $ax^2 + c = 0$ using the easiest method.

To solve a quadratic equation with form $ax^2 + c = 0$, you should find it easier to use the square-root rule method rather than the quadratic-formula method.

EXAMPLE 16-15: Show that the square-root rule method is easier than the quadratic-formula method to solve $3y^2 - 2 = 0$.

Solution:

<table>
<tr><th>Square-Root Rule Method</th><th>Quadratic-Formula Method</th></tr>
<tr><td>

$3y^2 - 2 = 0$

$3y^2 = 2$

$y^2 = \dfrac{2}{3}$

$y = \pm\sqrt{\dfrac{2}{3}}$

$y = \pm\dfrac{\sqrt{2}}{\sqrt{3}} \cdot \dfrac{\sqrt{3}}{\sqrt{3}}$

$y = \pm\dfrac{\sqrt{6}}{3}$

</td><td>

In $3y^2 - 2 = 0$, $a = 3$, $b = 0$, and $c = -2$.

$y = \dfrac{-b \pm \sqrt{b^2 - 4ac}}{2a}$

$y = \dfrac{-(0) \pm \sqrt{(0)^2 - 4(3)(-2)}}{2(3)}$

$y = \dfrac{0 \pm \sqrt{24}}{6}$

$y = \dfrac{\pm 2\sqrt{6}}{2(3)}$

$y = \pm\dfrac{\sqrt{6}}{3}$

</td></tr>
</table>

Note: The two solutions of $ax^2 + c = 0$ are $x = \pm\sqrt{\dfrac{-c}{a}}$.

RAISE YOUR GRADES

Can you ... ?

☑ identify quadratic equations

☑ write quadratic equations in standard form

☑ write quadratic equations in simplest form and then identify a, b, and c

☑ solve quadratic equations that factor using the zero-product property

☑ solve quadratic equations with form $x^2 = d$ using the Square Root Rule

☑ solve quadratic equations with form $ax^2 + h = k$ using the Square Root Rule

☑ solve quadratic equations with form $(x + h)^2 = k$ using the Square Root Rule

☑ complete the square for a binomial with form $x^2 + bx$

☑ solve quadratic equations by completing the square

☑ solve quadratic equations using the quadratic formula

☑ approximate solutions that contain radicals

☑ solve $ax^2 + bx + c = 0$ [$b \neq 0$ and $c \neq 0$] using the easiest method when $ax^2 + bx + c$ does not factor

☑ solve $ax^2 + bx + c = 0$ [$b \neq 0$ and $c \neq 0$] using the easiest method when $ax^2 + bx + c = 0$ does factor

☑ solve $ax^2 + bx = 0$ [$b \neq 0$] using the easiest method

☑ solve $ax^2 + c = 0$ using the easiest method

SUMMARY

1. If a, b, and c are real numbers [$a \neq 0$] and x is any variable, then any equation that can be written in the form $ax^2 + bx + c = 0$ is called a quadratic equation in one variable or just a quadratic equation by agreement.

2. If $a > 0$, then the quadratic equation $ax^2 + bx + c = 0$ is in standard form.

3. A quadratic equation in standard form $ax^2 + bx + c = 0$ [$a > 0$] is in simplest form if the greatest common factor [GCF] of a, b, and c is 1 or -1.

4. For a quadratic equation to be in simplest form, all five of the following conditions must be satisfied:
   (a) The right member is zero [0].
   (b) The left member is in descending powers [$ax^2 + bx + c$].
   (c) In $ax^2 + bx + c = 0$, a, b, and c are all integers.
   (d) In $ax^2 + bx + c = 0$, a is positive [$a > 0$].
   (e) In $ax^2 + bx + c = 0$, the GCF of a, b, and c is 1 or -1.

5. To solve a quadratic equation that factors in simplest form, you can use the Zero-Product Property Method:
   (a) Write the given quadratic equation in simplest form.
   (b) Factor the left member of the quadratic equation from Step (a).
   (c) Use the zero-product property to set both factors from Step (b) equal to zero.
   (d) Solve both linear equations from Step (c) to find the proposed solutions of the original quadratic equation.
   (e) Check by substituting each proposed solution from Step (d) in the original equation to see if you get a true number sentence.

6. To solve quadratic equations with form $x^2 = d$, you can use the Square Root Rule:
   If $x^2 = d$, then $x = \pm\sqrt{d}$.

7. $x = \pm\sqrt{d}$ means $x = \sqrt{d}$ or $x = -\sqrt{d}$.

8. To solve a quadratic equation with form $ax^2 + h = k$, you can use the Square-Root Rule Method:
   (a) Isolate x^2 in one member to get $x^2 = d$ form.
   (b) Solve the equation from Step (a) using the Square Root Rule.

(c) Check by substituting each proposed solution from Step (b) in the original equation to see if you get a true number sentence.

9. If a and c are both positive [or both negative], then $ax^2 + c = 0$ has no real-number solutions.

10. To solve quadratic equations with form $(x + h)^2 = k$, you
   (a) Use the Square Root Rule to get: $x + h = \pm\sqrt{k}$.
   (b) Simplify the radical $\sqrt{k}$ when possible.
   (c) Add the opposite of h to both members to find the proposed solutions of $(x + h)^2 = k$.
   (d) Check by substituting each proposed solution from Step (c) in $(x + h)^2 = k$ to see if you get a true number sentence.

11. Perfect Square Trinomial Rule for $x^2 + bx + c$:
   If $x^2 + bx + c$ is a perfect square trinomial, then the square of one-half b is equal to c:
   $$c = \left(\frac{1}{2}b\right)^2 \text{ or } \left(\frac{b}{2}\right)^2.$$

12. To make a binomial with form $x^2 + bx$ a perfect square trinomial with form $x^2 + bx + c$, you complete the square for $x^2 + bx$ by adding $c = \left(\frac{1}{2}b\right)^2$ or $\left(\frac{b}{2}\right)^2$ to it.

13. Every quadratic equation can be solved using the Complete-the-Square Method:
   (a) Write the given quadratic equation in $Ax^2 + Bx + C = 0$ form.
   (b) If $A \neq 1$ in $Ax^2 + Bx + C = 0$ from Step (a), then divide both members by A to get $x^2 + bx + c$ form $[a = 1]$.
   (c) Isolate $x^2 + bx$ in one member of the equation from Step (b).
   (d) Complete the square for $x^2 + bx$ by adding $\left(\frac{1}{2}b\right)^2$ or $\left(\frac{b}{2}\right)^2$ to both members of the equation from Step (c).
   (e) Rename the equation from Step (d) to get $(x + h)^2 = k$ form.
   (f) Solve $(x + h)^2 = k$ from Step (e) for x using the Square Root Rule to find the proposed solutions of the original quadratic equation.
   (g) Check by substituting each proposed solution from Step (f) in the original equation to see if you get a true number sentence.

14. The solutions of a quadratic equation in $ax^2 + bx + c = 0$ form are given by the quadratic formula: $x = \dfrac{-b \pm \sqrt{b^2 - 4ac}}{2a}$.

15. Every quadratic equation can be solved using the Quadratic-Formula Method:
   (a) Write the given quadratic equation in simplest form $ax^2 + bx + c = 0$.
   (b) Identify a, b, and c in $ax^2 + bx + c = 0$ from Step (a).
   (c) Evaluate the quadratic formula $x = \dfrac{-b \pm \sqrt{b^2 - 4ac}}{2a}$ for a, b, and c from Step (b) to find the proposed solutions of the original quadratic equation.
   (d) Check by substituting each proposed solution from Step (c) in the original equation to see if you get a true number sentence.

16. When the discriminant $b^2 - 4ac$ is negative in $\dfrac{-b \pm \sqrt{b^2 - 4ac}}{2a}$, the given quadratic equation has no real-number solutions.

17. To find the solutions of $ax^2 + bx + c = 0$ when the positive GCF of a, b, and c is not 1, you should divide both members by the positive GCF before substituting into the quadratic formula to avoid computing with such large numbers.

18. To approximate a solution that contains a radical, you can use Appendix Table 2 or a calculator.

19. Every quadratic equation can be identified as one of the following three types:
   Type 1: $ax^2 + bx + c = 0$ $[b \neq 0 \text{ and } c \neq 0]$
   Type 2: $ax^2 + bx = 0$ $[b \neq 0]$
   Type 3: $ax^2 + c = 0$

20. The easiest method to solve a quadratic equation with form:
   (a) $ax^2 + bx + c = 0$ $[b \neq 0 \text{ and } c \neq 0]$ when $ax^2 + bx + c$ does not factor as the product of two binomials is the quadratic-formula method.
   (b) $ax^2 + bx + c = 0$ $[b \neq 0 \text{ and } c \neq 0]$ when $ax^2 + bx + c$ does factor as the product of two binomials is the zero-product property method.

(c) $ax^2 + bx = 0 \ [b \neq 0]$ is the zero-product property method.

(d) $ax^2 + c = 0$ is the square-root rule method.

21. The two solutions of $ax^2 + bx = 0$ are $x = 0$ and $x = -\dfrac{b}{a}$.

22. The two solutions of $ax^2 + c = 0$ are $x = \pm\sqrt{\dfrac{-c}{a}}$.

SOLVED PROBLEMS

PROBLEM 16-1 Write each quadratic equation in standard form: (a) $2 - x^2 = -x$
(b) $3 = 2y^2$

Solution: Recall that if $a > 0$, then the quadratic equation $ax^2 + bx + c = 0$ is in standard form [see Example 16-2]:

(a)
$$2 - x^2 = -x$$
$$-x^2 + x + 2 = 0$$
$$x^2 - x - 2 = 0 \longleftarrow \text{standard form}$$

(b)
$$3 = 2y^2$$
$$0 = 2y^2 - 3$$
$$2y^2 - 3 = 0 \longleftarrow \text{standard form}$$

PROBLEM 16-2 Write each quadratic equation in simplest form and then identify a, b, and c [see Example 16-3]: (a) $2x^2 + 4x - 6 = 0$ (b) $12 = 18y^2$ (c) $2w^2 + \dfrac{3w}{5} = \dfrac{1}{3}$

(d) $3m^2 = \dfrac{3 - 9m}{2}$ (e) $\dfrac{2}{n^2} + 2 = \dfrac{5}{n}$ (f) $z = \sqrt{z} - 1$

Solution: Recall that a quadratic equation in standard form $ax^2 + bx + c = 0 \ [a > 0]$ is in simplest form if a, b, and c are all integers and the greatest common factor [GCF] of a, b, and c is 1 or -1 [see Example 16-3]:

(a) The GCF of 2, 4, and -6 is 2 or -2.
$$\frac{2x^2 + 4x - 6}{2} = \frac{0}{2}$$
$$x^2 + 2x - 3 = 0 \longleftarrow \text{simplest form}$$
In $x^2 + 2x - 3 = 0$: $a = 1$, $b = 2$, $c = -3$.

(b)
$$12 = 18y^2$$
$$0 = 18y^2 - 12$$
$$18y^2 - 12 = 0 \longleftarrow \text{standard form}$$
The GCF of -12 and 18 is 6 or -6.
$$\frac{18y^2 - 12}{6} = \frac{0}{6}$$
$$3y^2 - 2 = 0 \longleftarrow \text{simplest form}$$
In $3y^2 - 2 = 0$: $a = 3$, $b = 0$, $c = -2$.

(c) The LCD of $\dfrac{3w}{5}$ and $\dfrac{1}{3}$ is 15.
$$15\left(2w^2 + \frac{3w}{5}\right) = 15 \cdot \frac{1}{3}$$
$$15(2w^2) + 15 \cdot \frac{3w}{5} = 15 \cdot \frac{1}{3}$$
$$30w^2 + 9w = 5 \longleftarrow \text{fractions cleared}$$
$$30w^2 + 9w - 5 = 0 \longleftarrow \text{simplest form}$$
In $30w^2 + 9w - 5 = 0$: $a = 30$, $b = 9$,
$$c = -5.$$

(d) The LCD of $\dfrac{3m^2}{1}$ and $\dfrac{3 - 9m}{2}$ is 2.
$$2(3m^2) = 2 \cdot \frac{3 - 9m}{2}$$
$$6m^2 = 3 - 9m \longleftarrow \text{fractions cleared}$$
$$6m^2 + 9m - 3 = 0 \longleftarrow \text{standard form}$$
The GCF of 6, 9, and -3 is 3 or -3.
$$\frac{6m^2 + 9m - 3}{3} = \frac{0}{3}$$
$$2m^2 + 3m - 1 = 0 \longleftarrow \text{simplest form}$$
In $2m^2 + 3m - 1 = 0$: $a = 2$, $b = 3$,
$$c = -1.$$

(e) The LCD of $\dfrac{2}{n^2}$ and $\dfrac{5}{n}$ is n^2.

$$n^2\left(\dfrac{2}{n^2} + 2\right) = n^2 \cdot \dfrac{5}{n}$$

$$n^2 \cdot \dfrac{2}{n^2} + n^2(2) = n^2 \cdot \dfrac{5}{n}$$

$$2 + 2n^2 = 5n \quad \longleftarrow \text{ fractions cleared}$$

$$2n^2 - 5n + 2 = 0 \quad \longleftarrow \text{ simplest form}$$

In $2n^2 - 5n + 2 = 0$: $a = 2$, $b = -5$, $c = 2$.

(f)

$$z = \sqrt{z} - 1 \qquad \text{Clear radicals.}$$

$$z + 1 = \sqrt{z}$$

$$(z + 1)^2 = (\sqrt{z})^2$$

$$(z + 1)^2 = z \quad \longleftarrow \text{ radical cleared}$$

$$z^2 + z + 1 = 0 \quad \longleftarrow \text{ simplest form}$$

In $z^2 + z + 1 = 0$: $a = 1$, $b = 1$, $c = 1$.

PROBLEM 16-3   Solve using the zero-product property method: **(a)** $3x^2 = 0$ **(b)** $6y^2 = 9y$
(c) $w^2 - 1 = 0$ **(d)** $4m^2 + 14m - 30 = 0$

Solution: Recall that to solve a quadratic equation that factors in simplest form, you can use the following method:

Zero-Product Property Method:

1. Write the given quadratic equation in simplest form.
2. Factor the left member of the quadratic equation from Step 1.
3. Use the zero-product property to set both factors from Step 2 equal to zero.
4. Solve both linear equations from Step 3 to find the proposed solutions of the original quadratic equation.
5. Check by substituting each proposed solution from Step 4 in the original equation to see if you get a true number sentence [see Example 16-4]:

(a)

$$3x^2 = 0$$

$$\dfrac{3x^2}{3} = \dfrac{0}{3}$$

$$x^2 = 0 \quad \longleftarrow \text{ simplest form}$$

$$x(x) = 0 \qquad \text{Factor.}$$

$$x = 0 \text{ or } x = 0 \qquad \text{Use the zero-product property}$$

$$x = 0 \quad \longleftarrow \text{ solution [check as before]}$$

(b)

$$6y^2 = 9y$$

$$6y^2 - 9y = 0$$

$$\dfrac{6y^2 - 9y}{3} = \dfrac{0}{3}$$

$$2y^2 - 3y = 0$$

$$y(2y - 3) = 0$$

$$y = 0 \text{ or } 2y - 3 = 0$$

$$y = 0 \text{ or } \qquad 2y = 3$$

$$y = 0 \text{ or } \qquad y = \dfrac{3}{2}$$

(c)

$$w^2 - 1 = 0$$

$$(w + 1)(w - 1) = 0$$

$$w + 1 = 0 \quad \text{ or } w - 1 = 0$$

$$w = -1 \text{ or } \qquad w = 1$$

(d)

$$\dfrac{4m^2 + 14m - 30}{2} = \dfrac{0}{2}$$

$$2m^2 + 7m - 15 = 0$$

$$(2m - 3)(m + 5) = 0$$

$$2m - 3 = 0 \text{ or } m + 5 = 0$$

$$2m = 3 \text{ or } \qquad m = -5$$

$$m = \dfrac{3}{2} \text{ or } \qquad m = -5$$

PROBLEM 16-4   Solve using the square root rule: **(a)** $x^2 = 9$ **(b)** $y^2 = 12$ **(c)** $w^2 = \dfrac{5}{2}$
(d) $m^2 = -2$

Solution: Recall that to solve quadratic equations with form $x^2 = d$, you can use the following rule:

Square Root Rule:

If $x^2 = d$, then $x = \pm\sqrt{d}$ where $\pm\sqrt{d}$ means $\sqrt{d}$ or $-\sqrt{d}$ [see Example 16-5]:

(a) $x^2 = 9 \longleftarrow x^2 = d$ form

    $x = \pm\sqrt{9}$ Use the Square Root Rule.

    $x = \pm 3 \longleftarrow$ solutions [check as before]

(b) $y^2 = 12$

    $y = \pm\sqrt{12}$

    $y = \pm\sqrt{4}\sqrt{3}$ Simplify the radical.

    $y = \pm 2\sqrt{3}$

(c) $w^2 = \dfrac{5}{2}$

    $w = \pm\sqrt{\dfrac{5}{2}}$

    $w = \pm\dfrac{\sqrt{5}}{\sqrt{2}}$ Simplify.

    $w = \pm\dfrac{\sqrt{5}}{\sqrt{2}}\cdot\dfrac{\sqrt{2}}{\sqrt{2}}$ Rationalize the denominator.

    $w = \pm\dfrac{\sqrt{10}}{2}$

(d) $m^2 = -2$ has no real-number solutions because the square of a real number is never negative.

PROBLEM 16-5   Solve using the square-root rule method:

(a) $x^2 - 18 = 0$ (b) $3y^2 + 2 = 4$ (c) $2m^2 + 5 = 1$

Solution: Recall that to solve a quadratic equation with form $ax^2 + h = k$, you can use the Square-Root Rule Method:

1. Isolate x^2 in one member to get $x^2 = d$ form.
2. Solve the equation from Step 1 using the Square Root Rule.
3. Check by substituting each proposed solution from Step 2 in the original equation to see if you get a true number sentence [see Example 16-6]:

(a) $x^2 - 18 = 0 \longleftarrow ax^2 + h = k$ form

    $x^2 = 18 \longleftarrow x^2 = d$ form

    $x = \pm\sqrt{18}$ Use the Square Root Rule.

    $x = \pm\sqrt{9}\sqrt{2}$ Simplify the radical.

    $x = \pm 3\sqrt{2} \longleftarrow$ solutions [check as before]

(b) $3y^2 + 2 = 4$

    $3y^2 = 2$

    $y^2 = \dfrac{2}{3}$

    $y = \pm\sqrt{\dfrac{2}{3}}$

    $y = \pm\dfrac{\sqrt{2}}{\sqrt{3}}$ Simplify.

    $y = \pm\dfrac{\sqrt{2}}{\sqrt{3}}\cdot\dfrac{\sqrt{3}}{\sqrt{3}}$ Rationalize.

    $= \pm\dfrac{\sqrt{6}}{3}$

(c) $2m^2 + 5 = 1$

    $2m^2 = -4$

    $m^2 = -2$ Stop! The square of a real number can never be negative.

  $2m^2 + 5 = 1$ has no real-number solutions.

PROBLEM 16-6   Solve using the Square Root Rule: (a) $(x - 2)^2 = 9$ (b) $(y - 2)^2 = 3$
(c) $(w + \frac{3}{4})^2 = \frac{25}{16}$ (d) $(m + \frac{1}{2})^2 = \frac{5}{2}$ (e) $(n - \frac{2}{3})^2 = -\frac{8}{3}$

Solution: Recall that to solve equations with form $(x + h)^2 = k$, you

1. Use the Square Root Rule to get $x + h = \pm\sqrt{k}$.
2. Simplify the radical $\sqrt{k}$ when possible.
3. Add the opposite of h to both members to find the proposed solutions of $(x + h)^2 = k$.
4. Check by substituting each proposed solution from Step 3 in $(x + h)^2 = k$ to see if you get a true number sentence [see Example 16-7]:

(a) $(x - 2)^2 = 9$ ⟵ $(x + h)^2 = k$ form.

$\qquad x - 2 = \pm\sqrt{9}$ Use the Square Root Rule.

$\qquad x - 2 = \pm 3$ Simplify the radical when possible.

$\qquad\quad x = 2 \pm 3$ Add 2 [the opposite of -2] to both members.

$\qquad\quad x = 2 + 3 \text{ or } 2 - 3$ *Think:* 2 ± 3 means $2 + 3$ or $2 - 3$.

$\qquad\quad x = 5 \qquad \text{ or } -1$ ⟵ solutions [check as before]

(b) $(y - 2)^2 = 3$

$y - 2 = \pm\sqrt{3}$

$y = 2 \pm \sqrt{3}$

(c) $\left(w + \dfrac{3}{4}\right)^2 = \dfrac{25}{16}$

$w + \dfrac{3}{4} = \pm\sqrt{\dfrac{25}{16}}$

$w + \dfrac{3}{4} = \pm\dfrac{\sqrt{25}}{\sqrt{16}}$

$w + \dfrac{3}{4} = \pm\dfrac{5}{4}$

$w = -\dfrac{3}{4} \pm \dfrac{5}{4}$

$w = -\dfrac{3}{4} + \dfrac{5}{4} \text{ or } -\dfrac{3}{4} - \dfrac{5}{4}$

$w = \dfrac{1}{2} \qquad \text{ or } -2$

(d) $\left(m + \dfrac{1}{2}\right)^2 = \dfrac{5}{2}$

$m + \dfrac{1}{2} = \pm\sqrt{\dfrac{5}{2}}$

$m + \dfrac{1}{2} = \pm\dfrac{\sqrt{5}}{\sqrt{2}} \cdot \dfrac{\sqrt{2}}{\sqrt{2}}$

$m + \dfrac{1}{2} = \pm\dfrac{\sqrt{10}}{2}$

$m = -\dfrac{1}{2} \pm \dfrac{\sqrt{10}}{2}$

$m = \dfrac{-1 \pm \sqrt{10}}{2}$

(e) $\left(n - \dfrac{2}{3}\right)^2 = -\dfrac{8}{3}$ has no real-number solutions because the square of a real number can never be negative.

PROBLEM 16-7   Complete the square for: (a) $x^2 + 10x$ (b) $y^2 - 12y$ (c) $m^2 + m$

(d) $w^2 - \dfrac{1}{4}w$

Solution: Recall that to make a binomial with form $x^2 + bx$ a perfect square trinomial with form $x^2 + bx + c$, you complete the square for $x^2 + bx$ by adding $c = \left(\dfrac{1}{2}b\right)^2$ or $\left(\dfrac{b}{2}\right)^2$ to it [see Example 16-8]:

(a) $x^2 + 10x + \left(\dfrac{b}{2}\right)^2 = x^2 + 10x + \left(\dfrac{10}{2}\right)^2$ (b) $y^2 - 12y + \left(\dfrac{b}{2}\right)^2 = y^2 - 12y + \left(\dfrac{-12}{2}\right)^2$

$\qquad\qquad\qquad\qquad = x^2 + 10x + (5)^2$ $\qquad\qquad\qquad\qquad = y^2 - 12y + (-6)^2$

$\qquad\qquad\qquad\qquad = x^2 + 10x + 25$ $\qquad\qquad\qquad\qquad = y^2 - 12y + 36$

Check: $x^2 + 10x + 25 = (x + 5)(x + 5)$ *Check:* $y^2 - 12y + 36 = (y - 6)(y - 6)$

$\qquad\qquad\qquad\qquad = (x + 5)^2$ $\qquad\qquad\qquad\qquad = (y - 6)^2$

(c) $m^2 + m + \left(\dfrac{b}{2}\right)^2 = m^2 + 1m + \left(\dfrac{1}{2}\right)^2$

$$= m^2 + m + \dfrac{1^2}{2^2}$$

$$= m^2 + m + \dfrac{1}{4}$$

Check: $m^2 + m + \dfrac{1}{4} = \left(m + \dfrac{1}{2}\right)\left(m + \dfrac{1}{2}\right)$

$$= \left(m + \dfrac{1}{2}\right)^2$$

(d) $w^2 - \dfrac{1}{4}w + \left(\dfrac{1}{2}b\right)^2 = w^2 - \dfrac{1}{4}w + \left[\dfrac{1}{2}\left(-\dfrac{1}{4}\right)\right]^2$

$$= w^2 - \dfrac{1}{4}w + \left(\dfrac{-1}{8}\right)^2$$

$$= w^2 - \dfrac{1}{4}w + \dfrac{1}{64}$$

Check: $w^2 - \dfrac{1}{4}w + \dfrac{1}{64} = \left(w - \dfrac{1}{8}\right)\left(w - \dfrac{1}{8}\right)$

$$= \left(w - \dfrac{1}{8}\right)^2$$

PROBLEM 16-8 Solve using the complete-the-square method: **(a)** $x^2 + 10 = 7x$
(b) $-y^2 + 10y - 22 = 0$ **(c)** $8w^2 - 6w - 2 = 0$ **(d)** $2m + 2 = 3m^2$ **(e)** $n^2 + n + 1 = 0$

Solution: Recall that every quadratic equation can be solved using the following method:

Complete-the-Square Method:

1. Write the given quadratic equation in $Ax^2 + Bx + C = 0$ form.
2. If $A \neq 1$ in $Ax^2 + Bx + C = 0$ from Step 1, then divide both members by A to get $x^2 + bx + c = 0$ form $[a = 1]$.
3. Isolate $x^2 + bx$ in one member of the equation from Step 2.
4. Complete the square for $x^2 + bx$ by adding $\left(\dfrac{1}{2}b\right)^2$ or $\left(\dfrac{b}{2}\right)^2$ to both members of the equation from Step 3.
5. Rename the equation from Step 4 to get $(x + h)^2 = k$ form.
6. Solve $(x + h)^2 = k$ from Step 5 using the Square Root Rule to find the proposed solutions of the original quadratic equation.
7. Check by substituting each proposed solution from Step 6 in the original equation to see if you get a true number sentence [see Example 16-9]:

(a)

$x^2 + 10 = 7x$ ⟵ given equation

$x^2 - 7x + 10 = 0$ ⟵ $x^2 + bx + c$ form

$x^2 - 7x = -10$ Isolate $x^2 + bx$.

$x^2 - 7x + \left(\dfrac{-7}{2}\right)^2 = -10 + \left(\dfrac{-7}{2}\right)^2$ Complete the square.

$x^2 - 7x + \dfrac{49}{4} = -\dfrac{40}{4} + \dfrac{49}{4}$ Simplify.

$\left(x - \dfrac{7}{2}\right)^2 = \dfrac{9}{4}$ ⟵ $(x + h)^2 = k$ form

$x - \dfrac{7}{2} = \pm\sqrt{\dfrac{9}{4}}$ Use the Square Root Rule.

$x - \dfrac{7}{2} = \pm\dfrac{3}{2}$ Simplify.

$x = \dfrac{7}{2} \pm \dfrac{3}{2}$ Solve for x.

$x = \dfrac{7}{2} + \dfrac{3}{2}$ or $\dfrac{7}{2} - \dfrac{3}{2}$ Simplify.

$x = \quad$ 5 or 2 ⟵ solutions [check as before]

(b) $-y^2 + 10y - 22 = 0$

$y^2 - 10y + 22 = 0$

$y^2 - 10y = -22$

$y^2 - 10y + \left(\dfrac{-10}{2}\right)^2 = -22 + \left(\dfrac{-10}{2}\right)^2$

$y^2 - 10y + (-5)^2 = -22 + (-5)^2$

$y^2 - 10y + 25 = -22 + 25$

$(y - 5)^2 = 3$

$y - 5 = \pm\sqrt{3}$

$y = 5 \pm\sqrt{3}$

(c)
$$8w^2 - 6w - 2 = 0$$

$$\frac{8w^2 - 6w - 2}{8} = \frac{0}{8}$$

$$w^2 - \frac{3}{4}w - \frac{1}{4} = 0$$

$$w^2 - \frac{3}{4}w = \frac{1}{4}$$

$$w^2 - \frac{3}{4}w + \left[\frac{1}{2}\left(-\frac{3}{4}\right)\right]^2 = \frac{1}{4} + \left[\frac{1}{2}\left(-\frac{3}{4}\right)\right]^2$$

$$w^2 - \frac{3}{4}w + \left(-\frac{3}{8}\right)^2 = \frac{1}{4} + \left(-\frac{3}{8}\right)^2$$

$$w^2 - \frac{3}{4}w + \frac{9}{64} = \frac{16}{64} + \frac{9}{64}$$

$$\left(w - \frac{3}{8}\right)^2 = \frac{25}{64}$$

$$w - \frac{3}{8} = \pm\sqrt{\frac{25}{64}}$$

$$w - \frac{3}{8} = \pm\frac{\sqrt{25}}{\sqrt{64}}$$

$$w - \frac{3}{8} = \pm\frac{5}{8}$$

$$w = \frac{3}{8} \pm \frac{5}{8}$$

$$w = \frac{3}{8} + \frac{5}{8} \text{ or } \frac{3}{8} - \frac{5}{8}$$

$$w = \qquad 1 \text{ or } -\frac{1}{4}$$

(d)
$$2m + 2 = 3m^2$$

$$-3m^2 + 2m + 2 = 0$$

$$\frac{-3m^2 + 2m + 2}{-3} = \frac{0}{-3}$$

$$m^2 - \frac{2}{3}m - \frac{2}{3} = 0$$

$$m^2 - \frac{2}{3}m = \frac{2}{3}$$

$$m^2 - \frac{2}{3}m + \left[\frac{1}{2}\left(-\frac{2}{3}\right)\right]^2 = \frac{2}{3} + \left[\frac{1}{2}\left(-\frac{2}{3}\right)\right]^2$$

$$m^2 - \frac{2}{3}m + \left(-\frac{1}{3}\right)^2 = \frac{2}{3} + \left(-\frac{1}{3}\right)^2$$

$$m^2 - \frac{2}{3}m + \frac{1}{9} = \frac{6}{9} + \frac{1}{9}$$

$$\left(m - \frac{1}{3}\right)^2 = \frac{7}{9}$$

$$m - \frac{1}{3} = \pm\sqrt{\frac{7}{9}}$$

$$m - \frac{1}{3} = \pm\frac{\sqrt{7}}{3}$$

$$m = \frac{1}{3} \pm \frac{\sqrt{7}}{3} \text{ or } \frac{1 \pm \sqrt{7}}{3}$$

(e)
$$n^2 + n + 1 = 0$$

$$n^2 + n = -1$$

$$n^2 + 1n + \left(\frac{1}{2}\right)^2 = -1 + \left(\frac{1}{2}\right)^2$$

$$n^2 + n + \frac{1}{4} = -1 + \frac{1}{4}$$

$$\left(n + \frac{1}{2}\right)^2 = -\frac{3}{4} \qquad \text{Stop! The square of a real number is never negative.}$$

$n^2 + n + 1 = 0$ has no real-number solutions.

PROBLEM 16-9   Solve using the quadratic-formula method: **(a)** $x^2 + 10 = 7x$
(b) $-y^2 + 10y - 22 = 0$ **(c)** $8w^2 - 6w - 2 = 0$ **(d)** $2m + 2 = 3m^2$

(e) $n^2 + n + 1 = 0$ **(f)** $\frac{1}{3}x^2 = x + \frac{1}{4}$ **(g)** $\frac{y}{2} = \frac{2y^2 + 1}{5}$

Solution: Recall that every quadratic equation can be solved using the Quadratic-Formula Method:

1. Write the given quadratic equation in simplest form $ax^2 + bx + c = 0$.
2. Identify a, b, and c in $ax^2 + bx + c = 0$ from Step 1.
3. Evaluate the quadratic formula $x = \dfrac{-b \pm \sqrt{b^2 - 4ac}}{2a}$ for a, b, and c from Step 2 to find the proposed solutions of the original quadratic equation.
4. Check by substituting each proposed solution from Step 3 in the original equation to see if you get a true number sentence [see Example 16-10]:

(a) $\qquad x^2 + 10 = 7x$

$\qquad x^2 - 7x + 10 = 0 \longleftarrow$ simplest form

$\qquad$ In $x^2 - 7x + 10 = 0$: $a = 1, b = -7, c = 10.$ $\qquad$ Identify a, b, and c.

$\qquad x = \dfrac{-b \pm \sqrt{b^2 - 4ac}}{2a} \longleftarrow$ quadratic equation

$\qquad x = \dfrac{-(-7) \pm \sqrt{(-7)^2 - 4(1)(10)}}{2(1)} \qquad$ Evaluate for $a = 1, b = -7$, and $c = 10.$

$\qquad x = \dfrac{7 \pm \sqrt{9}}{2}$

$\qquad x = \dfrac{7 \pm 3}{2} \qquad\qquad\qquad$ Simplify.

$\qquad x = \dfrac{7 + 3}{2}$ or $\dfrac{7 - 3}{2}$

$\qquad x = 5 \qquad$ or $2 \longleftarrow$ solutions [check as before]

(b) $-y^2 + 10y - 22 = 0$

$\dfrac{-y^2 + 10y - 22}{-1} = \dfrac{0}{-1}$

$y^2 - 10y + 22 = 0$

In $y^2 - 10y + 22$: $a = 1, b = -10, c = 22.$

$y = \dfrac{-b \pm \sqrt{b^2 - 4ac}}{2a}$

$y = \dfrac{-(-10) \pm \sqrt{(-10)^2 - 4(1)(22)}}{2(1)}$

$y = \dfrac{10 \pm \sqrt{12}}{2}$

$y = \dfrac{2(5) \pm 2\sqrt{3}}{2}$

$y = \dfrac{2(5 \pm \sqrt{3})}{2}$

$y = 5 \pm \sqrt{3}$

(c) $8w^2 - 6w - 2 = 0$

$\dfrac{8w^2 - 6w - 2}{2} = \dfrac{0}{2}$

$4w^2 - 3w - 1 = 0$

In $4w^2 - 3w - 1$: $a = 4, b = -3, c = -1.$

$w = \dfrac{-b \pm \sqrt{b^2 - 4ac}}{2a}$

$w = \dfrac{-(-3) \pm \sqrt{(-3)^2 - 4(4)(-1)}}{2(4)}$

$w = \dfrac{3 \pm \sqrt{25}}{8}$

$w = \dfrac{3 \pm 5}{8}$

$w = \dfrac{3 + 5}{8}$ or $\dfrac{3 - 5}{8}$

$w = 1 \qquad$ or $-\dfrac{1}{4}$

(d) $2m + 2 = 3m^2$

$$0 = 3m^2 - 2m - 2$$

In $3m^2 - 2m - 2 = 0$: $a = 3, b = -2,$ $c = -2.$

$$m = \frac{-b \pm \sqrt{b^2 - 4ac}}{2a}$$

$$m = \frac{-(-2) \pm \sqrt{(-2)^2 - 4(3)(-2)}}{2(3)}$$

$$m = \frac{2 \pm \sqrt{28}}{6}$$

$$m = \frac{2 \pm 2\sqrt{7}}{6}$$

$$m = \frac{2(1 \pm \sqrt{7})}{2(3)}$$

$$m = \frac{1 \pm \sqrt{7}}{3}$$

(e) $n^2 + n + 1 = 0$

In $n^2 + n + 1 = 0$: $a = 1, b = 1, c = 1.$

$$n = \frac{-b \pm \sqrt{b^2 - 4ac}}{2a}$$

$$n = \frac{-(1) \pm \sqrt{(1)^2 - 4(1)(1)}}{2(1)}$$

$$n = \frac{-1 \pm \sqrt{-3}}{2} \qquad \text{Stop! } \sqrt{-3} \text{ is not a real number.}$$

$n^2 + n + 1 = 0$ has no real-number solutions.

(f) The LCD of $\frac{1}{3}$ and $\frac{1}{4}$ is 12.

$$12\left(\frac{1}{3}x^2\right) = 12\left(x + \frac{1}{4}\right)$$

$$12 \cdot \frac{1}{3}x^2 = 12(x) + 12 \cdot \frac{1}{4}$$

$$4x^2 = 12x + 3$$

$$4x^2 - 12x - 3 = 0 \longleftarrow \text{ simplest form}$$

In $4x^2 - 12x - 3 = 0$: $a = 4, b = -12,$ $c = -3$

$$x = \frac{-b \pm \sqrt{b^2 - 4ac}}{2a}$$

$$x = \frac{-(-12) \pm \sqrt{(-12)^2 - 4(4)(-3)}}{2(4)}$$

$$x = \frac{12 \pm \sqrt{192}}{8}$$

$$x = \frac{12 \pm 8\sqrt{3}}{8}$$

$$x = \frac{4(3) \pm 4(2)\sqrt{3}}{8}$$

$$x = \frac{4(3 \pm 2\sqrt{3})}{4(2)}$$

$$x = \frac{3 \pm 2\sqrt{3}}{2}$$

(g) The LCD of $\frac{y}{2}$ and $\frac{2y^2 + 1}{5}$ is 10.

$$10 \cdot \frac{y}{2} = 10 \cdot \frac{2y^2 + 1}{5}$$

$$5y = 2(2y^2 + 1)$$

$$5y = 4y^2 + 2$$

$$0 = 4y^2 - 5y + 2$$

In $4y^2 - 5y + 2 = 0$: $a = 4, b = -5,$ $c = 2.$

$$y = \frac{-b \pm \sqrt{b^2 - 4ac}}{2a}$$

$$y = \frac{-(-5) \pm \sqrt{(-5)^2 - 4(4)(2)}}{2(4)}$$

$$y = \frac{5 \pm \sqrt{-7}}{8} \qquad \text{Stop! } \sqrt{-7} \text{ is not a real number.}$$

$\frac{y}{2} = \frac{2y^2 + 1}{5}$ has no real-number solutions.

PROBLEM 16-10 Approximate the following solutions: **(a)** $\dfrac{1 \pm \sqrt{7}}{3}$ **(b)** $\pm \sqrt{5}$ **(c)** $5 \pm \sqrt{3}$

Solution: Recall that to approximate a solution that contains a radical, you can use Appendix Table 2 or a calculator [see Example 16-11]:

(a) $\dfrac{1 \pm \sqrt{7}}{3} = \dfrac{1 + \sqrt{7}}{3}$ or $\dfrac{1 - \sqrt{7}}{3}$ ← exact solutions

$\approx \dfrac{1 + 2.646}{3}$ or $\dfrac{1 - 2.646}{3}$ Approximate the radical using Appendix Table 2 or a calculator.

$= \dfrac{3.646}{3}$ or $\dfrac{-1.646}{3}$ Simplify when possible.

≈ 1.215 or -0.549 ← approximate solutions [round to the nearest thousandth when necessary]

(b) $\pm \sqrt{5} = \sqrt{5}$ or $-\sqrt{5}$

≈ 2.236 or -2.236

(c) $5 \pm \sqrt{3}$ $= 5 + \sqrt{3}$ or $5 - \sqrt{3}$

$\approx 5 + 1.732$ or $5 - 1.732$

$= 6.732$ or 3.268

PROBLEM 16-11   Solve using the easiest method: **(a)** $2y^2 - 3 = 0$ **(b)** $2w^2 + 5w = 0$
(c) $2m^2 + 5m - 3 = 0$ **(d)** $n^2 + 5n - 3 = 0$

Solution: Recall that the easiest method to solve a quadratic equation with form

1. $ax^2 + bx + c = 0$ [$b \neq 0$ and $c \neq 0$] when $ax^2 + bx + c$ does not factor as the product of two binomials is the quadratic-formula method [see Example 16-12]:
2. $ax^2 + bx + c = 0$ [$b \neq 0$ and $c \neq 0$] when $ax^2 + bx + c$ does factor as the product of two binomials is the zero-product property method [see Example 16-13]:
3. $ax^2 + bx = 0$ [$b \neq 0$] is the zero-product property method [see Example 16-14]:
4. $ax^2 + c = 0$ is the square-root rule method [see Example 16-15]:

(a) $2y^2 - 3 = 0$ ← $ax^2 + c = 0$ form

$2y^2 = 3$ Use the square-root rule method.

$y^2 = \dfrac{3}{2}$

$y = \pm \sqrt{\dfrac{3}{2}}$

$y = \pm \dfrac{\sqrt{3}}{\sqrt{2}} \cdot \dfrac{\sqrt{2}}{\sqrt{2}}$

$y = \pm \dfrac{\sqrt{6}}{2}$

(b) $2w^2 + 5w = 0$ ← $ax^2 + bx = 0$ form

$w(2w + 5) = 0$ Use the zero-product property method.

$w = 0$ or $2w + 5 = 0$

$w = 0$ or $2w = -5$

$w = 0$ or $w = -\dfrac{5}{2}$

(c) $2m^2 + 5m - 3 = 0$ [$ax^2 + bx + c = 0$ form]
Use the zero-product property method because $2m^2 + 5m - 3$ does factor.

$(2m - 1)(m + 3) = 0$

$2m - 1 = 0$ or $m + 3 = 0$

$2m = 1$ or $m = -3$

$m = \dfrac{1}{2}$ or $m = -3$

(d) $n^2 + 5n - 3 = 0$ [$ax^2 + bx + c = 0$ form]
Use the quadratic-formula method because $n^2 + 5n - 3$ does not factor.

In $n^2 + 5n - 3 = 0$: $a = 1$, $b = 5$, $c = -3$.

$n = \dfrac{-b \pm \sqrt{b^2 - 4ac}}{2a}$

$n = \dfrac{-(5) \pm \sqrt{(5)^2 - 4(1)(-3)}}{2(1)}$

$n = \dfrac{-5 \pm \sqrt{37}}{2}$

Supplementary Exercises

PROBLEM 16-12   Write each equation in standard form: **(a)** $3 + x^2 = 0$ **(b)** $2y + 5y^2 = 0$
(c) $2w^2 = 5$ **(d)** $3 = m + 5m^2$ **(e)** $-2n^2 - n = 0$ **(f)** $-5z^2 + 4 = -3z$

PROBLEM 16-13   Write each equation in simplest form and then identify a, b, and c:

(a) $3x + 15 + 12x^2 = 0$ **(b)** $12y - 30y^2 = 0$ **(c)** $w^2 + \dfrac{1}{2} = \dfrac{3}{2}w$ **(d)** $2m^2 = \dfrac{10m - 4}{3}$

(e) $\dfrac{3n}{4} + 5 = \dfrac{n^2}{6}$ **(f)** $z^2 - z - 2 = 4z^2 + 3z - 7$ **(g)** $x(x + 2) = 5$ **(h)** $\dfrac{1}{w} + w = \dfrac{3}{2}$

(i) $\sqrt{2y} + y = 5$

PROBLEM 16-14   Solve using the zero-product property method: **(a)** $5x^2 = 0$ **(b)** $4y - 5y^2 = 0$
(c) $4w^2 = 9$ **(d)** $m^2 - 2m = 15$ **(e)** $6n^2 = n + 15$ **(f)** $2z^2 + 2z - 40 = 0$
(g) $36r^2 + 30 = 69r$

PROBLEM 16-15   Solve using the square-root rule method: **(a)** $x^2 = 16$ **(b)** $y^2 = 27$
(c) $w^2 = \frac{2}{3}$ **(d)** $z^2 = -1$ **(e)** $m^2 - 32 = 0$ **(f)** $2n^2 - 3 = 5$ **(g)** $3r^2 + 5 = 2$
(h) $(a - 1)^2 = 1$ **(i)** $(b + \frac{2}{3})^2 = \frac{4}{9}$ **(j)** $(s - \frac{2}{3})^2 = \frac{5}{3}$ **(k)** $(c + \frac{1}{2})^2 = -\frac{1}{2}$

PROBLEM 16-16   Solve using the complete-the-square method: **(a)** $x^2 = 2x + 3$
(b) $y^2 = y + 12$ **(c)** $m^2 - 2m = 4$ **(d)** $n^2 + 5n = 7$ **(e)** $2w^2 + w - 5 = 0$
(f) $2z^2 - 8z + 3 = 0$ **(g)** $6x - 1 = -x^2$ **(h)** $9 - 8y = -y^2$ **(i)** $-w^2 - 4w - 7 = 0$
(j) $-z^2 + 6z - 10 = 0$ **(k)** $3 = 10m - 5m^2$ **(l)** $4 - 2n^2 = -3n$

PROBLEM 16-17   Solve using the quadratic-formula method: **(a)** $x^2 - 4x + 3 = 0$

(b) $y^2 = 5y + \dfrac{11}{4}$ **(c)** $2w = 6 - w^2$ **(d)** $4m^2 + 24m = 12$ **(e)** $3n^2 + 1 = 5n$

(f) $z^2 - 3z = 9$ **(g)** $\dfrac{1}{8}x^2 - x - 7 = 0$ **(h)** $y^2 = \dfrac{5y - 1}{3}$ **(i)** $\dfrac{1}{2}w^2 = \dfrac{w + 1}{3}$

(j) $\dfrac{1}{3}m^2 - \dfrac{2}{3}m - \dfrac{1}{2} = 0$

PROBLEM 16-18   Approximate the solutions of each quadratic equation using Appendix Table 2
or a calculator: **(a)** $x^2 - 5x + 1 = 0$ **(b)** $y^2 = 6y + 3$ **(c)** $w^2 - 2w = 5$

(d) $z^2 - 11 = 2z$ **(e)** $m^2 = \dfrac{6m + 2}{5}$ **(f)** $\dfrac{1}{5}n^2 - n - \dfrac{3}{20} = 0$

PROBLEM 16-19   Solve using the easiest method: **(a)** $x^2 = 0$ **(b)** $\frac{3}{4}y^2 = 0$
(c) $w^2 + w = 0$ **(d)** $2z^2 = 5z$ **(e)** $m^2 = 36$ **(f)** $25n^2 - 9 = 0$ **(g)** $r^2 = 5$
(h) $4s^2 - 1 = 5$ **(i)** $x^2 + 6 = 5x$ **(j)** $y^2 + 3y - 2 = 0$ **(k)** $7w = w^2 + 1$
(l) $10z^2 - 25z + 10 = 0$ **(m)** $30m^2 + 20m = 10$ **(n)** $6a^2 - 1 + a = 0$ **(o)** $4b^2 = 4b + 3$
(p) $16c^2 - 8c + 1 = 0$ **(q)** $6d^2 + 6 = 13d$ **(r)** $h^2 = 4h + 2$ **(s)** $k^2 = 5k + 5$
(t) $x^2 + 6x = 2$ **(u)** $y^2 + 3 = 5y$ **(v)** $w^2 - 3 = 6w$ **(w)** $\frac{1}{6}m^2 = m + \frac{2}{3}$

(x) $n^2 + \frac{8}{3}n + 1 = 0$ **(y)** $2 = \dfrac{3a - a^2}{2}$ **(z)** $\dfrac{b^2}{10} = \dfrac{2 - 6b}{5}$

Answers to Supplementary Exercises

(16-12) (a) $x^2 + 3 = 0$ (b) $5y^2 + 2y = 0$ (c) $2w^2 - 5 = 0$ (d) $5m^2 + m - 3 = 0$
(e) $2n^2 + n = 0$ (f) $5z^2 - 3z - 4 = 0$

(16-13) (a) $4x^2 + x + 5 = 0$; $a = 4$, $b = 1$, $c = 5$
(b) $5y^2 - 2y = 0$; $a = 5$, $b = -2$, $c = 0$
(c) $2w^2 - 3w + 1 = 0$; $a = 2$, $b = -3$, $c = 1$
(d) $3m^2 - 5m + 2 = 0$; $a = 3$, $b = -5$, $c = 2$
(e) $2n^2 - 9n - 60 = 0$; $a = 2$, $b = -9$, $c = -60$
(f) $3z^2 + 4z - 5 = 0$; $a = 3$, $b = 4$, $c = -5$
(g) $x^2 + 2x - 5 = 0$; $a = 1$, $b = 2$, $c = -5$
(h) $2w^2 - 3w + 2 = 0$; $a = 2$, $b = -3$, $c = 2$
(i) $y^2 - 12y + 25 = 0$; $a = 1$, $b = -12$, $c = 25$

(16-14) (a) 0 (b) $0, \frac{4}{5}$ (c) $\frac{3}{2}, -\frac{3}{2}$ (d) $5, -3$ (e) $\frac{5}{3}, -\frac{3}{2}$ (f) $4, -5$ (g) $\frac{2}{3}, \frac{5}{4}$

(16-15) (a) ± 4 (b) $\pm 3\sqrt{3}$ (c) $\pm\dfrac{\sqrt{6}}{3}$ (d) no real-number solutions   (e) $\pm 4\sqrt{2}$

(f) ± 2 (g) no real-number solutions   (h) $0, 2$ (i) $0, -\dfrac{4}{3}$ (j) $\dfrac{2 \pm \sqrt{15}}{3}$

(k) no real-number solutions

(16-16) (a) $3, -1$ (b) $4, -3$ (c) $1 \pm \sqrt{5}$ (d) $\dfrac{-5 \pm \sqrt{53}}{2}$ (e) $\dfrac{-1 \pm \sqrt{41}}{4}$

(f) $\dfrac{4 \pm \sqrt{10}}{2}$ (g) $-3 \pm \sqrt{10}$ (h) $4 \pm \sqrt{7}$ (i) no real-number solutions

(j) no real-number solutions   (k) $\dfrac{5 \pm \sqrt{10}}{5}$ (l) $\dfrac{3 \pm \sqrt{41}}{4}$

(16-17) (a) $1, 3$ (b) $\dfrac{11}{2}, -\dfrac{1}{2}$ (c) $-1 \pm \sqrt{7}$ (d) $-3 \pm 2\sqrt{3}$ (e) $\dfrac{5 \pm \sqrt{13}}{6}$

(f) $\dfrac{3 \pm 3\sqrt{5}}{2}$ (g) $4 \pm 6\sqrt{2}$ (h) $\dfrac{5 \pm \sqrt{13}}{6}$ (i) $\dfrac{1 \pm \sqrt{7}}{3}$ (j) $\dfrac{2 \pm \sqrt{10}}{2}$

(16-18) (a) $4.792, 0.209$ (b) $6.464, -0.464$ (c) $3.449, -1.449$ (d) $4.464, -2.464$
(e) $1.472, -0.272$ (f) $5.146, -0.146$

(16-19) (a) 0 (b) 0 (c) $0, -1$ (d) $0, \dfrac{5}{2}$ (e) ± 6 (f) $\pm\dfrac{3}{5}$ (g) $\pm\sqrt{5}$ (h) $\pm\dfrac{\sqrt{6}}{2}$

(i) $2, 3$ (j) $\dfrac{-3 \pm \sqrt{17}}{2}$ (k) $\dfrac{7 \pm 3\sqrt{5}}{2}$ (l) $2, \dfrac{1}{2}$ (m) $\dfrac{1}{3}, -1$ (n) $\dfrac{1}{3}, -\dfrac{1}{2}$

(o) $\dfrac{3}{2}, -\dfrac{1}{2}$ (p) $\dfrac{1}{4}$ (q) $\dfrac{3}{2}, \dfrac{2}{3}$ (r) $2 \pm \sqrt{6}$ (s) $\dfrac{5 \pm 3\sqrt{5}}{2}$ (t) $-3 \pm \sqrt{11}$

(u) $\dfrac{5 \pm \sqrt{13}}{2}$ (v) $3 \pm 2\sqrt{3}$ (w) $3 \pm \sqrt{13}$ (x) $\dfrac{-4 \pm \sqrt{7}}{3}$

(y) no real-number solutions   (z) $-6 \pm 2\sqrt{10}$

17 APPLICATIONS OF QUADRATIC EQUATIONS

THIS CHAPTER IS ABOUT

- ☑ Solving Number Problems Using Quadratic Equations
- ☑ Solving Geometry Problems Using Quadratic Equations
- ☑ Solving Work Problems Using Quadratic Equations
- ☑ Solving Uniform Motion Problems Using Quadratic Equations $[d = rt]$

17-1. Solving Number Problems Using Quadratic Equations

Agreement: In this chapter, the phrase "quadratic equation" will mean *a quadratic equation in one variable.*

To **solve a number problem using a quadratic equation,** you

1. *Read* the problem very carefully several times.
2. *Identify* the unknown numbers.
3. *Decide* how to represent the unknown numbers using one variable.
4. *Translate* the problem to a quadratic equation.
5. *Simplify* the quadratic equation.
6. *Solve* the quadratic equation.
7. *Interpret* the solutions of the quadratic equation with respect to each represented unknown number to find the proposed solutions of the original problem.
8. *Check* to see if the proposed solutions satisfy all the conditions of the original problem.

EXAMPLE 17-1: Solve the following number problem using a quadratic equation.

Solution:

1. *Read:* The sum of two whole numbers is 20. The positive difference of their squares is 16 less than their product. What are the two numbers?

2. *Identify:* The unknown numbers are $\begin{Bmatrix} \text{the larger whole number} \\ \text{the smaller whole number} \end{Bmatrix}$.

3. *Decide:* Let $n =$ the larger whole number
   then $20 - n =$ the smaller whole number [see the following *Notes 1* and *2*].

4. *Translate:* The positive difference of their squares is 16 less than their product.

$$(n)^2 - (20 - n)^2 = n(20 - n) - 16$$

5. *Simplify:*
$$n^2 - (400 - 40n + n^2) = 20n - n^2 - 16$$
$$n^2 - 400 + 40n - n^2 = 20n - n^2 - 16$$
$$-400 + 40n = 20n - n^2 - 16$$
$$n^2 + 20n - 384 = 0 \longleftarrow \text{simplest form}$$

6. *Solve:* $\qquad\qquad\qquad (n + 32)(n - 12) = 0 \qquad$ Factor or use the quadratic formula.

$$n + 32 = 0 \qquad \text{or } n - 12 = 0$$

$$n = -32 \text{ or} \qquad n = 12$$

7. *Interpret:* $\;n = -32$ is not a solution of the original problem because -32 is not a whole number.

$\qquad\qquad\quad n = 12$ means the larger whole number is 12.

$\qquad\qquad\quad 20 - n = 20 - 12 = 8$ means the smaller whole number is 8.

8. *Check:* $\qquad$ Did you find two whole numbers? Yes: 8 and 12.

$\qquad\qquad\quad$ Is the sum of the two whole numbers 20? Yes: $8 + 12 = 20$

$\qquad\qquad\quad$ Is the difference of their squares 16 less than their product? Yes:

$$12^2 - 8^2 = 144 - 64 = 80 \text{ and } 8(12) - 16 = 96 - 16 = 80$$

Note 1: $\;$ If you let $\qquad\qquad\quad n = $ the larger whole number

$\qquad\qquad$ then $\qquad\qquad 20 - n = $ the smaller whole number

$\qquad\qquad$ and $(n)^2 - (20 - n)^2 = $ the positive difference of their squares

$\qquad$ because the larger number comes first when writing a positive difference.

Note 2: $\;$ If you let $\qquad\qquad\quad n = $ the smaller whole number

$\qquad\qquad$ then $\qquad\qquad 20 - n = $ the larger whole number

$\qquad\qquad$ and $(20 - n)^2 - (n)^2 = $ the positive difference of their squares

$\qquad$ because the larger number comes first when writing a positive difference.

EXAMPLE 17-2: Let $\qquad n = $ the smaller number

$\qquad\qquad\qquad$ and $20 - n = $ the larger number for the problem in Example 17-1.

Solution:

Translate: The positive difference of their squares is 16 less than their product.

$$(20 - n)^2 - (n)^2 \qquad = \qquad (20 - n)n - 16$$

Simplify: $\qquad\qquad 400 - 40n + n^2 - n^2 = 20n - n^2 - 16$

$\qquad\qquad\qquad\qquad\quad 400 - 40n = 20n - n^2 - 16$

$\qquad\qquad\qquad\qquad n^2 - 60n + 416 = 0$

Solve: $\qquad\qquad\qquad\quad (n - 8)(n - 52) = 0$

$$n - 8 = 0 \quad \text{or} \quad n - 52 = 0$$

$$n = 8 \quad \text{or} \qquad n = 52$$

Interpret: $\;n = 52$ is not a solution of the original problem because

$\qquad\qquad 20 - n = 20 - 52 = -32$ is not a whole number.

$\qquad\qquad n = 8$ means the smaller whole number is 8. $\longleftarrow$ same solutions as

$\qquad\qquad 20 - n = 20 - 8 = 12$ means the larger whole number is 12. $\swarrow$ found in Example 17-1 [check as before].

17-2. Solve Geometry Problems Using Quadratic Equations

A. Identify the parts of a right triangle.

An **angle** that has a perfectly square corner is called a **right angle**:

A triangle that contains a right angle is called a **right triangle:**

In a right triangle, the side opposite the right angle is called the **hypotenuse** and is usually labeled with the lower-case letter c. In a right triangle, each side that forms the right angle is called a **leg** and the two legs are usually labeled with the lower-case letters a and b.

EXAMPLE 17-3: Identify the parts of the following right triangle.

Solution:

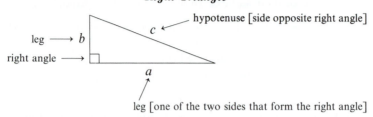

Right Triangle

hypotenuse [side opposite right angle]

leg ⟶ b

right angle ⟶

c

a

leg [one of the two sides that form the right angle]

B. Identify right triangles using the Pythagorean Theorem.

To identify whether a triangle is a right triangle given the lengths of its sides, you can use the following theorem:

Pythagorean Theorem

If a and b are the lengths of the two legs of a right triangle and c is the length of the hypotenuse, then $c^2 = a^2 + b^2$.

Note: The hypotenuse is always the longest side in any right triangle.

EXAMPLE 17-4: Identify which of the following are right triangles using the Pythagorean Theorem:

(a) 8 ft, 15 ft, 17 ft   **(b)** 3 m, 5 m, 6 m

Solution: **(a)** For 8 ft, 15 ft, and 17 ft:

$$c^2 = a^2 + b^2 \longleftarrow \text{Pythagorean Theorem}$$

hypotenuse [longest side] ⟶
$(17)^2$	$(8)^2 + (15)^2$ ⟵ legs [shorter sides]
289	64 + 225
289	289 ⟵ a right triangle [289 = 289]

(b) For 3 m, 5 m, and 6 m:

$$c^2 = a^2 + b^2 \longleftarrow \text{Pythagorean Theorem}$$

hypotenuse [longest side] ⟶
$(6)^2$	$(3)^2 + (5)^2$ ⟵ legs [shorter sides]
36	9 + 25
36	34 ⟵ not a right triangle [36 ≠ 34]

C. Solve a geometry problem using a quadratic equation.

To **solve a geometry problem using a quadratic equation,** you

1. *Read* the problem very carefully several times.
2. *Draw a picture* to help visualize the problem.
3. *Identify* the unknown measures.
4. *Decide* how to represent the unknown measures using one variable.
5. *Translate* the problem to a quadratic equation using the correct geometry formula from Appendix Table 7.
6. *Simplify* the quadratic equation.
7. *Solve* the quadratic equation.
8. *Interpret* the solutions of the quadratic equation with respect to each represented unknown measure to find the proposed solutions of the original problem.
9. *Check* to see if the proposed solutions satisfy all the conditions of the original problem.

EXAMPLE 17-5: Solve this geometry problem using a quadratic equation.

Solution:

1. *Read:*   The hypotenuse of a right triangle is 1 ft longer than one leg and 2 ft longer than the other leg. How long is each side of the right triangle?

2. *Draw a picture:*

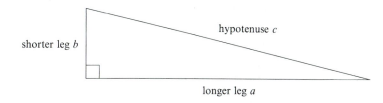

3. *Identify:*   The unknown measures are $\left\{\begin{array}{l}\text{the length of the hypotenuse}\\\text{the length of the longer leg}\\\text{the length of the shorter leg}\end{array}\right\}$.

4. *Decide:*   Let c = the length of the hypotenuse c
then $c - 1$ = the length of the longer leg a
and $c - 2$ = the length of the shorter leg b.

5. *Translate:* $c^2 = a^2 + b^2$ ⟵ Pythagorean Theorem

$c^2 = (c - 1)^2 + (c - 2)^2$ Substitute: $a = c - 1$ and $b = c - 2$

6. *Simplify:* $c^2 = c^2 - 2c + 1 + c^2 - 4c + 4$

$c^2 - 6c + 5 = 0$ ⟵ simplest form

7. *Solve:* $(c - 5)(c - 1) = 0$ Factor or use the quadratic formula.

$c - 5 = 0$ or $c - 1 = 0$

$c = 5$ or $c = 1$

8. *Interpret:* $c = 1$ is not a solution of the original problem because:
$c - 1 = 1 - 1 = 0$ [ft] and
$c - 2 = 1 - 2 = -1$ [ft] are not the required positive measures.

$c = 5$ means the length of the hypotenuse is 5 ft.
$c - 1 = 4$ means the length of the longer leg is 4 ft.
$c - 2 = 3$ means the length of the shorter leg is 3 ft.

9. *Check:*   Is a triangle with sides of 3, 4, and 5 a right triangle?   Yes:

c^2	$= a^2 + b^2$
$(5)^2$	$(4)^2 + (3)^2$
25	16 + 9
25	25

Is the hypotenuse "1 ft longer than one leg?"   Yes: $5 = 1 + 4$
Is the hypotenuse "2 ft longer than the other leg?"   Yes: $5 = 2 + 3$

17-3. Solving Work Problems Using Quadratic Equations

Recall: The unit rate for a person who completes a certain job [1 job] in n hours completes $\dfrac{1}{n}$ job per hour [see Example 14-10].

To find the amount of work completed with respect to the whole job, you evaluate the work formula: work = (unit rate)(time period), or $w = rt$ [see Example 14-11].

The total work is the whole completed job, or: total work = 1 [completed job].

To **solve a work problem using a quadratic equation,** you

1. *Read* the problem very carefully several times.
2. *Identify* the unknown time periods.
3. *Decide* how to represent the unknown time periods using one variable.
4. *Make a table* to help represent individual amounts of work using:
   work = (unit rate)(time period), or $w = rt$ [see Example 14-11].
5. *Translate* the problem to a quadratic equation.
6. *Simplify* the quadratic equation.
7. *Solve* the quadratic equation.
8. *Interpret* the solutions of the quadratic equation with respect to each represented unknown time period to find the proposed solutions of the original problem.
9. *Check* to see if the proposed solutions satisfy all the conditions of the original problem.

EXAMPLE 17-6: Solve this work problem using a quadratic equation.

1. *Read:*   Bobbie can mow a certain lawn in 2 hours less time than it takes Warren. Working together, they can mow the same lawn in $2\frac{2}{5}$ hours. How long does it take each to mow the lawn alone?

2. *Identify:*   The unknown time periods are $\left\{\begin{array}{l}\text{the time for Bobbie to mow the lawn alone}\\\text{the time for Warren to mow the lawn alone}\end{array}\right\}$.

3. *Decide:*   Let $x =$ the time for Bobbie to mow the lawn alone
             then $x + 2 =$ the time for Warren to mow the lawn alone.

4. *Make a table:*

	unit rate r [per hour]	time period t [in hours]	work $[w = rt]$ [portion of whole job]
Bobbie	$\dfrac{1}{x}$	$2\dfrac{2}{5}$ or $\dfrac{12}{5}$	$\dfrac{1}{x} \cdot \dfrac{12}{5}$ or $\dfrac{12}{5x}$
Warren	$\dfrac{1}{x + 2}$	$2\dfrac{2}{5}$ or $\dfrac{12}{5}$	$\dfrac{1}{x + 2} \cdot \dfrac{12}{5}$ or $\dfrac{12}{5(x + 2)}$

5. *Translate:*   Bobbie's work plus Warren's work equals the total work.

$$\frac{12}{5x} + \frac{12}{5(x + 2)} = 1$$

6. *Simplify:*   The LCD of $\dfrac{12}{5x}$ and $\dfrac{12}{5(x + 2)}$ is $5x(x + 2)$.   Clear fractions.

$$5x(x + 2) \cdot \frac{12}{5x} + 5x(x + 2) \cdot \frac{12}{5(x + 2)} = 5x(x + 2)(1)$$

$$(x + 2)12 + x(12) = 5x(x + 2) \longleftarrow \text{fractions cleared}$$

$$12x + 24 + 12x = 5x^2 + 10x$$

$$5x^2 - 14x - 24 = 0 \longleftarrow \text{simplest form}$$

7. *Solve:* $(5x + 6)(x - 4) = 0$ Factor or use the quadratic formula.

$$5x + 6 = 0 \quad \text{or } x - 4 = 0$$

$$5x = -6 \text{ or} \qquad x = 4$$

$$x = \frac{-6}{5} \text{ or} \qquad x = 4$$

8. *Interpret:* $x = -\frac{6}{5}$ is not a solution of the original problem because $-\frac{6}{5}$ hours is a negative time period.

x = 4 means the time for Bobbie to mow the lawn alone is 4 hours.
x + 2 = 4 + 2 = 6 means the time for Warren to mow the lawn alone is 6 hours.

9. *Check:*   Does Bobbie's unit rate [per hour] plus Warren's unit rate equal their unit rate together?   Yes:

$$\frac{1}{4} + \frac{1}{6} = \frac{3}{12} + \frac{2}{12} = \frac{5}{12} = \frac{1}{\frac{12}{5}} = \frac{1}{2\frac{2}{5}}$$

17-4. Solve Uniform Motion Problems Using Quadratic Equations [*d* = *rt*]

Recall 1: To represent an unknown time *t* given the distance *d* and the rate *r*, you solve the distance formula [*d* = *rt*] for *t* to get the time formula $t = \dfrac{d}{r}$ [see Section 14-4].

Recall 2: To represent an unknown rate *r* given the distance *d* and the time *t*, you solve the distance formula [*d* = *rt*] for *r* to get the rate formula $r = \dfrac{d}{t}$ [see Section 14-4].

Recall 3: To represent the groundspeed [rate with respect to the ground] of a plane flying
(a) with the wind [tailwind], you add the wind velocity to the airspeed [rate in still air];
(b) against the wind [headwind], you subtract the wind velocity from the airspeed [see Example 12-7].

Recall 4: To represent the landspeed [rate with respect to land or shore] of a boat moving
(a) with the current [downstream], you add the current rate to the waterspeed [rate in still water];
(b) against the current [upstream], you subtract the current rate from the waterspeed [see Example 12-9].

To **solve a uniform motion problem using a quadratic equation,** you

1. *Read* the problem very carefully several times.
2. *Identify* the unknown distances, rates, and times.
3. *Decide* how to represent the unknown distances, rates, and times using one variable.
4. *Make a table* to help represent the unknown rates [or times] using $r = \dfrac{d}{t} \left[\text{or } t = \dfrac{d}{r} \right]$.
5. *Translate* the problem to a quadratic equation.
6. *Simplify* the quadratic equation.
7. *Solve* the quadratic equation.
8. *Interpret* the solutions of the quadratic equation with respect to each unknown distance, rate, and time to find the proposed solutions of the original problem.
9. *Check* to see if the proposed solutions satisfy all the conditions of the original problem.

EXAMPLE 17-7: Solve this uniform motion problem using a quadratic equation.

Solution:

1. *Read:*   An airplane took 1 hour more to fly 800 km against a headwind than it did on the return trip with a tailwind of the same velocity. If the constant airspeed of the plane was 180 km/h, how long did the round trip take?

2. *Identify:*   The unknown rates and times are $\begin{cases} \text{the constant velocity of the wind} \\ \text{the groundspeed against the wind} \\ \text{the groundspeed with the wind} \\ \text{the time flying against the wind} \\ \text{the time flying with the wind} \end{cases}$.

3. *Decide:*   Let   x = the velocity of the wind
            then 180 − x = the groundspeed against the wind
            and 180 + x = the groundspeed with the wind.

	distance d [in kilometers]	rate r [groundspeed in km/h]	time $\left[t = \dfrac{d}{r} \right]$ [in hours]
4. *Make a table:* against the wind	800	$180 - x$	$\dfrac{800}{180 - x}$
with the wind	800	$180 + x$	$\dfrac{800}{180 + x}$

5. *Translate:*   The flight against the wind took 1 hour more than the flight with the wind.

$$\frac{800}{180 - x} \qquad = \qquad 1 \qquad + \qquad \frac{800}{180 + x}$$

6. *Simplify:*   The LCD of $\dfrac{800}{180 - x}$ and $\dfrac{800}{180 + x}$ is $(180 - x)(180 + x)$.   Clear fractions.

$$(180 - x)(180 + x) \cdot \frac{800}{180 - x} = (180 - x)(180 + x)(1)$$
$$+ (180 - x)(180 + x) \cdot \frac{800}{180 + x}$$

$$(180 + x)800 = (180 - x)(180 + x)$$
$$+ (180 - x)800 \longleftarrow \text{ fractions cleared}$$

$$144{,}000 + 800x = 32{,}400 - x^2 + 144{,}000 - 800x$$

$$800x = 32{,}400 - x^2 - 800x$$

$$x^2 + 1600x - 32{,}400 = 0 \longleftarrow \text{ simplest form}$$

7. *Solve:* $(x - 20)(x + 1620) = 0$ Factor or use the quadratic formula.

$$x - 20 = 0 \ \text{ or } \ x + 1620 = 0$$

$$x = 20 \ \text{or} \qquad x = -1620$$

8. *Interpret:* $x = -1620$ [km/h] is not a solution of the original problem because rates are never negative.

$x = 20$ means the velocity of the wind is 20 km/h.

$180 - x = 180 - 20 = 160$ means the groundspeed against the wind is 160 km/h.

$180 + x = 180 + 20 = 200$ means the groundspeed with the wind is 200 km/h.

$$\frac{800}{180 - x} = \frac{800}{160} = 5 \text{ means the time flying against the wind is 5 hours.}$$

$$\frac{800}{180 + x} = \frac{800}{200} = 4 \text{ means the time flying with the wind is 4 hours.}$$

9. *Check:*   Is 5 hours one hour more than 4 hours?   Yes: $5 = 1 + 4$

Does it take 5 hours to fly 800 km against the wind if the groundspeed is 160 km/h?

Yes: $d = rt$ or $800 = (160)(5)$

Does it take 4 hours to fly 800 km with the wind if the groundspeed is 200 km/h?

Yes: $d = rt$ or $800 = (200)(4)$

RAISE YOUR GRADES

Can you . . . ?

☑ solve a number problem using a quadratic equation

☑ identify the parts of a right triangle

☑ identify right triangles using the Pythagorean Theorem
☑ solve a geometry problem using a quadratic equation
☑ solve a work problem using a quadratic equation
☑ solve a uniform motion problem using a quadratic equation

SUMMARY

1. To solve a number problem, geometry problem, work problem, or uniform motion problem using a quadratic equation, you
   (a) *Read* the problem very carefully several times
   (b) *Draw a picture* to help visualize the problem when appropriate.
   (c) *Identify* the unknown values.
   (d) *Decide* how to represent the unknown values using one variable.
   (e) *Make a table* to help represent the unknown values when appropriate.
   (f) *Translate* the problem to a quadratic equation.
   (g) *Simplify* the quadratic equation.
   (h) *Solve* the quadratic equation.
   (i) *Interpret* the solutions of the quadratic equation with respect to each represented unknown to find the proposed solutions of the original problem.
   (j) *Check* to see if the proposed solutions satisfy all the conditions of the original problem.

2. An angle that has a perfectly square corner is called a right angle:

3. A triangle that contains a right angle is called a right triangle:

4. In a right triangle, the side opposite the right angle is called the hypotenuse and is usually labeled with the lower-case letter c.
5. In a right triangle, each side that forms the right angle is called a leg and the two legs are usually labeled with the lower-case letters a and b.
6. To identify whether a triangle is a right triangle given the lengths of its sides, you can use the Pythagorean Theorem.
7. Pythagorean Theorem:
   If a and b are the lengths of the two legs of a right triangle and c is the length of the hypotenuse, then $c^2 = a^2 + b^2$.
8. The hypotenuse is always the longest side in any right triangle.

9. The unit rate for a person who completes a certain job [1 job] in n hours is $\frac{1}{n}$ job per hour.

10. To find the amount of work completed with respect to the whole job, you evaluate the work formula: work = (unit rate)(time period), or $w = rt$.
11. The total work is the whole completed job, or: total work = 1 [completed job].

12. To represent the time t given the distance d and the rate r, you use $t = \frac{d}{r}$.

13. To represent the rate r given the distance d and the time t, you use $r = \frac{d}{t}$.

14. To represent the groundspeed of a plane flying
    (a) with the wind [tailwind], you add the wind velocity to the airspeed;
    (b) against the wind [headwind], you subtract the wind velocity from the airspeed.

15. To represent the landspeed of a boat moving
    (a) with the current [downstream], you add the current rate to the waterspeed;
    (b) against the current [upstream], you subtract the current rate from the waterspeed.

SOLVED PROBLEMS

PROBLEMS 17-1   Solve each number problem using a quadratic equation:

(a) The sum of a positive number and its square is 1. Find the number.

(b) The difference between a negative number and its reciprocal is 1. What is the number?

Solution: Recall that to solve a number problem using a quadratic equation, you

1. *Read* the problem very carefully several times.
2. *Identify* the unknown numbers.
3. *Decide* how to represent the unknown numbers using one variable.
4. *Translate* the problem to a quadratic equation.
5. *Simplify* the quadratic equation.
6. *Solve* the quadratic equation.
7. *Interpret* the solutions of the quadratic equation with respect to each represented unknown number to find the proposed solutions of the original problem.
8. *Check* to see if the proposed solutions satisfy all the conditions of the original problem.
    [See Example 17-1.]

(a) *Identify:*   The unknown numbers are $\begin{cases} \text{the positive number} \\ \text{the square of the positive number} \end{cases}$.

      Decide: Let $n =$ the positive number
               then $n^2 =$ the square of the positive number.

    Translate: The sum of a positive number and its square is 1.

$$\underbrace{n + n^2}_{} \qquad \underset{=}{=} 1$$

    Simplify: $n^2 + n - 1 = 0$ ⟵ simplest form [$n^2 + n - 1$ does not factor over the integers.]

    Solve: In $n^2 + n - 1 = 0$: $a = 1, b = 1, c = -1$.

$$n = \frac{-b \pm \sqrt{b^2 - 4ac}}{2a} \quad \longleftarrow \text{ quadratic formula}$$

$$n = \frac{-(1) \pm \sqrt{(1)^2 - 4(1)(-1)}}{2(1)} \qquad \text{Evaluate for } a = 1, b = 1, \text{ and } c = -1.$$

$$n = \frac{-1 \pm \sqrt{5}}{2}$$

    Interpret: $n = \dfrac{-1 - \sqrt{5}}{2}$ is a negative number [≈ -1.618] means $\dfrac{-1 - \sqrt{5}}{2}$ is not the solution of the original problem.

             $n = \dfrac{-1 + \sqrt{5}}{2}$ is a positive number [≈ 0.618] means $\dfrac{-1 + \sqrt{5}}{2}$ is the proposed solution of the original problem.

    Check: Is the sum of the positive number $\dfrac{-1 + \sqrt{5}}{2}$ and its square equal to 1? Yes:

$$n + n^2 = \left(\frac{-1 + \sqrt{5}}{2}\right) + \left(\frac{-1 + \sqrt{5}}{2}\right)^2$$

$$= \frac{-1 + \sqrt{5}}{2} + \frac{1 - 2\sqrt{5} + 5}{4}$$

$$= \frac{-2 + 2\sqrt{5} + 6 - 2\sqrt{5}}{4}$$

$$= 1$$

(b) *Identify:* The unknown numbers are $\begin{cases} \text{the negative number} \\ \text{the reciprocal of the negative number} \end{cases}$.

Decide: Let n = the negative number

then $\dfrac{1}{n}$ = the reciprocal of the negative number.

Translate: The difference between a negative number and its reciprocal is 1.

$$\underbrace{}$$
$$n - \frac{1}{n} \qquad\qquad\qquad = 1$$

Simplify: The LCD of $\dfrac{n}{1}, \dfrac{1}{n},$ and $\dfrac{1}{1}$ is n.   Clear fractions.

$$n(n) - n\left(\frac{1}{n}\right) = n(1)$$

$$n^2 - 1 = n \quad\longleftarrow\quad \text{fractions cleared}$$

$$n^2 - n - 1 = 0 \quad\longleftarrow\quad \text{simplest form } [n^2 - n - 1 \text{ does not factor over the integers.}]$$

Solve: In $n^2 - n - 1 = 0$: $a = 1, b = -1, c = -1$.

$$n = \frac{-b \pm \sqrt{b^2 - 4ac}}{2a} \quad\longleftarrow\quad \text{quadratic formula}$$

$$n = \frac{-(-1) \pm \sqrt{(-1)^2 - 4(1)(-1)}}{2(1)} \qquad \text{Evaluate for } a = 1, b = -1, \text{ and } c = -1.$$

$$n = \frac{1 \pm \sqrt{5}}{2}$$

Interpret: $n = \dfrac{1 + \sqrt{5}}{2}$ is a positive number $[\approx 1.618]$ means $\dfrac{1 + \sqrt{5}}{2}$ is not the solution of the original problem.

$n = \dfrac{1 - \sqrt{5}}{2}$ is a negative number $[\approx -0.618]$ means $\dfrac{1 - \sqrt{5}}{2}$ is the proposed solution of the original problem.

Check: Is the difference between the negative number $\dfrac{1 - \sqrt{5}}{2}$ and its reciprocal 1? Yes:

$$n - \frac{1}{n} = \left(\frac{1 - \sqrt{5}}{2}\right) - \frac{1}{\dfrac{1 - \sqrt{5}}{2}} = \frac{1 - \sqrt{5}}{2} - \frac{2}{(1 - \sqrt{5})}$$

$$= \frac{1 - \sqrt{5}}{2} - \frac{2 + 2\sqrt{5}}{-4} = \frac{2 - 2\sqrt{5} + 2 + 2\sqrt{5}}{4} = 1$$

$$\underline{}\quad \text{rationalize the denominator}$$

PROBLEM 17-2 Identify which are right triangles using the Pythagorean Theorem:
(a) 2 in., 3 in., $2\sqrt{3}$ in. $[\approx 3.464]$ (b) 63 cm, 84 cm, 105 cm

Solution: Recall that if *a* and *b* are the lengths of the two legs of a right triangle and *c* is the hypotenuse [longest side], then, by the Pythagorean Theorem, $c^2 = a^2 + b^2$ [see Example 17-4]:

(a) For 2 in., 3 in., and $2\sqrt{3}$ in.: $c^2 = a^2 + b^2$ ←——— Pythagorean Theorem

hypotenuse [longest side] $\longrightarrow$ $(2\sqrt{3})^2$ | $(2)^2 + (3)^2$ ←——— legs [shorter sides]

$4(3)$ | $4 + 9$

12 | 13 ←——— not a right triangle $[12 \neq 13]$

(b) For 63 cm, 84 cm, and 105 cm: $c^2 = a^2 + b^2$ ←——— Pythagorean Theorem

hypotenuse [longest side] $\longrightarrow$ $(105)^2$ | $(63)^2 + (84)^2$ ←——— legs [shorter sides]

$11{,}025$ | $3969 + 7056$

$11{,}025$ | $11{,}025$ ←——— a right triangle $[11{,}025 = 11{,}025]$

PROBLEM 17-3 Solve each geometry problem using a quadratic equation:

(a) The diagonal of a rectangle is 3 ft longer than its length and 5 ft longer than its width. Find the area of the rectangle.

(b) A rectangular swimming pool is surrounded by a concrete walkway of uniform width. The outside dimensions of the walkway are 13 m by 18 m. The area of the water surface of the pool is 150 m². What are the dimensions of the pool?

Solution: Recall that to solve a geometry problem using a quadratic equation, you

1. *Read* the problem very carefully several times.
2. *Draw a picture* to help visualize the problem.
3. *Identify* the unknown measures.
4. *Decide* how to represent the unknown measures using one variable.
5. *Translate* the problem to a quadratic equation using the correct geometry formula from Appendix Table 7.
6. *Simplify* the quadratic equation.
7. *Solve* the quadratic equation.
8. *Interpret* the solutions of the quadratic equation with respect to each represented unknown measure to find the proposed solutions of the original problem.
9. *Check* to see if the proposed solutions satisfy all the conditions of the original problem [see Example 17-5]:

(a) *Draw a picture:*

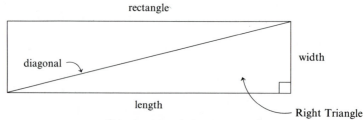

Identify: The unknown measures are $\begin{cases} \text{the length of the diagonal} \\ \text{the length of the rectangle} \\ \text{the width of the rectangle} \end{cases}$.

Decide: Let $d =$ the length of the diagonal [hypotenuse]
then $d - 3 =$ the length of the rectangle [longer leg]
and $d - 5 =$ the width of the rectangle [shorter leg].

Translate: $c^2 = a^2 + b^2$ ←——— Pythagorean Theorem

$(d)^2 = (d - 3)^2 + (d - 5)^2$ Substitute.

Simplify: $$d^2 = d^2 - 6d + 9 + d^2 - 10d + 25$$

$d^2 - 16d + 34 = 0$ ⟵ simplest form [$d^2 - 16d + 34$ does not factor over the integers.]

Solve: $$d = \frac{-b \pm \sqrt{b^2 - 4ac}}{2a}$$ ⟵ quadratic formula

$$d = \frac{-(-16) \pm \sqrt{(-16)^2 - 4(1)(34)}}{2(1)}$$ *Think:* $a = 1, b = -16,$ and $c = 34.$

$$d = \frac{16 \pm \sqrt{120}}{2}$$

$$d = 8 \pm \sqrt{30}$$

Interpret: $d = (8 - \sqrt{30})$ ft $[\approx 2.523 \text{ ft}]$ is not the solution to the original problem because the diagonal d must be 3 ft longer than the length of the rectangle and 5 ft longer than the width.

$d = 8 + \sqrt{30}$ $[\approx 13.477]$ means the length l of the diagonal is $(8 + \sqrt{30})$ ft.
$d - 3 = 5 + \sqrt{30}$ $[\approx 10.477]$ means the length l of the rectangle is $(5 + \sqrt{30})$ ft.
$d - 5 = 3 + \sqrt{30}$ $[\approx 8.477]$ means the width w of the rectangle is $(3 + \sqrt{30})$ ft.
$A = lw = (5 + \sqrt{30})(3 + \sqrt{30}) = 15 + 8\sqrt{30} + 30 = 45 + 8\sqrt{30}$ $[\approx 88.818]$ means the area of the rectangle is $(45 + 8\sqrt{30})$ ft².

Check: Do sides of $3 + \sqrt{30}$, $5 + \sqrt{30}$, and $8 + \sqrt{30}$ form a right triangle?
Yes:

$c^2 = a^2 + b^2$ ⟵ Pythagorean Theorem	
hypotenuse [longest side] ⟶ $(8 + \sqrt{30})^2$	$(3 + \sqrt{30})^2 + (5 + \sqrt{30})^2$ ⟵ legs [shorter-sides]
$64 + 16\sqrt{30} + 30$	$(9 + 6\sqrt{30} + 30) + (25 + 10\sqrt{30} + 30)$
$94 + 16\sqrt{30}$	$94 + 16\sqrt{30}$ ⟵ a right triangle

Is the diagonal 3 ft longer than the length?   Yes: $8 + \sqrt{30} = 3 + (5 + \sqrt{30})$
Is the diagonal 5 ft longer than the width?   Yes: $8 + \sqrt{30} = 5 + (3 + \sqrt{30})$

(b) *Draw a picture:*

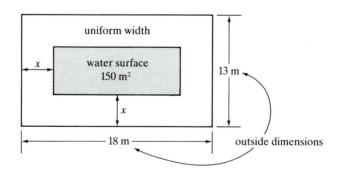

Identify: The unknown measures are $\begin{cases} \text{the uniform width of the walkway} \\ \text{the length } l \text{ of the pool} \\ \text{the width } w \text{ of the pool} \end{cases}$.

Decide: Let $x = $ the uniform width of the walkway
then $18 - 2x = $ the length l of the pool because $18 - x - x = 18 - 2x$
and $13 - 2x = $ the width w of the pool because $13 - x - x = 13 - 2x$.

Translate: $A = lw$ ⟵ area formula for a rectangle

$150 = (18 - 2x)(13 - 2x)$ Substitute the area and dimensions of the pool.

Simplify:

$$150 = 234 - 62x + 4x^2$$

$$4x^2 - 62x + 84 = 0$$

$$2x^2 - 31x + 42 = 0 \longleftarrow \text{simplest form}$$

Solve:

$$x = \frac{-b \pm \sqrt{b^2 - 4ac}}{2a} \longleftarrow \text{quadratic formula}$$

$$x = \frac{-(-31) \pm \sqrt{(-31)^2 - 4(2)(42)}}{2(2)} \qquad \textit{Think: } a = 2, b = -31, \text{ and } c = 42.$$

$$x = \frac{31 \pm \sqrt{625}}{4}$$

$$x = \frac{31 \pm 25}{4}$$

$$x = \frac{31 + 25}{4} \text{ or } \frac{31 - 25}{4}$$

$$x = 14 \text{ or } 1.5$$

Interpret: $x = 14$ m is not the solution of the original problem because the outside width of the pool $[13 - 2x]$ is a negative number for $x = 14$.
$x = 1.5$ means the uniform width of the walkway is 1.5 m.
$18 - 2x = 18 - 2(1.5) = 15$ means the length l of the pool is 15 m.
$13 - 2x = 13 - 2(1.5) = 10$ means the width w of the pool is 10 m.

Check: Is the outside length of the walkway 18 m? Yes: 15 m + 1.5 m + 1.5 m = 18 m
Is the outside width of the walkway 13 m? Yes: 10 m + 1.5 m + 1.5 m = 13 m
Is the area of the pool 150 m²? Yes: (10 m)(15 m) = 150 m²

PROBLEM 17-4 Solve each work problem using a quadratic equation:

(a) Bert agreed to do a certain job for $216. Because it took him 3 hours longer than planned, he earned $6 less an hour than expected. How long did the job actually take?

(b) It takes one outlet pipe 10 days longer to empty a city water tank than it does a larger outlet pipe. If both outlet pipes are used, they can empty the tank in 12 days. How long does it take each outlet pipe to empty the tank alone?

Solution: To solve a work problem using a quadratic equation, you

1. *Read* the problem very carefully several times.
2. *Identify* the unknown time periods.
3. *Decide* how to represent the unknown time periods using one variable.
4. *Make a table* to help represent individual amounts of work using work = (unit rate)(time period), or $w = rt$ [see Example 14-11].
5. *Translate* the problem to a quadratic equation.
6. *Simplify* the quadratic equation.
7. *Solve* the quadratic equation.
8. *Interpret* the solutions of the quadratic equations with respect to each represented unknown time period to find the proposed solutions of the original problem.
9. *Check* to see if the proposed solutions satisfy all the conditions of the original problem [see Example 17-6]:

(a) *Identify:* The unknown time periods are $\begin{cases} \text{the planned time to complete the job} \\ \text{the actual time to complete the job} \end{cases}$.

Decide: Let x = the planned time to complete the job
then $x + 3$ = the actual time to complete the job.

	work w [in dollars]	time period t [in hours]	unit rate $\left[r = \dfrac{w}{t}\right]$ [in dollars]
Make a table: planned	216	x	$\dfrac{216}{x}$
actual	216	$x + 3$	$\dfrac{216}{x + 3}$

Translate: The planned unit rate was $6 an hour more than the actual unit rate.

$$\underbrace{\frac{216}{x}} \quad \underbrace{=} \quad \underbrace{6} \quad \underbrace{+} \quad \underbrace{\frac{216}{x + 3}}$$

Simplify: The LCD of $\dfrac{216}{x}$ and $\dfrac{216}{x + 3}$ is $x(x + 3)$.   Clear fractions.

$$x(x + 3) \cdot \frac{216}{x} = x(x + 3)6 + x(x + 3) \cdot \frac{216}{x + 3}$$

$$(x + 3)(216) = 6x(x + 3) + x(216) \longleftarrow \text{fractions cleared}$$

$$216x + 648 = 6x^2 + 18x + 216x$$

$$6x^2 + 18x - 648 = 0 \longleftarrow \text{standard form [The GCF of 6, 18, and} -648 \text{ is 6 or } -6.]$$

$$x^2 + 3x - 108 = 0 \longleftarrow \text{simplest form}$$

Solve: $(x - 9)(x + 12) = 0$ Factor or use the quadratic formula.

$$x - 9 = 0 \text{ or } x + 12 = 0$$

$$x = 9 \text{ or} \qquad x = -12$$

Interpret: $x = -12$ [hours] cannot be the planned time to complete the job.
$x = 9$ means the planned time to complete the job was 9 hours.
$x + 3 = 12$ means the actual time to complete the job was 12 hours.

Check: Is $216 for 9 hours $6 more an hour than $216 for 12 hours?   Yes:
$$\frac{216}{9} - \frac{216}{12} = 24 - 18 = 6$$

(b) *Identify:* The unknown time periods are $\begin{cases} \text{the time for the smaller outlet pipe} \\ \text{the time for the larger outlet pipe} \end{cases}$.

Decide: Let $x =$ the time for the smaller outlet pipe.
then $x - 10 =$ the time for the larger outlet pipe.

	unit rate r [per day]	time period t [in days]	work $[w = rt]$ [portion of the whole job]
Make a table: smaller pipe	$\dfrac{1}{x}$	12	$\dfrac{1}{x} \cdot 12$ or $\dfrac{12}{x}$
larger pipe	$\dfrac{1}{x - 10}$	12	$\dfrac{1}{x - 10} \cdot 12$ or $\dfrac{12}{x - 10}$

Translate: The smaller pipe's work plus the larger pipe's work equals the total work.

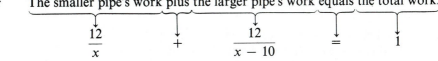

$$\underbrace{\frac{12}{x}} \quad \underbrace{+} \quad \underbrace{\frac{12}{x - 10}} \quad \underbrace{=} \quad \underbrace{1}$$

Simplify: The LCD of $\dfrac{12}{x}$ and $\dfrac{12}{x-10}$ is $x(x-10)$.   Clear fractions.

$$x(x-10)\cdot\frac{12}{x} + x(x-10)\cdot\frac{12}{x-10} = x(x-10)(1)$$

$$(x-10)12 + x(12) = x(x-10) \longleftarrow \text{fractions cleared}$$

$$12x - 120 + 12x = x^2 - 10x$$

$$x^2 - 34x + 120 = 0 \longleftarrow \text{simplest form}$$

Solve: $(x-4)(x-30) = 0$ Factor or use the quadratic formula.

$$x - 4 = 0 \text{ or } x - 30 = 0$$

$$x = 4 \text{ or } \qquad x = 30$$

Interpret: $x = 4$ [days] cannot be the time for the smaller outlet pipe because the larger outlet pipe takes 10 days less than the smaller outlet pipe: $4 - 10 = -6$ [days]
$x = 30$ means the time for the smaller outlet pipe is 30 days.
$x - 10 = 30 - 10 = 20$ means the time for the larger outlet pipe is 20 days.

Check: Does the smaller pipe's unit rate [per day] plus the larger pipe's unit rate equal their unit rate together?   Yes: $\dfrac{1}{30} + \dfrac{1}{20} = \dfrac{2}{60} + \dfrac{3}{60} = \dfrac{5}{60} = \dfrac{1}{12}$

PROBLEM 17-5   Solve each uniform motion problem using a quadratic equation:

(a) If George had driven 5 mph faster than he actually did over a 660-mile trip, he would have saved one hour. How fast did George drive on the trip? How long did the trip take?

(b) Two cars start at the same time from the same place and travel at constant speeds that differ by 10 mph, on roads that are at right angles to each other. In two hours, the cars are 100 miles apart. What is the constant speed of the slower car? How far did the faster car travel?

Solution: To solve a uniform motion problem using a quadratic equation, you

1. *Read* the problem very carefully several times.
2. *Identify* the unknown distances, rates, and times.
3. *Decide* how to represent the unknown distances, rates, and times using one variable.
4. *Make a table* to help represent the unknown rates [or times] using $r = \dfrac{d}{t}\left[\text{or } t = \dfrac{d}{r}\right]$.
5. *Translate* the problem to a quadratic equation.
6. *Simplify* the quadratic equation.
7. *Solve* the quadratic equation.
8. *Interpret* the solutions of the quadratic equation with respect to each unknown distance, rate, and time to find the proposed solutions of the original problem.
9. *Check* to see if the proposed solutions satisfy all the conditions of the original problem. [See Example 17-7.]

(a) *Identify:*   The unknown rates and times are $\left\{\begin{array}{l}\text{the actual rate}\\\text{the faster rate}\\\text{the time for the actual rate}\\\text{the time for the faster rate}\end{array}\right\}$.

Decide: Let $r =$ the actual rate
then $r + 5 =$ the faster rate.

	distance d [in miles]	rate r [in mph]	time $\left[t = \dfrac{d}{r} \right]$ [in hours]
Make a table: actual rate	660	r	$\dfrac{660}{r}$
faster rate	660	$r + 5$	$\dfrac{660}{r + 5}$

Translate: The time for the actual rate **is** one hour longer than the time for the faster rate.

$$\frac{660}{r} \quad = \quad 1 \quad + \quad \frac{660}{r + 5}$$

Simplify: The LCD of $\dfrac{660}{r}$ and $\dfrac{660}{r + 5}$ is $r(r + 5)$.   Clear fractions.

$$r(r + 5) \cdot \frac{660}{r} = r(r + 5)(1) + r(r + 5) \cdot \frac{660}{r + 5}$$

$$(r + 5)660 = r(r + 5) + r(660) \longleftarrow \text{fractions cleared}$$

$$660r + 3300 = r^2 + 5r + 660r$$

$$r^2 + 5r - 3300 = 0 \longleftarrow \text{simplest form}$$

Solve: $(r + 60)(r - 55) = 0$ Factor or use the quadratic equation.

$$r + 60 = 0 \quad \text{or} \quad r - 55 = 0$$

$$r = -60 \quad \text{or} \quad r = 55$$

Interpret: $r = -60$ [mph] cannot be the actual rate because car rates are not negative.
$r = 55$ means the actual rate is 55 mph.
$r + 5 = 55 + 5 = 60$ means the faster rate is 60 mph.

$$\frac{660}{r} = \frac{660}{55} = 12 \text{ means the time for the actual rate is 12 hours.}$$

$$\frac{660}{r + 5} = \frac{660}{60} = 11 \text{ means the time for the faster rate is 11 hours.}$$

Check: Is the time for the actual rate one hour more than the time for the faster rate?   Yes:
12 hours = 1 hour + 11 hours

(b) *Draw a picture:*

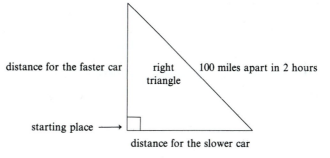

Identify: The unknown rates and distances are $\left\{ \begin{array}{l} \text{the rate for the faster car} \\ \text{the rate for the slower car} \\ \text{the distance for the faster car} \\ \text{the distance for the slower car} \end{array} \right\}$.

Decide: Let r = the rate for the slower car.
then $r + 10$ = the rate for the faster car.

	rate r [in mph]	time t [in hours]	distance $[d = rt]$ [in miles]
Make a table: slower car	r	2	$2r$
faster car	$r + 10$	2	$2(r + 10)$

Translate:

$$c^2 = a^2 + b^2 \longleftarrow \text{Pythagorean Theorem [see Example 17-5]}$$

hypotenuse $\longrightarrow (100)^2 = (2r)^2 + [2(r + 10)]^2 \longleftarrow$ legs [see picture and table]

Simplify:

$$100^2 = 2^2 r^2 + 2^2 (r + 10)^2$$

$$10{,}000 = 4r^2 + 4(r^2 + 20r + 100)$$

$$10{,}000 = 4r^2 + 4r^2 + 80r + 400$$

$$8r^2 + 80r - 9600 = 0 \longleftarrow \text{standard form [The positive GCF is 8.]}$$

$$r^2 + 10r - 1200 = 0 \longleftarrow \text{simplest form}$$

Solve:

$$(r + 40)(r - 30) = 0 \quad \text{Factor or use the quadratic formula.}$$

$$r + 40 = 0 \quad \text{or} \quad r - 30 = 0$$

$$r = -40 \quad \text{or} \quad r = 30$$

Interpret: $r = -40$ [mph] cannot be the rate of the slower car because car rates are never negative.
$r = 30$ means the rate for the slower car is 30 mph.
$r + 10 = 30 + 10 = 40$ means the rate for the faster car is 40 mph.
$2r = 2(30) = 60$ means the distance for the slower car is 60 miles.
$2(r + 10) = 2(40) = 80$ means the distance for the faster car is 80 miles.

Check: Is a triangle with sides of 60, 80, and 100 a right triangle?
Yes:

$$c^2 = a^2 + b^2 \longleftarrow \text{Pythagorean Theorem}$$

hypotenuse [longest side] $\longrightarrow$
$(100)^2$	$(60)^2 + (80)^2$	$\longleftarrow$ legs [shorter sides]
$10{,}000$	$3600 + 6400$	
$10{,}000$	$10{,}000$	$\longleftarrow$ a right triangle

Does it take 2 hours to travel 60 miles at 30 mph? Yes: $t = \dfrac{d}{r} = \dfrac{60}{30} = 2$ [hours]

Does it take 2 hours to travel 80 miles at 40 mph? Yes: $t = \dfrac{d}{r} = \dfrac{80}{40} = 2$ [hours]

Supplementary Exercises

PROBLEM 17-6 Solve each number problem using a quadratic equation:

(a) The square of a positive number is 12 more than the number itself. What is the number?

(b) One negative number is three times another number. The sum of their squares is 360. Find the numbers.

(c) One negative number is 7 more than than 3 times another number. The product of the numbers is 48. What are the numbers?

(d) The product of two consecutive positive integers is 72. Find the integers.

(e) One positive number is 2 more than another number. The sum of their reciprocals is 3. What are the numbers?

(f) The sum of a number and its reciprocal is 4. Find the two possible solutions.

PROBLEM 17-7 Identify which of the following are right triangles using the Pythagorean Theorem: **(a)** 1 yd, 2 yd, $\sqrt{3}$ yd [≈ 1.732] **(b)** 1 km, 1 km, $\sqrt{2}$ km [≈ 1.414] **(c)** 9 ft, 12 ft, 15 ft **(d)** 5 m, 12 m, 13 m **(e)** 3 in., $\sqrt{7}$ in. [≈ 2.646], 4 in. **(f)** 2 cm, 3 cm, $\sqrt{13}$ cm [≈ 3.606] **(g)** 5 mi, 6 mi, $2\sqrt{15}$ mi [≈ 7.746] **(h)** $\sqrt{3}$ mm, [≈ 1.732], $\sqrt{5}$ mm [≈ 2.236], $2\sqrt{2}$ mm [≈ 2.828]

PROBLEM 17-8 Solve each geometry problem using a quadratic equation:

(a) The length of a rectangle is 3 ft longer than the width. The area of the rectangle is 154 ft^2. Find the perimeter of the rectangle.

(b) The perimeter of a rectangle is 32 m. The area of the rectangle is 48 m^2. Find the dimensions of the rectangle.

(c) The side of one square is 3 in. longer than twice the length of the side of another square. The difference between the area of the two squares is 105 in.^{2} Find the perimeter of each square.

(d) A guy wire reaches from the top of a vertical 84-ft TV antenna to a ground anchor. The base of the TV antenna is 63 ft from the anchor. How long is the guy wire?

(e) A square has a perimeter of 12 cm. Find the length of the diagonal.

(f) A square has a 12-cm diagonal. Find the perimeter.

(g) The hypotenuse of a right triangle is 5 m longer than one leg and 2 m longer than the other leg. Find the perimeter and area of the right triangle.

(h) A strip of uniform width is mowed around the outside edge of a rectangular lawn that is 80 m by 120 m. What is the uniform width if the lawn is one-fourth mowed?

PROBLEM 17-9 Solve each work problem using a quadratic equation:

(a) Working alone, John can complete a certain job in 3 hours less time than Agnes. Working together, they can complete the same job in 2 hours. How long does it take each to do the job alone?

(b) It takes an outlet pipe 6 hours longer to empty a swimming pool than it does an inlet pipe to fill the pool. If both pipes are left open, the pool can be filled in 20 hours. How long does it take the inlet pipe to fill the pool if the outlet pipe is closed?

(c) Elizabeth took a job for $192. It took her 4 hours longer than expected and so she earned $2.40 less an hour than planned. How long does Elizabeth expect the job to take?

(d) Serena can paint a car in 3 hours less time than her competitor. If they work together, they can paint the same car in 4 hours. How long does it take Serena to paint the car?

PROBLEM 17-10 Solve each uniform motion problem using a quadratic equation:

(a) A car made a 400-mile trip at 10 mph faster than on the return trip over the same 400 miles which took 2 hours longer. Find the rates of both trips.

(b) Brett can row 16 miles downstream and then make the return trip upstream in a total time of 6 hours. The rate of the current is 2 mph. Find the landspeed downstream. How long did it take him to row upstream?

(c) Byron wanted to fly 300 miles due north. However, he took the wrong course and flew in a straight line to end up 50 miles due west of his planned destination. How far did Byron fly?

(d) An airplane flies between two cities that are 3200 km apart. It takes 20 minutes longer to make the trip against a headwind of 40 km/h than it does normally in still air. What is the normal airspeed of the plane? How long does the trip take with a 40 km/h tailwind?

Answers to Supplementary Exercises

(17-6) **(a)** 4   **(b)** $-6, -18$ **(c)** $-\dfrac{16}{3}, -9$ **(d)** 8, 9   **(e)** $\dfrac{-2 + \sqrt{10}}{3}, \dfrac{4 + \sqrt{10}}{3}$

(f) $2 \pm \sqrt{3}$

(17-7) **(a)** not a right triangle   **(b)** right triangle   **(c)** right triangle   **(d)** right triangle
(e) right triangle   **(f)** right triangle   **(g)** not a right triangle   **(h)** right triangle

(17-8) **(a)** 50 ft   **(b)** 4 m by 12 m   **(c)** 16 in., 44 in.   **(d)** 105 ft   **(e)** $3\sqrt{2}$ cm
(f) $24\sqrt{2}$ cm   **(g)** $(14 + 6\sqrt{5})$ m, $(15 + 7\sqrt{5})$ m^2 **(h)** $(50 - 10\sqrt{19})$ m

(17-9) **(a)** John [3 hr], Agnes [6 hr]   **(b)** $(-3 + \sqrt{129})$ hr   **(c)** 16 hr   **(d)** $\left(\dfrac{5 + \sqrt{73}}{2}\right)$ hr

(17-10) **(a)** 50 mph going, 40 mph returning   **(b)** 8 mph, 4 hr   **(c)** $(50\sqrt{37})$ mi
(d) 640 km/h, 4.71 hr

18 SETS AND INEQUALITIES IN ONE VARIABLE

THIS CHAPTER IS ABOUT

☑ **Graphing Simple Inequalities in One Variable**
☑ **Writing Set Notation**
☑ **Solving Linear Inequalities in One Variable**
☑ **Solving Simple Absolute Value Inequalities in One Variable**

18-1. Graphing Simple Inequalities in One Variable

A. Identify the relation symbols used in algebra.

Recall: The relation symbols used for arithmetic are $<$, $>$, $=$, $\neq$, and $\approx$ [see Example 1-3].

For algebra, you will need two more relation symbols.

EXAMPLE 18-1: List the relation symbols used in algebra.

Solution: $<$ is read as "is less than" [9 is less than 10 or $9 < 10$]. ⎫
$>$ is read as "is greater than" [10 is greater than 9 or $10 > 9$]. ⎬ the three basic relation symbols
$=$ is read as "is equal to" [9 is equal to 9 or $9 = 9$]. ⎭
$\neq$ is read as "is not equal to" [9 is not equal to 10 or $9 \neq 10$].
$\approx$ is read as "is approximately equal to" [9.9 is approximately equal to 10 or $9.9 \approx 10$].
$\leq$ is read as "is less than or equal to" [$9 \leq 10$ because $9 < 10$ and $9 \leq 9$ because $9 = 9$].
$\geq$ is read as "is greater than or equal to"
[$10 \geq 9$ because $10 > 9$ and $9 \geq 9$ because $9 = 9$].

Note 1: $a \leq b$ means $a < b$ or $a = b$ [$x \leq 9$ means $x < 9$ or $x = 9$].

Note 2: $a \geq b$ means $a > b$ or $a = b$ [$x \geq 9$ means $x > 9$ or $x = 9$].

Note 3: $a < b$ means the same as $b > a$ [$9 < 10$ means the same as $10 > 9$].

Note 4: $a \leq b$ means the same as $b \geq a$ [$9 \leq 10$ means the same as $10 \geq 9$].

B. Identify inequalities.

When an equality symbol ($=$) in an equation is replaced by an inequality symbol [$<$, $>$, $\leq$, $\geq$] the resulting algebraic sentence is called an **inequality.**

EXAMPLE 18-2: Identify which are inequalities: (a) $2x + 3 = 5$ (b) $2x + 3 < 5$
(c) $2x + 3y \geq 5$ (d) $-5 > x$ (e) $|x| \leq 2$

Solution: (a) $2x + 3 = 5$ [an equation] is not an inequality because $=$ is an equality symbol.
(b) $2x + 3 < 5$ is an inequality because $<$ is an inequality symbol.
(c) $2x + 3y \geq 5$ is an inequality because $\geq$ is an inequality symbol.
(d) $-5 > x$ is an inequality because $>$ is an inequality symbol.
(e) $|x| \leq 2$ is an inequality because $\leq$ is an inequality symbol.

C. **Identify simple inequalities in one variable.**

Every inequality has three distinct parts.

EXAMPLE 18-3: Identify the three distinct parts of $2x + 3 < 5$.

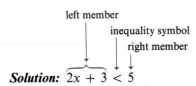

Solution: $2x + 3 < 5$

If x is any variable and a is a real number, then $x < a$, $x \leq a$, $x > a$, and $x \geq a$ are called **simple inequalities in one variable** and a is called the **boundary point.**

EXAMPLE 18-4: Identify which are simple inequalities in one variable:
(a) $2x + 3 = 5$ **(b)** $2x + 3 < 5$ **(c)** $2x + 3y \geq 5$ **(d)** $x < -5$ **(e)** $|x| \leq 2$

Solution:
(a) $2x + 3 = 5$ is not a simple inequality in one variable because it is not an inequality.
(b) $2x + 3 < 5$ is not a simple inequality in one variable because the variable is not alone in one member.
(c) $2x + 3y \geq 5$ is not a simple inequality in one variable because there are two variables.
(d) $x < -5$ is a simple inequality in one variable with boundary point -5.
(e) $|x| \leq 2$ is not a simple inequality in one variable because there are absolute value symbols around the variable.

Note: A simple inequality in one variable always has the variable isolated in the left member and a real number in the right member.

D. **Graph a simple inequality in one variable.**

To **graph a simple inequality in one variable,** you:

1. Graph the boundary point on a number line using:
   (a) an unshaded circle (◯) when the inequality symbol is $<$ or $>$.
   (b) a shaded circle (●) when the inequality symbol is $\leq$ or $\geq$.
2. Graph the **half-line** from the boundary point in Step 1 in the same direction that the inequality symbol points.

EXAMPLE 18-5: Graph each simple inequality in one variable: **(a)** $x < 2$ **(b)** $y \geq -3$

Solution:
(a)

Graph the boundary point $x = 2$ as an open circle to show that $x = 2$ is not part of the graph of $x < 2$. Graph the half-line to the left of $x = 2$ because $<$ in $x < 2$ points to the left.

(b)

Graph the boundary point $y = -3$ as a closed circle to show that $y = -3$ is part of the graph for $y \geq -3$. Graph the half-line to the right of $y = -3$ because $\geq$ in $y \geq -3$ points to the right.

Note 1: The graph in Example 18-5(**a**) is called an **open half-line** because the boundary point is not included in the graph.

Note 2: The graph in Example 18-5(**b**) is called a **closed half-line** because the boundary point is included in the graph.

Note 3: Always label the graph of an inequality with the original inequality.

18-2. Writing Set Notation

A. Write a set using roster notation.

A **set** is a collection of things. The objects in a set are called **members** or **elements** of the set. A set is represented by listing its members between **set braces** { }. A set with no members is called the **empty set** or **null set** and denoted by { } or ∅. To write a set using **roster notation,** you list each member of the set, separated by commas, between set braces.

EXAMPLE 18-6: Write each set using roster notation: **(a)** The set of all whole numbers less than 5. **(b)** The set of all real numbers less than 5. **(c)** The set of all counting numbers less than 1.

Solution:

roster notation

(a) The set of all whole numbers less than 5 = $\{0, 1, 2, 3, 4\}$.
(b) The set of all real numbers less than 5 cannot be written using roster notation because there is no way of listing *each* real number less than 5.
(c) The set of all counting numbers less than 1 = { } [or ∅].

B. Write a set using set-builder notation.

To write a set that cannot be written using roster notation, you can use **set-builder notation.**

EXAMPLE 18-7: Write "the set of all real numbers less than 5" in set builder notation.

set-builder notation

Solution: The set of all real numbers less than 5 = $\{x \,|\, x < 5\}$.

↑
"such that" symbol

Note: Read $\{x \,|\, x < 5\}$ as "the set of all x such that x is less than 5."

18-3. Solving Linear Inequalities in One Variable

A. Identify linear inequalities in one variable.

Recall: An equation that can be written in standard form as $Ax + B = C$, where A, B, and C are real numbers $[A \neq 0]$ and x is any variable, is called a linear equation in one variable [see Section 5-1, part *A*].

When the equality symbol $(=)$ in a linear equation in one variable is replaced by an inequality symbol $[<, >, \leq, \geq]$, the resulting algebraic sentence is called a **linear inequality in one variable.**

EXAMPLE 18-8: Identify which are linear inequalities in one variable:
(a) $2x + 3 = 5$ **(b)** $2x + 3 < 5$ **(c)** $2x + 3y \geq 5$ **(d)** $x < -5$ **(e)** $|x| \leq 2$

Solution:
(a) $2x + 3 = 5$ is a linear equation in one variable, not a linear inequality in one variable.
(b) $2x + 3 < 5$ is a linear inequality in one variable.
(c) $2x + 3y \geq 5$ is a linear inequality in two variables, not a linear inequality in one variable.
(d) $x < -5$ is a linear inequality in one variable [see the following *Note*].
(e) $|x| \leq 2$ is not a linear inequality in one variable because of the absolute value symbols.

Note: Every simple inequality in one variable is a linear inequality in one variable.

B. Solve a linear inequality in one variable using the Addition Rule for Inequalities.

Every linear inequality in one variable has infinitely many solutions [a half-line of solutions]. A **solution of an inequality in one variable** is any real number that can replace the variable to make a true number sentence. The set of all solutions is called the **solution set.**   To **solve a linear inequality in one variable** like $y - 2 \geq 1$, you add to get the variable alone in one member using the following rule:

Addition Rule for Inequalities

If $a < b$, then $a + c < b + c$
where it is understood that $<$ can be any inequality symbol $[<, >, \leq, \geq]$.

Note: The Addition Rule for Inequalities works in the same way as the Addition Rule for Equations [see Example 5-6].

EXAMPLE 18-9: Solve $y - 2 \geq 1$.

Solution: $y - 2 \geq 1$ ⟵── original inequality

$y - 2 + 2 \geq 1 + 2$ Add the subtrahend 2 to both members to
isolate the variable y in one member.

$y + 0 \geq 1 + 2$

$y \geq 1 + 2$

$y \geq 3$ ⟵── solution [simple inequality in one variable]

$y - 2 \geq 1$

⟵── graph [See the following *Note 1*].

$\{y \mid y \geq 3\}$ ⟵── solution set [See the following *Note 2.*]

Note 1: To graph the original inequality $y - 2 \geq 1$, you use the solution $y \geq 3$ because $y - 2 \geq 1$ and $y \geq 3$ are **equivalent inequalities.** That is, $y - 2 \geq 1$ and $y \geq 3$ have the same solutions.

Note 2: $y \geq 3$, 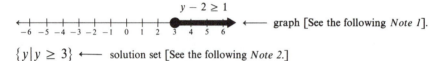, and $\{y \mid y \geq 3\}$ are all equivalent ways of writing the solution of $y - 2 \geq 1$.

C. Solve a linear inequality in one variable using the Subtraction Rule for Inequalities.

To solve a linear inequality in one variable like $x + 3 < 2$, you subtract to get the variable alone in one member using the following rule:

Subtraction Rule for Inequalities

If $a < b$, then $a - c < b - c$
where it is understood that $<$ can be any inequality symbol $[<, >, \leq, \geq]$.

Note: The Subtraction Rule for Inequalities works in the same way as the Subtraction Rule for Equations [see Example 5-5].

EXAMPLE 18-10: Solve $x + 3 < 2$.

Solution: $x + 3 < 2$ ⟵── original inequality

$x + 3 - 3 < 2 - 3$ Subtract the numerical addend 3 from both members to isolate
the variable x in one member.

$x + 0 < 2 - 3$

$x < 2 - 3$

$x < -1$ ⟵── solution

$x + 3 < 2$

⟵── graph

$\{x \mid x < -1\}$ ⟵── solution set

D. Solve a linear inequality in one variable using the Multiplication Rule for Inequalities.

To solve a linear inequality in one variable like $\dfrac{w}{2} > 1$, you multiply to get the variable alone in one member using the following rule:

Multiplication Rule for Inequalities

If $a < b$, then $ac < bc$ if $c > 0$ [See the following *Note*.]
and $ac > bc$ if $c < 0$ [See the following *Caution*.]
where it is understood that $<$ can be any inequality symbol $[<, >, \le, \ge]$.

Note: When both members of an inequality are multiplied by the same positive number, the Multiplication Rule for Inequalities works in the same way as the Multiplication Rule for Equations [see Example 5-8].

Caution: When both members of an inequality are multiplied by the same negative number, you must remember to reverse the inequality symbol to get the correct solution.

EXAMPLE 18-11: Solve: **(a)** $\dfrac{w}{2} > 1$ **(b)** $\dfrac{w}{-2} > 1$

Solution:

(a) $\dfrac{w}{2} > 1$ ⟵ original inequality

$2 \cdot \dfrac{w}{2} > 2(1)$ Multiply both members by the positive divisor 2 to isolate the variable w in one member.

$1w > 2(1)$

$w > 2(1)$

$w > 2$ ⟵ solution

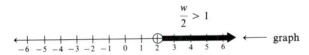

$\dfrac{w}{2} > 1$

⟵ graph

$\{w \mid w > 2\}$ ⟵ solution set

(b) $\dfrac{w}{-2} > 1$ ⟵ original inequality

$-2 \cdot \dfrac{w}{-2} < -2(1)$ Reverse the inequality symbol when you multiply both members by the same negative number [see the following *Note*].

$1w < -2(1)$

$w < -2(1)$

$w < -2$ ⟵ solution

$\dfrac{w}{-2} > 1$

⟵ graph

$\{w \mid w < -2\}$ ⟵ solution set

Note: To see that the inequality symbol must be reversed when you multiply both members by the same negative number, consider the following:

$2 > -1$ ⟵ true inequality

$-2(2) < -2(-1)$ ⟵ reverse inequality symbol to multiply both members by the same negative number

$-4 < 2$ ⟵ true inequality

E. Solve a linear inequality in one variable using the Division Rule for Inequalities.

To solve a linear inequality in one variable like $3m \le -12$, you divide to isolate the variable in one member using the following rule:

Division Rule for Inequalities

If $a < b$, then $\dfrac{a}{c} < \dfrac{b}{c}$ if $c > 0$ [See the following *Note.*]

and $\dfrac{a}{c} > \dfrac{b}{c}$ if $c < 0$ [See the following *Caution.*]

where it is understood that $<$ can be any inequality symbol $[<, >, \le, \ge]$.

Note: When both members of an inequality are divided by the same positive number, the Division Rule for Inequalities works in the same way as the Division Rule for Equations [see Example 5-7].

Caution: When both members of an inequality are divided by the same negative number, you must remember to reverse the inequality symbol to get the correct solution.

EXAMPLE 18-12: Solve **(a)** $3m \le -12$ **(b)** $-3m \le 12$

Solution:

(a) $3m \le -12 \longleftarrow$ original inequality

$\dfrac{3m}{3} \le \dfrac{-12}{3}$ Divide both members by the numerical factor 3 to isolate the variable m in one member.

$1m \le \dfrac{-12}{3}$

$m \le \dfrac{-12}{3}$

$m \le -4 \longleftarrow$ solution

$3m \le -12$

 $\longleftarrow$ graph

$\{m \,|\, m \le -4\} \longleftarrow$ solution set

(b) $-3m \le 12 \longleftarrow$ original inequality

$\dfrac{-3m}{-3} \ge \dfrac{12}{-3}$ Reverse the inequality symbol when you divide both members by the same negative number [see the following *Caution*].

$1m \ge \dfrac{12}{-3}$

$m \ge \dfrac{12}{-3}$

$m \ge -4 \longleftarrow$ solution

$-3m \le 12$

$\longleftarrow$ graph

$\{m \,|\, m \ge -4\} \longleftarrow$ solution set

Caution: To multiply or divide an inequality by the same negative number, you must remember to reverse the inequality symbol to get the correct solution.

F. Solve a linear inequality in one variable using the rules for inequalities.

The Addition, Subtraction, Multiplication, and Division Rules for Inequalities are collectively called the **rules for inequalities.** To solve a linear inequality in one variable like $2x + 3 < 9$, you use the rules for inequalities.

EXAMPLE 18-13: Solve: **(a)** $2x + 3 < 9$ **(b)** $\dfrac{y}{-3} - 4 \geq -5$ **(c)** $8 > 3w - 4 + w$

(d) $-m + 8 - 2m \leq 3m - 6 + m$

Solution:

(a) $2x + 3 < 9$ ⟵ original inequality

$2x + 3 - 3 < 9 - 3$ Use the rules for inequalities to isolate the variable x in one member.

$2x < 6$

$\dfrac{2x}{2} < \dfrac{6}{2}$

$x < 3$ ⟵ solution

$2x + 3 < 9$

⟵ graph

$\{x \mid x < 3\}$ ⟵ solution set

(b) $\dfrac{y}{-3} - 4 \geq -5$ ⟵ original inequality

$\dfrac{y}{-3} - 4 + 4 \geq -5 + 4$ Solve using the rules for inequalities.

$\dfrac{y}{-3} \geq -1$

$-3 \cdot \dfrac{y}{-3} \leq -3(-1)$ Remember to reverse the inequality symbol when you multiply or divide by a negative number.

$y \leq 3$ ⟵ solution

$\dfrac{y}{-3} - 4 \geq -5$

⟵ graph

$\{y \mid y \leq 3\}$ ⟵ solution set

(c) $8 > 3w - 4 + w$ ⟵ original inequality

$8 > 4w - 4$ Combine like terms.

$8 + 4 > 4w - 4 + 4$ Solve as before.

$12 > 4w$

$\dfrac{12}{4} > \dfrac{4w}{4}$

$3 > w \text{ or } w < 3$ ⟵ solution

$8 > 3w - 4 + w$

⟵ graph

$\{w \mid w < 3\}$ ⟵ solution set

(d) $-m + 8 - 2m \le 3m - 6 + m$ ←— original inequality

$-3m + 8 \le 4m - 6$ Combine like terms.

$-3m - 4m + 8 \le 4m - 4m - 6$ Collect like terms.

$-7m + 8 \le -6$ Solve as before.

$-7m \le -14$

$m \ge 2$ ←— solution

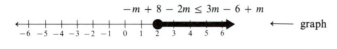

$-m + 8 - 2m \le 3m - 6 + m$

←— graph

$\{m \mid m \ge 2\}$ ←— solution set

G. Solve linear inequalities in one variable containing parentheses, fractions, or decimals.

To solve a linear inequality in one variable containing parentheses, fractions, or decimals, you first clear parentheses, fractions, and decimals.

EXAMPLE 18-14: Solve: **(a)** $2(3 - 5x) \le x - 4(2x - 3)$ **(b)** $\frac{2}{3}y - \frac{1}{2} > \frac{1}{4}$
(c) $-0.5w + 0.2 \le 0.3$

Solution:

(a) $2(3 - 5x) \le x - 4(2x - 3)$ ←— original inequality

$2(3) - 2(5x) \le x + (-4)(2x) - (-4)3$ Clear parentheses.

$6 - 10x \le x - 8x + 12$ ←— parentheses cleared

$6 - 10x \le -7x + 12$ Solve as before.

$6 - 3x \le 12$

$-3x \le 6$

$x \ge -2$ ←— solution

$2(3 - 5x) \le x - 4(2x - 3)$

←— graph

$\{x \mid x \ge -2\}$ ←— solution set

(b) The LCD of $\frac{2}{3}, \frac{1}{2},$ and $\frac{1}{4}$ is 12.   Clear fractions.

$12 \cdot \frac{2}{3}y - 12 \cdot \frac{1}{2} > 12 \cdot \frac{1}{4}$

$8y - 6 > 3$ ←— fractions cleared

$8y > 9$ Solve as before.

$y > \frac{9}{8}$ or $1\frac{1}{8}$ ←— solution

$\frac{2}{3}y - \frac{1}{2} > \frac{1}{4}$

←— graph

$\{y \mid y > \frac{9}{8}\}$ ←— solution set

(c) The LCD of $\frac{5}{10}$ (0.5), $\frac{2}{10}$ (0.2), and $\frac{3}{10}$ (0.3) is 10.   Clear decimals.

$$10(-0.5w) + 10(0.2) \le 10(0.3)$$

$$-5w + 2 \le 3 \longleftarrow \text{ decimals cleared}$$

$$-5w \le 1 \qquad \text{Solve as before.}$$

$$w \ge -\tfrac{1}{5} \longleftarrow \text{ solution}$$

$$-0.5w + 0.2 \le 0.3$$

graph

$\{w \,|\, w \ge -0.2\} \longleftarrow$ solution set

18-4. Solving Simple Absolute Value Inequalities in One Variable

A. Solve simple absolute value inequalities containing < or ≤.

Recall: The distance that a real number is from the origin (0) on a number line is called its absolute value [see Example 3-13].

If x is any variable and a is a real number, then $|x| < a$, $|x| \le a$, $|x| > a$, and $|x| \ge a$ are called **simple absolute value inequalities in one variable.**

Note: Because $|x|$ represents the distance that x is from the origin, $|x| < 2$ represents all the points that are less than 2 units from the origin and $|x| \le 2$ represents all the points that are less than or equal to 2 units from the origin.

The previous statement is generalized in the following rule:

Absolute Value Inequality Rule for < and ≤

If x is any variable and a is a positive real number [$a > 0$], then

1. The solutions of $|x| < a$ are $\{x \,|\, x > -a \text{ and } x < a\}$ [see the following *Note 1*].
2. The solutions of $|x| \le a$ are $\{x \,|\, x \ge -a \text{ and } x \le a\}$ [see the following *Note 2*].

Note 1: $|x| < a$ or $\{x \,|\, x > -a \text{ and } x < a\}$ represents all points that are less than a units from the origin [⟷].

Note 2: $|x| \le a$ or $\{x \,|\, x \ge -a \text{ and } x \le a\}$ represents all points that are less than or equal to a units from the origin [⟷].

EXAMPLE 18-15: Solve: **(a)** $|x| < 2$ **(b)** $|y| \le 4$ **(c)** $|w| < -1$

Solution: **(a)** The solutions of $|x| < 2$ are $\{x \,|\, x > -2 \text{ and } x < 2\}$.

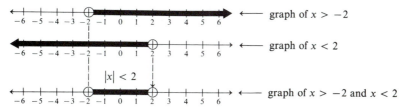

graph of $x > -2$

graph of $x < 2$

$|x| < 2$

graph of $x > -2$ and $x < 2$

(b) The solutions of $|y| \le 4$ are $\{y \,|\, y \ge -4 \text{ and } y \le 4\}$.

graph of $y \ge -4$ and $y \le 4$

(c) There are no solutions of $|w| < -1$ because $|w| \ge 0$ by definition.

B. Solve simple absolute value inequalities containing > or ≥.

Recall: Because $|x|$ represents the distance that x is from the origin, $|x| < 2$ represents all the points that are less than 2 units from the origin and $|x| \leq 2$ represents all the points that are less than or equal to 2 units from the origin.

Note: Because $|x|$ represents the distance that x is from the origin, $|x| > 2$ represents all the points that are more than 2 units from the origin and $|x| \geq 2$ represents all the points that are more than or equal to 2 units from the origin.

The previous statement is generalized in the following rule:

Absolute Value Inequality Rules for > and ≥

If x is any variable and a is a positive real number $[a > 0]$, then

1. The solutions of $|x| > a$ are $\{x \mid x < -a \text{ or } x > a\}$ [see the following *Note 1*].
2. The solutions of $|x| \geq a$ are $\{x \mid x \leq -a \text{ or } x \geq a\}$ [see the following *Note 2*].

Note 1: $|x| > a$ or $\{x \mid x < -a \text{ or } x > a\}$ represents all points that are more than a units from the origin [◄━━━⊕━━━━┼━━━━⊕━━━►].
$\qquad\qquad\qquad\qquad\quad -a \qquad 0 \qquad a$

Note 2: $|x| \geq a$ or $\{x \mid x \leq -a \text{ or } x \geq a\}$ represents all points that are more than or equal to a units from the origin [◄━━━●━━━━┼━━━━●━━━►].
$\qquad\qquad\qquad\qquad\qquad -a \qquad 0 \qquad a$

EXAMPLE 18-16: Solve: **(a)** $|x| > 2$ **(b)** $|y| \geq 1$ **(c)** $|w| > -1$

Solution: **(a)** The solutions of $|x| > 2$ are $\{x \mid x < -2 \text{ or } x > 2\}$.

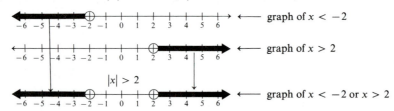

(b) The solutions of $|y| \geq 1$ are $\{y \mid y \leq -1 \text{ or } y \geq 1\}$.

$$|y| \geq 1$$

◄━━━●━━┼━●━━━━━━► ← graph of $y \leq -1$ or $y \geq 1$
$-6\ -5\ -4\ -3\ -2\ -1\ \ 0\ \ 1\ \ 2\ \ 3\ \ 4\ \ 5\ \ 6$

(c) Every real number is a solution of $|w| > -1$ because $|w|$ is never negative.

RAISE YOUR GRADES
Can you . . . ?

☑ identify the relation symbols used in algebra
☑ identify inequalities
☑ identify simple inequalities in one variable
☑ graph a simple inequality in one variable
☑ write a set using roster notation
☑ write a set using set builder notation
☑ identify linear inequalities in one variable
☑ solve a linear inequality in one variable using the Addition Rule for Inequalities
☑ solve a linear inequality in one variable using the Subtraction Rule for Inequalities
☑ solve a linear inequality in one variable using the Multiplication Rule for Inequalities

☑ solve a linear inequality in one variable using the Division Rule for Inequalities
☑ solve a linear inequality in one variable containing parentheses, fractions, or decimals
☑ solve simple absolute value inequalities containing < or ≤
☑ solve simple absolute value inequalities containing > or ≥

SUMMARY

1. The relation symbols used in algebra are < [is less than], > [is greater than], ≤ [is less than or equal to], ≥ [is greater than or equal to], = [is equal to], ≠ [is not equal to], and ≈ [is approximately equal to].
2. $a \leq b$ means $a < b$ or $a = b$.
3. $a \geq b$ means $a > b$ or $a = b$.
4. $a < b$ means the same as $b > a$.
5. $a \leq b$ means the same as $b \geq a$.
6. When an equality symbol (=) in an equation is replaced by an inequality symbol [<, >, ≤, ≥] the resulting algebraic sentence is called an inequality.
7. Every inequality has three distinct parts: left member, inequality symbol, right member.
8. If x is any variable and a is a real number, then $x < a$, $x > a$, $x \leq a$, and $x \geq a$ are called simple inequalities in one variable and a is called the boundary point.
9. To graph a simple inequality in one variable, you
   (a) Graph the boundary point on a number line using
      (i) an unshaded circle (◯) when the inequality symbol is < or >.
      (ii) a shaded circle (●) when the inequality symbol is ≤ or ≥.
   (b) Graph the half-line from the boundary point in Step (a) in the same direction that the inequality symbol points.
10. A set is a collection of things.
11. The objects in a set are called members or elements of the set.
12. A set is represented by listing its members between set braces { }.
13. A set with no members is called the empty set or null set.
14. The empty set is denoted by the symbol { } or ∅.
15. To write a set using roster notation, you list each member of the set, separated by commas, between set braces.
16. To write a set that cannot be written in roster notation, you can use set builder notation such as $\{x \mid x < 5\}$.
17. Read $\{x \mid x < 5\}$ as "set of all x such that x is less than 5."
18. When an equality symbol (=) in a linear equation in one variable is replaced by an inequality symbol [<, >, ≤, ≥] the resulting algebraic sentence is called a linear inequality in one variable.
19. Every linear inequality in one variable has infinitely many solutions [a half-line of solutions].
20. A solution of an inequality in one variable is any real number that can replace the variable to make a true number sentence.
21. The set of all solutions is called the solution set.
22. To solve a linear inequality in one variable, you use the following rules for inequalities:
   Addition Rule for Inequalities: If $a < b$, then $a + c < b + c$.
   Subtraction Rule for Inequalities: If $a < b$, then $a - c < b - c$.
   Multiplication Rule for Inequalities: If $a < b$, then $ac < bc$ if $c > 0$
   and $ac > bc$ if $c < 0$.

   Division Rule for Inequalities: If $a < b$, then $\dfrac{a}{c} < \dfrac{b}{c}$ if $c > 0$

   and $\dfrac{a}{c} > \dfrac{b}{c}$ if $c < 0$.

   It is understood in all four rules just stated that < can be any inequality symbol [<, >, ≤, ≥].

23. To multiply or divide an inequality by the same negative number, you must reverse the inequality symbol to get the correct solution.

24. To solve a linear inequality in one variable containing parentheses, fractions, or decimals, you first clear parentheses, fractions, and decimals.

25. If x is any variable and a is a real number, then $|x| < a$, $|x| > a$, $|x| \leq a$, and $|x| \geq a$ are called simple absolute value inequalities in one variable.

26. Absolute Value Inequality Rules for $<$ and $\leq$:
If x is any variable and a is a positive real number $[a > 0]$, then
(a) The solutions of $|x| < a$ are $\{x \mid x > -a \text{ and } x < a\}$.
(b) The solutions of $|x| \leq a$ are $\{x \mid x \geq -a \text{ and } x \leq a\}$.

27. $|x| < a$ or $\{x \mid x > -a \text{ and } x < a\}$ represents all points that are less than a units from the origin

$$[\overset{\longleftarrow\ \ \underset{-a}{\oplus}\ \ \underset{0}{|}\ \ \underset{a}{\oplus}\ \ \longrightarrow}{}].$$

28. $|x| \leq a$ or $\{x \mid x \geq -a \text{ and } x \leq a\}$ represents all points that are less than or equal to a units from the origin

$$[\overset{\longleftarrow\ \ \underset{-a}{\bullet}\ \ \underset{0}{|}\ \ \underset{a}{\bullet}\ \ \longrightarrow}{}].$$

29. Absolute Value Inequality Rules for $>$ and $\geq$:
If x is any variable and a is a positive real number $[a > 0]$, then
(a) The solutions of $|x| > a$ are $\{x \mid x < -a \text{ or } x > a\}$.
(b) The solutions of $|x| \geq a$ are $\{x \mid x \leq -a \text{ or } x \geq a\}$.

30. $|x| > a$ or $\{x \mid x < -a \text{ or } x > a\}$ represents all points that are more than a units from the origin

$$[\overset{\longleftarrow\ \ \underset{-a}{\oplus}\ \ \underset{0}{|}\ \ \underset{a}{\oplus}\ \ \longrightarrow}{}].$$

31. $|x| \geq a$ or $\{x \mid x \leq -a \text{ or } x \geq a\}$ represents all points that are more than or equal to a units from the origin

$$[\overset{\longleftarrow\ \ \underset{-a}{\bullet}\ \ \underset{0}{|}\ \ \underset{a}{\bullet}\ \ \longrightarrow}{}].$$

SOLVED PROBLEMS

PROBLEM 18-1 Graph each simple inequality in one variable: **(a)** $x < -2$ **(b)** $y \leq 0$ **(c)** $w > 2$ **(d)** $z \geq -1$

Solution: Recall that to graph a simple inequality in one variable, you

1. Graph the boundary point on a number line using
(a) an unshaded circle (○) when the inequality symbol is $<$ or $>$;
(b) a shaded circle (●) when the inequality symbol is $\leq$ or $\geq$.

2. Graph the half-line from the boundary in Step 1 in the same direction that the inequality symbol points [see Example 18-5]:

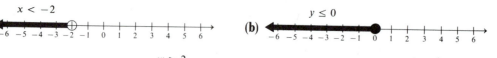

PROBLEM 18-2 Write each of the following sets using roster notation:

(a) The set of all natural numbers less than 6.
(b) The set of all positive integers greater than or equal to 6.
(c) The set of all rational numbers.
(d) The set of all positive integers that are also negative integers.

Solution: Recall that to write a set using roster notation, you list each member of the set, separated by commas, between set braces [see Example 18-6]:

(a) The set of all natural numbers less than $6 = \{1, 2, 3, 4, 5\}$.
(b) The set of all positive integers greater than or equal to $6 = \{6, 7, 8, \cdots\}$.
(c) The set of all rational numbers cannot be written in roster notation because the set of all rational numbers cannot be listed.
(d) The set of positive integers that are also negative integers $= \{ \ \}$ [or $\varnothing$].

PROBLEM 18-3 Write each of the following sets using set builder notation:

(a) The set of all real numbers less than 6.
(b) The set of all rational numbers.
(c) The set of all rational numbers that are also irrational numbers.

Solution: Recall that to write a set that cannot be written in roster notation, you can use set builder notation [see Example 18-7]:

(a) The set of all real numbers less than $6 = \{x \,|\, x < 6\}$.

(b) The set of all rational numbers $= \left\{ x \,\middle|\, x = \dfrac{a}{b} \text{ where } a \text{ and } b \text{ are integers } [b \neq 0] \right\}$.

(c) The set of all rational numbers that are also irrational numbers $= \{ \ \}$ [or $\varnothing$].

PROBLEM 18-4 Solve each linear inequality in one variable using the rules for inequalities:

(a) $x - 3 < -5$ **(b)** $y + 2 \geq 3$ **(c)** $\dfrac{m}{3} > 2$ **(d)** $\dfrac{n}{-4} \leq -2$ **(e)** $2u < 8$

(f) $-5v \leq 10$ **(g)** $2 - 3a > 5$ **(h)** $\dfrac{b}{2} - 3 \geq -1$ **(i)** $2w + 3 - 8w < 10$

(j) $z - 5 - 3z \leq 4 - 5z + 4z$ **(k)** $2c + 3 < 2c + 4$ **(l)** $3d - 3 \leq 3d - 4$

Solution: To solve a linear inequality in one variable, you use the following rules for inequalities:

Addition Rule for Inequalities: If $a < b$, then $a + c < b + c$.

Subtraction Rule for Inequalities: If $a < b$, then $a - c < b - c$.

Multiplication Rule for Inequalities: If $a < b$, then $ac < bc$ if $c > 0$
and $ac > bc$ if $c < 0$.

Division Rule for Inequalities: If $a < c$, then $\dfrac{a}{c} < \dfrac{b}{c}$ if $c > 0$

and $\dfrac{a}{c} > \dfrac{b}{c}$ if $c < 0$.

(a) $x - 3 < -5$
$x - 3 + 3 < -5 + 3$
$x < -2$
[See Example 18-9.]

(b) $y + 2 \geq 3$
$y + 2 - 2 \geq 3 - 2$
$y \geq 1$
[See Example 18-10.]

(c) $\dfrac{m}{3} > 2$
$3 \cdot \dfrac{m}{3} > 3(2)$
$m > 6$
[See Example 18-11(**a**).]

(d) $\dfrac{n}{-4} \leq -2$
$-4 \cdot \dfrac{n}{-4} \geq -4(-2)$
$n \geq 8$
[See Example 18-11(**b**).]

(e) $2u < 8$
$\dfrac{2u}{2} < \dfrac{8}{2}$
$u < 4$
[See Example 18-12(**a**).]

(f) $-5v \leq 10$
$\dfrac{-5v}{-5} \geq \dfrac{10}{-5}$
$v \geq -2$
[See Example 18-12(**b**).]

(g) $2 - 3a > 5$

$\qquad -3a > 3$

$\qquad\quad a < -1$

[See Example 18-13(a).]

(h) $\dfrac{b}{2} - 3 \geq -1$

$\qquad \dfrac{b}{2} \geq 2$

$\qquad\quad b \geq 4$

[See Example 18-13(b).]

(i) $2w + 3 - 8w < 10$

$\qquad 3 - 6w < 10$

$\qquad\quad -6w < 7$

$\qquad\qquad w > -\dfrac{7}{6}$

[See Example 18-13(c).]

(j) $z - 5 - 3z \leq 4 - 5z + 4z$

$\qquad -5 - 2z \leq 4 - z$

$\qquad\quad -5 - z \leq 4$

$\qquad\qquad -z \leq 9$

$\qquad\qquad\quad z \geq -9$

[See Example 18-13(d).]

(k) $2c + 3 < 2c + 4$

$\qquad 3 < 4 \longleftarrow$ true

Every real number is a solution of $2c + 3 < 2c + 4$.

(l) $3d - 3 \leq 3d - 4$

$\qquad -3 \leq -4 \longleftarrow$ false

There are no solutions of $3d - 3 \leq 3d - 4$.

PROBLEM 18-5 Solve linear inequalities in one variable that contain parentheses, fractions, or decimals: **(a)** $x - 2(3x - 4) < 2x + 3(5 - 2x)$ **(b)** $\frac{3}{4}y + \frac{1}{2} \geq \frac{1}{8}$ **(c)** $0.02m + 0.3 \leq 0.04$.

Solution: Recall that to solve linear inequalities in one variable that contain parentheses, fractions, or decimals, you first clear parentheses, fractions, and decimals [see Example 18-14]:

(a) $\qquad x - 2(3x - 4) < 2x + 3(5 - 2x)$

$\quad x + (-2)3x - (-2)4 < 2x + 3(5) - 3(2x)$

$\qquad\quad x - 6x + 8 < 2x + 15 - 6x$

$\qquad\qquad -5x + 8 < 15 - 4x$

$\qquad\qquad\quad -x + 8 < 15$

$\qquad\qquad\qquad -x < 7$

$\qquad\qquad\qquad\quad x > -7$

(b) The LCD of $\frac{3}{4}, \frac{1}{2}$, and $\frac{1}{8}$ is 8.

$\qquad 8 \cdot \frac{3}{4}y + 8 \cdot \frac{1}{2} \geq 8 \cdot \frac{1}{8}$

$\qquad\qquad 6y + 4 \geq 1$

$\qquad\qquad\quad 6y \geq -3$

$\qquad\qquad\quad\, y \geq -\frac{1}{2}$

(c) The LCD of $\frac{2}{100}$ (0.02), $\frac{3}{10}$ (0.3), and $\frac{4}{100}$ (0.04) is 100.

$\qquad 100(0.02)m + 100(0.3) \leq 100(0.04)$

$\qquad\qquad 2m + 30 \leq 4$

$\qquad\qquad\quad 2m \leq -26$

$\qquad\qquad\quad\, m \leq -13$

PROBLEM 18-6 Solve each simple absolute value inequality containing $<$ or $\leq$:

(a) $|x| < 3$ **(b)** $|y| \leq 1$ **(c)** $|w| \leq 0$ **(d)** $|z| < -2$

Solution: Recall the Absolute Value Inequality Rules for $<$ and $\leq$: If x is any variable and a is a positive real number [$a > 0$], then

1. The solutions of $|x| < a$ are $\{x \,|\, x > -a$ and $x < a\}$.
2. The solutions of $|x| \leq a$ are $\{x \,|\, x \geq -a$ and $x \leq a\}$ [see Example 18-15]:

(a) The solutions of $|x| < 3$ are $\{x \,|\, x > -3$ and $x < 3\}$ or 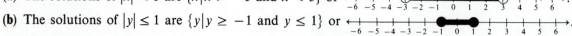.
(b) The solutions of $|y| \leq 1$ are $\{y \,|\, y \geq -1$ and $y \leq 1\}$ or
(c) The one and only solution of $|w| \leq 0$ is $w = 0$ because $|w| \geq 0$ by definition.
(d) There are no solutions of $|z| < -2$ because $|z| \geq 0$ by definition.

PROBLEM 18-7 Solve each simple absolute value inequality containing $>$ or $\geq$:

(a) $|x| > 3$ **(b)** $|y| \geq 2$ **(c)** $|w| > 0$ **(d)** $|z| \geq -2$

Solution: Recall the Absolute Value Inequality Rules for $>$ and $\geq$: If x is any variable and a is a positive real number $[a > 0]$, then

1. The solutions of $|x| > a$ are $\{x | x < -a \text{ or } x > a\}$.
2. The solutions of $|x| \geq a$ are $\{x | x \leq -a \text{ or } x \geq a\}$ [see Example 18-16]:

(a) The solutions of $|x| > 3$ are $\{x | x < -3 \text{ or } x > 3\}$ or .

(b) The solutions of $|y| \geq 2$ are $\{y | y \leq -2 \text{ or } y \geq 2\}$ or .

(c) The solutions of $|w| > 0$ are $\{w | w < 0 \text{ or } w > 0\}$ or or every real number except zero.

(d) Every real number is a solution of $|z| \geq -2$ because $|z|$ is never negative.

Supplementary Exercises

PROBLEM 18-8   Graph each simple inequality in one variable:

(a) $x < 0$ **(b)** $y \leq -1$ **(c)** $w > 0$ **(d)** $z \geq -2$

PROBLEM 18-9   Write each set using set notation:

(a) The set of zero.   **(b)** The set of counting numbers less than or equal to 3.   **(c)** The set of counting numbers greater than 3.   **(d)** The set of integers.   **(e)** The set of negative integers that are greater than 2.   **(f)** The set of real numbers that are greater than or equal to -3.   **(g)** The set of irrational numbers that can be written as $\dfrac{a}{b}$ when a and b are integers.

PROBLEM 18-10   Solve each linear equation in one variable: **(a)** $x + 2 < 5$ **(b)** $y - 3 \leq 2$

(c) $3w > 6$ **(d)** $\dfrac{z}{3} \geq -2$ **(e)** $-4m < 12$ **(f)** $\dfrac{n}{-2} \leq -3$ **(g)** $3a + 3 > -6$

(h) $\dfrac{b}{2} - 3 \geq -5$ **(i)** $5c + 3c < -24$ **(j)** $d - 3d + 5 \geq 3d + 1 - d$ **(k)** $2u \geq -2$

(l) $-5 \leq -2q + 3$ **(m)** $-2x + 13 < -7$ **(n)** $4w + 2 \geq 6w$ **(o)** $3(y + 2) < 15$

(p) $2(1 - v) + 3(2v - 2) \geq 12$ **(q)** $\dfrac{1}{3}u - \dfrac{1}{4} < \dfrac{5}{12}$ **(r)** $\dfrac{k + 4}{2} + \dfrac{k + 1}{4} \geq 3$

(s) $\dfrac{4}{3} > \dfrac{2}{3}(x + 1) + \dfrac{3}{4}(x - 1)$ **(t)** $0.3m + 0.1 < 2.5$ **(u)** $7.5 \geq 0.8(4 + 2v) + 1.1$

(v) $4 + 2(8 - 3y) \leq 9 - 4(1 - y)$ **(w)** $\dfrac{1}{3}(3w - 2) - \dfrac{1}{2}(5 - 2w) \geq -\dfrac{5}{6}(w - 3)$

(x) $0.5r - 0.3(60 - r) > 0.14$ **(y)** $8x + 5 - 2x \geq 3 + 6x + 2$ **(z)** $2 - x < 1 - x$

PROBLEM 18-11   Solve each simple absolute value inequality: **(a)** $|x| < 5$ **(b)** $|y| \leq 3$
(c) $|w| > 4$ **(d)** $|z| \geq 6$ **(e)** $|m| < 0$ **(f)** $|n| \geq 0$ **(g)** $|h| < -8$ **(h)** $|k| > -6$
(i) $|a| \leq 0$ **(j)** $|b| > 0$ **(k)** $|c| < \frac{1}{2}$ **(l)** $|d| > \frac{2}{3}$ **(m)** $|r| \leq 0.2$ **(n)** $|s| \geq 1.5$
(o) $|f| < 1$ **(p)** $|g| \geq -1$ **(q)** $|p| \leq 5$ **(r)** $|q| > -\frac{3}{4}$ **(s)** $|t| \geq 10$ **(t)** $|u| < -0.5$
(u) $|v| > 100$ **(v)** $|x| \leq 2\frac{1}{2}$ **(w)** $|y| > 16$ **(x)** $|w| \leq -0.03$ **(y)** $|m| \geq 3$ **(z)** $|n| < 12$

Answers to Supplementary Exercises

(18-8) **(a)**

$x < 0$

(b)

$y \leq -1$

(c)

$w > 0$

(d)

$z \geq -2$

(18-9) **(a)** $\{0\}$ **(b)** $\{1, 2, 3\}$ **(c)** $\{4, 5, 6, \cdots\}$ **(d)** $\{\cdots, -3, -2, -1, 0, 1, 2, 3, \cdots\}$
(e) $\{\ \}$ or $\varnothing$ **(f)** $\{x \mid x \geq -3\}$ **(g)** $\{\ \}$ or $\varnothing$

(18-10) **(a)** $x < 3$ **(b)** $y \leq 5$ **(c)** $w > 2$ **(d)** $z \geq -6$ **(e)** $m > -3$ **(f)** $n \geq 6$
(g) $a > -3$ **(h)** $b \geq -4$ **(i)** $c < -3$ **(j)** $d \leq 1$ **(k)** $u \geq -1$ **(l)** $q \leq 4$
(m) $x > 10$ **(n)** $w \leq 1$ **(o)** $y < 3$ **(p)** $v \geq 4$ **(q)** $u < 2$ **(r)** $k \geq 1$
(s) $x < 1$ **(t)** $m < 8$ **(u)** $v \leq 2$ **(v)** $y \geq \frac{3}{2}$ **(w)** $w \geq 2$ **(x)** $r > 22.675$
(y) all real numbers **(z)** no solutions

(18-11) **(a)** $\{x \mid x > -5 \text{ and } x < 5\}$ **(b)** $\{y \mid y \geq -3 \text{ and } y \leq 3\}$ **(c)** $\{w \mid w < -4 \text{ or } w > 4\}$
(d) $\{z \mid z \leq -6 \text{ or } z \geq 6\}$ **(e)** no solutions **(f)** all real numbers **(g)** no solutions
(h) all real numbers **(i)** $a = 0$ **(j)** all real numbers except 0
(k) $\{c \mid c > -\frac{1}{2} \text{ and } c < \frac{1}{2}\}$ **(l)** $\{d \mid d < -\frac{2}{3} \text{ or } d > \frac{2}{3}\}$ **(m)** $\{r \mid r \geq -0.2 \text{ and } r \leq 0.2\}$
(n) $\{s \mid s \leq -1.5 \text{ or } s \geq 1.5\}$ **(o)** $\{f \mid f > -1 \text{ and } f < 1\}$ **(p)** all real numbers
(q) $\{p \mid p \geq -5 \text{ and } p \leq 5\}$ **(r)** all real numbers **(s)** $\{t \mid t \leq -10 \text{ or } t \geq 10\}$
(t) no solutions **(u)** $\{v \mid v < -100 \text{ or } v > 100\}$ **(v)** $\{x \mid x \geq -2\frac{1}{2} \text{ and } x \leq 2\frac{1}{2}\}$
(w) $\{y \mid y < -16 \text{ or } y > 16\}$ **(x)** no solutions **(y)** $\{m \mid m \leq -3 \text{ or } m \geq 3\}$
(z) $\{n \mid n > -12 \text{ and } n < 12\}$

FINAL EXAMINATION

Chapters 10–18

Part 1: Skills and Concepts (24 questions)

1. The graph $y = x$ is

(a)

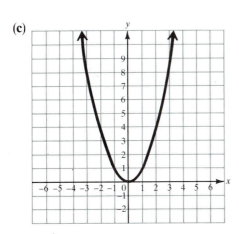

(b)

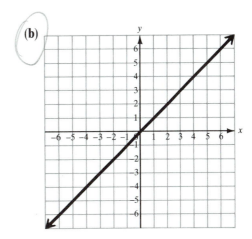

(c)

(d)

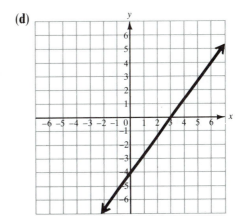

(e) none of these.

2. The graph of $y = x^2$ is

(a)

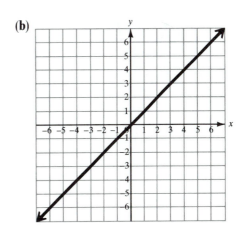

(b)

(c)

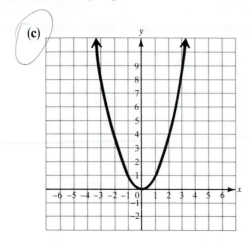

(d)

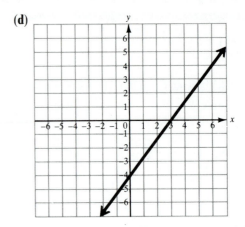

(e) none of these.

---

3. The graph of $y = |x|$ is

(a)

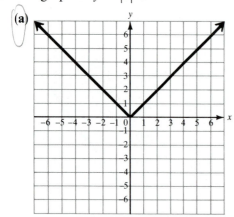

(b)

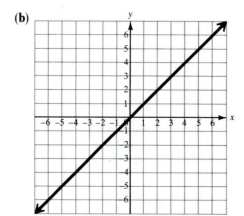

(c)

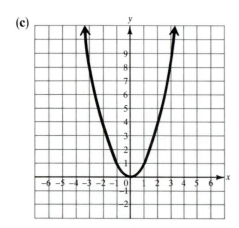

(d)

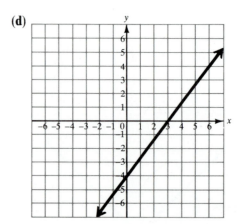

(e) none of these.

4. The graph of $4x - 3y = 12$ is

(a)
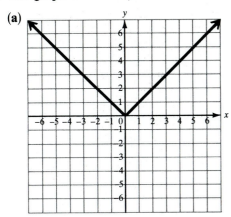

(b)
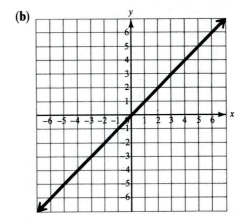

(c)
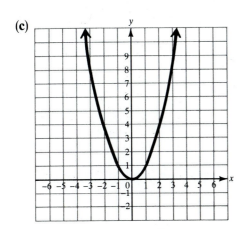

(d)
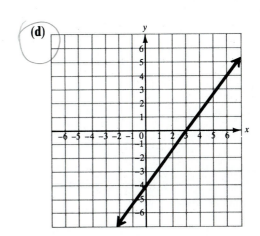

(e) none of these.

5. The solution of $\begin{cases} x - y = 5 \\ x = 2y + 1 \end{cases}$ is

(a) $(9, 4)$ **(b)** $(0, -5)$ **(c)** no solutions   **(d)** infinitely many solutions
(e) none of these.

6. The solution of $\begin{cases} 2x + 3y = -5 \\ 3x - 2y = 12 \end{cases}$ is

(a) $(-\frac{5}{2}, 0)$ **(b)** $(2, -3)$ **(c)** no solutions   **(d)** infinitely many solutions
(e) none of these.

7. The solution of $\begin{cases} 4y = 2 + 3x \\ 9x - 12y - 6 = 0 \end{cases}$ is

(a) $(0, \frac{1}{2})$ **(b)** $(\frac{2}{3}, 0)$ **(c)** no solutions   **(d)** infinitely many solutions   **(e)** none of these.

8. The result of $\dfrac{6 - 4x}{x^2 - 9} + \dfrac{2x}{x - 3} - \dfrac{x}{x + 3}$ in simplest form is

(a) $\dfrac{x^2 + 5x + 6}{x^2 - 9}$ **(b)** $\dfrac{x^2 - x + 6}{x^2 - 9}$ **(c)** $\dfrac{x + 2}{x + 3}$ **(d)** $\dfrac{x + 2}{x - 3}$ **(e)** none of these.

9. The product of $\dfrac{m^3n}{n^2 - n - 6} \cdot \dfrac{n^2 + n - 2}{m^2n^3}$ in simplest form is

(a) $\dfrac{m^3n(n^2 + n - 2)}{m^2n^3(n^2 - n - 6)}$ (b) $\dfrac{m(m + 1)}{n^2(m - 3)}$ (c) $\dfrac{m + 1}{n^2(m - 3)}$ (d) all of these

(e) none of these.

10. The quotient of $\dfrac{x^2 + 5x + 4}{x^2 + 12x + 32} \div \dfrac{x^2 - 12x + 35}{x^2 + 3x - 40}$ in simplest form is

(a) $\dfrac{(x + 4)(x + 1)}{(x + 4)(x - 7)}$ (b) $\dfrac{x + 1}{x - 7}$ (c) $\dfrac{(x + 1)(x - 5)}{(x - 7)(x - 5)}$ (d) all of these   (e) none of these.

11. $\dfrac{\dfrac{1}{x} + \dfrac{1}{2}}{\dfrac{1}{x^2} - \dfrac{1}{4}}$ simplified is

(a) $\dfrac{2x}{2 - x}$ (b) $\dfrac{(x + 2)(4 - x^2)}{8x^3}$ (c) $\dfrac{2x(4 - x^2)}{x + 2}$ (d) all of these   (e) none of these.

12. The solution of $\dfrac{5}{3x} = \dfrac{3}{2x} - \dfrac{1}{x^2}$ is

(a) 0   (b) 1   (c) -6 (d) no solutions   (e) none of these.

13. The result of $\sqrt{24m} - \sqrt{54m^3} + \sqrt{150\, m^5}$ in simplest form is

(a) $(5m^2 - 3m + 2)\sqrt{6m}$ (b) $5m^2 - 3m + 2\sqrt{6m}$ (c) $\sqrt{6m}$ (d) all of these

(e) none of these.

14. The product of $2\sqrt{x}\sqrt{xy}$ in simplest form is

(a) $2\sqrt{x^2y}$ (b) $2xy$ (c) $2x\sqrt{y}$ (d) all of these   (e) none of these.

15. The quotient of $\dfrac{\sqrt{x^3y^5}}{\sqrt{x^2y}}$ in simplest form is

(a) $\sqrt{\dfrac{x^3y^5}{x^2y}}$ (b) $\sqrt{xy^4}$ (c) $y\sqrt{x}$ (d) all of these   (e) none of these.

16. The quotient of $\sqrt{\dfrac{5}{2}}$ in simplest form is

(a) $\dfrac{\sqrt{5}}{2}$ (b) $\sqrt{\dfrac{25}{4}}$ (c) $\sqrt{\dfrac{10}{4}}$ (d) $\dfrac{\sqrt{10}}{2}$ (e) none of these.

17. The solution of $\sqrt{x} + x = 0$ is

(a) 0, 1   (b) 0   (c) 1   (d) no solutions   (e) none of these.

18. The solution of $w^2 = 2w$ is

(a) 0, 2   (b) 0   (c) 2   (d) no solutions   (e) none of these.

19. The solution of $3x^2 - 2 = 0$ in simplest form is

(a) $\dfrac{\pm\sqrt{6}}{3}$ (b) $\dfrac{2}{3}$ (c) $\pm\sqrt{\dfrac{2}{3}}$ (d) $\dfrac{\sqrt{6}}{3}$ (e) none of these.

20. The solution of $x^2 = 4x + 1$ in simplest form is

(a) $\dfrac{4 \pm \sqrt{20}}{2}$ (b) $\dfrac{4 \pm 2\sqrt{5}}{2}$ (c) $2 \pm \sqrt{5}$ (d) no solutions   (e) none of these.

21. The solution of $2x^2 + 3 = 8x$ in simplest form is
    (a) $\dfrac{8 \pm \sqrt{40}}{4}$ (b) $2 - \sqrt{10}$ (c) $\dfrac{4 - \sqrt{10}}{2}$ (d) no solutions   (e) none of these.

22. The solution of $3 - 4x > -5$ is
    (a) $x < 2$ (b) ![number line from -6 to 6 with open circle at 2, shaded to left]   (c) $\{x \mid x < 2\}$ (d) all of these
    (e) none of these.

23. The solution of $|x| < 2$ is
    (a) $\{x \mid x > -2 \text{ or } x > 2\}$ (b) $\{x \mid x > -2 \text{ and } x < 2\}$ (c) $\{x \mid x < -2 \text{ or } x > 2\}$
    (d) all of these   (e) none of these.

24. The solution of $|x| > 2$ is
    (a) $\{x \mid x > -2 \text{ or } x < 2\}$ (b) $\{x \mid x > -2 \text{ and } x < 2\}$ (c) $\{x \mid x < -2 \text{ or } x > 2\}$
    (d) all of these   (e) none of these.

Part 2: Problem Solving (26 questions)

25. The sum of two numbers is 24. The difference between the numbers is 7.5. What are the numbers?

26. The difference between two numbers is 50. One number is 3 times the other number. Find the numbers.

27. Peter is one year younger than Sue. The sum of their ages is 85. How old is each?

28. In 3 years, Richard will be twice as old as Margie is now. Two years ago, the sum of their ages was 20. How old is each now?

29. Twice the sum of the digits of a 2-digit number is 34. The tens digit is one more than the units digit. Find the number.

30. The sum of the digits of a 2-digit number is 14. When the reversed number is subtracted from the original number, the result is 36. What is the original number?

31. A nurse wants to strengthen 20 cc of a 50% alcohol solution to an 80% alcohol solution. How much pure alcohol should he add?

32. A store owner mixed two different kinds of coffee that sold for $6/lb each and $4.50/lb each to get a mixture that would be worth $5.50/lb. How many pounds of each kind of coffee were used if the coffee in the mixture is worth $825 in all?

33. Joe can row downstream at 8 km/h. He can row upstream at 2 km/h. What is the rate of the current and his waterspeed?

34. At a constant airspeed against a headwind, it took Stephanie 7 hours to fly 1050 miles to San Jose. At the same constant airspeed, it took her 5 hours to return flying with a tailwind of the same velocity. What is the wind velocity and her airspeed?

35. The denominator of a fraction is 7 more than the numerator. If 1 is added to the numerator and 2 is subtracted from the denominator, the resulting fraction will equal $\frac{3}{4}$. Find the original fraction.

36. One number is three times another number. The sum of their reciprocals is 1. What are the numbers?

37. A certain car can travel 315 km on 15 gallons of gas. At that rate, how far can the car travel on a full tank of 20 gallons of gas?

38. It takes 2 minutes to run around a certain track at 10 mph. How long will it take to walk around the track at 4 mph?

39. Jennifer can unload a truck in 3 hours. It takes Jaclyn 6 hours to unload the same truck. How long would it take to unload the truck if they worked together?

40. A bathtub can be filled with both faucets on in 20 minutes. It takes 30 minutes to drain the bathtub. How long will it take to fill the bathtub with both faucets on if the drain is accidentally left open?

41. At the same rate of speed, it takes one truck 2 hours longer to travel 375 miles than it takes a car to travel 250 miles. How long does the truck travel? What is that same rate of speed?

42. An airplane travels 600 km with a tailwind of 25 km/h. On the return trip against the same wind velocity, it takes the airplane the same amount of time to travel only 450 km. What is the airspeed of the airplane? How long is that amount of time?

43. The sum of a positive number and its square is 3. What is the number?

44. The sum of a number and its reciprocal is 4. Find the two possible solutions.

45. One leg of a right triangle is 1 ft less than the other leg. The hypotenuse is 3 ft. Find the exact perimeter.

46. A rectangular pool is surrounded by a walkway of uniform width. The outside dimensions of the walkway are 20 m by 17 m. The water surface area is 108 m². What is the uniform width of the walkway?

47. Working alone, Paul can complete a certain job in 4 hours less time than Ernie. Working together, they can complete the same job in 3 hours. Exactly how long would it take each to complete the job alone?

48. Florence agreed to do a job for $216. Because it took her 6 hours longer than expected, she earned $6 less an hour than anticipated. How long did the job actually take? What was the original expected hourly wage?

49. A bus traveled 560 km at a constant rate. A truck traveled the same distance 10 km/h faster than the bus in 1 hour less time. Find the constant rate of the bus. How long did it take the truck to travel the 560 km?

50. Two cars start at the same place and travel on roads that are at right angles to each other. The difference between their constant rates is 35 mph. At the end of 4 hours, the cars are 260 miles apart. What is the speed of the slower car? How far did the faster car travel?

Final Examination Answers

Part 1

1. (b)	**2.** (c)	**3.** (a)	**4.** (d)	**5.** (a)
6. (b)	**7.** (c)	**8.** (d)	**9.** (e)	**10.** (b)
11. (a)	**12.** (c)	**13.** (a)	**14.** (c)	**15.** (e)
16. (d)	**17.** (b)	**18.** (a)	**19.** (a)	**20.** (c)
21. (e)	**22.** (d)	**23.** (b)	**24.** (c)	

Part 2

25. 8.25, 15.75 **26.** 25, 75 **27.** 42 yr, 43 yr **28.** 9 yr, 15 yr

29. 98 **30.** 95 **31.** 30 cc

32. 100 lb of $6, 50 lb of $4.50 **33.** 3 km/h [current], 5 km/h [waterspeed]

34. 30 mph [wind], 180 mph [airspeed] **35.** $\dfrac{11}{18}$ **36.** $\dfrac{4}{3}$, 4

37. 420 km **38.** 5 min **39.** 2 hr **40.** 1 hr

41. 6 hr, 62.5 mph **42.** 175 km/h, 3 hr **43.** $\dfrac{-1 + \sqrt{13}}{2}$ **44.** $2 \pm \sqrt{3}$

45. $(3 + \sqrt{17})$ ft **46.** 4 m **47.** $(1 + \sqrt{13})$ hr, $(5 + \sqrt{13})$ hr

48. 18 hr, $18 **49.** 70 km/h, 7 hr **50.** 25 mph, 240 mi

APPENDIX TABLE 1: *Numbers and Properties*

Numbers

Natural Numbers or *Counting Numbers:* $1, 2, 3, \cdots$

Whole Numbers: $0, 1, 2, 3, \cdots$

Integers: $\cdots, -3, -2, -1, 0, 1, 2, 3, \cdots$

Rational Numbers: All numbers that can be written as $\dfrac{a}{b}$ where a and b are integers ($b \neq 0$).

Irrational Numbers: All numbers that equal decimals which neither terminate nor repeat.

Real Numbers: All rational and irrational numbers.

Properties

Commutative Properties:	$a + b = b + a$	$ab = ba$
Associative Properties:	$(a + b) + c = a + (b + c)$	$(ab)c = a(bc)$
Distributive Properties:	$a(b + c) = ab + ac$	$a(b - c) = ab - ac$
Zero-Product Property:	$ab = 0$ means $a = 0$ or $b = 0$	
Identity Properties:	$a + 0 = 0 + a = a$	$a(1) = 1(a) = a$
Zero Property:	$a(0) = 0(a) = 0$	
Subtraction Properties:	$a - b = a + (-b)$	$a - 0 = a + 0 = a$
	$0 - a = 0 + (-a) = -a$	$a - b = 0$ means $a = b$
Division Properties:	$\dfrac{a}{b} = \dfrac{1}{b} \cdot a = a \cdot \dfrac{1}{b}$	$0 \div a = \dfrac{0}{a} = 0 \ (a \neq 0)$
	$a \div 0$ or $\dfrac{a}{0}$ is not defined	$a \div 1 = \dfrac{a}{1} = a$
	$\dfrac{a}{b} = 1$ means $a = b \ (a \neq 0$ and $b \neq 0)$	
Inverse Properties:	$a + (-a) = -a + a = 0$	$a \cdot \dfrac{1}{a} = \dfrac{1}{a} \cdot a = 1 \ (a \neq 0)$
Negative Properties:	$-(-a) = a$	$-1(a) = a(-1) = -a$
	$a(-b) = (-a)b = -(ab) = -ab$	$(-a)(-b) = ab$
	$\dfrac{a}{b} = \dfrac{-a}{-b} = -\dfrac{-a}{b} = -\dfrac{a}{-b}$	$\dfrac{-a}{b} = \dfrac{a}{-b} = -\dfrac{a}{b} = -\dfrac{-a}{-b}$

APPENDIX TABLE 2: Squares and Square Roots [for whole numbers 1–100]

Number N	Square N^2	Square Root $\sqrt{N}$	Number N	Square N^2	Square Root $\sqrt{N}$	Number N	Square N^2	Square Root $\sqrt{N}$
0	0	0	35	1225	5.916	70	4900	8.367
1	1	1	36	1296	6	71	5041	8.426
2	4	1.414	37	1369	6.083	72	5184	8.485
3	9	1.732	38	1444	6.164	73	5329	8.544
4	16	2	39	1521	6.245	74	5476	8.602
5	25	2.236	40	1600	6.325	75	5625	8.660
6	36	2.449	41	1681	6.403	76	5776	8.718
7	49	2.646	42	1764	6.481	77	5929	8.775
8	64	2.828	43	1849	6.557	78	6084	8.832
9	81	3	44	1936	6.633	79	6241	8.888
10	100	3.162	45	2025	6.708	80	6400	8.944
11	121	3.317	46	2116	6.782	81	6561	9
12	144	3.464	47	2209	6.856	82	6724	9.055
13	169	3.606	48	2304	6.928	83	6889	9.110
14	196	3.742	49	2401	7	84	7056	9.165
15	225	3.873	50	2500	7.071	85	7225	9.220
16	256	4	51	2601	7.141	86	7396	9.274
17	289	4.123	52	2704	7.211	87	7569	9.327
18	324	4.243	53	2809	7.280	88	7744	9.381
19	361	4.359	54	2916	7.348	89	7921	9.434
20	400	4.472	55	3025	7.416	90	8100	9.487
21	441	4.583	56	3136	7.483	91	8281	9.539
22	484	4.690	57	3249	7.550	92	8464	9.592
23	529	4.796	58	3364	7.616	93	8649	9.644
24	576	4.899	59	3481	7.681	94	8836	9.695
25	625	5	60	3600	7.746	95	9025	9.747
26	676	5.099	61	3721	7.810	96	9216	9.798
27	729	5.196	62	3844	7.874	97	9409	9.849
28	784	5.292	63	3969	7.937	98	9604	9.899
29	841	5.385	64	4096	8	99	9801	9.950
30	900	5.477	65	4225	8.062	100	10,000	10
31	961	5.568	66	4356	8.124			
32	1024	5.657	67	4489	8.185			
33	1089	5.745	68	4624	8.246			
34	1156	5.831	69	4761	8.307			

APPENDIX TABLE 3: Product of Primes
[for whole numbers 1–200]

N	Product of Primes	N	Product of Primes	N	Product of Primes	N	Product of Primes	N	Product of Primes
1		41	p	81	3^4	121	11^2	161	$7 \cdot 23$
2	$p*$	42	$2 \cdot 3 \cdot 7$	82	$2 \cdot 41$	122	$2 \cdot 61$	162	$2 \cdot 3^4$
3	p	43	p	83	p	123	$3 \cdot 41$	163	p
4	2^2	44	$2^2 \cdot 11$	84	$2^2 \cdot 3 \cdot 7$	124	$2^2 \cdot 31$	164	$2^2 \cdot 41$
5	p	45	$3^2 \cdot 5$	85	$5 \cdot 17$	125	5^3	165	$3 \cdot 5 \cdot 11$
6	$2 \cdot 3$	46	$2 \cdot 23$	86	$2 \cdot 43$	126	$2 \cdot 3^2 \cdot 7$	166	$2 \cdot 83$
7	p	47	p	87	$3 \cdot 29$	127	p	167	p
8	2^3	48	$2^4 \cdot 3$	88	$2^3 \cdot 11$	128	2^7	168	$2^3 \cdot 3 \cdot 7$
9	3^2	49	7^2	89	p	129	$3 \cdot 43$	169	13^2
10	$2 \cdot 5$	50	$2 \cdot 5^2$	90	$2 \cdot 3^2 \cdot 5$	130	$2 \cdot 5 \cdot 13$	170	$2 \cdot 5 \cdot 17$
11	p	51	$3 \cdot 17$	91	$7 \cdot 13$	131	p	171	$3^2 \cdot 19$
12	$2^2 \cdot 3$	52	$2^2 \cdot 13$	92	$2^2 \cdot 23$	132	$2^2 \cdot 3 \cdot 11$	172	$2^2 \cdot 43$
13	p	53	p	93	$3 \cdot 31$	133	$7 \cdot 19$	173	p
14	$2 \cdot 7$	54	$2 \cdot 3^3$	94	$2 \cdot 47$	134	$2 \cdot 67$	174	$2 \cdot 3 \cdot 29$
15	$3 \cdot 5$	55	$5 \cdot 11$	95	$5 \cdot 19$	135	$3^3 \cdot 5$	175	$5^2 \cdot 7$
16	2^4	56	$2^3 \cdot 7$	96	$2^5 \cdot 3$	136	$2^3 \cdot 17$	176	$2^4 \cdot 11$
17	p	57	$3 \cdot 19$	97	p	137	p	177	$3 \cdot 59$
18	$2 \cdot 3^2$	58	$2 \cdot 29$	98	$2 \cdot 7^2$	138	$2 \cdot 3 \cdot 23$	178	$2 \cdot 89$
19	p	59	p	99	$3^2 \cdot 11$	139	p	179	p
20	$2^2 \cdot 5$	60	$2^2 \cdot 3 \cdot 5$	100	$2^2 \cdot 5^2$	140	$2^2 \cdot 5 \cdot 7$	180	$2^2 \cdot 3^2 \cdot 5$
21	$3 \cdot 7$	61	p	101	p	141	$3 \cdot 47$	181	p
22	$2 \cdot 11$	62	$2 \cdot 31$	102	$2 \cdot 3 \cdot 17$	142	$2 \cdot 71$	182	$2 \cdot 7 \cdot 13$
23	p	63	$3^2 \cdot 7$	103	p	143	$11 \cdot 13$	183	$3 \cdot 61$
24	$2^3 \cdot 3$	64	2^6	104	$2^3 \cdot 13$	144	$2^4 \cdot 3^2$	184	$2^3 \cdot 23$
25	5^2	65	$5 \cdot 13$	105	$3 \cdot 5 \cdot 7$	145	$5 \cdot 29$	185	$5 \cdot 37$
26	$2 \cdot 13$	66	$2 \cdot 3 \cdot 11$	106	$2 \cdot 53$	146	$2 \cdot 73$	186	$2 \cdot 3 \cdot 31$
27	3^3	67	p	107	p	147	$3 \cdot 7^2$	187	$11 \cdot 17$
28	$2^2 \cdot 7$	68	$2^2 \cdot 17$	108	$2^2 \cdot 3^3$	148	$2^2 \cdot 37$	188	$2^2 \cdot 47$
29	p	69	$3 \cdot 23$	109	p	149	p	189	$3^3 \cdot 7$
30	$2 \cdot 3 \cdot 5$	70	$2 \cdot 5 \cdot 7$	110	$2 \cdot 5 \cdot 11$	150	$2 \cdot 3 \cdot 5^2$	190	$2 \cdot 5 \cdot 19$
31	p	71	p	111	$3 \cdot 37$	151	p	191	p
32	2^5	72	$2^3 \cdot 3^2$	112	$2^4 \cdot 7$	152	$2^3 \cdot 19$	192	$2^6 \cdot 3$
33	$3 \cdot 11$	73	p	113	p	153	$3^2 \cdot 17$	193	p
34	$2 \cdot 17$	74	$2 \cdot 37$	114	$2 \cdot 3 \cdot 19$	154	$2 \cdot 7 \cdot 11$	194	$2 \cdot 97$
35	$5 \cdot 7$	75	$3 \cdot 5^2$	115	$5 \cdot 23$	155	$5 \cdot 31$	195	$3 \cdot 5 \cdot 13$
36	$2^2 \cdot 3^2$	76	$2^2 \cdot 19$	116	$2^2 \cdot 29$	156	$2^2 \cdot 3 \cdot 13$	196	$2^2 \cdot 7^2$
37	p	77	$7 \cdot 11$	117	$3^2 \cdot 13$	157	p	197	p
38	$2 \cdot 19$	78	$2 \cdot 3 \cdot 13$	118	$2 \cdot 59$	158	$2 \cdot 79$	198	$2 \cdot 3^2 \cdot 11$
39	$3 \cdot 13$	79	p	119	$7 \cdot 17$	159	$3 \cdot 53$	199	p
40	$2^3 \cdot 5$	80	$2^4 \cdot 5$	120	$2^3 \cdot 3 \cdot 5$	160	$2^5 \cdot 5$	200	$2^3 \cdot 5^2$

* p means prime.

APPENDIX TABLE 4: *Systems of Measure* *[U.S./Metric]*

U.S. Customary System of Measures	Metric System of Measures

Length

1 ft	= 12 in.	in.: inch(es)
1 yd	= 3 ft	ft: foot (feet)
1 min	= 1760 yd	yd: yard(s)
1 min	= 5280 ft	mi: mile(s)

Length

1 cm	= 10 mm	mm: millimeter(s)
1 m	= 100 cm	cm: centimeter(s)
1 km	= 1000 m	m: meter(s)
		km: kilometer(s)

Capacity

1 tsp	= 80 gtt	gtt: drop(s)
1 tbsp	= 3 tsp	tsp: teaspoon(s)
1 fl oz	= 2 tbsp	tbsp: tablespoon(s)
1 c	= 8 fl oz	fl oz: fluid ounce(s)
1 pt	= 2 c	c: cup(s)
1 qt	= 2 pt	pt: pint(s)
1 gal	= 4 qt	qt: quart(s)
		gal: gallon(s)

Capacity

1 L	= 1000 mL	mL: milliliter(s)
1 kL	= 1000 L	L: liter(s)
		kL: kiloliter(s)

Weight

1 lb	= 16 oz	oz: ounce(s)
1 T	= 2000 lb	lb: pound(s)
		T: ton(s) or short ton(s)

Mass

1 g	= 1000 mg	mg: milligram(s)
1 kg	= 1000 g	g: gram(s)
1 t	= 1000 kg	kg: kilogram(s)
		t: tonne(s) or metric ton(s)

Area

$1\ ft^2$	$= 144\ in.^2$	$in.^2$: square inch(es)
$1\ yd^2$	$= 9\ ft^2$	ft^2: square foot (feet)
1 A	$= 4840\ yd^2$	yd^2: square yard(s)
$1\ mi^2$	= 640 A	A: acre(s)
		mi^2: square mile(s)

Area

$1\ cm^2$	$= 100\ mm^2$	mm^2: = square millimeter(s)
$1\ m^2$	$= 10{,}000\ cm^2$	cm^2: = square centimeter(s)
1 ha	$= 10{,}000\ m^2$	m^2: = square meter(s)
$1\ km^2$	= 100 ha	ha: = hectare(s)
		km^2: = square kilometer(s)

Volume

$1\ ft^3$	$= 1728\ in.^3$	$in.^3$: cubic inch(es)
$1\ yd^3$	$= 27\ ft^3$	ft^3: cubic foot (feet)
		yd^3: cubic yard(s)

Volume

$1\ cm^3$	$= 1000\ mm^3$	mm^3: cubic millimeter(s)
$1\ m^3$	$= 1{,}000{,}000\ cm^3 (cc)$	$cm^3(cc)$: cubic centimeter(s)
		m^3: cubic meter(s)

Temperature

Water boils at 212°F.   °F: degrees Fahrenheit
The normal human body temperature is 98.6°F.
Water freezes at 32°F.

Temperature

Water boils at 100°C.   °C: degrees Celsius
The normal human body temperature is 37°C.
Water freezes at 0°C.

Time

1 min	= 60 sec	sec: second(s)	1 hr	= 60 min	min: minute(s)
1 da	= 24 hr	hr: hour(s)	1 wk	= 7 da	wk: week(s)
1 yr	= 12 mo	mo: month(s)	1 yr	≈ 365 da	da: day(s)
					yr: year(s)

APPENDIX TABLE 5: *Conversion Factors* [U.S./Metric]

	U.S. Customary/Metric		Paper-and-Pencil Conversion Factors	Calculator Conversion Factors
	From	*To*	*Multiply By*	*Multiply By*
Length	inches (in.)	millimeters (mm)	25	**25.4**
	inches	centimeters (cm)	2.5	**2.54**
	feet (ft)	meters (m)	0.3	**0.3048**
	yards (yd)	meters	0.9	**0.9144**
	miles (mi)	kilometers (km)	1.6	1.609
Capacity	drops (gtt)	milliliters (mL)	16	16.23
	teaspoons (tsp)	milliliters	5	4.929
	tablespoons (tbsp)	milliliters	15	14.79
	fluid ounces (fl oz)	milliliters	30	29.57
	cups (c)	liters (L)	0.24	0.2366
	pints (pt)	liters	0.47	0.4732
	quarts (qt)	liters	0.95	0.9464
	gallons (gal)	liters	3.8	3.785
Weight (Mass)	ounces (oz)	grams (g)	28	28.35
	pounds (lb)	kilograms (kg)	0.45	0.4536
	tons (T)	tonnes (t)	0.9	0.9072
Area	square inches (in.^{2})	square centimeters (cm^2)	6.5	6.452
	square feet (ft^2)	square meters (m^2)	0.09	0.09290
	square yards (yd^2)	square meters	0.8	0.8361
	square miles (mi^2)	square kilometers (km^2)	2.6	2.590
	acres (A)	hectares (ha)	0.4	0.4047
Volume	cubic inches (in.^{3})	cubic centimeters (cm^3 or cc)	16	16.39
	cubic feet (ft^3)	cubic meters (m^3)	0.03	0.02832
	cubic yards (yd^3)	cubic meters	0.8	0.7646
Temperature	degrees Fahrenheit (°F)	degrees Celsius (°C)	$\frac{5}{9}$ (after subtracting 32)	0.5556 (after subtracting 32)

Note: All conversion factors in bold type are exact. All others are rounded.

TABLE 5: [*Continued*]

	Metric/U.S. Customary		*Paper-and-Pencil Conversion Factors*	*Calculator Conversion Factors*
	From	*To*	*Multiply By*	*Multiply By*
Length	millimeters (mm)	inches (in.)	0.04	0.03937
	centimeters (cm)	inches	0.4	0.3937
	meters (m)	feet (ft)	3.3	3.280
	meters	yards (yd)	1.1	1.094
	kilometers (km)	miles (mi)	0.6	0.6214
Capacity	milliliters (mL)	drops (gtt)	0.06	0.06161
	milliliters	teaspoons (tsp)	0.2	0.2029
	milliliters	tablespoons (tbsp)	0.07	0.06763
	milliliters	fluid ounces (fl oz)	0.03	0.03381
	liters (L)	cups (c)	4.2	4.227
	liters	pints (pt)	2.1	2.113
	liters	quarts (qt)	1.1	1.057
	liters	gallons (gal)	0.26	0.2642
Mass (Weight)	grams (g)	ounces (oz)	0.035	0.03527
	kilograms (kg)	pounds (lb)	2.2	2.205
	tonnes (t)	tons (T)	1.1	1.102
Area	square centimeters (cm^2)	square inches (in.^{2})	0.16	0.1550
	square meters (m^2)	square feet (ft^2)	11	10.76
	square meters	square yards (yd^2)	1.2	1.196
	square kilometers (km^2)	square miles (mi^2)	0.4	0.3861
	hectares (ha)	acres (A)	2.5	2.471
Volume	cubic centimeters (cm^3)	cubic inches (in.^{3})	0.06	0.06102
	cubic meters (m^3)	cubic feet (ft^3)	35	35.31
	cubic meters	cubic yards (yd^3)	1.3	1.308
Temperature	degrees Celsius (°C)	degrees Fahrenheit (°F)	$\frac{9}{5}$ (then add 32)	**1.8** (then add 32)

Note: All conversion factors in bold type are exact. All others are rounded.

APPENDIX TABLE 6: *Rules and Formulas*

Rules for Equations

If $a = b$, then: $a + c = b + c$

If $a = b$, then: $ac = bc$

If $a = b$, then: $a^2 = b^2$

If $a^2 = b$, then: $a = \pm\sqrt{b}$

Rules for Inequalities

If $a < b$, then: $a + c < b + c$

If $a < b$ and $c > 0$, then: $ac < bc$

If $a < b$ and $c < 0$, then: $ac > bc$

Rules for Exponents

If m and n are integers and r and s are terms, then:

$r^n = \overbrace{r \cdot r \cdot r \cdots r}^{n \text{ repeated factors of } r}$

$10^n = 1\overbrace{000 \cdots 0}^{n \text{ zeros}}$

$10^{-n} = 0.\overbrace{000 \cdots 01}^{n \text{ zeros}}$

$r^0 = 1 \ (r \neq 0)$

$0^n = 0 \ (n \neq 0)$

0^0 is not defined

$1^n = 1$

$-1^n = -1$

$(-1)^n = \begin{cases} 1 \text{ if } n \text{ is even} \\ -1 \text{ if } n \text{ is odd} \end{cases}$

$r^m r^n = r^{m+n}$

$(rs)^n = r^n s^n$

$(r^m)^n = r^{mn}$

$\dfrac{r^m}{r^n} = r^{m-n} \ (r \neq 0)$

$\left(\dfrac{r}{s}\right)^n = \dfrac{r^n}{s^n} \ (s \neq 0)$

$r^{-n} = \dfrac{1}{r^n} \ (r \neq 0)$

Slopes and Straight Lines

$\text{slope} = m = \dfrac{\text{change in } y}{\text{change in } x} = \dfrac{y_2 - y_1}{x_2 - x_1} \ (x_1 \neq x_2)$

slope-intercept form: $y = mx + b$

point-slope form: $y - y_1 = m(x - x_1)$

Products of Binomials

F   O   I   L

$(a + b)(c + d) = ac + ad + bc + bd$

$(a + b)(a - b) = a^2 - b^2$

$(a + b)(a + b) = a^2 + 2ab + b^2$

$(a - b)(a - b) = a^2 - 2ab + b^2$

Rules for Rational Expressions

If P, Q, R, and S are polynomials, then:

$\dfrac{P}{Q} = \dfrac{P \cdot R}{Q \cdot R} \ (Q \neq 0 \text{ and } R \neq 0)$

$\dfrac{P}{Q} \cdot \dfrac{R}{S} = \dfrac{P \cdot R}{Q \cdot S} \ (Q \neq 0 \text{ and } S \neq 0)$

$\dfrac{P}{Q} \div \dfrac{R}{S} = \dfrac{P}{Q} \cdot \dfrac{S}{R} \ (Q \neq 0, R \neq 0, \text{ and } S \neq 0)$

$\dfrac{P}{Q} + \dfrac{R}{Q} = \dfrac{P + R}{Q} \ (Q \neq 0)$

$\dfrac{P}{Q} - \dfrac{R}{Q} = \dfrac{P - R}{Q} \ (Q \neq 0)$

$\dfrac{\frac{P}{Q}}{\frac{R}{S}} = \dfrac{P}{Q} \div \dfrac{R}{S} \ (Q \neq 0, R \neq 0, \text{ and } S \neq 0)$

Rules for Radical Expressions

If r and s are rational expressions, then:

$\sqrt{rs} = \sqrt{r}\sqrt{s} \ (r \geq 0 \text{ and } s \geq 0)$

$\sqrt{\dfrac{r}{s}} = \dfrac{\sqrt{r}}{\sqrt{s}} \ (r \geq 0 \text{ and } s > 0)$

$\sqrt{r^2} = |r| = \begin{cases} r \text{ if } r \geq 0 \\ -r \text{ if } r < 0 \end{cases}$

$\sqrt{r^n} = r^{\frac{n}{2}} \text{ if } n \text{ is even } (r \geq 0)$

$\sqrt{r^n} = r^{\frac{n-1}{2}}\sqrt{r} \text{ if } n \text{ is odd } (r \geq 0)$

$(\sqrt{r})^2 = r \ (r \geq 0)$

Quadratic Formula

If $ax^2 + bx + c = 0 \ (a \neq 0)$, then: $x = \dfrac{-b \pm \sqrt{b^2 - 4ac}}{2a}$

APPENDIX TABLE 7: *Geometry Formulas*

	Figure	Perimeter (P)	Area (A)
Square		$P = 4s$	$A = s^2$
Rectangle		$P = 2(l + w)$	$A = lw$
Parallelogram		$P = 2(a + b)$	$A = bh$
Triangle		$P = a + b + c$	$A = \dfrac{1}{2}bh$

	Figure	Circumference (C)	Area (A)
Circle		$C = \pi d$ $C = 2\pi r$	$A = \pi r^2$

	Figure	Volume (V)	Surface Area (SA)
Cube		$V = e^3$	$SA = 6e^2$
Rectangular Prism (box)		$V = lwh$	$SA = 2(lw + lh + wh)$
Cylinder		$V = \pi r^2 h$	$SA = 2\pi r(r + h)$
Sphere		$V = \dfrac{4}{3}\pi r^3$	$SA = 4\pi r^2$

INDEX

Page numbers in bold type indicate solved problems.